T0217828

Algorithms and Computation in Mathematics

Volume 26

With this forward-thinking series Springer recognizes that the prevailing trend in mathematical research towards algorithmic and constructive processes is one of long-term importance. This series is intended to further the development of computational and algorithmic mathematics. In particular, Algorithms and Computation in Mathematics emphasizes the computational aspects of algebraic geometry, number theory, combinatorics, commutative, non-commutative and differential algebra, geometric and algebraic topology, group theory, optimization, dynamical systems and Lie theory. Proposals or manuscripts that center on content in non-computational aspects of one of these fields will also be regarded if the presentation gives consideration to the contents' usefulness in algorithmic processes.

More information about this series at https://link.springer.com/bookseries/3339

Oleg N. Karpenkov

Geometry of Continued Fractions

Second Edition

 Springer

Oleg N. Karpenkov
Department of Mathematical Sciences
The University of Liverpool
Liverpool, UK

ISSN 1431-1550
Algorithms and Computation in Mathematics
ISBN 978-3-662-65279-4   ISBN 978-3-662-65277-0   (eBook)
https://doi.org/10.1007/978-3-662-65277-0

This Springer imprint is published by the registered company Springer-Verlag GmbH, DE part of Springer Nature.
The registered company address is: Heidelberger Platz 3, 14197 Berlin, Germany

Preface to the Second Edition

The idea of the second edition was originally motivated by improvement of certain notation within the chapters and correcting various typos suggested by the readers. However during this work I decided to add several interesting theorems that were missing in the first edition. For the convenience of the readers who are familiar with the first edition I would like to underline here the main changes that were made. My intention was not to overload the book with new topics but rather to improve the exposition of the existing ones.

- First of all in Section 1.3 we relate partial numerators and partial denominators to the classical notion of continuants. We supplement numerous formulae via expressions in terms of continuants further in the text.
- We have added a criterion of rational angles congruence in Subsection 2.1.8 and of integer triangle convergence (Proposition 6.7).
- In the new Section 2.5 and Section 18.6 we show the classification of integer-regular polygons and polyhedra respectively.
- We have included an explicit expression for LLS sequences of adjacent angles in terms of certain long continued fractions (see Section 5.5).
- Two algorithms to compute LLS sequences are added to Chapter 4 (see Section 4.5).

Finally, the chapter on Gauss Reduction Theory (Chapter 7 of the first edition of the book) was a subject of the major metamorphose. It was substantially revised and split into several new chapters:

- Markov numbers are discussed in a separate Chapter 7 now.
- The section on geometry of continued fractions is substantially extended to new Chapter 8. In particular we have added a new technique of computation of LLS sequence periods for $GL(2, \mathbb{Z})$ matrices.
- Chapter 9 on continuant representations of $GL(2, \mathbb{Z})$ matrices is new. It is very much in the spirit of Gauss Reduction Theory.

- The semigroup of reduced matrices is discussed separately in Chapter 10.
- The remaining material (of Chapter 7 of the first edition of the book) is now placed in Chapter 11: here we have added proofs for elliptic and parabolic matrices and revised the main case of the hyperbolic matrices. Additionally we have extended the exposition to the group $GL(2, \mathbb{Z})$ (originally it was mostly regarding $SL(2, \mathbb{Z})$).

Further examples and exercises were added to different chapters of the book.

University of Liverpool   *Oleg Karpenkov*
                                                                  February 2022

Acknowledgements

First, I would like to thank Vladimir Arnold, who introduced the subject of continued fractions to me and who provided me with all necessary remarks and discussions for many years. Second, I am grateful to many people who helped me with remarks and corrections related to particular subjects discussed in this book. Among them are F. Aicardi, T. Garrity, V. Goryunov, I. Pak, E.I. Pavlovskaya, C.M. Series, M. Skopenkov, A.B. Sossinski, A.V. Ustinov, A.M. Vershik, and J. Wallner. Especially I would like to express my gratitude to Thomas Garrity for exhaustively reading through the manuscript and giving some suggestions to improve the book. My special thanks for the amazing support from Martin Peters in particular and from the Publisher Team in general. Finally, I am grateful to my wife, Tanya, who encouraged and inspired me during the years of working on both editions of this book.

The major part of this book was written at the Technische Universität Graz. The work was completed at the University of Liverpool. I am grateful to the Technische Universität Graz for hospitality and excellent working conditions. Work on this book was supported by the Austrian Science Fund (FWF), grant M 1273-N18.

Finally I would like to thank G. R. Gerardo, R. Janssen, S. Kristensen, G. Panti, M. Peters, M. H. Tilijese, J. Wattis, M. van-Son for corrections, comments, and remarks that were implemented in the second edition.

Preface to the First Edition

Continued fractions appear in many different branches of mathematics: the theory of Diophantine approximations, algebraic number theory, coding theory, toric geometry, dynamical systems, ergodic theory, topology, etc. One of the metamathematical explanations of this phenomenon is based on an interesting structure of the set of real numbers endowed with two operations: addition $a + b$ and inversion $1/b$. This structure appeared for the first time in the Euclidean algorithm, which was known several thousand years ago. Similarly to the structures of fields and rings (with operations of addition $a + b$ and multiplication $a * b$), structures with addition and inversion can be found in many branches of mathematics. That is the reason why continued fractions can be encountered far away from number theory. In particular, continued fractions have a geometric interpretation in terms of integer geometry, which we place as a cornerstone for this book.

The main goal of the first part of the book is to explore geometric ideas behind regular continued fractions. On the one hand, we present geometrical interpretation of classical theorems, such as the Gauss—Kuzmin theorem on the distribution of elements of continued fractions, Lagrange's theorem on the periodicity of continued fractions, and the algorithm of Gaussian reduction. On the other hand, we present some recent results related to toric geometry and the first steps of integer trigonometry of lattices. The first part is rather elementary and will be interesting for both students in mathematics and researchers. This part is a result of a series of lecture courses at the Graz University of Technology (Austria). The material is appropriate for master's and doctoral students who already have basic knowledge of linear algebra, algebraic number theory, and measure theory. Several chapters demand certain experience in differential and algebraic geometry. Nevertheless, I believe that it is possible for strong bachelor's students as well to understand this material.

In the second part of the book we study an integer geometric generalization of continued fractions to the multidimensional case. Such a generalization was first considered by F. Klein in 1895. Later, this subject was almost completely abandoned due to the computational complexity of the structure involved in the calculation of the generalized continued fractions. The interest in Klein's generalization was revived by V.I. Arnold approximately one hundred years after its invention, when computers became strong enough to overcome the computational complexity. After a brief introduction to multidimensional integer geometry, we study essentially new questions for the multidimensional cases and questions arising as extensions of the classical ones (such as Lagrange's theorem and Gauss—Kuzmin statistics). This part is an exposition of recent results in this area. We emphasize that the majority of examples and even certain statements of this part are on two-dimensional continued fractions. The situation in higher dimensions is more technical and less studied, and in many cases we formulate the corresponding problems and conjectures. The second part is intended mostly for researchers in the fields of algebraic number theory, Diophantine equations and approximations, and algebraic geometry. Several chapters of this part can be added to a course for master's or doctoral students.

Finally, I should mention many other interesting generalizations of continued fractions, coming from algorithmic, dynamical, and approximation properties of continued fractions. These generalizations are all distinct in higher dimensions. We briefly describe the most famous of them in Chapter 27.

University of Liverpool   *Oleg Karpenkov*
                                                                     February 2013

Contents

Part I
Regular Continued Fractions

In the first part of this book we study geometry of continued fractions in the plane. As usually happens for low dimensions, this case is the richest, and many interesting results from different areas of mathematics show up here. While trying to prove analogous multidimensional statements, several research groups have invented completely different multidimensional analogues of continued fractions (see, e.g., Chapter 27). This is due to the fact that many relations and regularities that are similar in the planar case become completely different in higher dimensions. So we start with the classical planar case.

The first two chapters are preliminary. Nevertheless, we strongly recommend that the reader look through them to get the necessary notation. We start in Chapter 1 with a brief introduction of the notions and definitions of the classical theory of continued fractions. Further, in Chapter 2, we shed light on the geometry of the planar integer lattice. We present this geometry from the classical point of view, where one has a set of objects and a group of congruences acting on the set of objects. This approach leads to natural definitions of various integer invariants, such as integer lengths and integer areas.

In Chapter 3 we present the main geometric construction of continued fractions. It is based on the notion of sails, which are attributes of integer angles in integer geometry.

In the next three chapters we come to the problem of the description of integer angles in terms of integer parameters of their sails. We start in Chapter 4 with a construction of a complete invariant of integer angles (it is written in terms of integer characteristics of the corresponding sails). Further, in Chapter 5, we use these invariants to define trigonometric functions for integer angles in integer geometry. These functions have many nice properties similar to Euclidean trigonometric functions, while the others are totally different. For instance, there is an integer analogue of the formula $\alpha + \beta + \gamma = \pi$ for the angles in a triangle in the Euclidean plane, which we prove in Chapter 6.

Further in Chapter 7 we briefly discuss discrete Markov spectrum and introduce the notion of Markov—Davenport characteristics for matrices.

In Chapter 8 we introduce a notion of geometric continued fractions related to arrangements of pairs of lines passing through the origin in the plane. One can consider these arrangements to be invariant lines of certain real two-dimensional matrices. So geometric continued fractions give rise to powerful invariants of conjugacy classes of matrices. They are especially useful for the study of $GL(2, \mathbb{Z})$ matrices when the geometric continued fractions have a rich structure of automorphisms. We discuss both theoretical and computational aspects in the chapter.

We slightly touch the subject of the structure of the groups $SL(2, \mathbb{Z})$ and $GL(2, \mathbb{Z})$ in Chapter 9. First we discuss the classic representation, when the group $SL(2, \mathbb{Z})$ is generated by two elements. Further we introduce a continuant description of $GL(2, \mathbb{Z})$ matrices and show how to use it in order to write the elements of $SL(2, \mathbb{Z})$ in terms of these two generators.

Further in Chapter 10 we introduce a semigroup of reduced matrices and study its basic properties. In this chapter we collect some preliminary material for Gauss Reduction Theory discussed in the next chapter.

Chapter 11 is dedicated to the study of integer conjugacy classes of $GL(2, \mathbb{Z})$ matrices. Here we separate the following three distinct cases: elliptic (complex spectra) matrices; parabolic (real rational spectra) matrices; and hyperbolic (real irrational spectra) matrices. The last case is a little more advanced. It gives rise to a famous Gauss Reduction Theory, which is could be seen as a certain Euclidean algorithm on the space of $GL(2, \mathbb{Z})$ matrices. We discuss the geometric approach to the Gauss Reduction Algorithm. We conclude this chapter with several statistical etudes.

In Chapter 12 we focus on the classical question related to Lagrange's theorem on the periodicity of continued fractions for quadratic irrationalities, which is based on the structure of the set of solutions of Pell's equations. In this chapter we also discuss a few questions related to Dirichlet groups of matrices, in particular a problem of solving the equation $X^n = A$ in $GL(n, \mathbb{Z})$.

We continue in Chapter 13 with a classical question of Gauss—Kuzmin statistics for the distribution of elements of continued fractions for arbitrary integers. After the introduction of the classical approach via ergodic theory, we present a geometric meaning of the Gauss—Kuzmin statistics in terms of cross ratios in projective geometry.

Regular continued fractions are very good approximations to real numbers. Since there is a vast number of publications on this subject, we do not attempt to cover all interesting problems on continued fractions related to approximations. In Chapter 14 we discuss two questions related to best approximations of real numbers and of arrangements of lines in the plane.

In the last three chapters of the first part we introduce applications of continued fractions to differential and algebraic geometry. In Chapter 15 we introduce an "infinitesimal" version of continued fractions arising as sails of planar curves. We show the relation of these fractions to the motion of a body along this curve satisfying Kepler's second law. Further, in Chapter 16, we give a supplementary definition of extended integer angles, which we use to prove the formula on the sum of integer angles in a triangle. In addition, we introduce a definition of the sum of integer angles, which is a good candidate for the correct addition operation on continued fractions. Finally, in Chapter 17, we describe the global relations on singular points of toric surfaces via integer tangents of the integer angles of polygons associated to these surfaces.

Chapter 1
Classical Notions and Definitions

In this chapter we bring together some basic definitions and facts on continued fractions. After a small introduction we prove a convergence theorem for infinite regular continued fractions. Further, we prove existence and uniqueness of continued fractions for a given number (odd and even continued fractions in the rational case). Finally, we discuss approximation properties of continued fractions. For more details on the classical theory of continued fractions we refer the reader to [116].

1.1 Continued fractions

1.1.1 Definition of a continued fraction

We start with basic definitions in the theory of regular continued fractions.

Definition 1.1. Let $\alpha, a_0, \ldots, a_n$ be real numbers satisfying

$$\alpha = a_0 + \cfrac{1}{a_1 + \cfrac{1}{a_2 + \cfrac{1}{\ddots + \cfrac{1}{a_n}}}} \qquad (1.1)$$

The expression in the right-hand side of the equality is called a *continued fraction expansion of a given number* α (or just a *continued fraction of* α), and denoted by $[a_0; a_1 : \cdots : a_n]$. The numbers $a_0, \ldots, a_n$ are the *elements* of this continued fraction.

Definition 1.2. A continued fraction is *odd* (*even*) if it has an odd (even) number of elements.

© Springer-Verlag GmbH Germany, part of Springer Nature 2022
O. N. Karpenkov, *Geometry of Continued Fractions*,
Algorithms and Computation in Mathematics 26,
https://doi.org/10.1007/978-3-662-65277-0_1

Definition 1.3. An *infinite* continued fraction with an infinite sequence of elements $a_0, a_1, \ldots$ is the following limit if it exists

$$\lim_{k \to \infty} [a_0; a_1 : \cdots : a_k].$$

We denote it by $[a_0; a_1 : \cdots]$.

Definition 1.4. The number $[a_0; a_1 : \cdots : a_{k-1}]$ is called the *k-convergent* (or just *convergent*) to the finite or infinite continued fraction $[a_0; a_1 : \cdots]$.

Definition 1.5. When the first element a_0 of a continued fraction is an integer and all other elements are positive integers, we say that this continued fraction is *regular*.

1.1.2 Regular continued fractions for rational numbers

Let us show how to construct a regular continued fraction for a rational number α.

Algorithm to construct a regular continued fraction.

Input data. Consider a rational number α.

Goal of the algorithm. We are constructing a continued fraction for α. At each step we define a pair of integers (a_{k-1}, r_k).

Step 1. If α is an integer, then $\alpha = [\alpha]$, and the algorithm terminates. Suppose now that α is not an integer. Let us subtract the integer part $\lfloor \alpha \rfloor$ and invert the remainder, i.e.,

$$\alpha = \lfloor \alpha \rfloor + \frac{1}{1/(\alpha - \lfloor \alpha \rfloor)}.$$

Write $a_0 = \lfloor \alpha \rfloor$, and continue with the remaining part $r_1 = 1/(\alpha - a_0)$.

Inductive Step k. Suppose we have completed $k - 1$ steps and get the numbers a_{k-2} and r_{k-1}. Let us find a_{k-1} and r_k:

$$r_{k-1} = \lfloor r_{k-1} \rfloor + \frac{1}{1/(r_{k-1} - \lfloor r_{k-1} \rfloor)}.$$

Then we set $a_{k-1} = \lfloor r_{k-1} \rfloor$ and $r_k = 1/(r_{k-1} - a_{k-1})$.

Output. The algorithm for noninteger numbers stops at Step n when r_n is an integer. We get

$$\alpha = [a_0; a_1 : \cdots : a_{n-1} : a_n],$$

where $a_n = r_n$.

Remark 1.6. Let $r_i = p_i/q_i$ with positive relatively prime integers p_i, q_i for $i \geq 1$. Then for every $k \geq 1$ we have

$$\frac{p_{k+1}}{q_{k+1}} = r_{k+1} = \frac{1}{r_k - \lfloor r_k \rfloor} = \frac{1}{\frac{p_k}{q_k} - \lfloor p_k/q_k \rfloor} = \frac{q_k}{p_k - q_k \lfloor p_k/q_k \rfloor}.$$

Since $r_k > 1$, its denominator is less than its numerator (i.e., $q_{k+1} < p_{k+1} = q_k$). Hence the sequence of denominators q_k decreases with the growth of k. Therefore, the algorithm stops in a finite number of steps.

Example 1.7. Let us study one example:

$$\frac{9}{7} = 1 + \frac{1}{\left(\frac{7}{2}\right)} = 1 + \frac{1}{3 + \frac{1}{2}}.$$

So we get the continued fraction $[1; 3 : 2]$.

1.1.3 Regular continued fractions and the Euclidean algorithm

The term *continued fraction* was used for the first time by J. Wallis in 1695, but the story of continued fractions starts much earlier with the Euclidean algorithm, named after the Greek mathematician Euclid, who described it in his *Elements* (Books VII and X). Actually the algorithm was known before Euclid; it was mentioned in the *Topics* of Aristotle. For further historical details we refer to the book [63] by D.H. Fowler. The goal of the algorithm is *to find the greatest common divisor of a pair of integers*. Let us briefly describe the algorithm.

Euclidean algorithm.

Input data. Consider two positive integers p and q.

Goal of the algorithm. To find the greatest common divisor of p and q (usually it is denoted by $\gcd(p, q)$). We do this in several iterative steps.

Step 1. Set $q_0 = q$. Find integers a_0 and q_1 with $q > r_1 \geq 0$ such that

$$p = a_0 q + q_1.$$

Inductive Step k. Suppose that we have completed $k - 1$ steps and get the integers a_{k-2} and q_{k-1}. Find integers a_{k-1} and q_k where $q_{k-1} > q_k \geq 0$ such that

$$q_{k-2} = a_{k-1} q_{k-1} + q_k.$$

Output. The algorithm stops at Step n when $r_n = 0$. It is clear that

$$\gcd(p, q) = \gcd(q, q_1) = \gcd(q_1, q_2) = \cdots = \gcd(q_{n-2}, q_{n-1}) = q_{n-1}.$$

Remark 1.8. Since the sequence (q_k) is a decreasing sequence of positive integers, the algorithm always stops.

Remark 1.9. The integers a_{k-1} and r_k at each step are uniquely defined. The Euclidean algorithm actually generates the elements of a continued fraction. We always have

$$\frac{p}{q} = [a_0; a_1 : \cdots : a_n].$$

1.1.4 Continued fractions with arbitrary elements

In general one can consider continued fractions with arbitrary real elements. In Chapter 6 and in Chapter 17 we use continued fractions with arbitrary integer elements to describe integer triangles and later to generalize to the case of toric varieties. In Chapter 15 we present some of the geometric ideas behind continued fractions with arbitrary integer and real coefficients. In the rest of the book we shall deal with regular continued fractions.

In the case of continued fractions with arbitrary elements, we could encounter infinity while calculating the value of a continued fraction. For instance, this is the case for the continued fraction $[1; 1 : 0]$. To avoid the annoying consideration of different cases of infinities, we propose to add an element ∞ to the field of real numbers $\mathbb{R}$ and define the following operations:

$$\infty + a = a + \infty = \infty, \quad \frac{1}{0} = \infty, \quad \frac{1}{\infty} = 0.$$

Denote the resulting set by $\overline{\mathbb{R}}$. Notice that we do not add two elements: $+\infty$ and $-\infty$, but only one: ∞. In some sense we are considering the projectivization of the real line.

In this notation, for the continued fraction $[1 : 1; 0]$ we have

$$[1; 1 : 0] = 1 + \frac{1}{1 + \frac{1}{0}} = 1 + \frac{1}{1 + \infty} = 1 + \frac{1}{\infty} = 1 + 0 = 1.$$

1.2 Convergence of infinite regular continued fractions

In this section we show that a sequence of k-convergents of any infinite regular continued fraction converges to some real number. First, we give a formula for numerators and denominators of rational numbers in terms of elements of continued fractions of these numbers. Second, we introduce partial numerators and denominators and prove some relations between them. Finally, we formulate and prove a theorem on convergence of infinite continued fractions.

Let us say a few words about the numerators and denominators of rational numbers and the elements of their continued fractions. It turns out that the numerator and the denominator can be expressed as polynomial functions of the elements of the continued fraction. Namely, there exists a unique pair of polynomials P_k and Q_k in the variables $x_0, \ldots, x_k$ with nonnegative integer coefficients such that

$$\frac{P_k(x_0, \ldots, x_k)}{Q_k(x_0, \ldots, x_k)} = [x_0; x_1 : \cdots : x_k], \quad \text{and} \quad P_k(0, \ldots, 0) + Q_k(0, \ldots, 0) = 1.$$

The first condition defines the polynomials only up to a multiplicative constant, so the second condition is a necessary normalization condition.

Example 1.10. Let us calculate the above polynomials for the continued fractions $[x_0], [x_0; x_1], [x_0; x_1 : x_2], \ldots$. The polynomials are as follows:

$$\begin{array}{ll}
P_0(x_0) = x_0, & Q_0(x_0) = 1; \\
P_1(x_0, x_1) = x_0 x_1 + 1, & Q_1(x_0, x_1) = x_1; \\
P_2(x_0, x_1, x_2) = x_0 x_1 x_2 + x_2 + x_0, & Q_2(x_0, x_1, x_2) = x_1 x_2 + 1;.
\end{array}$$

$\cdots$

Now the numerator and the denominator are described via polynomials P and Q as follows. Consider a finite (or infinite) continued fraction $[a_0; a_1 : \cdots : a_n]$ with $n \geq k$ (or $[a_0; a_1 : \cdots]$ respectively). We set

$$p_k = P_k(a_0, \ldots, a_k) \quad \text{and} \quad q_k = Q_k(a_0, \ldots, a_k).$$

In some sense the numbers p_k and q_k may be considered *partial numerators* and *partial denominators* of rational numbers. Namely (by Proposition 1.13 below) these numbers are the numerators and denominators of k-convergents. If $k = n$, we get the numerator and the denominator of $[a_0; a_1 : \cdots : a_n]$.

Let us prove several interesting facts and relations on the numbers p_k and q_k (and hence that they are true numerators and denominators for $k = n$). In Proposition 1.12 below we prove that the integers p_k and q_k are relatively prime for every k. Then in Proposition 1.13 we prove an important recurrence relation on both numerators and denominators, which we will use in several further chapters. After that we deduce one more useful relation (see Proposition 1.15) and prove Theorem 1.16 on the convergence of infinite continued fractions.

For the next several propositions we use some additional notation:

$$\hat{p}_k = P_{k-1}(a_1, \ldots, a_k) \quad \text{and} \quad \hat{q}_k = Q_{k-1}(a_1, \ldots, a_k).$$

We start with the following lemma.

Lemma 1.11. *The following holds:*

$$\begin{cases} p_k = a_0 \hat{p}_k + \hat{q}_k, \\ q_k = \hat{p}_k. \end{cases}$$

Proof. Since

$$\frac{\hat{p}_k}{\hat{q}_k} = [a_1; a_2 : \cdots : a_k].$$

we get

$$\frac{p_k}{q_k} = a_0 + \frac{1}{\hat{p}_k/\hat{q}_k} = \frac{a_0\hat{p}_k + \hat{q}_k}{\hat{p}_k}.$$

Therefore,

$$\begin{cases} p_k = \lambda(a_0\hat{p}_k + \hat{q}_k), \\ q_k = \lambda\hat{p}_k. \end{cases}$$

Now we rewrite the second condition for polynomials P_k and Q_k:

$$\begin{aligned} 1 &= P_k(0,\ldots,0) + Q_k(0,\ldots,0) \\ &= \lambda(a_0 P_{k-1}(0,\ldots,0) + Q_{k-1}(0,\ldots,0)) + \lambda P_{k-1}(0,\ldots,0). \end{aligned}$$

It is clear that the last expression represents a positive integer that is divisible by λ. Hence 1 is divisible by λ, and therefore $\lambda = 1$.

$\square$

Proposition 1.12. *Let $[a_0; a_1 : \cdots : a_n]$ be a continued fraction with integer elements. Then the corresponding integers p_k and q_k are relatively prime.*

Proof. We prove the statement by induction on k.

It is clear that $p_0 = a_0$ and $q_0 = 1$ are relatively prime.

Suppose that the statement holds for $k-1$. Then $\hat{p}_k$ and $\hat{q}_k$ are relatively prime by the induction assumption. Now the statement holds directly from the equalities of Lemma 1.11.

$\square$

Let us now present a recursive definition of partial numerators and denominators.

Proposition 1.13. *For every integer k we get*

$$\begin{cases} p_k = a_k p_{k-1} + p_{k-2}, \\ q_k = a_k q_{k-1} + q_{k-2}. \end{cases}$$

Proof. We prove the statement by induction on k.

For $k = 2$ the statement holds, since

$$\frac{p_0}{q_0} = \frac{a_0}{1} \quad \text{and} \quad \frac{p_1}{q_1} = \frac{a_0 a_1 + 1}{a_1},$$

and therefore

$$\frac{p_2}{q_2} = \frac{a_2 + a_0 a_1 a_2 + a_0}{a_1 a_2 + 1} = \frac{a_2 p_1 + p_0}{a_2 q_1 + q_0}.$$

Suppose the statement holds for $k - 1$. Let us prove it for k:

$$\frac{p_k}{q_k} = a_0 + \frac{1}{\hat{p}_k/\hat{q}_k} = a_0 + \frac{1}{\frac{a_k\hat{p}_{k-1}+\hat{p}_{k-2}}{a_k\hat{q}_{k-1}+\hat{q}_{k-2}}} = \frac{a_0(a_k\hat{p}_{k-1}+\hat{p}_{k-2})+a_k\hat{q}_{k-1}+\hat{q}_{k-2}}{a_k\hat{p}_{k-1}+\hat{p}_{k-2}}$$

$$= \frac{a_k(a_0\hat{p}_{k-1}+\hat{q}_{k-1})+(a_0\hat{p}_{k-2}+\hat{q}_{k-2})}{a_k\hat{p}_{k-1}+\hat{p}_{k-2}} = \frac{a_kp_{k-1}+p_{k-2}}{a_kq_{k-1}+q_{k-2}}.$$

(The last equality holds by Lemma 1.11.) Therefore, the relations of the system hold. $\qquad\square$

From Proposition 1.13 it follows that the partial numerators and denominators are bounded by the Fibonacci numbers F_i (defined inductively as follows: $F_1 = F_2 = 1$, and $F_n = F_{n-1} + F_{n-2}$).

Corollary 1.14. *For regular continued fractions the following estimates hold:*

$$|p_k| \geq F_k \quad \text{and} \quad q_k \geq F_{k+1}.$$

Proof. We prove the statement by induction on k. Direct calculations show that

$$|p_0| \geq 0, \quad |p_1| \geq 1, \quad \text{and} \quad |p_2| \geq 1,$$
$$|q_0| \geq 1, \quad |q_1| \geq 1, \quad \text{and} \quad |q_2| \geq 2.$$

Suppose that the statement holds for $k-2$ and $k-1$, we prove it for k. Notice that the q_k are all positive and the p_k are either all negative or all nonnegative. We apply Proposition 1.13:

$$|p_k| = a_k|p_{k-1}| + |p_{k-2}| \geq 1 \cdot F_{k-1} + F_{k-2} = F_k$$

and

$$q_k = a_kq_{k-1} + q_{k-2} \geq 1 \cdot F_k + F_{k-1} = F_{k+1}.$$

This concludes the proof. $\qquad\square$

In the next proposition we present another useful expression for partial numerators and denominators.

Proposition 1.15. *For every $k \geq 1$, the following holds:*

$$\frac{p_{k-1}}{q_{k-1}} - \frac{p_k}{q_k} = \frac{(-1)^k}{q_{k-1}q_k}.$$

Proof. Let us multiply both sides by $q_{k-1}q_k$. We get

$$p_{k-1}q_k - p_kq_{k-1} = (-1)^k.$$

We prove this by induction on k.
For $k = 1$ we have

$$p_0q_1 - p_1q_0 = a_1a_0 - (a_1a_0 + 1) = -1.$$

Suppose the statement holds for $k-1$. Let us prove it for k. By Proposition 1.13 we get

$$p_{k-1}q_k - p_kq_{k-1} = p_{k-1}(a_kq_{k-1}+q_{k-2}) - (a_kp_{k-1}+p_{k-2})q_{k-1}$$
$$= -(p_{k-2}q_{k-1} - p_{k-1}q_{k-2}) = (-1)^k.$$

Therefore, the statement holds.   □

Now we are ready to prove the following fundamental theorem.

Theorem 1.16. *The sequence of k-convergents of an arbitrary regular continued fraction converges.*

Notice that the theorem does not always hold for continued fractions with arbitrary real elements.

Proof. Let $[a_0; a_1 : \cdots]$ be an infinite regular continued fraction. From Proposition 1.15 and Corollary 1.14 we have

$$\left| \frac{p_{k-1}}{q_{k-1}} - \frac{p_k}{q_k} \right| \le \frac{1}{F_kF_{k+1}}.$$

Since the sum

$$\sum_{k=1}^{\infty} \frac{1}{F_kF_{k+1}}$$

converges (we leave this statement as an exercise for the reader), the sequence $\left(\frac{p_k}{q_k} \right)$ is a Cauchy sequence. Therefore, $\left(\frac{p_k}{q_k} \right)$ converges.   □

1.3 Continuants

Let us briefly mention a general approach to partial numerators and partial denominators via continuants.

Definition 1.17. Let n be a positive integer. A *continuant* K_n is a polynomial with integer coefficients defined recursively by

$$K_{-1}() = 0;$$
$$K_0() = 1;$$
$$K_1(a_1) = a_1;$$
$$K_n(a_1, a_2, \ldots, a_n) = a_nK_{n-1}(a_1, a_2, \ldots, a_{n-1}) + K_{n-2}(a_1, a_2, \ldots, a_{n-2}).$$

Remark 1.18. Directly by construction of continuants we have

$$P_n(a_0, \ldots, a_n) = K_{n+1}(a_0, a_1, a_2, \ldots, a_n) \quad \text{and} \quad Q_n(a_0, \ldots, a_n) = K_n(a_1, a_2, \ldots, a_n).$$

Therefore,

$$\frac{K_{n+1}(a_0,a_1,\ldots,a_n)}{K_n(a_1,a_2,\ldots,a_n)} = [a_0;a_1 : \cdots : a_n].$$

1.4 Existence and uniqueness of a regular continued fraction for a given real number

In the next theorem we show that for every real number there exists a regular continued fraction representing it. For an irrational number, the corresponding regular continued fraction is unique and infinite. For a rational number there are exactly two continued fractions,

$$[a_0;a_1 : \cdots : a_n] = [a_0;a_1 : \cdots : a_n-1 : 1],$$

with one of them odd, the other even.

Theorem 1.19. (*i*) *For every rational number there exist a unique odd and a unique even regular continued fraction.*
(*ii*) *For every irrational number* α *there exists a unique infinite regular continued fraction whose k-convergents converge to* α.

For instance, $\frac{9}{7} = [1;3:2] = [1;3:1:1]$ and $\pi = [3;7:15:1:292:1:1:1:2: \cdots]$.

Proof. **Existence.** In Section 1.1 we have shown how to construct a regular continued fraction $[a_0;a_1 : \cdots a_n]$ for a rational number α. Notice that if α is rational but not an integer, then $a_n > 1$, and therefore,

$$\alpha = [a_0;a_1 : \cdots : a_n] = [a_0;a_1 : \cdots : a_n-1 : 1].$$

One of these continued fractions is odd and the other is even. For an integer α we always get $\alpha = [\alpha] = [\alpha-1;1]$.

In the case of an irrational number α, the algorithm never terminates, generating the regular continued fraction $\alpha' = [a_0;a_1 : a_2 : \cdots]$ and the sequence of remainders $r_k > 1$ such that

$$\alpha = [a_0;a_1 : \cdots : a_{k-1} : r_k].$$

Let us show that $\alpha = \alpha'$. From Proposition 1.13 for $[a_0;a_1 : \cdots : a_{k-1} : r_k]$ and $[a_0;a_1 : \cdots : a_{k-1} : \cdots]$ we have

$$\alpha = \frac{p_{n-1}r_n + p_{n-2}}{q_{n-1}r_n + q_{n-2}} \quad \text{and} \quad \frac{p_n}{q_n} = \frac{p_{n-1}a_n + p_{n-2}}{q_{n-1}a_n + q_{n-2}}.$$

Using these expressions and the fact that $a_n = \lfloor r_n \rfloor$, we have

$$\left| \alpha - \frac{p_n}{q_n} \right| = \left| \frac{(p_{n-1}q_{n-2} - p_{n-2}q_{n-1})(r_n - a_n)}{(q_{n-1}r_n + q_{n-2})(q_{n-1}a_n + q_{n-2})} \right|$$

$$< \left| \frac{1}{(q_{n-1}r_n + q_{n-2})(q_{n-1}a_n + q_{n-2})} \right|$$

$$< \frac{1}{q_{n-1}q_n} \le \frac{1}{F_n F_{n+1}}.$$

The last inequality follows from Corollary 1.14.

Therefore, the sequence $\left(\frac{p_n}{q_n}\right)$ converges to α, and hence $\alpha = \alpha'$.

Uniqueness. Consider a rational number α. Let us prove the uniqueness of the finite regular continued fraction $\alpha = [a_0; a_1 : \cdots : a_n]$, where $a_n \ne 1$. We prove this by reductio ad absurdum.

Suppose

$$\alpha = [a_0; a_1 : \cdots : a_k : a_{k+1} : \cdots : a_n] = [a_0; a_1 : \cdots : a_k : a'_{k+1} : \cdots : a'_m],$$

where $a_{k+1} \ne a'_{k+1}$. Then we have

$$\alpha = \frac{p_k r_{k+1} + p_{k-1}}{q_k r_{k+1} + q_{k-1}} = \frac{p'_k r'_{k+1} + p'_{k-1}}{q'_k r'_{k+1} + q'_{k-1}} = \frac{p_k r'_{k+1} + p_{k-1}}{q_k r'_{k+1} + q_{k-1}}.$$

Therefore, $r_{k+1} = r'_{k+1}$, and thus $a_{k+1} = \lfloor r_{k+1} \rfloor = \lfloor r'_{k+1} \rfloor = a'_{k+1}$. We arrived at a contradiction.

The proof of uniqueness for continued fractions of irrational numbers is the same as in the case of rational numbers.   □

1.5 Monotone behavior of convergents

In this section we prove two statements on the monotone behavior of convergents. We start with a statement on the sequences of odd and even convergents.

Proposition 1.20. (*i*) *The sequence of even convergents* (p_{2k}/q_{2k}) *is increasing, and the sequence of odd convergents* (p_{2k+1}/q_{2k+1}) *is decreasing.*
(*ii*) *For every real* α *and a nonnegative integer k we have*

$$\frac{p_{2k}}{q_{2k}} \le \alpha \quad and \quad \frac{p_{2k+1}}{q_{2k+1}} \ge \alpha.$$

Equality holds only for the last convergent in case of rational α.

Proof. (*i*). By Proposition 1.13 we have

$$\frac{p_{m-2}}{q_{m-2}} - \frac{p_m}{q_m} = \left(\frac{p_{m-2}}{q_{m-2}} - \frac{p_{m-1}}{q_{m-1}} \right) - \left(\frac{p_{m-1}}{q_{m-1}} - \frac{p_m}{q_m} \right) = \frac{(-1)^{m-1}}{q_{m-1}} \left(\frac{1}{q_{m-2}} - \frac{1}{q_m} \right).$$

Since the sequence of the denominators (q_k) is increasing (by Proposition 1.15), we have that p_{m-2}/q_{m-2} is greater than p_m/q_m for all even m, and it is less for all odd m. This concludes the proof of (i).

(ii). The sequence of even (odd) convergents is increasing (decreasing) and tends to α in the irrational case, and we end up with some $p_n/q_n = \alpha$ in the rational case. This implies the second statement of the proposition. □

In the second statement we show that the larger k is, the better a k-convergent approximates α.

Proposition 1.21. *The sequence of real numbers*

$$\left| \alpha - \frac{p_k}{q_k} \right|, \quad k = 0, 1, 2, \ldots,$$

is strongly decreasing, except for the case of $\alpha = [a_0, 1, 1]$, where this sequence consists of the following three elements: $(1/2, 1/2, 0)$.

Proof. Recall that r_k denotes the reminder $[a_{k+1}; a_{k+2} : \cdots]$.

Let us first prove that

$$|\alpha - \lfloor \alpha \rfloor| > \left| \alpha - \frac{p_1}{q_1} \right|.$$

This inequality is equivalent to

$$\left| a_0 + \frac{1}{a_1 + 1/r_2} - a_0 \right| > \left| a_0 + \frac{1}{a_1 + 1/r_2} - a_0 + \frac{1}{a_1} \right|,$$

which is equivalent to

$$\frac{1}{a_1 + 1/r_2} > \frac{1/r_2}{a_1(a_1 + 1/r_2)},$$

and further to

$$a_1 r_2 > 1.$$

The only case in which $a_1 r_2 \leq 1$ is $a_1 = r_2 = 1$. This is exactly the exceptional case mentioned in the formulation of the theorem.

Now we proceed with the general case of $k \geq 2$. From Proposition 1.13 we have

$$\left| \alpha - \frac{p_{k-1}}{q_{k-1}} \right| = \frac{|p_{k-1}q_{k-2} - q_{k-1}p_{k-2}|}{q_{k-1}(q_{k-1}r_k + q_{k-2})} = \frac{1}{q_{k-1}(q_{k-1}r_k + q_{k-2})}. \tag{1.2}$$

Similarly, we have

$$\left| \alpha - \frac{p_k}{q_k} \right| = \frac{1}{q_k(q_k r_{k+1} + q_{k-1})}. \tag{1.3}$$

Let us estimate the expression on the right side of the last equality:

$$\frac{1}{q_k(q_k r_{k+1} + q_{k-1})} \leq \frac{1}{q_k(q_k + q_{k-1})} = \frac{1}{q_k(q_{k-1}a_k + q_{k-2} + q_{k-1})} \tag{1.4}$$
$$= \frac{1}{q_k(q_{k-1}(a_k+1) + q_{k-2})} \leq \frac{1}{q_k(q_{k-1}r_k + q_{k-2})} < \frac{1}{q_{k-1}(q_{k-1}r_k + q_{k-2})}.$$

In the last inequality we used the fact that $q_k > q_{k-1}$, which follows directly from Proposition 1.13 for $k \geq 2$:

$$q_k = q_{k-1}a_k + q_{k-2} > q_{k-1}.$$

Let us substitute the first and the last expressions of inequality (1.4) by equalities (1.2) and (1.3). We get

$$\left| \alpha - \frac{p_k}{q_k} \right| > \left| \alpha - \frac{p_{k-1}}{q_{k-1}} \right|.$$

Therefore, the sequence is strictly decreasing.   □

1.6 Approximation rates of regular continued fractions

It is interesting to compare advantages and disadvantages of continued fractions with respect to decimal expansions. It is clear that it is very easy algorithmically to add and to multiply two numbers if you know their decimal expansions, while for continued fractions these operations are hard. For instance, there is no algebraic expression for the elements of sums and products via elements of summands or factors. Probably that is the reason why in real life, decimal expansions are widely known and continued fractions are not. On the other hand, quadratic irrationalities have periodic continued fractions, while their decimal expansions are not distinguishable from an arbitrary transcendental number (see Chapter 12). In addition, continued fractions give exact expressions for certain expressions involving exponents and logarithms (see Exercise 1.4 below). They play a key role in Gauss's reduction theory for describing $SL(2,\mathbb{Z})$ matrices (see Chapter 11). It turns out that convergents of continued fractions are good as rational approximations of real numbers. In this section we say a few words about the approximation rates of convergents. Later, in Chapter 14, we study questions related to best approximations of real numbers by rational numbers.

Theorem 1.22. *Consider the inequality*

$$\left| \alpha - \frac{p}{q} \right| < \frac{c}{q^2}. \tag{1.5}$$

Let $c \geq \frac{1}{\sqrt{5}}$. Then for every α, the inequality has an infinite number of integer solutions (p,q).

Let us reformulate the essence of Theorem 1.22.

Lemma 1.23. *Theorem 1.22 is true if for every irrational α, there are infinitely many integer solutions (p,q) of*

$$\left| \alpha - \frac{p}{q} \right| \leq \frac{1}{\sqrt{5}q^2}.$$

Proof. It is clear that for rational $\alpha = p/q$ the integer pairs (np, nq) are solutions. In the irrational case it is enough to prove the theorem for $c = \frac{1}{\sqrt{5}}$. Let us prove that the equality can occur at most twice. Suppose we have

$$\alpha - \frac{p}{q} = \pm \frac{1}{\sqrt{5}q^2} \qquad \text{and} \qquad \alpha - \frac{\hat{p}}{\hat{q}} = \pm \frac{1}{\sqrt{5}\hat{q}^2}$$

for some choice of signs. Then

$$\frac{p}{q} - \frac{\hat{p}}{\hat{q}} = \frac{1}{\sqrt{5}}\left(\pm \frac{1}{\hat{q}^2} - \pm \frac{1}{q^2} \right).$$

Since 1 and $\frac{1}{\sqrt{5}}$ are linearly independent over $\mathbb{Q}$, we have

$$\frac{p}{q} = \frac{\hat{p}}{\hat{q}} \qquad \text{and} \qquad |q| = |\hat{q}|.$$

Hence equality holds at most for two pairs of integers (p,q) and $(-p,-q)$. □

Our next step in the proof of Theorem 1.22 is to give an estimate on the growth of denominators.

Lemma 1.24. *Consider a real number α with regular continued fraction $[a_0; a_1 : \cdots]$. Let p_k/q_k be its k-convergents. Then for an infinite sequence of integers we have*

$$\frac{q_k}{q_{k-1}} \geq \frac{1+\sqrt{5}}{2}.$$

Proof. By Proposition 1.13 we have

$$\frac{q_k}{q_{k-1}} = a_k + \frac{q_{k-2}}{q_{k-1}}.$$

Hence if $a_k \geq 2$, then

$$\frac{q_k}{q_{k-1}} \geq a_k \geq 2 > \frac{1+\sqrt{5}}{2}.$$

Suppose now that $a_k = 1$ and let

$$\frac{q_k}{q_{k-1}} < \frac{1+\sqrt{5}}{2}.$$

Then

$$\frac{q_{k-1}}{q_{k-2}} = \frac{1}{\frac{q_k}{q_{k-1}} - 1} > \frac{1}{\frac{1+\sqrt{5}}{2} - 1} = \frac{1+\sqrt{5}}{2}.$$

Therefore, at least one of every two sequential numbers q_{k-1}/q_{k-2} and q_k/q_{k-1} satisfies the condition of the lemma, and hence there are infinitely many q_k/q_{k-1} satisfying the condition.

$\square$

Proof of Theorem 1.22. From Lemma 1.23 it is enough to show that for every irrational α there exist infinitely many pairs of integers (p, q) satisfying

$$\left| \alpha - \frac{p}{q} \right| \le \frac{1}{\sqrt{5}q^2}.$$

Consider two consecutive convergents. From Proposition 1.15 we have

$$\left| \frac{p_{k-1}}{q_{k-1}} - \frac{p_k}{q_k} \right| = \frac{1}{q_{k-1}q_k}.$$

In addition, from Proposition 1.20 it follows that α is contained in the segment with endpoints p_{k-1}/q_{k-1} and p_k/q_k. Hence if the distance between endpoints is small enough, namely

$$\frac{1}{q_{k-1}q_k} \le \frac{1}{\sqrt{5}q_{k-1}^2} + \frac{1}{\sqrt{5}q_k^2},$$

then either (p_{k-1}, q_{k-1}) or (p_k, q_k) is a solution of the above equation. The last inequality is equivalent to the following:

$$\left(\frac{q_k}{q_{k-1}} \right)^2 - \sqrt{5}\left(\frac{q_k}{q_{k-1}} \right) + 1 > 0.$$

The quadratic polynomial on the left side has two roots: $\frac{\pm 1 + \sqrt{5}}{2}$. Hence the inequality holds when

$$\frac{q_k}{q_{k-1}} \ge \frac{1+\sqrt{5}}{2}.$$

From Lemma 1.24 we have infinitely many k satisfying the last inequality. Therefore, the number of solutions of the above inequality is also infinite. This concludes the proof of Theorem 1.22.

$\square$

It turns out that for certain numbers the estimate of the approximation rate cannot be essentially improved.

Proposition 1.25. *Let α be the golden ratio, i.e.,*

$$\alpha = \frac{1+\sqrt{5}}{2} = [1; 1:1:1:1:\cdots]$$

If $c < \frac{1}{\sqrt{5}}$, then inequality (1.5) has only finitely many solutions.

Denote the golden ratio by θ and its conjugate $(1 - \sqrt{5})/2$ by $\overline{\theta}$.

Proof. First of all, let us show that it is enough to check only all the convergents p_k/q_k. From Theorem 1.16 it follows that best approximations are convergents to a number. Let p/q be a rational number such that $q_k \leq q < q_{k+1}$. Therefore,

$$q^2 \left| \alpha - \frac{p}{q} \right| \geq q^2 \left| \alpha - \frac{p_k}{q_k} \right| \geq q_k^2 \left| \alpha - \frac{p_k}{q_k} \right|.$$

Hence if p/q is a solution of inequality (1.5), then p_k/q_k is a solution of inequality (1.5) as well. Therefore, if there are infinitely many solutions of inequality (1.5) then there are infinitely many convergents to the golden ratio among them.

Second, we prove the statement for the convergents. From Proposition 1.13 it follows that the k-convergent to the golden ratio equals F_{k+1}/F_k. Recall *Binet's formula* for Fibonacci numbers via the golden ratio and its conjugate:

$$F_k = \frac{\theta^k - \overline{\theta}^k}{\sqrt{5}}$$

(for more details see [221]). We have

$$\left| \theta - \frac{p_{k-1}}{q_{k-1}} \right| = \left| \theta - \frac{F_{k+1}}{F_k} \right| = \left| \theta - \frac{\theta^{k+1} - \overline{\theta}^{k+1}}{\theta^k - \overline{\theta}^k} \right| = \left| \frac{\overline{\theta}^k}{\theta^k - \overline{\theta}^k} \right| = \left| \frac{1 - \overline{\theta}^{2k}}{(\theta^k - \overline{\theta}^k)^2} \right|$$

$$= \frac{1}{\sqrt{5}F_k^2} \left| 1 - \overline{\theta}^{2k} \right|.$$

Since $|\overline{\theta}| < 1$, we have $\left| 1 - \overline{\theta}^{2k} \right| = 1 + o(1)$ and hence

$$\left| \theta - \frac{p_{k-1}}{q_{k-1}} \right| = \frac{1}{\sqrt{5}q_{k-1}^2} + o\left(\frac{1}{q_{k-1}^2} \right).$$

This implies the statement for convergents and concludes the proof of Proposition 1.25. $\qquad\qquad\square$

1.7 Exercises

Exercise 1.1. Prove that for every $k \geq 2$ we get

$$\frac{p_{k-2}}{q_{k-2}} - \frac{p_k}{q_k} = \frac{(-1)^{k-1}a_k}{q_k q_{k-2}}.$$

Exercise 1.2. Prove that for every $k \geq 1$ we get

$$\frac{q_k}{q_{k-1}} = [a_k; a_{k-1} : \cdots : a_1].$$

Exercise 1.3. Consider the convergents of a regular continued fraction. Prove that for every $k \geq 0$ we get

$$\frac{1}{q_k q_{k+1}} \geq \left| \alpha - \frac{p_k}{q_k} \right| > \frac{1}{q_k(q_{k+1} + q_k)}.$$

Exercise 1.4. Prove that (a) $\sqrt{2} = [1; (2)]$;
(b) $\exp(1) = [2; 1 : 2 : 1 : 1 : 4 : 1 : 1 : 6 : 1 : 1 : 8 : 1 : 1 : 10 : \cdots]$.

Exercise 1.5. Consider an irrational number α.
(a) Suppose that we know that $\alpha \approx 4,17$. Is it true that $\frac{417}{100}$ is its best approximation?
(b) Find the set of all real numbers for which the rational number $[1; 2 : 3 : 4]$ is one of the best approximations.

Exercise 1.6. Construct an infinite continued fraction with real coefficients that has exactly two limit points: 1 and -1.

Chapter 2
On Integer Geometry

In many questions, the geometric approach gives an intuitive visualization that leads to a better understanding of a problem and sometimes even to its solution. In the next chapters we give an interpretation of the elements of continued fractions in terms of integer geometry, with the continued fractions being associated to certain invariants of integer angles. The geometric viewpoint on continued fractions also gives key ideas for generalizing Gauss—Kuzmin statistics to studying multidimensional Gauss's reduction theory, leading to several results in toric geometry.

The notion of geometry in general can be interpreted in many different ways. In this book we think of a geometry as of a *set of objects* and a *congruence relation*, which is normally defined by some group of transformations. For instance, in Euclidean geometry in the plane, we study points, lines, segments, polygons, circles, etc., with the congruence relation being defined by the group of all length-preserving transformations of the plane (this group is generated by all symmetries about the lines in the plane). The aim of this chapter is to introduce basic ideas of integer geometry. In order to have a general impression we start with the simplest, two-dimensional, case. We will need multidimensional integer geometry only in the second part of the book, so we describe it later, in Chapters 18 and 19.

We start in Section 2.1 with general definitions of integer geometry, and in particular, define integer lengths, distances, areas of triangles, and indexes of angles. Further, in Sections 2.2 and 2.3 we extend the notion of integer area to the case of arbitrary polygons whose vertices have integer coordinates. In Section 2.4 we formulate and prove the famous Pick's formula that shows how to find areas of polygons simply by counting points with integer coordinates contained in them. Further in Section 2.5 we show that there are only 6 distinct elementary integer-regular polygons in the plane. Finally, in Section 2.6 we formulate one theorem in the spirit of Pick's theorem: it is the so-called twelve-point theorem.

© Springer-Verlag GmbH Germany, part of Springer Nature 2022
O. N. Karpenkov, *Geometry of Continued Fractions*,
Algorithms and Computation in Mathematics 26,
https://doi.org/10.1007/978-3-662-65277-0_2

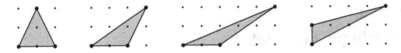

Fig. 2.1 The following triangles are integer congruent.

2.1 Basic notions and definitions

2.1.1 Objects and congruence relation of integer geometry

We consider the integer lattice $\mathbb{Z}^2$ in $\mathbb{R}^2$. The objects of *integer geometry* are *integer points* with integer coordinates, *integer segments and polygons* having all vertices in the integer lattice, *integer lines* passing through pairs of integer points, *integer angles* with vertices at integer points. The congruence relation is defined by the group of affine transformations preserving the integer lattice, i.e. $\mathrm{Aff}(2,\mathbb{Z})$. This group is a semidirect product of $\mathrm{GL}(2,\mathbb{Z})$ and the group of translations on integer vectors. We use $\cong$ to indicate that two objects are integer congruent.

The congruence relation is slightly different from that in the Euclidean case. We illustrate this with integer congruent triangles in Fig. 2.1.

Remark 2.1. To avoid confusion we add prefixes for triangles and angles, writing $\triangle ABC$ and $\angle ABC$ respectively.

2.1.2 Invariants of integer geometry

As long as we have objects and a group of transformations that acts on the objects, we get *invariants* — quantities that are preserved by transformations of the objects. Usually the study of these invariants is the main subject of the corresponding geometry. For instance, in Euclidean geometry, the invariants are lengths of segments, areas of polyhedra, measures of angles, etc.

So *what are the invariants in integer geometry?* There are two different types of invariants in integer geometry: affine invariants and lattice invariants. Affine invariants are intrinsic to affine geometries in general. Affine transformations preserve

– the set of lines;
– the property that a point on a line is between two other points;
– the property that two points are in one half-plane with respect to a line.

Lattice invariants are induced by the group structure of the vectors in the lattice. The majority of lattice invariants are indices of certain subgroups (i.e., sublattices).

2.1.3 Index of sublattices

Recall several notions from group theory. Let H be a subgroup of a group G. Consider an element $g \in G$. The sets

$$gH = \{gh \mid h \in H\} \quad \text{and} \quad Hg = \{hg \mid h \in H\}$$

are called *left* and *right cosets* of H in G.

The *index* of the subgroup H in the group G is the number of left cosets of H in G, denoted by $|G : H|$. (For example, for a positive integer n, we have $|\mathbb{Z} : n\mathbb{Z}| = n$.) Note that within this book we work with abelian groups $\mathbb{Z}^k$, where for every g left cosets of g coincide with right cosets, i.e., $gH = Hg$.

In integer geometry we consider the additive group of integer vectors $\mathbb{Z}^n$, which is called the *integer lattice*. A subgroup of an integer lattice is called an *integer sublattice* of the integer lattice. Actually, any integer sublattice is isomorphic to $\mathbb{Z}^k$ for $k \leq n$, with the sublattice contained in some k-dimensional linear subspace of $\mathbb{R}^n$. The number k is the *dimension* of the sublattice.

Let A be an integer point and G be an integer sublattice. We say that the set of points

$$L = \{A + g \mid g \in G\}$$

is the *integer affine sublattice*. Basically, the group G does not depend on the choice of the point A in L. The invariant definition for G is as follows: the lattice G is the set of all vectors with endpoints in L.

Consider two integer affine sublattices $L_1 \subset L_2$ whose integer sublattices G_1 and G_2 have the same dimension. It is clear that $G_1 \subset G_2$, so the index $|G_2 : G_1|$ is well defined. Since $|G_2 : G_1|$ is an invariant of the pair (G_1, G_2) under the action of $GL(n, \mathbb{Z})$, it is also an invariant of the pair (L_1, L_2) under the action of $Aff(n, \mathbb{Z})$.

One of the simple ways to calculate the index of a two-dimensional sublattice in $\mathbb{Z}^2$ is by counting integer points in a basis parallelogram for the sublattice.

Proposition 2.2. *The index of a sublattice generated by a pair of integer vectors v and w in $\mathbb{Z}^2$ equals the number of all integer points P satisfying*

$$AP = \alpha v + \beta w \quad \text{with} \quad 0 \leq \alpha < 1; \quad 0 \leq \beta < 1,$$

where A is an arbitrary integer point.

Proof. Let H be the subgroup of $\mathbb{Z}^2$ generated by v and w. Define

$$Par(v, w) = \{\alpha v + \beta w \mid 0 \leq \alpha, \beta < 1\}.$$

First, we show that for every integer vector g there exists an integer point $P \in Par(v, w)$ such that $AP \in gH$. The point P is constructed as follows. Let

$$g = \lambda_1 v + \lambda_2 w.$$

Consider

$$P = (\lambda_1 - \lfloor \lambda_1 \rfloor)v + (\lambda_2 - \lfloor \lambda_2 \rfloor)w + A.$$

Since

$$0 \le \lambda_1 - \lfloor \lambda_1 \rfloor, \lambda_2 - \lfloor \lambda_2 \rfloor < 1,$$

the point P is inside the parallelogram and $AP \in gH$.

Second, we prove the uniqueness of P. Suppose that for two integer points $P_1, P_2 \in \mathrm{Par}(v, w)$ we have $AP_1 \in gH$ and $AP_2 \in gH$. Hence the vector $P_1 P_2$ is an element in H. The only element of H of type

$$\alpha v + \beta w \quad \text{with} \quad 0 \le \alpha, \beta < 1$$

is the zero vector. Hence $P_1 = P_2$.

Therefore, the integer points of $\mathrm{Par}(v, w)$ are in one-to-one correspondence to the cosets of H in $\mathbb{Z}^2$.   □

Example 2.3. Consider the points A, B, and C as follows:

Then we have six points satisfying the above condition. Therefore, the index of the sublattice generated by AB and AC is 6.

2.1.4 Integer length of integer segments

Now we are ready to define several integer invariants. We start with integer length.

Definition 2.4. The *integer length* of an integer segment AB is the number of integer points in the interior of AB plus one. We denote it by $1\ell(AB)$.

For example, the integer lengths of the segments AB and CD are both equal to 2.

$$1\ell(AB) = 2, \quad 1\ell(CD) = 2$$

an alternative invariant definition is as follows: the integer length of an integer segment AB is the index of the sublattice generated by the vector $B - A$ in the lattice of integer points in the line containing AB.

In Euclidean geometry the length is a complete invariant of congruence classes of segments. The same is true in lattice geometry.

Proposition 2.5. *Two integer segments are congruent if and only if they have the same integer length.*

Proof. Let AB be an integer segment of length k. Let us prove that it is integer congruent to the integer segment with endpoints $O(0,0)$ and $K(k,0)$. Consider an integer translation sending A to the origin O. Let this translation send B to an integer point $B'(x,y)$.

Since $l\ell(OB') = l\ell(AB) = k$, we have $x = kx'$ and $y = ky'$, where x' and y' are relatively prime. Hence there exists an integer pair (a,b) such that

$$bx' - ay' = 1.$$

Hence the matrix

$$\begin{pmatrix} x' & y' \\ a & b \end{pmatrix}$$

is invertible, and therefore, it is in $SL(2,\mathbb{Z})$. This matrix takes the segment OK to OB', and thus $OK \cong OB' \cong OB$. Hence all the segments of integer length k are congruent to the segment OK. Therefore, they are all integer congruent. $\qquad\square$

2.1.5 Integer distance to integer lines

There is no natural simple definition of orthogonal vectors in integer geometry. Still, it is possible to give a natural definition of an integer distance from a point to a line.

Definition 2.6. Consider integer points A, B, and C that do not lie in one line. An *integer distance* from the point A to the integer segment (line) BC is the index of the sublattice generated by all integer vectors AV, where V is an integer point of the line BC in the integer lattice. We denote it by $\mathrm{ld}(A,BC)$.
The points A, B, and C are collinear, we agree to say that the integer distance from A to BC is zero.

One of the geometric interpretations of integer distance from a point A to BC is as follows. Draw all integer lines parallel to BC. One of them contains the point A (let us call it AA'). The integer distance $\mathrm{ld}(A,BC)$ is the number of lines in the region bounded by the lines AB and AA' plus one. For the example of Fig. 2.2, we have the following:

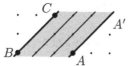

There are two integer lines parallel to AB and lying in the region with boundary lines AB and AA'. Hence, $\mathrm{ld}(A,BC) = 2 + 1 = 3$.

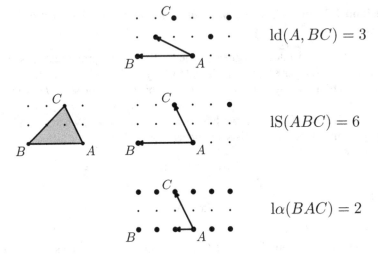

Fig. 2.2 A triangle $\triangle ABC$ and the sublattices for calculation of $\mathrm{ld}(A,BC)$, $\mathrm{lS}(\triangle ABC)$, and $\mathrm{l}\alpha(\angle BAC)$.

2.1.6 Integer area of integer triangles

Let us start with the notion of integer area for integer triangles. We will define the integer areas for polygons later, in the next subsection.

Definition 2.7. The *integer area*, denoted by $\mathrm{lS}(\triangle ABC)$, of an integer triangle $\triangle ABC$ is the index of the sublattice generated by the vectors AB and AC in the integer lattice.
For the points A, B, and C in one line we say that $\mathrm{lS}(\triangle ABC) = 0$.

For instance, the triangle in Fig. 2.2 has integer area equal to 6.

As in Euclidean geometry, integer area does not uniquely determine the congruence class of triangles. Nevertheless, all integer triangles of unit integer area are congruent, since the vectors of the edges of such triangles generate the integer lattice.

2.1.7 Index of rational angles

An angle is called *rational* if its vertex is an integer point and both its edges contain integer points other than the vertex.

Definition 2.8. The *index* of a rational angle $\angle BAC$, denoted by $\mathrm{l}\alpha(\angle BAC)$, is the index of the sublattice generated by all integer vectors of the lines AB and AC in the integer lattice.
In addition, if the points A, B, and C are collinear, we say that $\mathrm{l}\alpha(\angle BAC) = 0$.

Geometrically, the index of an angle $\angle BAC$ is the integer area of the smallest triangle $\triangle AB'C'$ whose edges AB' and AC' generate the sublattices of lines containing the corresponding edges. For instance, the angle $\angle BAC$ in Fig. 2.2 has index equal to 2:

Let us conclude this subsection with a general remark.

Remark 2.9. We write $l\ell$, ld, lS, and $l\alpha$ (starting with the letter "l") to indicate that all these notions are well defined for any lattice, and not just for the integer lattice.

2.1.8 Congruence of rational angles

Let us formulate a simple necessary and sufficient condition of integer congruence of two integer angles.

Proposition 2.10. *Consider four integer points*

$$A_i = (p_i, q_i), \quad and \quad B_i = (r_i, s_i) \quad for\ i = 1, 2.$$

Then the angle $\angle A_1OB_1$ is integer congruent to the angle $\angle A_2OB_2$ if and only if the below matrix is in $\mathrm{GL}(2, \mathbb{Z})$:

$$T = \begin{pmatrix} \dfrac{p_2}{\gcd(p_2,q_2)} & \dfrac{r_2}{\gcd(r_2,s_2)} \\ \dfrac{q_2}{\gcd(p_2,q_2)} & \dfrac{s_2}{\gcd(r_2,s_2)} \end{pmatrix} \cdot \begin{pmatrix} \dfrac{p_1}{\gcd(p_1,q_1)} & \dfrac{r_1}{\gcd(r_1,s_1)} \\ \dfrac{q_1}{\gcd(p_1,q_1)} & \dfrac{s_1}{\gcd(r_1,s_1)} \end{pmatrix}^{-1}.$$

Remark 2.11. If all the coordinates of the points in the above proposition are relatively prime to each other then

$$T = \begin{pmatrix} p_2 & r_2 \\ q_2 & s_2 \end{pmatrix} \cdot \begin{pmatrix} p_1 & r_1 \\ q_1 & s_1 \end{pmatrix}^{-1}.$$

Proof of Proposition 2.10. Denote by A_i' and B_i' the nonzero integer points on the rays OA_i and OB_i closest to O ($i = 1, 2$). Then the matrix T is the transition matrix from the basis (OA_1', OB_1') to the basis (OA_2', OB_2'). Hence the statement of the proposition follows directly from the definition of the integer congruence for the angles. $\qquad\square$

Remark 2.12. To verify that a matrix is in $\mathrm{GL}(2, \mathbb{Z})$ one checks if the determinant of the matrix is equal to ± 1 and that all the elements of the matrix are integers.

Example 2.13. Consider

$$A = (3, -4) \quad \text{and} \quad B = (3, 2).$$

Let us check if the angles $\angle AOB$ and $\angle BOA$ are congruent. We compute

$$T = \begin{pmatrix} 3 & 3 \\ 2 & -4 \end{pmatrix} \cdot \begin{pmatrix} 3 & 3 \\ -4 & 2 \end{pmatrix}^{-1} = \begin{pmatrix} 1 & 0 \\ -2/3 & -1 \end{pmatrix}.$$

Although the value of the determinant of T is -1, one of the elements of T is not an integer. Therefore the angles $\angle AOB$ and $\angle BOA$ are not integer congruent.

2.2 Empty triangles: their integer and Euclidean areas

Let us completely study the case of the following "smallest" triangles.

Definition 2.14. An integer triangle is called *empty* if it does not contain integer points other than its vertices.

We show that empty triangles are exactly the triangles of integer area 1 and Euclidean area $1/2$. In particular, this means that all empty triangles are integer congruent (as we show later, this is not true in the multidimensional case for empty tetrahedra).

Denote by $S(\triangle ABC)$ the Euclidean area of the triangle $\triangle ABC$. Recall that

$$S(\triangle ABC) = \frac{1}{2} \left| \det(AB, AC) \right|.$$

(Here by $\det(v, w)$ we denote the determinant of the (2×2) matrix whose first and second columns contain the coordinates of the vectors v and w respectively.)

Proposition 2.15. *Consider an integer triangle $\triangle ABC$. Then the following statements are equivalent:*
(a) *$\triangle ABC$ is empty;*
(b) *$\mathrm{l}S(\triangle ABC) = 1$;*
(c) *$S(\triangle ABC) = 1/2$.*

Proof. **(a)** $\Rightarrow$ **(b).** Let an integer triangle $\triangle ABC$ be empty. Then by symmetry, a parallelogram with edges AB and AC does not contain integer points except for its vertices as well. Therefore, by Proposition 2.2 there is only one coset for the subgroup generated by AB and AC. Hence, AB and AC generate the integer lattice, and $\mathrm{l}S(\triangle ABC) = 1$.

(b) $\Rightarrow$ **(c).** Let $\mathrm{l}S(\triangle ABC) = 1$. Hence the vectors AB and AC generate the integer lattice. Thus any integer point is an integer combination of them, so

$$(1,0) = \lambda_1 AB + \lambda_2 AC \quad \text{and} \quad (0,1) = \mu_1 AB + \mu_2 AC,$$

with integers $\lambda_1, \lambda_2, \mu_1, \mu_2$. Let also

$$AB = b_1(1,0) + b_2(0,1) \quad \text{and} \quad AC = c_1(1,0) + c_2(0,1)$$

for some integers a_1, a_2, b_1, b_2. Therefore,

$$\begin{pmatrix} b_1 & c_1 \\ b_2 & c_2 \end{pmatrix} \begin{pmatrix} \lambda_1 & \mu_1 \\ \lambda_2 & \mu_2 \end{pmatrix} = \begin{pmatrix} 1 & 0 \\ 0 & 1 \end{pmatrix}.$$

Since these matrices are integer, both their determinants equal either 1 or -1. Hence the Euclidean area of $\triangle ABC$ coincides with the Euclidean area of the triangle with vertices $(0,0)$, $(1,0)$, and $(0,1)$ and equals $1/2$.

(c) $\Rightarrow$ (a). Consider an integer triangle of Euclidean area $1/2$. Suppose that it has an integer point in the interior or on its sides. Then there exists an integer triangle with Euclidean area smaller than $1/2$, which is impossible (since the determinant of two integer vectors is an integer). □

Remark 2.16. Proposition 2.15 implies that all empty integer triangles are congruent.

2.3 Integer area of polygons

In this section we show that every integer polygon is decomposable into empty triangles. Then we define the integer area of polygons to be the number of empty triangles in the decomposition. Additionally, we show that this definition of integer area is well defined and coincides with the definition of the integer area of triangles.

Consider a closed broken line $A_0A_1 \ldots A_{n-1}A_n$ $(A_n = A_0)$ with finitely many vertices and without self-intersections in $\mathbb{R}^2$. By the polygonal Jordan curve theorem, this broken line separates the plane into two regions: one of them is homeomorphic to a disk and the other to an annulus. An *n-gon* (or just a *polygon*) $A_1 \ldots A_n$ is the closure of the region homeomorphic to a disk.

Proposition 2.17. (*i*) *Every integer polygon admits a decomposition into empty triangles.*
(*ii*) *Two decompositions of the same polygon into empty triangles have the same number of empty triangles.*
(*iii*) *The number of empty triangles in any decomposition of a triangle $\triangle ABC$ equals* $\text{lS}(\triangle ABC)$.

Proposition 2.17 introduces a natural extension of the integer area to the case of integer polygons.

Definition 2.18. The *integer area of an integer polygon* is the number of empty triangles in its decomposition into empty triangles.

Proof of Proposition 2.17. (*i*). We prove the statement for integer *n*-gons by induction on *n*.

Base of induction. We start with triangles. To prove the statement for triangles we use another induction on the number of integer points in the closure triangle.

If there are exactly three integer points in the closure of a triangle, then they are the vertices of the triangle (since the triangle is integer). Hence this triangle is empty; the decomposition consists of one empty triangle.

Suppose that every triangle with $k' < k$ integer points ($k \geq 3$) is decomposable into empty triangles. Consider an arbitrary integer triangle $\triangle ABC$ with k integer points. Since $k \geq 3$, there exists an integer point P in the triangle distinct from the vertices. Decompose the triangle into $\triangle ABP$, $\triangle BCP$, and $\triangle CAP$ (excluding triangles of zero area). This decomposition consists of at least two triangles, and each of these triangles has at most $k-1$ integer points. By the induction assumption each of these triangles admits a decomposition into empty triangles. Hence $\triangle ABC$ is decomposable as well.

Therefore, every integer triangle admits a decomposition into integer triangles.

Induction step. Suppose that every integer k'-gon for $k' < k$ is decomposable into empty triangles. Let us find a decomposition for an arbitrary integer k-gon $A_1 \ldots A_k$. If there exists an integer index s such that A_s, A_{s+1}, A_{s+2} are collinear then we just remove one of the points and reduce the number of vertices. Suppose that this is not the case. Consider the ray with vertex at A_1 and containing $A_1 A_2$ and start to rotate it in the direction of the vertex A_3. We stop when we reach A_3 (then $A_s = A_3$) or at the moment when the ray contains some vertex A_s at the same half-plane as A_1 with respect to the line $A_2 A_3$ for the first time. Now we decompose the polygon $A_1 \ldots A_k$ into polygons $A_1 \ldots A_s$ and $A_1 A_s \ldots A_k$. Both of these polygons have fewer than k vertices. By the induction assumption each of these polygons admits a decomposition into empty triangles. Hence $A_1 \ldots A_k$ is decomposable as well.

This concludes the proof of (*i*).

(*ii*). Let a polygon P have two decompositions, one into n_1 empty triangles and the other into n_2 empty triangles. Then by Proposition 2.15 we have

$$\frac{n_1}{2} = S(P) = \frac{n_2}{2},$$

since Euclidean area is additive. Hence, $n_1 = n_2$.

(*iii*). Consider an integer triangle $\triangle ABC$. Define

$$\mathrm{Par}(AB, AC) = \{A + \alpha v + \beta w | 0 \leq \alpha, \beta < 1\}.$$

Denote by $\#(\triangle ABC)$ the number of empty triangles in the decomposition (by (i) and (ii) this is well defined). Let us prove that if $\mathrm{lS}(\triangle ABC) = n$, then $\#(\triangle ABC) = n$, by induction on n.

Base of induction. Suppose that $\mathrm{lS}(\triangle ABC) = 1$. Then by Proposition 2.15 we have $\#(\triangle ABC) = 1$.

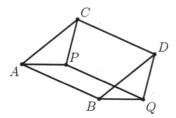

Fig. 2.3 Integer area is additive.

Induction step. Suppose that the statement holds for every $k' < k$, for $k > 1$. Let us prove it for k. Consider a triangle $\triangle ABC$ such that $\mathrm{IS}(\triangle ABC) = k$. By Proposition 2.2, there exists an integer point P in the parallelogram $\mathrm{Par}(AB,AC)$ distinct from the vertices. Without loss of generality, we suppose that P is not on the edge AB. Set $Q = P + AB$ and $D = C + AB$.

The triangle $\triangle BQD$ is obtained from $\triangle APC$ by shifting by the integer vector AB (see Fig. 2.3). Hence the total number of orbits of integer points with respect to the shift on vector multiples of AB in the parallelograms $\mathrm{Par}(AB,AP)$ and $\mathrm{Par}(PQ,PC)$ equals the number of such orbits in the parallelogram $\mathrm{Par}(AB,AC)$. Therefore, from Proposition 2.2, we get

$$\mathrm{IS}(\triangle ABC) = \mathrm{IS}(\triangle APB) + \mathrm{IS}(\triangle CPD).$$

By Proposition 2.15 we have

$$\#(\triangle ABC) = 2S(\triangle ABC) = S(ABCD) = S(ABQP) + S(CPQD)$$
$$= 2S(\triangle ABP) + 2S(\triangle CPD) = \#(\triangle APB) + \#(\triangle CPD).$$

Since P is not on AB, we have $\#(\triangle APB) < k$ and $\#(\triangle CPD) < k$. Therefore, by the inductive hypothesis, we get

$$\mathrm{IS}(\triangle ABC) = \mathrm{IS}(\triangle APB) + \mathrm{IS}(\triangle CPD) = \#(\triangle APB) + \#(\triangle CPD) = \#(\triangle ABC).$$

The proof is completed by induction. □

Corollary 2.19. *The integer area of polygons in the plane is twice the Euclidean area.*

Proof. The statement holds for empty triangles by Proposition 2.15. Now the corollary follows directly from Proposition 2.17 and the definition of integer area. □

Remark 2.20. Corollary 2.19 implies that the index of an integer sublattice generated by vectors AB and AC in the integer lattice equals

$$|\det(AB,AC)|.$$

2.4 Pick's formula

We conclude this section with a nice formula that describes a relation between integer points in a polygon and its Euclidean area.

Theorem 2.21. (Pick's formula.) *The Euclidean area S of an integer polygon satisfies the following relation:*

$$S = I + E/2 - 1,$$

where I is the number of integer points in the interior of the polygon and E is the number of integer points in the edges.

For the integer area one should multiply the right part of the formula by two.

Example 2.22. For instance, for the following pentagon

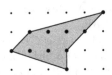

the number of inner integer points equals 4, the number of integer points on the edges equals 6, and the area equals 6. So,

$$6 = 4 + 6/2 - 1.$$

Proof. We prove Pick's formula by the induction on the area.
Base of induction. If the Euclidean area of an integer polygon is $1/2$, then by Proposition 2.15 the polygon is an empty triangle. For the empty triangle we have

$$S = 0 + 3/2 - 1 = 1/2.$$

Induction step. Fix k. Suppose that the statement holds for every integer polygon of area $k'/2$, where $k' < k$. Let us prove the statement for polygons of area $k/2$. Let P be an integer polygon of area $k/2$. Then it can be decomposed into two integer polygons P_1 and P_2 (for instance as in the proof of Proposition 2.17 (i)) intersecting in an integer segment A_iA_j. Suppose that P_i has I_i interior points and E_i boundary points. Let A_iA_j contain $\hat{E}$ integer points. Since the areas of P_1 and P_2 are less than $k/2$, by the induction assumption induction we have

$$S(P) = S(P_1) + S(P_2) = I_1 + E_1/2 - 1 + I_2 + E_2/2 - 1$$
$$= (I_1 + I_2 + \hat{E}) + (E_1 + E_2 - 2\hat{E} - 2)/2 - 1.$$

Since the number of inner integer points of P is $I_1 + I_2 + \hat{E}$ and the number of boundary integer points is $E_1 + E_2 - 2\hat{E} - 2$, Pick's formula holds for P. This concludes the induction step.   □

We have formulated Pick's theorem traditionally in Euclidean geometry. It is clear that this theorem is not true for all lattices. Still, the lattice analogue of the theorem holds for all lattices.

2.5 Integer-regular polygon

Let us first introduce several general notions from the theory of convex polygons. Recall that a *convex polygon* in the plane is the convex hull of a finite set of points that are not contained in one line. A point v of a polygon P is a *vertex* if for every pair of distinct to v points (v_1, v_2) of the polygon P the segment $v_1 v_2$ does not contain v.

Definition 2.23. A *flag* $F = (v, \ell)$ is a collection of a point v and a line ℓ satisfying the condition $v \in \ell$.
We say that a flag $F = (v, \ell)$ is a *flag of a polygon* P if the following three conditions hold:
— the point v is a vertex of P;
— the line ℓ contains another vertex of P distinct to v;
— the polygon P is contained in one of the half-planes with respect to ℓ.

A polygon is *integer* if all its vertices are integer points.

Now we are in position to define integer-regular polygons.

Definition 2.24. An integer polygon is said to be *integer-regular* if for any pair of its flags there exists an integer affine transformation of the plane sending the first flag to the second one.

Finally let us define elementary integer polygons.

Definition 2.25. We say that a point $v = (tx, ty)$ is a *t-dilate* of the point $w = (x, y)$, we write $v = tw$.
An integer polygon P_1 with vertices $(tv_1, \ldots, tv_n)$ is said to be the *t-dilate* of the polygon P_2 with vertices $(v_1, \ldots, v_n)$, denote it by $P_1 = tP_2$.
An integer polygon is *elementary* if it is not the *t*-dilate of an integer polygon for every integer $t > 1$.

Despite the situation in the Euclidean geometry with regular *n*-gons for every integer $n \geq 3$, there are only finitely many elementary integer-regular polyhedra in the integer lattice case. They are all either triangles, or quadrangles, or hexagons. Let us list all of them.

Proposition 2.26. *Any elementary integer-regular polygon is integer congruent to one of the following 6 polygons:*

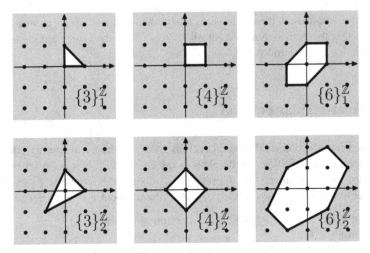

We formally denote them by $\{3\}_1^\mathbb{Z}$, $\{3\}_2^\mathbb{Z}$, $\{4\}_1^\mathbb{Z}$, $\{4\}_2^\mathbb{Z}$, $\{6\}_1^\mathbb{Z}$, and $\{6\}_2^\mathbb{Z}$ following Schläfli notation for the Euclidean case (for more details see Section 18.6).   □

We refer to [100] for the proof of Proposition 2.26.

Later in Section 18.6 we show the classification of the elementary integer-regular polyhedra in higher dimensions.

2.6 The twelve-point theorem

Let us also mention here an interesting theorem coming from the theory of toric varieties (see [66] and [117]; several interesting proofs of this theorem can be found in [181] and [185]).

First we give a definition of a dual polygon.

Definition 2.27. Let $P = A_1 \ldots A_n$ be a convex integer polygon. Suppose that O is the only integer point in the interior of P. Draw the integer vectors $OB_1, \ldots, OB_n$ of unit integer length and parallel to $A_1A_2, \ldots, A_nA_1$ respectively. The polygon $P^* = B_1 \ldots B_n$ is said to be *dual* to P.

The following statement holds.

Theorem 2.28. (The Twelve-Point Theorem.) *Suppose that P is a convex integer polygon containing a single integer point in its interior. Denote by E and E^* the number of integer points in the boundary of P and P^*. Then we have*

$$E + E^* = 12.$$

□

We leave the proof of this theorem as an exercise for the reader.

2.7 Exercises

Exercise 2.1. For any triangle $\triangle ABC$, show that

$$\mathrm{lS}(\triangle ABC) = \mathrm{l}\ell(AB)\,\mathrm{ld}(C,AB).$$

Exercise 2.2. Let ℓ_1, ℓ_2, and ℓ_3 be three parallel integer lines and let $A_i \in l_i$ be integer points. Suppose also that the lines ℓ_1 and ℓ_3 are in different half-planes with respect to the line ℓ_2. Prove that

$$\mathrm{ld}(A_1,\ell_3) = \mathrm{ld}(A_1,\ell_2) + \mathrm{ld}(A_2,\ell_3).$$

Exercise 2.3. Geometric interpretation of integer distance. Let $\mathrm{ld}(O,AB) = k$. Prove that there are exactly $k-1$ integer lines parallel to AB such that the point O and the segment AB are in different half-planes with respect to these lines.

Exercise 2.4. Find an example of two integer noncongruent triangles that have the same integer area.

Exercise 2.5. Consider an empty tetrahedron in $\mathbb{R}^3$ with vertices in $\mathbb{Z}^3$. Suppose that it does not contain other points of the lattice $\mathbb{Z}^3$. Is it true that the edges of the tetrahedron generate the whole lattice $\mathbb{Z}^3$?

Exercise 2.6. For any triangle $\triangle ABC$, show that

$$\mathrm{lS}(\triangle ABC) = \mathrm{l}\ell(AB)\,\mathrm{l}\ell(BC)\,\mathrm{l}\alpha(\angle ABC).$$

Exercise 2.7. Prove that the index of a subgroup generated by v and w in the integer lattice is the absolute value of $\det(v,w)$.

Exercise 2.8. Find a proof of the twelve-point theorem.

Exercise 2.9. Classify all integer-regular polygons (namely, prove Proposition 2.26).

Chapter 3
Geometry of Regular Continued Fractions

Continued fractions play an important role in the geometry of numbers. In this chapter we describe a classical geometric interpretation of regular continued fractions in terms of integer lengths of edges and indices of angles for the boundaries of convex hulls of all integer points inside certain angles. In the next chapter we will extend this construction to construct a complete invariant of integer angles. For the geometry of continued fractions with arbitrary elements see Chapter 15.

3.1 Classical construction

Let us start with the classical geometric construction of continued fractions.

Consider an arbitrary number $\alpha \geq 1$. Denote a ray $\{y = \alpha x \mid x \geq 0\}$ by r_α. This ray divides the first quadrant $\{(x,y) \mid x,y \geq 0\}$ into two angles. Consider one of them and take the convex hull of all the integer points except the origin inside this angle. The boundary of the convex hull is a broken line, which is called the *sail* of the angle. The same construction is applied to the second angle. See Fig. 3.1 for an example when $\alpha = 7/5$.

If α is a rational number, then both sails of the angles consist of finitely many segments and two rays, as in the example of $\alpha = 7/5$. Denote the vertices of the broken line in the sail of the angle containing $(1,0)$ by $A_0, \ldots, A_n$ starting with $A_0 = (1,0)$. In the same way we denote the vertices of the broken line in the sail of the angle containing $(0,1)$ by $B_0, \ldots, B_m$, starting with $B_0 = (0,1)$. We call these broken lines the *principal part* of the sails. As we will show later, $m = n - 1$ or $m = n$.

In the case of an irrational α, each of the sails is a union of one ray and an infinite broken line. Denote the broken line starting from $(1,0)$ by $A_0 A_1 \ldots$, and the broken line starting from $(0,1)$ by $B_0 B_1 \ldots$ respectively, and call them *principal parts* of the sail.

It is interesting to note that in the case of $\alpha = 7/5$, shown in Fig. 3.1, the ratios of the coordinates of extremal points of the convex hulls

© Springer-Verlag GmbH Germany, part of Springer Nature 2022
O. N. Karpenkov, *Geometry of Continued Fractions*,
Algorithms and Computation in Mathematics 26,
https://doi.org/10.1007/978-3-662-65277-0_3

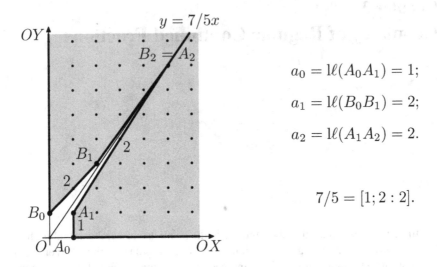

$$a_0 = 1\ell(A_0 A_1) = 1;$$
$$a_1 = 1\ell(B_0 B_1) = 2;$$
$$a_2 = 1\ell(A_1 A_2) = 2.$$

$$7/5 = [1; 2 : 2].$$

Fig. 3.1 The sails for $7/5$.

$$A_1 = (1,1), \quad B_1 = (2,3), \quad B_2 = A_2 = (5,7)$$

are convergents to the number $7/5$. This does not happen by chance for $7/5$; such a situation is typical. In the next theorem we show a general relation between A_i and B_i and the convergents for an arbitrary real number $\alpha \geq 1$ (we discuss the case of $0 \leq \alpha < 1$ later, in Theorem 3.5).

Theorem 3.1. *Consider $\alpha \geq 1$. Let $A_0 A_1 A_2 \ldots$ and $B_0 B_1 B_2 \ldots$ be the principal parts of the corresponding sails (finite or infinite). Then*

$$A_i = (q_{2i-2}, p_{2i-2}) \quad and \quad B_i = (q_{2i-1}, p_{2i-1}), \quad i = 1, 2, \ldots \quad ,$$

where p_k/q_k are convergents. The only exception is for the rational case for the following reason: the last vertices of the principal parts for both sails coincide with (q_n, p_n), where p_n/q_n is the last convergent, i.e., $\alpha = p_n/q_n$.

Remark 3.2. In terms of continuants the expressions of Theorem 3.1 where $\alpha = [a_0; a_1 : \cdots]$ is the regular continued fraction are as follows:

$$A_i = (K_{2i-2}(a_1, \ldots, a_{2i-2}), K_{2i-1}(a_0, a_1 \ldots, a_{2i-2})) \quad \text{and}$$
$$B_i = (K_{2i-1}(a_1, \ldots, a_{2i-1}), K_{2i}(a_0, a_1 \ldots, a_{2i-1})),$$

$i = 1, 2, \ldots$ (with a similar exception for the last value for rational α).

We start the proof with the following lemma.

Lemma 3.3. (a) *The segment with endpoints (q_{2k-2}, p_{2k-2}) and (q_{2k}, p_{2k}) is in the sail $A_0 A_1 A_2 \ldots$.*

(b) *The segment with endpoints* (q_{2k-1}, p_{2k-1}) *and* (q_{2k+1}, p_{2k+1}) *is in the sail* $B_0 B_1 B_2 \ldots$.

Proof. We start with item (a). First, notice that both points (q_{2k-2}, p_{2k-2}) and (q_{2k}, p_{2k}) are in the convex hull of the sail including $A_0 A_1 A_2 \ldots$, since

$$\alpha > \frac{p_{2k-2}}{q_{2k-2}} \quad \text{and} \quad \alpha > \frac{p_{2k}}{q_{2k}}.$$

Consider the line l passing through the points (q_{2k-2}, p_{2k-2}) and (q_{2k}, p_{2k}). From Proposition 1.13 we know that all integer points on l are as follows:

$$(q_{2k-2}, p_{2k-2}) + \lambda(q_{2k-1}, pp_{2k-1}), \quad \lambda \in \mathbb{Z}.$$

Let us prove that the line l is at unit integer distance to the origin. The integer distance is equivalent to the index of the sublattice generated by the vectors (q_{2k-2}, p_{2k-2}) and (q_{2k-1}, p_{2k-1}). From Proposition 1.15 it follows that

$$|p_{2k-2} q_{2k-1} - p_{2k-1} q_{2k-2}| = 1,$$

i.e., the Euclidean area of the corresponding triangle is $1/2$. Hence by Proposition 2.15 these vectors generate the lattice, and $\mathrm{ld}\left((0,0), l\right) = 1$.

Therefore, there is no integer point in the interior of the band between l and the line parallel to l and passing through the origin.

It is clear that the point $(q_{2k-2}, p_{2k-2}) - (q_{2k-1}, p_{2k-1})$ has nonpositive coordinates, and therefore it is not in the cone.

Let us show that the point

$$(q_{2k-2}, p_{2k-2}) + (a_{2k} + 1)(q_{2k-1}, p_{2k-1}) = (q_{2k}, p_{2k}) + (q_{2k-1}, p_{2k-1})$$

is not in the cone. Notice that the point $(q_{2k}, p_{2k}) + (q_{2k-1}, p_{2k-1})$ belongs to the segment with endpoints

$$(q_{2k-1}, p_{2k-1}) \quad \text{and} \quad (q_{2k+1}, p_{2k+1}) = (q_{2k-1}, p_{2k-1}) + a_{2k+1}(q_{2k}, p_{2k})$$

which is contained in the other cone of the quadrant.

Therefore, the segment with endpoints (q_{2k-2}, p_{2k-2}) and (q_{2k}, p_{2k}) is contained in the sail $A_0 A_1 A_2 \ldots$ (see Fig. 3.2).

The proof for the second item is similar. $\qquad \square$

Proof of Theorem 3.1. Actually, Lemma 3.3 almost proves the theorem. We have to check only the endpoints of broken lines.

First, note that $A_1 = (1, \lfloor \alpha \rfloor)$, i.e., $A_1 = (q_0, p_0)$.

Second, (if $\alpha \notin \mathbb{Z}$) we have

$$B_1 = (0, 1) + a_1(1, \lfloor \alpha \rfloor) = (q_1, p_1).$$

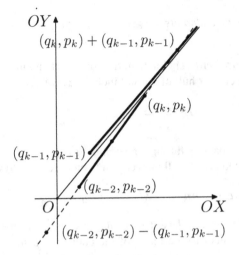

Fig. 3.2 The segment with endpoints (q_{2k-2}, p_{2k-2}) and (q_{2k}, p_{2k}) is in the sail.

Finally, we have to check in the case of rational α whether the segment with endpoints (q_{n-1}, p_{n-1}) and (q_n, p_n) for the last two convergents is in one of the two sails. The point (q_{n-1}, p_{n-1}) is in one of the two sails by Lemma 3.3. The point (q_n, p_n) is in the intersection of the sails. The last thing we have to show is that the triangle with vertices $(0,0)$, (p_{n-1}, q_{n-1}), and (q_n, p_n) is empty. From Proposition 1.15 it follows that the Euclidean area of the triangle is $1/2$; hence by Proposition 2.15 the triangle is empty. Therefore, the segment with endpoints (q_{n-1}, p_{n-1}) and (q_n, p_n) is in the sail.

We have found all the segments of the principal parts of both sails, concluding the proof.   □

Remark 3.4. In the case of rational α with the principle parts of the corresponding angles $A_1 \ldots A_n$ and $B_1 \ldots B_m$, we have the following. If the last element of the odd continued fraction is 1, then $n = m + 1$, otherwise $n = m$.

Let us formulate a similar theorem for the case $0 < \alpha < 1$.

Theorem 3.5. *Consider $0 < \alpha < 1$. Let $A_0 A_1 A_2 \ldots$ and $B_0 B_1 B_2 \ldots$ be the principal parts of the corresponding sails (finite or infinite). Then*

$$A_i = (q_{2i}, p_{2i}) \quad and \quad B_i = (q_{2i-1}, p_{2i-1}), \quad i = 1, 2, \ldots$$

where p_k/q_k are convergents. The only exception is when α is rational, in which case the last vertex of the principal parts for both sails coincide with (q_n, p_n), where p_n/q_n is the last convergent, i.e. $\alpha = p_n/q_n$.

Proof. The proof repeats the proof of Theorem 3.1 except for the following difference. Since $a_0 = 0$, the point A_0 should coincide with A_1. This is the explanation for the shift in indices.   □

Remark 3.6. Similarly to Theorem 3.1 we have the following continuant expression for Theorem 3.5:

$$A_i = (K_{2i}(a_1,\dots,a_{2i}), K_{2i+1}(0,a_1\dots,a_{2i})) \quad \text{and}$$
$$B_i = (K_{2i-1}(a_1,\dots,a_{2i-1}), K_{2i}(0,a_1\dots,a_{2i-1})),$$

$i = 1,2,\dots$, where $\alpha = [0;a_1 : \cdots]$ is the regular continued fraction (with a similar exception for the last value for rational α).

3.2 Geometric interpretation of the elements of continued fractions

Let us first formulate a corollary to Theorem 3.1.

Corollary 3.7. *Consider $\alpha \geq 1$. Let $A_0A_1A_2\dots$ and $B_0B_1B_2\dots$ be the principal parts of the corresponding sails (finite or infinite). Then*

$$1\ell(A_iA_{i+1}) = a_{2i} \quad \text{and} \quad 1\ell(B_iB_{i+1}) = a_{2i+1}, \quad i = 0,1,2,\dots \quad ,$$

where $\alpha = [a_0; a_1 : a_2 : \cdots]$. The only exception occurs when α is rational, in which case the last edges of the principal parts for both sails coincide with (q_n, p_n), where p_n/q_n is the last convergent, i.e., $\alpha = p_n/q_n$. The integer length of this edge is 1.

In the case of $0 < \alpha < 1$, the corollary also holds, the only difference being that $1\ell(A_iA_{i+1}) = a_{2i+2}$ instead a_{2i}.

Proof. The corollary follows directly from the explicit formulas for A_k and B_k of Theorem 3.1, after applying Proposition 1.13. □

Theorem 3.1 and Corollary 3.7 lead to an interesting algorithm for constructing sails for regular continued fractions.

Algorithm to construct sails of real numbers.

Input data. Given a regular continued fraction $\alpha = [a_0; a_1 : a_2 : \cdots]$ with $a_0 > 0$. Set $O = (0,0)$.

Goal of the algorithm. To construct the sails for α.

Step 1. Assign $A_0 = (1,0)$, $B_0 = (0,1)$ and construct

$$A_1 = A_0 + a_0OB_0 \quad \text{and further} \quad B_1 = B_0 + a_1OA_1.$$

Inductive Step k. Suppose that we have already constructed $A_0\dots A_k$ and $B_0\dots B_k$. Then put

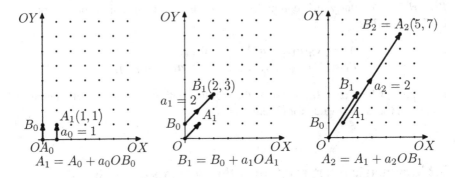

Fig. 3.3 Construction of the continued fraction for $7/5$.

$$A_{k+1} = A_k + a_{2k}OB_k \quad \text{and} \quad B_{k+1} = B_k + a_{2k+1}OA_{k+1}.$$

Output. If α is rational, then the algorithm constructs both sails in finite time. If α is irrational, then the algorithm also calculates both sails but in infinitely many steps.

We leave the case $0 < \alpha < 1$ as an easy exercise for the reader.

In Fig. 3.3 we show the steps of the algorithm for the particular case $\alpha = [1; 2 : 2]$.

3.3 Index of an angle, duality of sails

In this subsection we show that there exists a duality between edges and angles. An important consequence of this duality is that all the elements of the regular continued fraction can be read from one of the sails. We restrict ourselves to the case $\alpha \geq 1$. Reduction to the case $0 < \alpha < 1$ is straightforward, so we omit it.

Theorem 3.8. *Consider $\alpha \geq 1$. Let $A_0A_1A_2\ldots$ and $B_0B_1B_2\ldots$ be the principal parts of the corresponding sails (finite or infinite). Then*

$$l\alpha(\angle A_iA_{i+1}A_{i+2}) = a_{2i+1} \quad \text{and} \quad l\alpha(\angle B_iB_{i+1}B_{i+2}) = a_{2i+2}$$

for all admissible indices, where $\alpha = [a_0; a_1 : a_2 : \cdots]$.

Proof. Let us calculate the index of an integer angle at some vertex of the principal part of one of the two sails. By Theorem 3.1 it is equivalent to calculate the index of the angle between a pair of vectors (q_{i-1}, p_{i-1}) and (q_{i+1}, p_{i+1}):

$$l\alpha\left(\angle(q_{i-1}, p_{i-1})(0,0)(q_{i+1}, p_{i+1})\right) = |p_{i-1}q_{i+1} - p_{i+1}q_{i-1}| =$$
$$|p_{i-1}(a_{i+1}q_i + q_{i-1}) - (a_{i+1}p_i + p_{i-1})q_{i-1}| = a_{i+1}|p_{i-1}q_i - p_iq_{i-1}| = a_{i+1}.$$

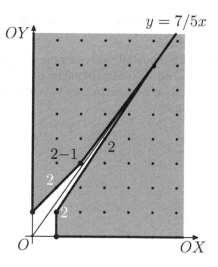

Fig. 3.4 The edge–angle duality for $7/5$.

The second equality follows from Proposition 1.13, while the last follows from Proposition 1.15. □

Edge–angle duality. From Corollary 3.7 and Theorem 3.8, we get that

$$l\alpha(\angle A_i A_{i+1} A_{i+2}) = l\ell(B_i B_{i+1}) \quad \text{and} \quad l\alpha(\angle B_i B_{i+1} B_{i+2}) = l\ell(A_{i+1} A_{i+2}).$$

The only exception is the very last angle (for rational slopes) where $l\alpha$ is smaller than the corresponding $l\ell$ by one.

For the example of $\alpha = 7/5$ (see Fig. 3.1 and 3.4), we get

$$l\alpha(\angle A_0 A_1 A_2) = l\ell(B_0 B_1) = a_1 = 2;$$
$$l\alpha(\angle B_0 B_1 B_2) = l\ell(A_1 A_2) - 1 = a_2 - 1 = 1.$$

Remark 3.9. We return to edge-angle duality a little later in the slightly different setting of LLS sequences (see Proposition 5.17).

3.4 Exercises

Exercise 3.1. Prove that every sail is homeomorphic to $\mathbb{R}^1$.

Exercise 3.2. Describe the edge–angle duality for the case $0 < \alpha < 1$.

Exercise 3.3. Construct the sails of continued fractions for rational numbers $3/2$, $10/7$, $18/13$, and $13/18$.

Exercise 3.4. Find geometrically all the convergents to the rational numbers $8/5$, $9/4$, and $17/8$.

Exercise 3.5. Is it sufficient to know all indices for the angles in the principal parts of the sail defined by $\alpha > 1$ and the indices of angles of its dual sail to reconstruct α in a unique way? The question is interesting for both of the following cases: if α is rational and if α is irrational.

Chapter 4
Complete Invariant of Integer Angles

In this chapter we generalize the classical geometric interpretation of regular continued fractions presented in Chapter 3 to the case of arbitrary integer angles, constructing a certain integer broken line called the *sail* of an angle. We combine the integer invariants of a sail into a sequence of positive integers called an *LLS sequence*. From one side, the notion of LLS sequence extends the notion of continued fraction (see Remark 4.8), about which we will say more in the next chapter. From another side, LLS sequences distinguish the integer angles. Sails and LLS sequences of angles play a central role in the geometry of numbers. In particular, we use LLS sequences in integer trigonometry and its relations to toric singularities and in Gauss's reduction theory. F. Klein generalized the notion of sail to the multidimensional case to study integer solutions of homogenous decomposable forms. We will study this generalization in the second part of this book.

In Section 4.1 we give a preliminary definition of an integer sine. Further, in Section 4.2, we define sails of integer angles and corresponding LLS sequences. Further, in Section 4.3 we prove that an LLS sequence is a complete invariant of an integer angle. We discuss angles sharing the same edge in Section 4.4. We conclude in Section 4.5 demonstration of two ways to compute the LLS sequence for a given angle.

4.1 Integer sines of rational angles

Recall the following general definitions. An angle is *integer* if its vertex is an integer point. If the integer angle $\angle ABC$ has integer points distinct from B on both edges AB and BC, we call it *rational*.

Let us introduce the notion of the integer sine function for rational angles. We put in the definition an integer analogue of the sine formula of the Euclidean case. In this section we consider an integer angle $\angle ABC$ with integer vertex B.

© Springer-Verlag GmbH Germany, part of Springer Nature 2022
O. N. Karpenkov, *Geometry of Continued Fractions*,
Algorithms and Computation in Mathematics 26,
https://doi.org/10.1007/978-3-662-65277-0_4

Definition 4.1. Let $\angle ABC$ be a rational angle, where A and C are lattice points distinct from B. The *integer sine* of the angle is the following number:

$$\frac{\text{lS}(\triangle ABC)}{\text{l}\ell(AB)\,\text{l}\ell(BC)},$$

denoted by $\text{lsin}\,\angle ABC$.

Remark 4.2. Notice that the integer sine is well defined, i.e., it does not depend on the choice of points A and C. We leave the reader to check this as an exercise.

Let us briefly compare sin and lsin. One difference is in the multiplicative constant $1/2$. This is due to the fact that an empty triangle has integer area 1 but Euclidean area $1/2$, as shown in Proposition 2.15. Yet the difference between sin and lsin is much stronger, as illustrated in the following proposition.

Proposition 4.3. *The integer sine of a rational angle coincides with the index of the angle.*

In particular, this implies that the integer sine takes all possible nonnegative integer values.

Proof. Consider a rational angle $\angle ABC$ with A, B, C not contained on one line. Let A' and C' be the nearest integer points to B in the open rays BA and BC, respectively. Then from the definition of integer length, we get

$$\text{l}\ell(BA') = \text{l}\ell(BC') = 1.$$

Hence,
$$\text{lsin}(\angle A'BC') = \text{lS}(\triangle A'BC').$$

The integer area of $\triangle A'BC'$ is the index of the sublattice generated by BA' and BC' in $\mathbb{Z}^2$. Since the vectors BA' and BC' generate all integer points of the lines AB and BC, the integer area of $\triangle A'BC'$ is equivalent to the index of the angle. Therefore, we get
$$\text{lsin}(\angle ABC) = \text{l}\alpha(\triangle ABC).$$

If A, B, and C are on one line, then

$$\text{lsin}(\angle ABC) = \text{l}\alpha(\triangle ABC) = 0.$$

This concludes the proof. $\square$

4.2 Sails for arbitrary angles and their LLS sequences

In Chapter 3 we studied sails for a certain subset of integer angles. Let us extend the notion of sails to the case of arbitrary integer angles.

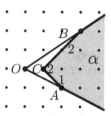

Fig. 4.1 An angle and its sail.

Definition 4.4. Let α be an arbitrary integer angle. Consider the convex hull of the set of all integer points in α except the origin. The boundary of this convex hull is called the *sail* of this angle.

Remark 4.5. Notice that the sail of a lattice angle is a broken line homeomorphic to $\mathbb{R}$. A sail is either a two-sided infinite broken line or a one-sided infinite broken line including also a ray at one side or a finite broken line that includes two rays at both sides. These rays appear when edges of angles contain integer points other than the vertex of the corresponding angle.

Let us now define a very important characteristic of the sail that we will use in the future.

Definition 4.6. Consider an arbitrary angle α with integer vertex. Let the sail for this angle be a broken line with sequence of vertices (A_i). Define:

$$a_{2k} = \mathrm{l}\ell A_k A_{k+1},$$
$$a_{2k-1} = \mathrm{lsin}(\angle A_{k-1} A_k A_{k+1})$$

for all admissible indices. The *lattice length sine sequence* (*LLS sequence* for short) for the sail is the sequence (a_n). Denote it by $\mathrm{LLS}(\alpha)$

The LLS sequence can be either finite or infinite on one or both sides.

Example 4.7. Consider an example of the integer angle α centered at the origin O with edges passing through points $A = (2,-1)$ and $B = (3,2)$ respectively. Denote by C the point $(1,0)$. Then the broken line ABC is the sail for α (see Fig. 4.1). Now we have

$$\mathrm{l}\ell(AC) = 1, \qquad \mathrm{lsin}(\angle ACB) = 2, \quad \text{and} \quad \mathrm{l}\ell(CB) = 2.$$

Hence

$$\mathrm{LLS}(\alpha) = (1,2,2).$$

Remark 4.8. Notice that LLS sequences contain only positive integer elements. Consider a finite to the left LLS sequence. Regular continued fractions define a one-to-one correspondence between such LLS sequences and real numbers not smaller than 1; namely, the sequence $(a_0, a_1, \ldots)$ corresponds to the real number $[a_0; a_1 : \cdots]$. This continued fraction is an important invariant of integer angles (we will call it the *integer tangent*), and we will study this invariant in the next chapter.

4.3 On complete invariants of angles with integer vertex

Proposition 4.9. *Integer length, integer area, index and integer sine of a rational angle are invariant under the action of the group of integer affine transformations* $\mathrm{Aff}(2,\mathbb{Z})$.

Proof. Every integer linear transformation sends subgroups to subgroups and the corresponding cosets to the corresponding cosets; hence all integer indices are invariant under integer linear transformations. Thus, every integer affine transformation preserves indexes of the corresponding integer lattice subgroups in $\mathbb{Z}^2$ as well. Therefore, the integer area and the index (and hence the integer sine) of a rational angle are also preserved. The integer length is preserved, since all the inner integer points map to inner integer points.   □

Corollary 4.10. *The LLS sequence is an invariant of lattice angles with respect to* $\mathrm{Aff}(2,\mathbb{Z})$.

Proof. First, note that convex hulls are preserved by the elements of $\mathrm{Aff}(2,\mathbb{Z})$ (we leave this as an easy exercise for the reader). Then the statement follows directly from the fact that the integer length and the index are invariants of $\mathrm{Aff}(2,\mathbb{Z})$.   □

It is interesting to note that the angles $\angle ABC$ and $\angle CBA$ are not necessarily integer congruent (i.e., there is no integer affine transformation taking one to another). For example, if we consider

$$A = (3,-2), \qquad B = (0,0), \quad \text{and} \quad C = (2,1),$$

then $\angle ABC$ and $\angle CBA$ are not integer congruent. We will say more about such pairs of angles in the next section.

Theorem 4.11. *Two angles with vertices at integer points are lattice congruent if and only if they have the same LLS sequences (here we assume that the subsequences of lengths and angles of one LLS sequence respectively coincide with the subsequences of lengths and angles of the other).*

Proof. From Corollary 4.10 we know that the LLS sequence is an integer invariant of angles. It remains to prove that if two LLS sequences coincide, then the corresponding angles are integer congruent.

Consider two angles α and β with sails (A_i) and (B_i). Let the corresponding LLS sequences coincide and be equivalent to (a_i). Let the vertices of the angles α and β be O_α and O_β respectively. Consider an affine transformation taking O_β to O_α, B_0 to A_0, and B_1 to A_1. This affine transformation is integer, since $\mathrm{l}\ell(A_0A_1) = \mathrm{l}\ell(B_0B_1) = a_0$ and $\mathrm{ld}(O_\alpha,A_0A_1) = \mathrm{ld}(O_\beta,B_0B_1)$. Suppose the angle β is taken to some angle γ with sail (C_i) (we already know that the vertex of γ is O_α, $C_0 = A_0$, and $C_1 = A_1$).

Let us prove that the broken lines $A_0A_1A_2\ldots$ and $C_0C_1C_2\ldots$ coincide by induction. Suppose $A_0A_1\ldots A_{k-1}$ coincides with $C_0C_1\ldots C_{k-1}$. Let us prove that $A_k = C_k$.

First, we know that

$$\text{lsin}(\angle A_{k-2}A_{k-1}A_k) = \text{lsin}(\angle C_{k-2}C_{k-1}C_k) = a_{2k-3}$$

and

$$\text{l}\ell(A_{k-1}A_k) = \text{l}\ell(C_{k-1}C_k) = a_{2k-2}.$$

Hence we have

$$\text{ld}(A_k, A_{k-2}A_{k-1}) = \text{ld}(C_k, C_{k-2}C_{k-1}) = a_{2k-2}a_{2k-3}.$$

This follows from the simple fact (see Exercise 4.6) that

$$\text{ld}(A, BC) = \frac{\text{lS}(\angle ABC)}{\text{l}\ell(BC)}.$$

Notice that $\angle A_{k-2}A_{k-1}A_k$ and $\angle C_{k-2}C_{k-1}C_k$ are two parts of convex sails, and hence the points A_k and C_k are on the other side of the point O_α with respect to the line $A_{k-2}A_{k-1} = C_{k-2}C_{k-1}$. Hence A_k and C_k are both on a line l_1 parallel to the line $A_{k-2}A_{k-1}$ containing points at integer distance $a_{2k-2}a_{2k-3}$ from the line $A_{k-2}A_{k-1}$.

Second, we have

$$\text{ld}(A_k, O_\alpha A_{k-1}) = \text{lS}(\angle O_\alpha A_{k-1}A_k) = a_{2k-1} = \text{lS}(\angle O_\alpha C_{k-1}C_k) = \text{ld}(C_k, O_\alpha C_{k-1}),$$

where $O_\alpha A_{k-1} = O_\alpha C_{k-1}$. Again, by convexity we know that the points A_k and C_k are in a different half space from the point $A_{k-2} = C_{k-2}$ with respect to the line $O_\alpha A_{k-1}$. Therefore, C_k and A_k are on a line l_2 parallel to $O_\alpha A_{k-1}$.

Since $O_\alpha A_{k-1}$ and $A_{k-2}A_{k-1}$ are not parallel, the intersection of lines l_1 and l_2 is a point coinciding with both A_k and C_k.

Therefore, the broken lines $A_0A_1A_2\ldots$ and $C_0C_1C_2\ldots$ coincide.

The fact that $\ldots A_{-1}A_0A_1$ and $\ldots C_{-1}C_0C_1$ coincide follows from the considered case after performing a GL$(2,\mathbb{Z})$ transformation taking A_0 to A_1 and A_1 to A_0, namely the transformation

$$\begin{pmatrix} 1 & 0 \\ a_0 & -1 \end{pmatrix}.$$

We apply this transformation to both sails (A_i) and (C_i) and get that the images of C_k and A_k coincide for every negative k. Therefore, the whole sails (A_i) and (B_i) coincide. $\qquad\square$

Theorem 4.12. *For every sequence of positive integers (odd finite or infinite on one or both sides) there exists an angle with vertex at the origin whose LLS sequence is this sequence.*

Proof. First we consider the case of a sequence $a_0, a_1, a_2, \ldots$ (odd finite or infinite to the right). Let $A = (1,0)$, $B = (0,0)$, and $C = (1, \alpha)$, where $[a_0; a_1 : a_2 : \cdots]$ is the continued fraction for α. The angle $\angle ABC$ has the desired LLS sequence. This follows directly from Corollary 3.7 and Theorem 3.8.

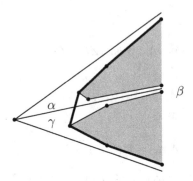

Fig. 4.2 The sails for α and β coincide starting from some vertex.

When the sequence is infinite to the left, we construct the angle $\angle ABC$ for the inverse sequence. Then the angle $\angle CBA$ is the angle with the prescribed LLS sequence.

The remaining case is the case of both-side infinite sequences. Here we construct the angle for the part with nonnegative indices. Then apply to this angle the transformation we used in the proof of Theorem 4.11 and construct the angle for the remaining part of the sail and get the angle with the prescribed LLS sequence.   □

4.4 Equivalent tails of the angles sharing an edge

Let us prove the following remarkable proposition (see also in [3], pp. 24–25).

Proposition 4.13. *Let* $\alpha_1 > \alpha_2 > \alpha_3$ *be distinct real numbers and let* α_1 *be irrational. Consider two angles* α *and* β *defined by pairs of rays in the lines* $(y = \alpha_1 x, y = \alpha_2 x)$ *and* $(y = \alpha_1 x, y = \alpha_3 x)$, *where* $x > 0$. *Then the LLS sequences of these two angles coincide from some element (up to a sequence shift).*

Proof. Denote by γ the angle defined by rays $(y = \alpha_2 x, y = \alpha_3 x)$, $x > 0$. It is clear that every integer point of the angle β is either in α or in γ. Then the convex hull of all integer points of β is the convex hull of the union of the convex hulls of all integer points for α and γ. Hence the sail for β is contained in the sails for α and γ except for one edge, which represents the common support line to the convex hulls for α and β (see Fig. 4.2) Therefore, the sails of α and β coincide starting from some vertex. Hence, the LLS sequences coincide as well.   □

From duality reasons we immediately have the following statement.

Corollary 4.14. *Let* $\alpha_2 > \alpha_1 > \alpha_3$ *be distinct real numbers and let* α_2 *be irrational. Consider two angles* α *and* β *defined by pairs of rays in the lines* $(y = \alpha_1 x, y = \alpha_2 x)$

and $(y = \alpha_1 x, y = \alpha_3 x)$, where $x > 0$. Then the LLS sequences of these two angles coincide from some element (up to a sequence shift). □

4.5 Two algorithms to compute the LLS sequence of an angle

In this section we describe two algorithms to write LLS sequences for a given rational angle. The first one requires a brute force study of rather small amount of cases. For the second method we provide explicit formulae for the LLS sequence in terms of coordinates of the integer vectors defining a given rational angle.

4.5.1 Brute force algorithm

Let us start with an algorithm that includes some brute force study of a curtain number of cases.

Brute force algorithm to find the LLS sequence for an angle

Input data. We are given an integer angle ABC. In particular we have the coordinates of the two integer nonzero vectors

$$BA = (p, q), \quad \text{and} \quad CA = (r, s).$$

Goal of the algorithm. To construct the LLS sequence of the angle $\angle ABC$.

Step 1. First of all we compute primitive integer vectors on the edges $BA' = (p', q')$ and $BC' = (r', s')$ where

$$(p', q') = \frac{(p, q)}{\gcd(p, q)} \quad \text{and} \quad (r', s') = \frac{(r, s)}{\gcd(r, s)}.$$

Step 2. Secondly we compute integer sine of the angle by the formula

$$\operatorname{lsin} \angle ABC = \operatorname{lsin} \angle A'BC' = \left| \det \begin{pmatrix} p' & r' \\ q' & s' \end{pmatrix} \right|.$$

(here we take the absolute value of the determinant). In the case where integer sine is 0 we have a degenerate angle (either 0 or π) and we do not construct an LLS sequence for it.

Step 3. Now let the integer sine be nonzero. Then

$$\angle ABC \cong \text{larctan} \, \frac{\text{lsin} \angle ABC}{k},$$

where k is one of the integer number satisfying

$$\gcd(k, \text{lsin} \angle ABC) = 1, \quad \text{and} \quad 1 \leq k \leq \text{lsin} \angle ABC.$$

In this step we use brute force method to find the value of k by comparing $A'BC'$ with

$$\text{larctan} \, \frac{\text{lsin} \angle ABC}{k}$$

for all admissible k. Here note that larctan $\frac{\text{lsin} \angle ABC}{k}$ is defined by the vectors

$$v_1 = (1,0), \quad \text{and} \quad v_2 = (\text{lsin} \angle ABC, k).$$

There will be precisely one value of k that will give a positive test.

Output. The sequence of elements of the odd regular continued fraction for the rational number

$$\frac{\text{lsin} \angle ABC}{k}$$

is the LLS sequence for the angle $\angle ABC$.

Example 4.15. Consider the angle $\alpha = \angle AOB$ with

$$A = (10, 4) \quad \text{and} \quad B = (12, 21).$$

Let us compute its LLS sequence.

First of all, we find the primitive vectors parallel to OA and OB. They are

$$OA' = (5,2) \quad \text{and} \quad OB' = (4,7).$$

Then we have

$$\text{lsin} \, \alpha = \left| \det \begin{pmatrix} 5 & 4 \\ 2 & 7 \end{pmatrix} \right| = 27.$$

Now we need to check all values of $0 < k \leq 27$ that are not divisible by 3.

For $k = 1$: the vectors defining larctan$(27/1)$ are $(1,0)$ and $(1,27)$ respectively. Hence the transition matrix between the angles is as follows:

$$\begin{pmatrix} 1 & 1 \\ 0 & 27 \end{pmatrix} \cdot \begin{pmatrix} 5 & 4 \\ 2 & 7 \end{pmatrix}^{-1} = \begin{pmatrix} 5/27 & 1/27 \\ -2 & 5 \end{pmatrix}.$$

It is clear that this matrix is not in $\text{GL}(2, \mathbb{Z})$, and hence by Proposition 2.10 the angles are not integer congruent. The same happens for all admissible k except for $k = 17$. (We omit computations for the cases $k \neq 17$.) If $k = 17$ then we compare with larctan$(27/17)$. The transition matrix is as follows:

$$\begin{pmatrix} 1 & 17 \\ 0 & 27 \end{pmatrix} \cdot \begin{pmatrix} 5 & 4 \\ 2 & 7 \end{pmatrix}^{-1} = \begin{pmatrix} -1 & 3 \\ -2 & 5 \end{pmatrix} \in GL(2, \mathbb{Z}).$$

Hence by Proposition 2.10

$$\alpha \cong \text{larctan} \frac{27}{17}.$$

Note that the odd regular continued fraction for $27/17$ is

$$\frac{27}{17} = [1; 1 : 1 : 2 : 3].$$

Therefore the LLS sequence for α is

$$(1, 1, 1, 2, 3).$$

4.5.2 Explicite formulae for LLS sequences via given coordinates of the angle

It turns out that it is possible to write down an explicit expression for the elements of the LLS sequence for angles due to the following remarkable formula.

Theorem 4.16. *Consider two linearly independent integer vectors*

$$OA = (p, q) \quad and \quad OB = (r, s)$$

such that none of them are proportional either to $(1, 0)$ or to $(0, 1)$. Further let two sequences of integers

$$(a_0, a_1, \ldots, a_{2m}) \quad and \quad (b_0, a_1, \ldots, b_{2n})$$

be defined as the sequences of elements of odd regular continued fractions of rational numbers

- $|q/p|$ *and* $|s/r|$ *if* $\det(OA, OB) \cdot \text{sign} \frac{p}{q} < 0$;
- $|p/q|$ *and* $|r/s|$ *if* $\det(OA, OB) \cdot \text{sign} \frac{p}{q} > 0$.

Set

$$\varepsilon = -\text{sign} \frac{p}{q} \quad and \quad \delta = \text{sign} \frac{r}{s}.$$

Denote

$$\alpha = [\varepsilon a_{2m} : \varepsilon a_{2m-1} : \cdots : \varepsilon a_1 : \varepsilon a_0 : 0 : \delta b_0 : \delta b_1 : \cdots : \delta b_{2n}]$$

and let

$$|\alpha| = [c_0; c_1 : \cdots : c_{2k}]$$

be the regular odd continued fraction for $|\alpha|$. Finally we set

- $S = (c_0, c_1, \ldots, c_{2k})$ *in case* $c_0 \neq 0$;
- $S = (c_2, \ldots, c_{2k})$ *in case* $c_0 = 0$.

Then S is the LLS sequence for the angle $\angle AOB$.

Remark 4.17. Currently we do not have all necessary tools for the proof of Theorem 4.16. We do it later in Subsection 15.2.4, when the LLS-sequence for broken lines is introduced.

Remark 4.18. (**A shorter formula that allows one to write integer tangents for the angles with irrational value of** s/t.) One can simplify the computation of the continued fraction for

$$\alpha = [\varepsilon a_{2m} : \varepsilon a_{2m-1} : \cdots : \varepsilon a_1 : \varepsilon a_0 : 0 : \delta b_0 : \delta b_1 : \cdots : \delta b_{2n}].$$

Here

$$\alpha = [\varepsilon a_{2m} : \varepsilon a_{2m-1} : \cdots : \varepsilon a_1 : \varepsilon a_0 : 0 : \delta \xi]$$

where

$$\xi = \begin{cases} q/p, & \text{if } \det(OA, OB) \cdot \text{sign} \frac{p}{q} < 0; \\ p/q, & \text{otherwise.} \end{cases}$$

It is interesting to note that the last expression gives an option to find integer tangent of the corresponding angle even if ξ is irrational, and the corresponding LLS sequence is infinite.

Remark 4.19. For the completeness of the observation let us consider the cases of vectors $(1,0)$ and $(0,1)$.

Example 4.20. As above in Example 4.15 we consider the angle $\alpha = \angle AOB$ with

$$A = (10, 4) \quad \text{and} \quad B = (12, 21).$$

Let us compute its LLS sequence using the techniques suggested by Theorem 4.16.
Note first that

$$\frac{10}{4} \cdot \det \begin{pmatrix} 10 & 12 \\ 4 & 21 \end{pmatrix} = \frac{10}{4} \cdot 162 = 405 > 0,$$

hence we consider $10/4$ and $12/21$ respectively. We have

$$\varepsilon = -\text{sign} \frac{10}{4} = -1 \quad \text{and} \quad \delta = \text{sign} \frac{12}{21} = 1.$$

Further we have

$$\frac{10}{4} = \frac{5}{2} = [1, 1, 2], \quad \text{and} \quad \frac{12}{21} = \frac{2}{7} = [0; 1 : 1 : 2 : 1].$$

So the expression for the long continued fraction is as follows:

$$[-1; -1 : -2 : 0 : 0 : 1 : 1 : 2 : 1] = -\frac{27}{17}.$$

Let us now write the odd continued fraction for $|-27/17|$:

$$\left|-\frac{27}{17}\right| = [1; 1 : 1 : 2 : 3].$$

Since the first element of the continued fraction is not equal to zero ($1 \neq 0$) the LLS sequence for α is

$$(1, 1, 1, 2, 3).$$

4.6 Exercises

Exercise 4.1. Show that the integer sine of a rational angle $\angle ABC$ does not depend on the choice of integer points B and C on the edges.

Exercise 4.2. Prove that convex sets are taken to convex sets and their boundaries to boundaries under affine transformations.

Exercise 4.3. Let $A = (3, -2)$, $B = (0,0)$, and $C = (2,1)$. Prove that $\angle ABC$ and $\angle CBA$ are not integer congruent.

Exercise 4.4. Prove that if the principal parts of the sails of two angles are integer congruent, then the angles themselves are integer congruent.

Exercise 4.5. Let d be a positive integer and l an integer line. Prove that all lattice points lying at integer distance are contained in two lines parallel to l at the same Euclidean distance from l.

Exercise 4.6. Prove that

$$\mathrm{ld}(A, BC) = \frac{\mathrm{IS}(\angle ABC)}{\mathrm{l}\ell(BC)}.$$

Exercise 4.7. Compute the LLS sequence for the angle AOB where O is the origin and the points A and B are as follows:
(i) $A = (5,7)$ and $B = (11, -6)$;
(ii) $A = (1,0)$ and $B = (4, -7)$;
(iii) $A = (8,3)$ and $B = (11, -1)$.

Chapter 5
Integer Trigonometry for Integer Angles

In this chapter we explain how to interpret regular continued fractions related to LLS sequences in terms of integer trigonometric functions. Integer trigonometry has many similarities to Euclidean trigonometry (for instance, integer arctangents coincide with real arctangents; the formulas for adjacent angles are similar). From another point of view they are totally different, since integer sines and cosines are positive integers; there are two right angles in integer trigonometry, etc. In this chapter we discuss basic properties of integer trigonometry. In Chapter 6 we use integer trigonometric functions to describe angles of integer triangles, which will further result in global relations for toric singularities on toric surfaces (see Chapter 17).

For rational angles we introduce definitions of integer sines, cosines, and tangents. In addition to rational integer angles, there are three types of *irrational integer angles*. If an integer angle $\angle ABC$ has an integer point distinct from the vertex B in AB but not in BC (in BC but not in AB), we call the angle *R-irrational* (or respectively *L-irrational*). In case both edges of an angle do not contain lattice points other than B, the angle is called *LR-irrational*. It is only for R-irrational angles that we have a definition of integer tangents. The trigonometric functions are not defined for L-irrational and LR-irrational angles. For more information on integer trigonometry we refer to [102] and [104].

5.1 Definition of trigonometric functions

We start with the definition of the integer tangent for rational and R-irrational angles.

Definition 5.1. Consider a rational or R-irrational angle $\angle ABC$. Let A, B, and C be noncollinear. Suppose the LLS sequence of $\angle ABC$ is $(a_0, a_1, a_2, \ldots)$ (finite or infinite). Then the *integer tangent* of the angle $\angle ABC$ equals

$$\operatorname{ltan}(\angle ABC) = [a_0; a_1 : a_2 : \cdots].$$

© Springer-Verlag GmbH Germany, part of Springer Nature 2022
O. N. Karpenkov, *Geometry of Continued Fractions*,
Algorithms and Computation in Mathematics 26,
https://doi.org/10.1007/978-3-662-65277-0_5

When the points A, B, and C are collinear (but the points A and C are distinct from B) we say that $\text{ltan}(\angle ABC) = 0$.

Consider a rational angle $\angle ABC$ and recall that

$$\text{lsin}(\angle ABC) = \frac{\text{lS}(\triangle ABC)}{\text{l}\ell(AB)\,\text{l}\ell(BC)}.$$

Definition 5.2. For a rational angle α we define

$$\text{lcos}\,\alpha = \frac{\text{lsin}\,\alpha}{\text{ltan}\,\alpha}.$$

It is clear that the integer sine, integer tangent, and therefore integer cosine are invariants of $\text{Aff}(2,\mathbb{Z})$.

Now we give the inverse function to the integer tangent.

Definition 5.3. Consider a real $\alpha \geq 1$ or $\alpha = 0$. The *integer arctangent* of α is the angle with vertex at the origin and edges

$$\{y = 0 \mid x \geq 0\} \qquad \text{and} \qquad \{y = \alpha x \mid x \geq 0\}.$$

We define the zero angle as $\text{larctan}\,0$. The angle π is the angle ABC, where $A = (1,0)$, $B = (0,0)$, and $C = (-1,0)$.

5.2 Basic properties of integer trigonometry

First we show that the integer tangent and arctangent are in fact inverse to each other.

Proposition 5.4. (*i*) *For every real $s \geq 1$, we have* $\text{ltan}(\text{larctan}\,s) = s$.
(*ii*) *For every rational or R-irrational angle α, the following holds:*

$$\text{larctan}(\text{ltan}\,\alpha) \cong \alpha.$$

Proof. (*i*) From Corollary 3.7 and Theorem 3.8 we have that the elements of the LLS sequence for $\text{larctan}\,s$ coincide with the elements of the regular continued fraction for s. Hence the statement holds by definition of the integer tangent.

(*ii*) Both angles $\text{larctan}(\text{ltan}\,\alpha)$ and α have the same LLS sequences. Therefore, they are congruent by Theorem 4.11.   □

In the following proposition we collect several trigonometric properties.

Proposition 5.5. (*i*) *The values of integer trigonometric functions for integer congruent angles coincide.*
(*ii*) *The lattice sine and cosine of any rational angle are relatively prime positive*

integers.

(*iii*) *For every angle* α *the following inequalities hold:*

$$\text{lsin}\,\alpha \geq \text{lcos}\,\alpha \quad and \quad \text{ltan}\,\alpha \geq 1.$$

Equality holds if and only if the lattice vectors of the angle rays generate the whole lattice.

(*iv*) (**Description of lattice angles.**) *Two integer angles* α *and* β *are congruent if and only if* $\text{ltan}\,\alpha = \text{ltan}\,\beta$.

Proof. (*i*) This is a direct corollary of the fact that integer sine and tangent are defined only by values of certain indices.

(*ii*) By Proposition 5.4 it is enough to prove the assertion for angles larctan α for rational $\alpha \geq 1$.

Consider $\frac{m}{n} \geq 1$, where m and n are relatively prime integers. Then the sail of the angle will contain the point (n,m). It is also clear that the lattice distance between the point (n,m) and the line $y=0$ is m, and hence $\text{lsin}\left(\text{larctan}\,\frac{m}{n}\right) = m$.

Since $\text{ltan}\left(\text{larctan}\,\frac{m}{n}\right) = \frac{m}{n}$ and $\text{lsin}\left(\text{larctan}\,\frac{m}{n}\right) = m$, we have $\text{lcos}\left(\text{larctan}\,\frac{m}{n}\right) = n$.

(*iii*) This is true for all integer arctangent angles. Therefore, it is true for all angles.

(*iv*) The LLS sequence is uniquely defined by the integer tangent. Therefore, the statement follows from Theorem 4.11.   □

5.3 Transpose integer angles

Let us give a definition of the integer angle transpose to a given one.

Definition 5.6. The integer angle $\angle BOA$ is said to be *transpose* to the integer angle $\angle AOB$. We denote it by $(\angle AOB)^t$.

It immediately follows from the definition that for any integer angle α, we have

$$(\alpha^t)^t \cong \alpha.$$

Further, we will use the following notion. Suppose that some arbitrary integers a, b, and c, where $c \geq 1$, satisfy $ab \equiv 1 (\text{mod}\, c)$. Then we write

$$a \equiv \left(b(\text{mod}\, c)\right)^{-1}.$$

For the trigonometric functions of transpose integer angles the following relations hold.

Theorem 5.7. Trigonometric relations for transpose angles. *Let an integer angle* α *be not contained in a line. Then*

(1) If $\alpha \cong \text{larctan}(1)$, then $\alpha^t \cong \text{larctan}(1)$.
(2) If $\alpha \ncong \text{larctan}(1)$, then

$$\text{lsin}(\alpha^t) = \text{lsin}\,\alpha, \quad \text{lcos}(\alpha^t) \equiv \left(\text{lcos}\,\alpha \,(\text{mod}\,\text{lsin}\,\alpha)\right)^{-1}.$$

Example 5.8. Let us check the statement of Theorem 5.7 for the simple case of

$$\alpha = \text{larctan}\,\frac{7}{5}$$

Since $10/7 = [1;2:2]$, the LLS sequence of α is $(1,2,2)$. Hence the LLS sequence of the transpose angle is $(2,2,1)$. Now

$$\alpha^t \cong \text{larctan}([2;2:1]) = \text{larctan}\,\frac{7}{3}.$$

In particular, $3 \cdot 5 - 1 = 14$ is divisible by $\text{lsin}\,\alpha = 7$, and hence we have

$$\text{lsin}(\alpha^t) = 7 = \text{lsin}\,\alpha; \quad \text{lcos}(\alpha^t) = 3 \equiv \left(5(\text{mod}\,7)\right)^{-1} = \left(\text{lcos}\,\alpha\,(\text{mod}\,\text{lsin}\,\alpha)\right)^{-1}.$$

(See Fig. 5.1.)

Proof. Consider an integer angle α. Let $\text{ltan}\,\alpha = p/q$, where $\gcd(p,q) = 1$. By Proposition 5.4(ii), we have $\alpha \cong \text{larctan}(p/q)$. The case $p/q = 1$ is trivial. Consider the case $p/q > 1$.

Let $A = (1,0)$, $B = (q,p)$, and $O = (0,0)$. Suppose that an integer point $C = (q',p')$ of the sail of the angle $\text{larctan}(p/q)$ is the closest integer point to the endpoint B. Both coordinates of C are positive integers, since $p/q > 1$. Since the triangle $\triangle BOC$ is empty and the orientation of the pair of vectors (OB, OC) does not coincide with the orientation of the pair of vectors (OA, OB), we have

$$\det\begin{pmatrix} q & q' \\ p & p' \end{pmatrix} = -1.$$

Consider a linear transformation ξ of the two-dimensional plane,

$$\xi = \begin{pmatrix} p - p' & q' - q \\ p & -q \end{pmatrix}.$$

Since $\det(\xi) = -1$, the transformation ξ is integer-linear and changes the orientation. Direct calculations show that the transformation ξ takes the angle $\text{larctan}'(p/q)$ to the angle $\text{larctan}(p/(p-p'))$. (See the example in Fig. 5.1.)

Since $\gcd(p,p') = 1$ and $p > p-p'$, the following holds:

$$\begin{cases} \text{lsin}(\alpha^t) = p, \\ \text{lcos}(\alpha^t) = p - p'. \end{cases}$$

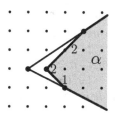

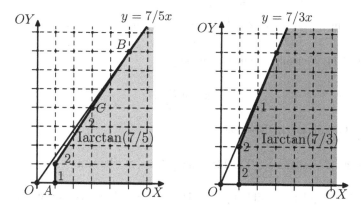

Fig. 5.1 If α is integer-congruent to larctan$(7/5)$, then α^t is integer-congruent to larctan$(7/3)$.

Since $qp' - pq' = -1$, we have $qp' \equiv -1 \pmod{p}$. Therefore,

$$\text{lcos}\,\alpha\,\text{lcos}(\alpha^t) = q(p - p') \equiv 1 \pmod{p}.$$

From that we have

$$\begin{cases} \text{lsin}(\alpha^t) = \text{lsin}\,\alpha, \\ \text{lcos}(\alpha^t) \equiv \left(\text{lcos}\,\alpha\,(\text{mod}\,\text{lsin}\,\alpha)\right)^{-1}. \end{cases}$$

This concludes the proof of Theorem 5.7. $\qquad\qquad\qquad\square$

5.4 Adjacent integer angles

Now we define an integer angle adjacent to a given one.

Definition 5.9. An integer angle $\angle BOA'$ is said to be *adjacent* to an integer angle $\angle AOB$ if the points A, O, and A' are contained in the same straight line. We denote the angle $\angle BOA'$ by $\pi - \angle AOB$.

For the trigonometric functions of adjacent integer angles the following relations hold.

Theorem 5.10. Trigonometric relations for adjacent angles. *Let α be some integer angle. Then one of the following holds:*
(1) If α is the zero angle, then $\pi - \alpha \cong \pi$.
(2) If α is the straight angle, then $\pi - \alpha \cong 0$.
(3) If $\alpha \cong \text{larctan}(1)$, then $\pi - \alpha \cong \text{larctan}(1)$.
(4) If α is neither zero nor straight nor integer-congruent to $\text{larctan}(1)$, then

$$\pi - \alpha \cong \text{larctan}' \left(\frac{\text{ltan}\,\alpha}{\text{ltan}(\alpha) - 1} \right),$$
$$\text{lsin}(\pi - \alpha) = \text{lsin}\,\alpha, \quad \text{lcos}(\pi - \alpha) \equiv (-\text{lcos}\,\alpha(\text{mod}\,\text{lsin}\,\alpha))^{-1}.$$

Example 5.11. Consider the integer angle

$$\text{larctan}\,\frac{225}{157}.$$

Then for the adjacent angle we have

$$\text{lsin}\left(\pi - \text{larctan}\,\frac{225}{157}\right) = 225 \quad \text{and} \quad \text{lcos}\left(\pi - \text{larctan}\,\frac{225}{157}\right) = 182,$$

since $157 \cdot 182 + 1$ is divisible by 225. This is precisely in accordance with Theorem 5.10. Hence

$$\pi - \text{larctan}\,\frac{225}{157} \cong \text{larctan}\,\frac{225}{182}.$$

Remark 5.12. Suppose that an integer angle α is neither zero nor straight. Then the conditions

$$\begin{cases} \text{lcos}(\pi - \alpha) \equiv (-\text{lcos}\,\alpha(\text{mod}\,\text{lsin}\,\alpha))^{-1}, \\ 0 < \text{lcos}(\pi - \alpha) \leq \text{lsin}\,\alpha, \end{cases}$$

uniquely determine the value $\text{lcos}(\pi - \alpha)$.

Proof of Theorem 5.10. Consider an integer angle α. Directly from the definitions it follows that if $\alpha \cong 0$, then $\pi - \alpha \cong \pi$, and, if $\alpha \cong \pi$, then $\pi - \alpha \cong 0$.

Suppose that $\text{ltan}\,\alpha = p/q > 0$, where $\gcd(p, q) = 1$. Then by Proposition 5.4(ii) we have $\alpha \cong \text{larctan}(p/q)$. Therefore,

$$\pi - \alpha \cong \pi - \text{larctan}(p/q).$$

It follows immediately that if $p/q = 1$, then $\pi - \alpha \cong \text{larctan}(1)$.

Now let $\alpha \ncong \text{larctan}(1)$, and hence $p/q > 1$. Consider a linear transformation ξ_1 of the two-dimensional plane,

$$\xi_1 = \begin{pmatrix} -1 & 1 \\ 0 & 1 \end{pmatrix}.$$

Since $\det(\xi_1) = -1$, the transformation ξ_1 is integer-linear and changes the orientation. Direct calculations show that the transformation ξ_1 takes the cone cor-

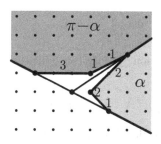

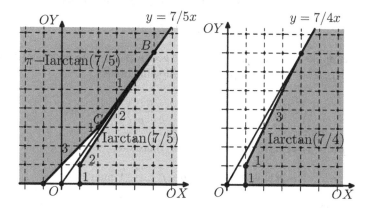

Fig. 5.2 If α is integer-congruent to larctan$(7/5)$, then $\pi-\alpha$ is integer-congruent to larctan$(7/4)$.

responding to the angle $\pi-\text{larctan}(p/q)$ to the cone corresponding to the angle larctan$'(p/(p-q))$. Since ξ_1 changes the orientation, we have to transpose larctan$'(p/(p-q))$. (See the example in Fig. 5.2). Therefore,

$$\pi-\alpha \cong \text{larctan}'\left(\frac{\text{ltan}\,\alpha}{\text{ltan}(\alpha)-1}\right).$$

Now we show that

$$\begin{cases} \text{lsin}(\pi-\alpha) = \text{lsin}\,\alpha, \\ \text{lcos}(\pi-\alpha) \equiv \left(-\text{lcos}\,\alpha\,(\text{mod}\,\text{lsin}\,\alpha)\right)^{-1}. \end{cases}$$

Let $A=(1,0)$, $B=(q,p)$, and $O=(0,0)$. Consider the sail of the angle

$$\pi-\text{larctan}(p/q),$$

which is integer-congruent to $\pi-\alpha$. Suppose that an integer point $C=(q',p')$ of the sail for $\pi-\text{larctan}(p/q)$ is the closest integer point to the endpoint of the sail B (or equivalently, the segment BC is in the sail and has the unit integer length). The coordinate p' is positive, since $p/q > 1$. Since the triangle $\triangle BOC$ is empty and the vectors OB and OC define the same orientation as OA and OB,

$$\det \begin{pmatrix} q & q' \\ p & p' \end{pmatrix} = 1.$$

Consider a linear transformation ξ_2 of the two-dimensional plane,

$$\xi_2 = \begin{pmatrix} p' - p & q - q' \\ -p & q \end{pmatrix}.$$

Since $\det(\xi_2) = 1$, the transformation ξ_2 is integer-linear and orientation-preserving. Direct calculations show that the transformation ξ_2 takes the integer angle

$$\pi - \text{larctan}(p/q)$$

to the integer angle $\text{larctan}(p/(p-p'))$. Since $\gcd(p, p') = 1$, we have $\text{lsin}(\pi-\alpha) = p$. Since $\gcd(p, p') = 1$ and $p > p-p'$, the following holds:

$$\text{lcos}(\pi-\alpha) = \text{lcos}\left(\text{larctan}\left(\frac{p}{p-p'}\right)\right) = p - p'.$$

Since $qp' - pq' = 1$, we have $qp' \equiv 1 \pmod{p}$. Therefore,

$$\text{lcos}\,\alpha\,\text{lcos}(\pi-\alpha) = q(p-p') \equiv -1 \pmod{p}.$$

From that we have

$$\begin{cases} \text{lsin}(\pi-\alpha) = \text{lsin}\,\alpha, \\ \text{lcos}(\pi-\alpha) \equiv \left(-\text{lcos}\,\alpha \,(\text{mod}\,\text{lsin}\,\alpha)\right)^{-1}. \end{cases}$$

This concludes the proof of Theorem 5.10.   □

Remark 5.13. The statements of Theorem 5.7 and Theorem 5.10 were first introduced in terms of regular continued fractions and so-called *zig-zags* by P. Popescu-Pampu [182].

The following statement is an easy corollary of Theorem 5.10.

Corollary 5.14. *For every integer angle α, the following holds:*

$$\pi - (\pi - \alpha) \cong \alpha.$$

                                                                              □

5.5 LLS sequences for adjacent angles

The formula for the LLS sequences of adjacent angles is a little more complicated than the formula for the LLS sequences of transpose angles. In this section we discuss it.

Let us define a flip on the set of all positive integer sequences of odd length.

Definition 5.15. Let s be a sequence of an odd number of positive integers. We say that $s^\perp$ is an *adjacency dual* sequence to s, where $s^\perp$ is defined as follows

- **Finite sequences:**

$$(1)^\perp = (1);$$
$$(2)^\perp = (2);$$
$$(k)^\perp = (1, k-2, 1);$$
$$(1, k-2, 1)^\perp = (k);$$
$$(1, a_1, a_2, \ldots, a_m, 1)^\perp = (a_m+1, a_{m-1}, \ldots, a_3, a_2, a_1+1);$$
$$(b, a_1, a_2, \ldots, a_n, 1)^\perp = (a_n+1, a_{n-1}, \ldots, a_2, a_1, b-1, 1);$$
$$(1, a_1, a_2, \ldots, a_n, c)^\perp = (1, c-1, a_n, \ldots, a_2, a_1+1);$$
$$(b, a_1, a_2 \ldots, a_n, c)^\perp = (1, c-1, a_n, \ldots, a_2, a_1, b-1, 1).$$

- **One side infinite sequences:**

$$(1, a_1, a_2, \ldots)^\perp = (\ldots, a_2, a_1 + 1);$$
$$(b, a_1, a_2, \ldots)^\perp = (\ldots, a_2, a_1, b-1, 1);$$
$$(\ldots, a_{-2}, a_{-1}, 1)^\perp = (a_{-1}+1, a_{-2}, \ldots);$$
$$(\ldots, a_{-2}, a_{-1}, c)^\perp = (1, c-1, a_{-1}+1, a_{-2}, \ldots).$$

- **Both sides infinite sequences:**

$$(\ldots, a_{-2}, a_{-1}, a_0, a_1, a_2, \ldots)^\perp = (\ldots, a_2, a_1, a_0, a_{-1}, a_{-2}, \ldots).$$

Here we assume that the indexes $m \geq 2$ and $n \geq 1$, and the integer elements satisfy $a_i > 0$; $b, c \geq 2$; $k \geq 3$.

Remark 5.16. The adjacency duality is a duality, i.e.,

$$((s)^\perp)^\perp = s.$$

Informally speaking the relation between s and $s^\perp$ is evocative of the relation between odd and even regular continued fractions, but applied to both ends of the sequence s.

Proposition 5.17. *The LLS sequences of a rational angle and its adjacent angle are adjacency dual to each other.*

Example 5.18. Consider the integer angle

$$\alpha = \operatorname{larctan} \frac{225}{157}.$$

Then according to Example 5.11 for the adjacent angle we have

$$\pi - \alpha = \pi - \mathrm{larctan}\,\frac{225}{157} \cong \mathrm{larctan}\,\frac{225}{182}.$$

The corresponding LLS sequences for α and $\pi - \alpha$

$$(1,2,3,4,5) \quad \text{and} \quad (1,4,4,3,3) \quad (=(1,5-1,4,3,2+1))$$

respectively. We have checked Proposition 5.17 for the angle $\mathrm{larctan}\,\frac{225}{157}$.

Proof of Proposition 5.17. Consider an integer angle α.

The cases when

$$\alpha \cong \mathrm{larctan}\,k \quad \text{or} \quad \alpha \cong \pi - \mathrm{larctan}\,k$$

for a positive integer k are straightforward and left for the reader to check.

Now let the LLS sequence for α be either

$$(b, a_1, \ldots, a_n, c)$$

with $b > 1$ and $c > 1$, or

$$(b, a_1, a_2, \ldots)$$

with $b > 1$.

Then the sail of the transposed adjacent angle will start with two vertices $(-1,0)$ and $(0,1)$; and the direction of the second vector is $(1,b)$. Therefore, the first two elements of $(\pi - \alpha)^t$ are

$$\mathrm{l}\ell((-1,0),(0,1)) = 1;$$
$$\mathrm{lsin}(\angle(-1,0)(0,1)(1,b+1)) = b-1.$$

Further elements of the LLS sequence of $(\pi - \alpha)^t$ are computed by edge-angle duality (see Corollary 3.7 and Theorem 3.8).

Finally for the case of finite sequences it remains to say that the LLS sequence for $\pi - \alpha$ is reversed to the LLS sequence for $(\pi - \alpha)^t$. This concludes the proof for the case of α having the LLS sequence

$$(b, a_1, \ldots, a_n, c)$$

with $b > 1$ and $c > 1$.

The proof of the cases starting with an element 1 is similar, for that reason they are omitted here.

The cases of LLS sequences bounded from the right is reduced to the considered cases by transposing all the angles involved.

Finally the case of doubly infinite sequences is a limiting case of one side infinite sequences. Here we consider angles $\angle A_i OB$ passing through vertices of the sail A_i

tending to the ray OA. The LLS sequences of $\angle A_iOB$ and $\pi - \angle A_iOB$ asymptotically coincide with $\angle AOB$ and $\pi - \angle AOB$ respectively, as the ray OA_i approaches the ray OA. (We return to doubly infinite sequences later, see Proposition 8.6.) □

Remark 5.19. It is not a surprise that the behavior of the formula for the LLS sequence of adjacent angles has a similar behavior on both its ends. In fact there is an informal reason for this symmetry. Indeed let $\alpha = \angle AOB$ and let us denote by A' and B' the points symmetric to A and B respectively with respect to the origin O. Then we have:

$$\angle AOB = \alpha;$$
$$\angle BOA' = \pi - \alpha;$$
$$\angle A'OB' \cong \alpha.$$

So in some sense the angle $\pi - \alpha$ is surrounded by two copies of α. That provides symmetric behavior of the LLS sequences.

5.6 Right integer angles

We define right integer angles by analogy with Euclidean angles, using their symmetric properties.

Definition 5.20. An integer angle is said to be *right* if it is integer-congruent to both its transpose and adjacent angles.

It turns out that in integer geometry there exist exactly two integer inequivalent right integer angles.

Proposition 5.21. *Every integer right angle is integer-congruent to either* larctan(1) *or* larctan(2).

Proof. Let α be an integer right angle.

Since $(\pi-0) \not\cong 0$ and $(\pi-\pi) \not\cong \pi$, we have $\operatorname{ltan}\alpha > 0$.

By the definition of integer right angles and Theorem 5.7, we obtain

$$\operatorname{lcos}(\alpha) \equiv \operatorname{lcos}(\alpha^t) \equiv \big(\operatorname{lcos}\alpha(\operatorname{mod}\operatorname{lsin}\alpha)\big)^{-1}.$$

By the definition of integer right angles and Theorem 5.10, we obtain

$$\operatorname{lcos}(\alpha) \equiv \operatorname{lcos}(\pi-\alpha) \equiv \big(-\operatorname{lcos}\alpha(\operatorname{mod}\operatorname{lsin}\alpha)\big)^{-1}.$$

Hence,

$$\big(\operatorname{lcos}\alpha(\operatorname{mod}\operatorname{lsin}\alpha)\big)^{-1} \equiv \big(-\operatorname{lcos}\alpha(\operatorname{mod}\operatorname{lsin}\alpha)\big)^{-1}.$$

Therefore, $\operatorname{lsin}\alpha = 1$ or $\operatorname{lsin}\alpha = 2$.

The integer angles with $\operatorname{lsin}\alpha = 1$ are integer-congruent to larctan(1). The integer angles with $\operatorname{lsin}\alpha = 2$ are integer-congruent to larctan(2). The proof is complete.

□

5.7 Opposite interior angles

First we give the definition of parallel lines. Two integer lines are said to be *parallel* if there exists an integer shift of the plane by an integer vector taking the first line to the second.

Definition 5.22. Consider two distinct integer parallel lines AB and CD, where A, B, C, and D are integer points. Let the points A and D be in different open half-planes with respect to the line BC. Then the integer angle $\angle ABC$ is called *opposite interior* to the integer angle $\angle DCB$.

We have the following proposition on opposite interior integer angles.

Proposition 5.23. *Two integer angles opposite interior to each other are integer-congruent.*

Proof. Consider two distinct integer parallel lines AB and CD. Let the points A and D be in distinct open half-planes with respect to the line BC. Let us prove that $\angle ABC \cong \angle DCB$.

Consider the central symmetry S of the two-dimensional plane at the midpoint of the segment BC. Let P be an arbitrary integer point. Note that $S(P) = C + PB$. Since the point C and the vector PB are both integer, $S(P)$ is also integer. Since the central symmetry is self-inverse, the inverse map S^{-1} also takes the integer lattice to itself.

Therefore, the central symmetry S is an integer-affine transformation. Since S takes the angle $\angle ABC$ to the angle $\angle DCB$, we obtain $\angle ABC \cong \angle DCB$ □

5.8 Exercises

Exercise 5.1. Find both algebraically and geometrically the integer trigonometric functions for the adjacent and transpose angles to the angles $\frac{7}{3}$, larctan $\frac{9}{4}$, larctan $\frac{13}{5}$.

Exercise 5.2. Let α be an arbitrary integer angle. Find expressions in the trigonometric functions for the angles $\pi - \alpha'$ and $(\pi - \alpha)'$. Are these two angles integer congruent?

Exercise 5.3. Can an integer triangle have two integer right angles? What about three integer right angles?

Chapter 6
Integer Angles of Integer Triangles

In this chapter we study integer triangles, based on the results from previous chapters. We start with the sine formula for integer triangles. Then we introduce integer analogues of classical Euclidean criteria for congruence for triangles and present several examples. Further, we verify which triples of angles can be taken as angles of an integer triangle; this generalizes the Euclidean condition $\alpha + \beta + \gamma = \pi$ for the angles of a triangle (this formula will be used in Chapter 17 to study toric singularities). Then we exhibit trigonometric relations for angles of integer triangles. Finally, we give examples of integer triangles with small area.

6.1 Integer sine formula

In this section we give the integer sine formula for angles and edges of integer triangles.

Let A, B, C be three distinct and noncollinear integer points. We denote the integer triangle with vertices A, B, and C by $\triangle ABC$.

Proposition 6.1. (The sine formula for integer triangles.) *For any integer triangle $\triangle ABC$ the following holds:*

$$\frac{\mathrm{l}\ell(AB)}{\mathrm{lsin}(\angle BCA)} = \frac{\mathrm{l}\ell(BC)}{\mathrm{lsin}(\angle CAB)} = \frac{\mathrm{l}\ell(CA)}{\mathrm{lsin}(\angle ABC)} = \frac{\mathrm{l}\ell(AB)\,\mathrm{l}\ell(BC)\,\mathrm{l}\ell(CA)}{\mathrm{lS}(\triangle ABC)}.$$

Proof. We have

$$\mathrm{lS}(\triangle ABC) = \mathrm{l}\ell(AB)\,\mathrm{l}\ell(AC)\,\mathrm{lsin}(\angle CAB) = \mathrm{l}\ell(BA)\,\mathrm{l}\ell(BC)\,\mathrm{lsin}(\angle CBA)$$
$$= \mathrm{l}\ell(CB)\,\mathrm{l}\ell(CA)\,\mathrm{lsin}(\angle ACB),$$

After inverting the expressions and multiplying by all three integer lengths, we get the statement of the proposition. $\square$

© Springer-Verlag GmbH Germany, part of Springer Nature 2022
O. N. Karpenkov, *Geometry of Continued Fractions*,
Algorithms and Computation in Mathematics 26,
https://doi.org/10.1007/978-3-662-65277-0_6

The following is an open problem.

Problem 1. Find an integer analogue of the cosine formula for triangles.

Recall that the cosine formula for Euclidean triangle $\triangle ABC$ with $a = |BC|$, $b = |AC|$, $c = |AB|$, and $\alpha = \angle BAC$ is

$$a^2 = b^2 + c^2 - 2bc\cos\alpha.$$

It seems that to write an integer cosine formula is a hard task. Currently there are no theorems related to addition in integer trigonometry.

6.2 On integer congruence criteria for triangles

We start with integer analogues for the three Euclidean critera of triangle congruence. It turns out that only the first criterion is valid in integer geometry. Further, we present an additional criterion of congruence: on three angles and the area.

Proposition 6.2. (The first criterion of integer triangle integer congruence.) *Consider integer triangles $\triangle ABC$ and $\triangle A'B'C'$. Suppose that*

$$AB \cong A'B', \quad AC \cong A'C', \quad \text{and} \quad \angle CAB \cong \angle C'A'B'.$$

Then $\triangle A'B'C' \cong \triangle ABC$. □

Proof. By definition there exists an integer affine transformation taking $\angle CAB$ to $\angle C'A'B'$. Since the integer lengths of the corresponding segments are the same, the transformation takes B and C to B' and C'. □

It turns out that the literal generalizations of the second and third criteria from Euclidean geometry to integer geometry do not hold. The following two examples illustrate these phenomena.

Example 6.3. The second criterion of triangle integer congruence does not hold in integer geometry. In Fig. 6.1 we show two integer triangles $\triangle ABC$ and $\triangle A'B'C'$. We have

$$AB \cong A'B', \quad \angle ABC \cong \angle A'B'C' \cong \text{larctan}(1),$$
$$\text{and} \quad \angle CAB \cong \angle C'A'B' \cong \text{larctan}(1).$$

The triangle $\triangle A'B'C'$ is not integer congruent to the triangle $\triangle ABC$, since

$$\text{IS}(\triangle ABC) = 4 \quad \text{and} \quad \text{IS}(\triangle A'B'C') = 8.$$

Example 6.4. The third criterion of triangle integer congruence does not hold in integer geometry. In Fig. 6.2 we show two integer triangles $\triangle ABC$ and $\triangle A'B'C'$.

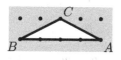
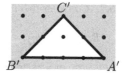

Fig. 6.1 A counterexample to the second criterion of integer congruence for triangles.

Fig. 6.2 A counterexample to the third criterion of integer congruence for triangles.

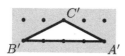

Fig. 6.3 The additional criterion of integer congruence is not improvable.

All edges of both triangles are integer congruent (of length one), but the triangles are not integer congruent, since $\mathrm{lS}(\triangle ABC) = 1$ and $\mathrm{lS}(\triangle A'B'C') = 3$.

Instead of the second and the third criteria, we have the following additional criterion.

Proposition 6.5. (An additional criterion of integer triangle integer congruence.) *Consider two integer triangles $\triangle ABC$ and $\triangle A'B'C'$ of the same integer area. Suppose that*

$$\angle ABC \cong \angle A'B'C', \quad \angle CAB \cong \angle C'A'B', \quad \angle BCA \cong \angle B'C'A'.$$

Then $\triangle A'B'C' \cong \triangle ABC$.

Proof. All the integer lengths of the corresponding edges are the same by the sine formula. Hence the triangles are integer congruent by the first criterion.   □

In the following example we show that the additional criterion of integer triangle integer congruence is not improvable.

Example 6.6. In Fig. 6.3 we give an example of two integer inequivalent triangles $\triangle ABC$ and $\triangle A'B'C'$ of the same integer area 4 and congruent angles $\angle ABC$, $\angle CAB$, and $\angle A'B'C'$, $\angle C'A'B'$ all integer-equivalent to the angle larctan(1), but $\triangle ABC \not\cong \triangle A'B'C'$.

Finally, we give a general formula for integer congruence of two triangles.

Proposition 6.7. *Consider two integer triangles $\triangle A_1 B_1 C_1$ and $\triangle A_2 B_2 C_2$ with*

$$B_i A_i = (p_i, q_i), \quad and \quad B_i C_i = (r_i, s_i) \quad for\ i = 1, 2.$$

Then these triangles are vertex-to-vertex integer congruent (i.e., there exists a $\mathrm{GL}(2, \mathbb{Z})$ *transformation sending A_1 to A_2, B_1 to B_2, and C_1 to C_2) if and only if the below matrix is in* $\mathrm{GL}(2, \mathbb{Z})$:

$$T = \begin{pmatrix} p_2 & r_2 \\ q_2 & s_2 \end{pmatrix} \cdot \begin{pmatrix} p_1 & r_1 \\ q_1 & s_1 \end{pmatrix}^{-1}.$$

Proof. Note that the matrix T is the transition matrix from the basis $(B_1 A_1, B_1 C_1)$ to the basis $(B_2 A_2, B_2 C_2)$. Now the statement of the proposition follows directly from the definition of the integer congruence for the triangles.   □

6.3 On sums of angles in triangles

In Euclidean geometry a triple of angles is a triple of angles in some triangle if and only if their sum equals π. Let us introduce a generalization of this statement to the case of integer geometry.

First, we reformulate the Euclidean criterion in the form of tan functions. A *triangle with angles* α, β, *and* γ *exists if and only if*

$$\begin{cases} \tan(\alpha + \beta + \gamma) = 0, \\ \tan(\alpha + \beta) \notin [0; \tan \alpha] \end{cases}$$

(without loss of generality, we suppose that α *is acute).* The next theorem is a translation of this condition into the integer case.

Second we give several preliminary definitions.

Let n be an arbitrary positive integer, and let $A = (x, y)$ be an arbitrary integer point. Denote by nA the point (nx, ny).

Definition 6.8. Consider an integer polygon or broken line with vertices $A_0, \ldots, A_k$. The polygon or broken line $nA_0 \ldots nA_k$ is called *n-multiple* (or *multiple*) to $A_0, \ldots, A_k$.

Let p_i (for $i = 1, \ldots, k$) be rational numbers and let $[a_{1,i}; a_{2,i} : \cdots : a_{n_i, i}]$ be their odd continued fractions. Define

$$]p_1, \ldots, p_k[= [a_{1,1}; a_{2,1} : \cdots : a_{n_1,1} : a_{1,2} : a_{2,2} : \cdots : a_{n_2,2} : \cdots : a_{1,k} : a_{2,k} : \cdots : a_{n_k,k}].$$

For instance,

$$\left] \frac{3}{2}, \frac{7}{5} \right[= [1; 1 : 1 : 1 : 2 : 2] = \frac{31}{19}.$$

Now we are ready to formulate the generalization of the Euclidean theorem on sum of angles in triangles.

Theorem 6.9. On sums of integer tangents of angles in integer triangles.

(*i*) *Let* $(\alpha_1, \alpha_2, \alpha_3)$ *be an ordered triple of angles. There exists a triangle with consecutive angles integer congruent to* α_1, α_2, *and* α_3 *if and only if there exists* $i \in \{1, 2, 3\}$ *such that the angles* $\alpha = \alpha_i$, $\beta = \alpha_{i+1(\mathrm{mod}\,3)}$, $\gamma = \alpha_{i+2(\mathrm{mod}\,3)}$ *satisfy the following conditions:*

(*a*) *for* $\xi =\,]\,\mathrm{ltan}\,\alpha, -1, \mathrm{ltan}\,\beta\,[$, *the following holds:* $\xi < 0$ *or* $\xi > \mathrm{ltan}\,\alpha$ *or* $\xi = \infty$;

(*b*) $]\,\mathrm{ltan}\,\alpha, -1, \mathrm{ltan}\,\beta, -1, \mathrm{ltan}\,\gamma\,[= 0$.

(*ii*) *Let* α, β, *and* γ *be the consecutive angles of some integer triangle. Then this triangle is multiple to the triangle with vertices* $A_0 = (0, 0)$, $B_0 = (\lambda_2 \,\mathrm{lcos}\,\alpha, \lambda_2 \,\mathrm{lsin}\,\alpha)$, *and* $C_0 = (\lambda_1, 0)$, *where*

$$\lambda_1 = \frac{\mathrm{lsin}\,\beta}{\gcd(\mathrm{lsin}\,\alpha, \mathrm{lsin}\,\beta, \mathrm{lsin}\,\gamma)} \quad and \quad \lambda_2 = \frac{\mathrm{lsin}\,\gamma}{\gcd(\mathrm{lsin}\,\alpha, \mathrm{lsin}\,\beta, \mathrm{lsin}\,\gamma)}.$$

We are not yet ready to prove the first statement of this theorem; we shall do it later, in Section 16. We prove the second statement of the theorem after a small remark.

Remark 6.10. The statement of Theorem 6.9(i) does not necessarily hold for even continued fractions of the tangents. For instance, consider an integer triangle with integer area equaling 7 and all angles integer congruent to $\mathrm{larctan}\,7/3$. For the odd continued fractions $7/3 = [2; 2 : 1]$ of all angles we have

$$[2; 2 : 1 : -1 : 2 : 2 : 1 : -1 : 2 : 2 : 1] = 0.$$

If instead we take the even continued fractions $7/3 = [2; 3]$, then we have

$$[2; 3 : -1 : 2 : 3 : -1 : 2 : 3] = \frac{35}{13} \neq 0.$$

Remark 6.11. In Chapter 17 we discuss the relation on integer angles of integer polygons similar to the case of triangles.

Example 6.12. Let us consider the triangle with vertices $A = (0, 0)$, $B = (1, 3)$, and $C = (3, 0)$. Set $\alpha = \angle CAB$, $\beta = \angle ABC$, and $\gamma = \angle BCA$:

Then we have:

$$\operatorname{ltan}\alpha = 3 = [3];$$
$$\operatorname{ltan}\beta = 9/7 = [1;3:2];$$
$$\operatorname{ltan}\gamma = 3/2 = [1;1:1].$$

Let us check the expressions for both conditions of Theorem 6.9 for the triple (α,β,γ). We have

(a)] $\operatorname{ltan}\alpha, -1, \operatorname{ltan}\beta \,[= [3;-1:1:3:2] = -3/2 \notin [0,3] = [0, \operatorname{ltan}\alpha]$;
(b)] $\operatorname{ltan}\alpha, -1, \operatorname{ltan}\beta, -1, \operatorname{ltan}\gamma \,[= [3;-1:1:3:2:-1:1:1:1] = 0$.

In our case both conditions are fulfilled.

Proof of the second statement of Theorem 6.9. Consider a triangle $\triangle ABC$ with rational angles α, β, and γ (at vertices at A, B, and C respectively).

Suppose that S is the integer area of $\triangle ABC$. Then by the definition of integer sine, the following holds:

$$\begin{cases} l\ell(AB)\,l\ell(AC) = S/\operatorname{lsin}\alpha, \\ l\ell(BC)\,l\ell(BA) = S/\operatorname{lsin}\beta, \\ l\ell(CA)\,l\ell(CB) = S/\operatorname{lsin}\gamma. \end{cases}$$

Let $l\ell(AC) = a\operatorname{lsin}\beta$, then from the above system of equations we have

$$l\ell(AB) = \frac{S}{\operatorname{lsin}\beta \cdot l\ell(BC)} = \frac{\operatorname{lsin}\gamma}{\operatorname{lsin}\beta} \cdot a\operatorname{lsin}\beta = a\operatorname{lsin}\gamma.$$

Similarly,

$$l\ell(BC) = a\operatorname{lsin}\alpha.$$

By the first criterion of integer congruence, the triangle $\triangle ABC$ is integer congruent to $\triangle A_1 B_1 C_1$, where

$$A_1 = (0,0) \quad B_1 = (a\operatorname{lsin}\gamma\operatorname{lcos}\alpha, a\operatorname{lsin}\gamma\operatorname{lsin}\alpha), \quad C_1 = (a\operatorname{lsin}\beta, 0).$$

It is also clear that the condition for the triangle $\triangle A_1 B_1 C_1$ to be integer is as follows:

$$q = \frac{k}{\gcd(\operatorname{lsin}\alpha, \operatorname{lsin}\beta, \operatorname{lsin}\gamma)}$$

where k should be a positive integer. Therefore, the triangle $\triangle A_1 B_1 C_1$ is multiple to the triangle $\triangle A_0 B_0 C_0$ of the theorem. $\square$

6.4 Angles and segments of integer triangles

Let us find the integer tangents of all angles and the integer lengths of all edges of any integer triangle, knowing the integer lengths of two edges and the integer tangent of the angle between them. Suppose that we know the integer lengths of the edges AB, AC and the integer tangent of the angle $\angle BAC$ in the triangle $\triangle ABC$. Let us show how to restore the integer length and the integer tangents for the remaining edge and the rational angles of the triangle.

For simplicity we fix some integer basis and use the system of coordinates OXY corresponding to this basis (denoted by $(*,*)$).

Theorem 6.13. *Consider some triangle $\triangle ABC$. Let*

$$1\ell(AB) = c, \qquad 1\ell(AC) = b, \quad \text{and} \quad \angle CAB \cong \alpha.$$

Then the angles $\angle BCA$ and $\angle ABC$ are defined in the following way:

$$\angle BCA \cong \begin{cases} \pi - \text{larctan}\left(\frac{c\,\text{lsin}\,\alpha}{c\,\text{lcos}\,\alpha - b}\right) & \text{if } c\,\text{lcos}\,\alpha > b, \\ \text{larctan}(1) & \text{if } c\,\text{lcos}\,\alpha = b, \\ \text{larctan}'\left(\frac{c\,\text{lsin}\,\alpha}{b - c\,\text{lcos}\,\alpha}\right) & \text{if } c\,\text{lcos}\,\alpha < b, \end{cases}$$

$$\angle ABC \cong \begin{cases} \pi - \text{larctan}'\left(\frac{b\,\text{lsin}(\alpha')}{b\,\text{lcos}(\alpha') - c}\right) & \text{if } b\,\text{lcos}(\alpha') > c, \\ \text{larctan}(1) & \text{if } b\,\text{lcos}(\alpha') = c, \\ \text{larctan}\left(\frac{b\,\text{lsin}(\alpha')}{c - b\,\text{lcos}(\alpha')}\right) & \text{if } b\,\text{lcos}(\alpha') < c. \end{cases}$$

Proof. We start with proving the formula for the angle $\angle BCA$. Let $\alpha \cong \text{larctan}(p/q)$, where $\gcd(p,q) = 1$. Then $\triangle CAB \cong \triangle DOE$, where $D = (b,0)$, $O = (0,0)$, and $E = (qc, pc)$. Let us express the angle $\angle DEO$. Denote by Q the point $(qc, 0)$. If $qc - b = 0$, then $\angle BCA = \angle DEO \cong \text{larctan}\,1$. If $qc - b \neq 0$, then we have

$$\angle QDE \cong \text{larctan}\left(\frac{cp}{cq - b}\right) \cong \text{larctan}\left(\frac{c\,\text{lsin}\,\alpha}{c\,\text{lcos}\,\alpha - b}\right).$$

The expression for the angle $\angle BCA$ follows directly from the above expression for $\angle QDE$, since $\angle BCA \cong \angle QDE$. (See Fig. 6.4; here $1\ell(OD) = b$, $1\ell(OQ) = c\,\text{lcos}\,\alpha$, and therefore $1\ell(DQ) = |c\,\text{lcos}\,\alpha - b|$.)

To obtain an expression for the angle $\angle ABC$, we consider the triangle $\triangle BAC$. Calculate the angle $\angle CBA$ and then transpose all angles in the expression. Finally, the integer length of CB is defined from the integer sine formula. $\qquad\square$

6.5 Examples of integer triangles

Let us define certain types of triangles occurring in integer geometry. Since dual angles are not necessarily congruent, we have more different types than in the Euclidean case.

Definition 6.14. An integer triangle $\triangle ACB$ is said to be *dual* to the triangle $\triangle ABC$. An integer triangle is said to be *self-dual* if it is integer congruent to the dual triangle.

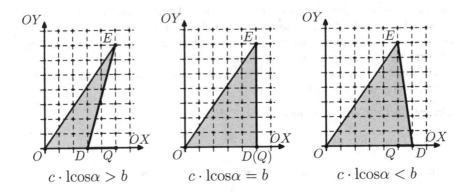

$$c \cdot \mathrm{lcos}\alpha > b \qquad\qquad c \cdot \mathrm{lcos}\alpha = b \qquad\qquad c \cdot \mathrm{lcos}\alpha < b$$

Fig. 6.4 Three possible configuration of points O, D, and Q.

An integer triangle is said to be *pseudo-isosceles* if it has at least two integer congruent angles.

An integer triangle is said to be *integer isosceles* if it is pseudo-isosceles and self-dual.

An integer triangle is said to be *pseudo-regular* if all its angles and all its edges are integer congruent.

An integer triangle is said to be *integer regular* if it is pseudo-regular and self-dual.

By the first criterion of integer congruence for integer triangles, the number of integer congruence classes for integer triangles with bounded integer area is always finite. In Fig. 6.5 we show the complete list of 33 triangles representing all integer congruence classes of integer triangles with integer areas not greater than 10. We enumerate the vertices of the triangle clockwise. Near each vertex of a triangle we write the tangent of the corresponding rational angle. Inside any triangle we write its area. We draw dual triangles on the same light gray region (if they are not self-dual). Integer regular triangles are colored in dark gray, integer isosceles but not integer regular triangles are white, and the others are light gray.

Integer triangles of small area. The above criteria allow us to enumerate all integer triangles of small integer area up to integer equivalence. In the following table we write down the numbers $N(d)$ of noncongruent integer triangles of integer area d for $d \leq 20$ (here the dual noncongruent triangles are counted as two).

d	1	2	3	4	5	6	7	8	9	10	11	12	13	14	15	16	17	18	19	20
$N(d)$	1	1	2	3	2	4	4	5	5	6	4	10	6	8	8	11	6	13	8	14

Let us prove an easy statement on the statistics of $N(d)$.

Proposition 6.15. *We have*

$$\frac{d}{3} \leq N(d) \leq \frac{d(d+1)}{2}.$$

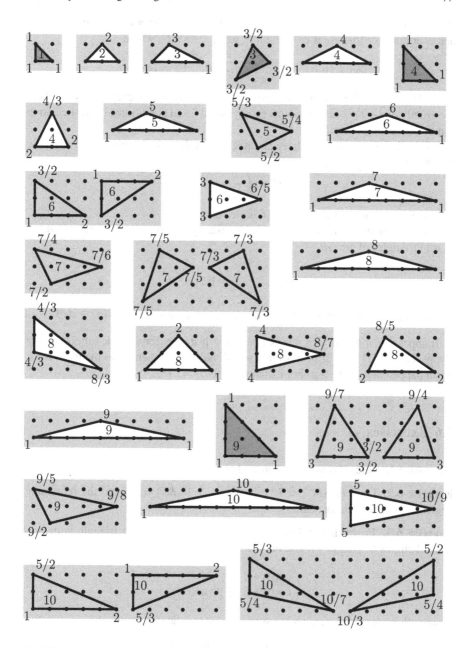

Fig. 6.5 List of integer triangles of integer area less than or equal to 10.

Proof. First, let us show that $N(d) \leq d(d+1)/2$. From Theorem 6.9(ii) we know that every integer triangle is equivalent to some $\triangle A_0 B_0 C_0$ where

$$A_0 = (0,0), \quad B_0 = (\lambda_2 q, \lambda_2 p), \quad C_0 = (\lambda_1, 0),$$

where $0 < q < p$. The area of the triangle is exactly $\lambda_1 \lambda_2 p$. Hence $\lambda_2 p \leq d$. There are exactly $d(d+1)/2$ integer points satisfying all listed conditions. Each such point can be chosen to construct B_0.

If d is divisible by $\lambda_2 q$, we have an integer triangle with C_0 constructed uniquely by setting

$$C_0 = \left(\frac{d}{\lambda_2 q}, 0 \right).$$

Hence there are at most $d(d+1)/2$ triangles of area d.

Let us show that $N(d) \geq d/3$. There are exactly d noncongruent angles that appear in triangles of area d:

$$\text{larctan}\, \frac{k}{d}, \qquad k = 1, 2, \dots, d.$$

Each triangle contains at most three noncongruent angles. Hence $N(d) \geq d/3$. $\quad\square$

It appears that the growth rate is linear. For instance, for a prime number d, we always have $N(d) \leq d$. Still, in some exceptional cases, when d has many divisors, it can happen that $N(d) > d$. For instance, $N(240) = 248$.

6.6 Exercises

Exercise 6.1. Find all triangles of area 11.

Exercise 6.2. Let $\text{ltan}(ABC) = 7/5$; $\text{l}\ell(AB) = 3$; $\text{l}\ell(BC) = 5$. Find the remaining angles and the remaining edge.

Exercise 6.3. Prove that in Euclidean geometry there exists a triangle with prescribed angles α, β, and γ if and only if

$$\begin{cases} \tan(\alpha+\beta+\gamma) = 0, \\ \tan(\alpha+\beta) \notin [0; \tan\alpha] \end{cases}$$

(here without loss of generality we suppose that α is acute).

Exercise 6.4. Is there a triangle with angles congruent to

$$\text{larctan}(24/7), \qquad \text{larctan}(24/13), \quad \text{and} \quad \text{larctan}\,4?$$

Exercise 6.5. Suppose q has n divisors. Find an upper estimate on $N(q)$ linear in d and n.

Chapter 7
Minima of Quadratic Forms, the Markov Spectrum and the Markov-Davenport Characteristics

The theory of Markov numbers studies minima of the absolute values of the quadratic decomposable forms in the plane over the lattice of integer points (excluding the origin). In this chapter we briefly discuss this classical subject, focusing on the discrete Markov spectrum that has the most relevant connection to geometry of continued fractions. We conclude this chapter with the notion of Markov—Davenport characteristic that we use later in the study of the conjugacy classes for matrices.

7.1 Calculation of minima of quadratic forms

Consider a quadratic form

$$f(x,y) = ax^2 + bxy + cy^2$$

with real coefficients and positive discriminant $\Delta(f) = b^2 - 4ac$. Set

$$m(f) = \inf_{(x,y) \in \mathbb{Z}^2 \setminus \{(0,0)\}} |f(x,y)|.$$

The set of all possible values of $\sqrt{\Delta(f)}/m(f)$ is called the *Markov spectrum*. For a general reference on the subject we recommend the book of T.W. Cusick [47].

Proposition 7.1. (*i*) *Let f be a quadratic form with positive discriminant. Consider the geometric continued fraction defined by two lines, all of whose points satisfy the equation*

$$f(x,y) = 0.$$

Let S be the set of all vertices in all four sails of this continued fraction. Then we have

$$m(f) = \inf_{(x,y) \in S} |f(x,y)|.$$

© Springer-Verlag GmbH Germany, part of Springer Nature 2022
O. N. Karpenkov, *Geometry of Continued Fractions*,
Algorithms and Computation in Mathematics 26,
https://doi.org/10.1007/978-3-662-65277-0_7

(ii) *Let the LLS sequence of the continued fraction be* $[\ldots, a_{-1}, a_0, a_1, \ldots]$. *Suppose that* $v_i = (x_i, y_i)$ *is a vertex of the sail corresponding to the element* a_i *of the sail (i.e., the integer sine at vertex* v_i *in the sail equals* a_i*). Then*

$$\sqrt{\Delta(f)}/m(f) = \sup_{i \in \mathbb{Z}} \left(a_i + [0; a_{i+1} : a_{i+2} : \cdots] + [0; a_{i-1} : a_{i-2} : \cdots] \right).$$

$\square$

The first statement of Proposition 7.1 is a straightforward corollary of the convexity of sails. For further details regarding the second statement of Proposition 7.1 we refer to the original papers of A. Markov [146] and [147] (see also [47]).

Remark 7.2. If the LLS sequence is periodic, the expression of Proposition 7.1(*ii*) is

$$\sqrt{\Delta(f)}/m(f) = \sqrt{\left(a_0 + \frac{1}{\operatorname{ltan} \alpha} + \frac{1}{\operatorname{ltan} \alpha^t} \right)^2 - \frac{4}{\operatorname{lsin}^2 \alpha}} \quad,$$

where the angle α is integer congruent to $\operatorname{larctan}([a_1; \cdots : a_{2n-1}])$, and where $(a_0, a_1, \ldots, a_{2n-1})$ is some minimal even period of the LLS sequence.

If f has integer coefficients, the corresponding continued fraction is either finite or periodic. The minimum is then attained at some vertex of the sail of the geometric continued fraction. So the calculation of this minimum is similar to the calculation of the complexity of minimal periods studied above in this chapter.

Remark 7.3. Instead of a quadratic form in two variables one can take a form of degree n in n variables corresponding to the product of n linear real forms. Any such form defines a point in the *multidimensional Markov spectrum*. Namely, for a form f one takes the infimum of the set of absolute values of f for all nonzero integer points divided by the n-th root of the discriminant of f. There is not much known about the multidimensional Markov spectrum. We refer the interested reader to [55], [56], [57], and [209] (see also in [79]).

7.2 Some properties of Markov spectrum

The Markov spectrum is a closed set that has a complicated structure. It does not contain points less than $\sqrt{5}$. On the segment $[\sqrt{5}, 3]$ the spectrum has a unique limit point, which is 3. It was described by A. Markov (see in [146] and [147]). The first elements in the spectrum in the increasing order are as follows:

$$\sqrt{5} = 1 + [0 : (1)] + [0 : (1)];$$

$$\sqrt{8} = 1 + [0 : (2)] + [0 : (2)];$$

$$\frac{\sqrt{221}}{5} = 2 + [0; (2 : 1 : 1 : 2)] + [0; (1 : 1 : 2 : 2)],$$

$$\frac{\sqrt{1517}}{13} = 2 + [0; (2 : 1 : 1 : 1 : 1 : 2)] + [0; (1 : 1 : 1 : 1 : 2 : 2)],$$

$$\frac{\sqrt{7565}}{29} = 2 + [0; (2 : 2 : 2 : 1 : 1 : 2)] + [0; (1 : 1 : 2 : 2 : 2 : 2)],$$

$$\frac{\sqrt{2600}}{17} = 2 + [0; (2 : 1 : 1 : 1 : 1 : 1 : 1 : 2)] + [0; (1 : 1 : 1 : 1 : 1 : 1 : 2 : 2)].$$

All these points are enumerated in Theorem 7.4. All real numbers greater than the so-called *Freiman's constant*

$$F = \frac{221564096 + 283748\sqrt{462}}{491993569} = 4.5278295661...$$

are in the Markov spectrum. The segment $[3, F]$ has not been completely studied. It has many gaps, i.e., open segments that belong to the complement to the Markov spectrum. The largest gap in the Markov spectrum below 3 is the open segment $(\sqrt{12}, \sqrt{13})$. The first gaps, including the segment $(\sqrt{12}, \sqrt{13})$, were calculated by O. Perron in [179] and [180]. Nowadays many other gaps are known, but a complete description of the gaps is unknown. For further information we refer the interested reader to [44] and [47] (see also [91]).

It is also known that in some neighborhood of 3 the Markov spectrum has Lebesgue measure zero (for instance, in the paper [35] by R.T. Bumby, it is shown that the spectrum is of measure zero for the the ray $x < 3.33440$).

The value 3 of the spectrum is interesting on its own. For example, as stated in [47] for any real number x and positive integer n we have that

$$2 + [0; (1 : 1)^n : x] + [0; 2 : (1 : 1)^{n-1} : x] = 3.$$

Hence, 3 is in fact represented by infinitely many forms.

Finally for the discrete Markov Spectrum we have the following remarkable theorem.

Theorem 7.4. (A. Markov [147]) (*i*) *The Markov spectrum below 3 consists of the numbers* $\sqrt{9m^2 - 4}/m$, *where m is a positive integer such that*

$$m^2 + m_1^2 + m_2^2 = 3mm_1m_2, \qquad m_2 \leq m_1 \leq m,$$

for some positive integers m_1 *and* m_2.
(*ii*) *Let the triple* (m, m_1, m_2) *fulfill the conditions of item* (*i*). *Suppose that u is the least positive residue satisfying*

$$m_2 u \equiv \pm m_1 \pmod m$$

and v is defined from

$$u^2 + 1 = vm.$$

Then the form

$$f_m(x,y) = mx^2 + (3m - 2u)xy + (v - 3u)y^2$$

represents the value $\sqrt{9m^2 - 4}/m$ in the Markov spectrum. The form representing this value is unique up to the $\mathrm{GL}(2,\mathbb{Z})$ group action and the multiplication by a non-zero constant.   □

Remark 7.5. It is interesting to observe that the theory of the Markov spectrum is based on the study of continued fractions with elements 1, 2, and 3. This is due to the fact that LLS sequences containing greater elements correspond to the quadratic forms that contribute to the Markov spectrum above Freiman's constant.

7.3 Markov numbers

Finally, let us say a few words about the integer solutions of the equation mentioned in Theorem 7.4. The equation

$$x^2 + y^2 + z^2 = 3xyz$$

is called the *Markov Diophantine equation*. A *Markov number* is a positive integer x for which there exist positive integers y and z such that the triple (x,y,z) is a solution of the Markov Diophantine equation. The first few Markov numbers are

$$1, 2, 5, 13, 29, 34, 89, 169, 194, 233, 433, 610, 985, 1325.$$

The corresponding solutions are *Markov triples*

$$(1,1,1), (1,1,2), (1,2,5), (1,5,13), (2,5,29), (1,13,34), (1,34,89), (2,29,169), \ldots$$

It turns out that Markov triples possess the following regularity.

Proposition 7.6. *Every Markov triple is obtained from $(1,1,1)$ by applying a sequence operations of the following two types:*
 — permute the numbers x, y, and z in the triple (x,y,z);
 — if (x,y,z) is a Markov triple, then $(x,y,3xy - z)$ is also a Markov triple.   □

Without loss of generality we consider only solutions with $x \le y \le z$. The structure of such solutions forms a tree, whose vertices are almost all of valence 3 (except for two vertices). An edge between two of vertices corresponds to the operation described in Proposition 7.6. We show several of the first vertices of the tree in Fig. 7.1.

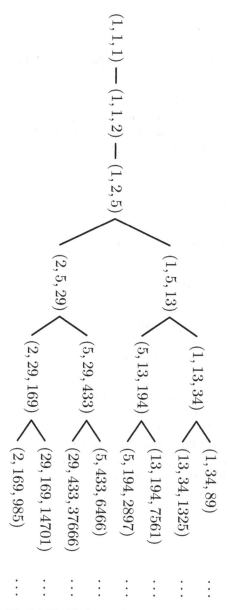

Fig. 7.1 The Markov number tree.

Finally we would like to mention one very interesting old conjecture related to number theoretical properties of triples.

Conjecture 2. Any Markov number occurs only once as the greatest element of a Markov triple.

This conjecture was introduced by G. Frobenius in 1913 (see [65], pp. 458–487) and studied very intensively since that due to its relations to modular groups and other questions of modern number theory. We refer to the book [3] by M. Aigner for a comprehensive description of history and development of the conjecture.

7.4 Markov—Davenport characteristic

In this section we relate matrices with integer coefficients to quadratic forms with integer coefficients.

Definition 7.7. The *Markov—Davenport characteristic* (or the *MD-characteristic*, for short) of any 2×2 matrix A is the functional

$$\Delta_A : \mathbb{R}^2 \to \mathbb{R} \qquad \text{defined by} \qquad \Delta_A(v) = 2S(O, v, A(v)),$$

where $S(O, v, A(v))$ is the nonoriented area of the triangle with vertices v, $A(v)$, and the origin O.

Remark 7.8. In the case when A has two real distinct eigenvalues, Δ_A is proportional to the absolute value of the product of two linear forms annulating different eigenlines of A (see Proposition 7.12 below). Such forms were considered by A. Markov in relation to Markov minima, (see Remark 7.14). Similar forms in the three-dimensional case were considered by H. Davenport, see, e.g., [55, 56, 57]. Later in Subsection 25.4.1 we return to the study of the MD-characteristic in the multidimensional settings.

Example 7.9. Let us consider an example of matrix

$$A = \begin{pmatrix} 1 & 3 \\ 2 & 4 \end{pmatrix}.$$

Consider $v = (x, y)$, then

$$A(v) = (x + 3y, 2x + 4y).$$

Then

$$\Delta_A(x, y) = \left| \det \begin{pmatrix} x & x + 3y \\ y & 2x + 4y \end{pmatrix} \right| = |2x^2 + 3xy - 3y^2|.$$

As we see the MD-characteristic is a quadratic function on the plane. In general we have the following formula.

Proposition 7.10. *Let*

$$A = \begin{pmatrix} a & c \\ b & d \end{pmatrix}.$$

Then

$$\Delta_A(x,y) = |bx^2 + (d-a)xy - cy^2|.$$

Proof. We have

$$\Delta_A(x,y) = \left| \det \begin{pmatrix} x & ax+cy \\ y & bx+dy \end{pmatrix} \right| = |bx^2 + (d-a)xy - cy^2|.$$

$\square$

One of the important properties of the MD-characteristic for a $GL(2,\mathbb{Z})$ matrix A is its invariance under the action of A. Namely, we have the following statement.

Proposition 7.11. *We have*

$$\Delta_A(v) = \Delta_A(A(v)).$$

Proof. The triangle with vertices $O, v, A(v)$ is integer congruent to the triangle with vertices $O = A(O), A(v), A^2(v) = A(A(v))$. Therefore,

$$\Delta_A(v) = \Delta_A(A(v)).$$

$\square$

The invariance of the MD-characteristic has the following geometric explanation in terms of eigenlines.

Proposition 7.12. *Assume that A has two distinct eigenvalues. Let l_1 and l_2 be two nonzero linear forms annulating eigenspaces of these eigenvectors (they might be complex here). Then there exists a non-zero constant α such that*

$$\Delta_A = |\alpha(l_1 \cdot l_2)|.$$

Proof. The statement is straightforward in the coordinates of the eigenbasis. $\square$

Remark 7.13. In particular from Proposition 7.12 one may conclude the following. If a matrix A has two distinct real eigenvalues then the level sets of Δ_A are hyperbolae whose asymptotes are eigenspaces. If a matrix A has two distinct complex (non-real) eigenvalues then the level sets of Δ_A are ellipses. (See Fig. 7.2.)

Remark 7.14. (**Markov minima for matrices.**) Consider an integer matrix with distinct real irrational eigenvalues. Definition 7.7 suggests that the elements of Markov spectrum can be associated to matrices via minima of the absolute values of their MD-characteristic on the integer lattice except the origin. Note that such minima are attained at vertices of one of the sails of four angles centered at the origin and with edges on inct eigenlines of the corresponding matrix. (For further information, see [110, 111].)

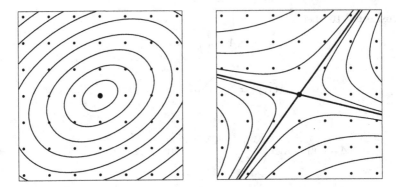

Fig. 7.2 Level sets of Δ_A in elliptic (Left) and hyperbolic (Right) cases.

7.5 Exercises

Exercise 7.1. Find Markov minima of the following forms
(a) $x^2 + xy - y^2$;   (b) $2x^2 - 4xy + y^2$;   (c) $2x^2 - 7y^2$.

Exercise 7.2. Find matrices whose MD-characteristics represent the first five elements in the Markov spectrum.

Chapter 8
Geometric Continued Fractions

In this chapter we set a more general definition of geometric continued fractions, which is related to the arrangements of pairs of distinct lines passing through the origin (see section 8.1 for basic definitions). Further in Subsection 8.2 we relate these continued fractions to $GL(2, \mathbb{R})$ matrices. In Section 8.3 we show that all four LLS sequences in the geometric continued fraction are the same for the case of irrational slopes of the lines. This property implies that the LLS sequence is well-defined with hyperbolic matrices (see in Section 8.4). We discuss periodic properties of algebraic sails related to $GL(2, \mathbb{Z})$ operators in Section 8.5. This chapter is concluded in Section 8.6 with the a small discussion on how to compute LLS cycles of $GL(2, \mathbb{Z})$ matrices.

8.1 Definition of a geometric continued fraction

In 1895, F. Klein introduced a generalization of the geometric interpretation of regular continued fractions to the multidimensional case (see his original papers [119] and [120]). We will discuss the general multidimensional case in the second part of the book. Now we give a definition for the one-dimensional Klein geometric continued fraction. It is slightly different from the definition of regular continued fractions.

Definition 8.1. Consider a pair of distinct lines l_1 and l_2 passing through an integer point O. The lines l_1 and l_2 divide the plane into four cones. A *geometric continued fraction* is the union of the sails of these angles corresponding to these cones.

Notice that the LLS sequences of the angles between the lines are lattice invariants of a geometric continued fraction. Since the LLS sequence of an angle is its integer complete invariant (see Theorem 4.11), we have the following corollary.

Corollary 8.2. *The LLS sequence is a complete invariant of lattice angles under the group of lattice affine transformations* $\mathrm{Aff}(2, \mathbb{Z})$.

© Springer-Verlag GmbH Germany, part of Springer Nature 2022
O. N. Karpenkov, *Geometry of Continued Fractions*,
Algorithms and Computation in Mathematics 26,
https://doi.org/10.1007/978-3-662-65277-0_8

8.2 Geometric continued fractions of hyperbolic $GL(2,\mathbb{R})$ matrices

Recall the following general definition.

Definition 8.3. A $GL(2,\mathbb{R})$ matrix is called *hyperbolic/parabolic/elliptic* if its eigenvalues are distinct and real/coincide/distinct and complex conjugate.

Let us associate continued fractions to hyperbolic matrices.

Definition 8.4. Consider a hyperbolic matrix A in $GL(2,\mathbb{R})$ with two distinct eigenlines. The geometric continued fraction defined by these two straight lines is said to be *associated* to A.

In Fig. 8.1 we show the geometric continued fraction for the matrix

$$\begin{pmatrix} 7 & 18 \\ 5 & 13 \end{pmatrix}.$$

Integer lengths of edges are denoted by black digits, and integer sines, by white. The LLS sequences of all four sails are equivalent to or inverse to

$$(\ldots,2,1,1,3,2,1,1,\ldots).$$

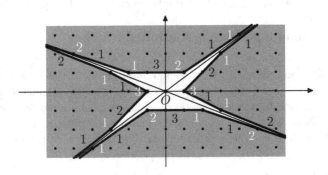

Fig. 8.1 Geometric continued fraction of the matrix $\begin{pmatrix} 7 & 18 \\ 5 & 13 \end{pmatrix}$.

8.3 Duality of sails

As we have already seen in Fig. 8.1, the continued fractions of adjacent angles are equivalent up to a reversal of order. This happens not by chance, such situation is general. Before showing this, we discuss the notion of sail duality.

Definition 8.5. Two sails are *dual* with respect to each other if the LLS sequence of one sail coincides (up to an index shift if the LLS sequences are infinite on both sides) with the inverted LLS sequence of the second sail such that the subsequences of lengths and angles of the first sail coincide respectively with the subsequences of angles and sails of the second one.

Proposition 8.6. *Let A be a hyperbolic matrix with no integer eigenvectors. Then the sails of the opposite octants are congruent. The sails of the adjacent octants are dual.*

Proof. The sails of opposite octants are congruent, since they are taken one to another by the symmetry about the origin.

Let us prove the duality. Let one of the sails for the geometric continued fraction be a broken line (A_i). Without loss of generality we may fix coordinates such that $A_0 = (1,0)$ and $A_1 = (1,a_0)$. Set $B_0 = (0,1)$. Notice that B_0 is on the dual sail (since the sail is defined by the lines $y = \alpha x$ and $y = \beta x$, where $a_0 < \alpha < a_0+1$ and $-1 < \beta < 0$ in the chosen coordinates). Denote the remaining points of the dual sail by B_i, starting from B_0.

Let (a_i) and (b_i) be the LLS sequences for the sails (A_i) and (B_i). In Section 3 we have already shown the edge—angle duality in the orthant of points with positive coordinates. In other words, we know that $a_i = b_{-i-1}$ for $i \geq 1$.

Let B_i' be the point symmetric to B_i with respect to the origin ($i \in \mathbb{Z}$). Recall that

$$B_{-1} = (a_1, a_0 a_1 + 1) \quad \text{and} \quad A_2 = (a_1 a_2 + 1, a_0 a_1 a_2 + a_0 + a_2).$$

Consider the linear transformation taking A_2 to $(1,0)$ and A_1 to $(1,a_2)$, namely

$$\begin{pmatrix} a_0 a_1 + 1 & -a_1 \\ a_0 a_1 a_2 + a_0 + a_2 & -a_1 a_2 - 1 \end{pmatrix}.$$

This transformation takes the point B_{-1}' to the point $(0,1)$. Now we have edge—angle duality of $A_2 A_1 A_0 \ldots$ and $B_{-1}' B_0' \ldots$. Therefore, $a_i = b_{-i-1}$ for $i \leq 1$. Hence the LLS sequences are inverse to each other. $\qquad\square$

Example 8.7. Let us consider the example of the matrix

$$\begin{pmatrix} 7 & 18 \\ 5 & 13 \end{pmatrix}$$

mentioned before (See Fig. 8.1). As we can see the LLS sequences for all four sails coincide and are periodic with period $(1,1,3,2)$. Note the following small difference between the sails and their dual sails: The sail containing the point $(1,0)$ and its opposite sail have integer sines with period $(1,3)$ and integer lengths with period $(1,2)$. However, two dual sails both have integer sines with period $(1,2)$ and integer lengths with period $(1,3)$.

8.4 LLS sequences for hyperbolic matrices

Let us define LLS sequences for hyperbolic matrices.

Definition 8.8. The *LLS sequence* of a matrix A in $GL(2, \mathbb{R})$ is the LLS sequence for any of its sails up to the choice of a direction of a sequence and zero element.

In the next proposition we list some basic properties of LLS sequences for hyperbolic matrices.

Proposition 8.9. (*i*) *For any sequence of integers infinite in two sides there exists a hyperbolic matrix whose LLS sequence coincides with the given sequence.*

(*ii*) *Let hyperbolic matrices A and B have the same LLS sequence. Then there exists a matrix C commuting with A and integer conjugate to B.*   □

8.5 Algebraic sails and their LLS cycles

8.5.1 Algebraic sails

Consider now the case of hyperbolic matrices in $GL(2, \mathbb{Z})$. It turns out that all such matrices have characteristic polynomials irreducible over the rational numbers. The sails of such matrices are called *algebraic*.

Proposition 8.10. *Every algebraic sail has a periodic LLS sequence.*

Remark 8.11. Later, in Corollary 11.10, we show that a sail with periodic LLS sequence is algebraic.

Proof. Let A be hyperbolic matrix in $SL(2, \mathbb{Z})$. The matrix A preserves its invariant lines and the lattice $\mathbb{Z}^2$. Hence it acts on each of the sails by shifting the vertices of the corresponding broken line along the broken line. Therefore, the LLS sequences of algebraic sails are periodic.   □

In Fig. 8.1, we show the sails of a hyperbolic algebraic matrix with the period of the LLS sequence equal to $(2, 1, 1, 3)$.

8.5.2 LLS periods and LLS cycles of $GL(2, \mathbb{Z})$ matrices

We start with the hyperbolic matrices with positive eigenvalues (here we require that these eigenvalues are irrational). Such matrices act on each of their sails as a shift, so we can give the following definition.

Definition 8.12. Let M be an $SL(2, \mathbb{Z})$ matrix with positive irrational eigenvalues. Then M acts on the sail for M as a shift. Hence M also defines a shift of the LLS sequence. The factor of the LLS sequence with respect to this shift is called the *LLS cycle* of M.

Any sequence of the corresponding LLS cycle (i.e., we mark a starting element of the LLS-cycle) is called an *LLS period* of M.

So the LLS cycle is a cyclically ordered sequence of an even number of integer elements. Notice that $(1, 2, 1, 2)$ and $(2, 1, 2, 1)$ are distinct LLS periods representing the same LLS cycle. Notice further that the LLS periods $(1, 2, 1, 2)$ and $(1, 2)$ represent distinct LLS sycles.

If one or both of the eigenvalues of M are negative, then we still can define the LLS cycles and LLS periods using the fact that the eigenvalues of M^2 are positive.

Definition 8.13. Let M be a hyperbolic $GL(2, \mathbb{Z})$ matrix. Then M^2 has the LLS cycle, which is a twice repeated cyclic sequence. We say that this cyclic sequence (once repeated) is the *LLS cycle* of M.

Any (non-cyclic) sequence of the corresponding LLS cycle is called an *LLS period* of M.

Remark 8.14. Note that inverse matrices to each other have reversed periods. It is interesting to note that for palindromic sequences the matrices are conjugate to their inverses.

For matrices with positive eigenvalues we have the following proposition.

Proposition 8.15. *Two hyperbolic $SL(2, \mathbb{Z})$ matrices M_1 and M_2 with positive irrational eigenvalues are integer conjugate if and only if their LLS cycles coincide.*

Proof. Let M_1 and M_2 be integer conjugate. Then they have integer congruent geometric continued fractions, and they define the same shift of these continued fractions. Therefore the LLS cycles of M_1 and M_2 coincide.

Suppose now that the LLS cycles of the matrices M_1 and M_2 coincide. This means that the corresponding continued fractions have the same LLS sequences, and therefore they are integer congruent. Therefore, there exists a matrix M_2' integer conjugate to M_2 whose geometric continued fraction coincides with the geometric continued fraction of M_1. Suppose v is a vertex of the geometric continued fraction of the operator M_1. Then

$$M_1(v) = M_2'(v) \quad \text{and} \quad M_1(M_1(v)) = M_2'(M_1(v)),$$

since the matrices M_1 and M_2' define the same shift of the geometric continued fraction. Since the vectors v and $M_1(v)$ form a basis of $\mathbb{R}^2$, we have

$$M_1 = M_2'.$$

Therefore, M_1 is integer conjugate to M_2. □

In the case of arbitrary $GL(2, \mathbb{Z})$ matrices we have a weaker statement.

Corollary 8.16. *Consider two hyperbolic* $SL(2,\mathbb{Z})$ *matrices* M_1 *and* M_2 *with irrational eigenvalues. Then the LLS cycles* M_1 *and* M_2 *coincide if and only if* M_1 *is integer conjugate either to* M_2 *or to* $-M_2$.

Remark 8.17. This statement is rather surprising as the matrix equation

$$X^2 = M$$

has four $GL(2,\mathbb{R})$ solutions in the case where the eigenvalues of M are positive and distinct.

Proof. It is clear that if M_1 is integer conjugate either to M_2 or to $-M_2$ then LLS cycles coincide.

Now let LLS cycles coincide. Then M_1^2 is integer conjugate to M_2^2. Let us consider two cases.

In the case where LLS cycles of M_1 and M_2 have even number of elements, then M does not swap lengths and sines in the LLS sequence, hence it maps the cones either to themselves or to the opposite ones. In both cases we have $\det M_1 \cdot \det M_2 = 1$, so the eigenvalues are either simultaneously positive or negative for both M_1 and M_2. Therefore, M_1 is integer conjugate either to M_2 or to $-M_2$.

In the case where LLS cycles of M_1 and M_2 have an odd number of elements, then both M_1 and M_2 swap lengths and sines in the LLS sequence, hence it maps the cones to the adjacent ones. In both cases we have $\det M_1 = \det M_2 = -1$, so one of the eigenvalues is positive and one is negative for both M_1 and M_2. Therefore, M_1 is integer conjugate either to M_2 or to $-M_2$.

In both cases M_1 is integer conjugate either to M_2 or to $-M_2$. This concludes the proof.

$\square$

We conclude this subsection with a statement on the form of LLS sequences for a given matrix.

Proposition 8.18. *Let* M *be a* $GL(2,\mathbb{Z})$ *hyperbolic matrix with irrational eigenvalues and let*

$$(a_1,\ldots,a_n)$$

be its LLS cycle. Let also m *be the minimal lengths of the periods of the LLS sequence for* M. *Then the list of all LLS periods for* M *consists of the following* m *sequences*

$$(a_{k+1},\ldots,a_{k+n}) \quad for\ k = 1,\ldots,m$$

(here the indices are taken modulo n*).*

Proof. Operator M^2 acts on all sails as shifts by n vertices. Hence we get all of the periods are obtained by applying M to vertices of both sails. $\square$

Remark 8.19. Note that all periods starting at elements of the LLS sequence with odd indexing correspond to the same sail. All periods starting at elements with even indexing correspond to the sails of the adjacent angles.

8.6 Computing LLS cycles of GL$(2, \mathbb{Z})$ matrices

In this section we address a question of computing LLS sequences for GL$(2, \mathbb{Z})$ matrices with irrational real eigenvalues (such eigenvalues will be automatically distinct).

8.6.1 Differences of sequences

Let us start with the following supplementary definition.

Definition 8.20. Let $m > n$ be non-negative integers. Consider two number sequences

$$S_a = (a_1, \ldots, a_m) \quad \text{and} \quad S_b = (b_1, \ldots, b_n).$$

We say that there exists a *difference* of the sequences S_a and S_b if there exists $k \leq m+1$ satisfying the following conditions
  (a) $b_i = a_i$ for $1 \leq i < k$;
  (b) either $k = m+1$ or $b_k \neq a_k$;
  (c) $b_{k+i} = a_{k+i+m-n}$ for $0 \leq i \leq n - k$.
In this case we write

$$S_a - S_b = (a_k, a_{k+1}, \ldots, a_{k+n-m-1}).$$

Example 8.21. We start with the following example:

$$(1, 2, 3, 4, 5, 6, 7, 8) - (1, 2, 3, 6, 7, 8) = (4, 5).$$

Example 8.22. The expression

$$(1, 2, 3, 4, 5, 6, 7, 8) - (1, 4, 8)$$

does not exist.

8.6.2 LLS cycles for SL$(2, \mathbb{Z})$ matrices with positive eigenvalues

In this subsection we show how to compute LLS cycles for SL$(2, \mathbb{Z})$ matrices with positive irrational eigenvalues (see also in [112]). The computation is based on the following statement.

Proposition 8.23. *Let an* SL$(2, \mathbb{Z})$ *matrix* M *have distinct irrational positive eigenvalues. Let also* P *be any non-zero integer point. Then there exists a difference*

$$\text{LLS}\left(\angle PO(M^3(P))\right) - \text{LLS}\left(\angle PO(M^2(P))\right),$$

and the resulting sequence is an LLS period of M.

Remark 8.24. It can happen that the obtained period of the LLS sequence are not of the minimal length.

Example 8.25. Consider an $SL(2,\mathbb{Z})$ matrix

$$M = \begin{pmatrix} 7 & -37 \\ -10 & -53 \end{pmatrix}.$$

Note that this matrix has two eigenvalues $30 \pm \sqrt{899}$ that are both positive irrational numbers. Hence we can apply Proposition 8.23.

Let us take $P = (1,1)$. Then we have

$$M^2(P) = (-1801, 2579) \quad \text{and} \quad M^3(P) = (-108030, 154697).$$

Further we compute LLS sequences of the corresponding angles (e.g., using one of the algorithms of Section 4.5):

$$\text{LLS}\big(\angle PO(M^2(P))\big) = (1,1,2,3,5,1,2,3,4);$$
$$\text{LLS}\big(\angle PO(M^3(P))\big) = (1,1,2,3,5,1,2,3,5,1,2,3,4).$$

Hence

$$\text{LLS}\big(\angle PO(M^3(P))\big) - \text{LLS}\big(\angle PO(M^2(P))\big) = (5,1,2,3).$$

Therefore, the LLS cycle of M is $(5,1,2,3)$.

Proof of Proposition 8.23. The complement to the union of eigenlines consists of four cones. Consider one of these cones containing the point P (denote it by C). Then the convex angle $\angle P_0 OP_1$ is the fundamental domain of C up to the action of the group of (integer) powers of M. Hence there is at least one vertex of the sail for C in the interior of $\angle P_0 OP_1$. Denote one of the sail vertices inside $\angle P_0 OP_1$ by v. We immediately get that the angle $\angle P_0 OP_3$ contains vertices

$$v_0 = v, \quad v_1 = M(v), \quad \text{and} \quad v_2 = M^2(v).$$

Thus by convexity reasons, the sail for the angle $\angle P_0 OP_3$ contains the part of the sail of C between the vertices v_0 and v_2.

Namely, the sail of $\angle P_0 OP_3$ consists of the following four parts:

- S_1: a part of the sail contained in $P_0 O v_0$;
- S_2: a part of the sail contained in $v_0 O v_1$;
- S_3: a part of the sail contained in $v_1 O v_2$;
- S_4: a part of the sail contained in $v_2 O P_3$.

Since $v_1 = M(v)$ and $v_2 = M^2(v)$ the parts S_2 and S_3 are periods of the sail for C.

Similarly the vertices v_0 and v_1 are in the sail of $\angle P_0 OP_2$. Therefore, the sail for $\angle P_0 OP_2$ consists of the following three parts

- S_1': a part of the sail contained in $P_0 O v_0$;

- S'_2: a part of the sail contained in $v_0 O v_1$;
- S'_3: a part of the sail contained in $v_1 O P_2$;

Let us observe that

$$\begin{cases} S'_1 = S_1; \\ S'_2 = S_2; \\ S'_2 \cong S_3; \\ S'_3 \cong S_4. \end{cases}$$

Therefore, the difference of the LLS sequences for the angles $\angle P_0 O P_3$ and $\angle P_0 O P_2$ is precisely a period of the LLS sequence for C. This period corresponds to the sequences between the points v_1 and v_2. This period is an LLS period for M as $M(v_1) = v_2$. After considering this period up to a cyclic shift we have the LLS cycle for M. This concludes the proof. □

Remark 8.26. Note that it is not sufficient to consider the difference of the LLS sequences for the angles $\angle P_0 O P_2$ and $\angle P_0 O P_1$. Here it is not possible to determine one of the integer sines of the periods. Let us illustrate this with the following example.
  Consider a matrix

$$M = \begin{pmatrix} 1 & 2 \\ 1 & 3 \end{pmatrix}$$

and set $P = (4, -1)$. Then

$$P_1 = M(P_0) = (2, 1), \quad P_2 = M^2(P_0) = (4, 5), \quad \text{and} \quad P_3 = M^3(P_0) = (14, 19).$$

The LLS sequences for the angles $\angle P_0 O P_1$, $\angle P_0 O P_2$ and $\angle P_0 O P_3$ are respectively

$$(1,4,1);$$
$$(1,3,1,3,1);$$
$$(1,3,1,2,1,3,1).$$

We have

$$\mathrm{LLS}\big(\angle POP_3)\big) - \mathrm{LLS}\big(\angle POP_2)\big) = (1,3,1,2,1,3,1) - (1,3,1,3,1) = (2,1)$$

which is a correct period for the LLS sequence of M, while the difference

$$\mathrm{LLS}\big(\angle POP_2)\big) - \mathrm{LLS}\big(\angle POP_1)\big) = (1,3,1,3,1) - (1,4,1)$$

is not even defined.

8.6.3 LLS cycles for GL$(2,\mathbb{Z})$ matrices

Let us extend the result of Proposition 8.23 to GL$(2,\mathbb{Z})$ matrices with distinct (not necessarily positive) irrational eigenvalues. (See also in [112].)

Corollary 8.27. *Let a* $GL(2,\mathbb{Z})$ *matrix M have distinct irrational eigenvalues. Let also P be any integer point distinct from the origin. Then there exists a difference*

$$LLS(\angle PO(M^6(P))) - LLS(\angle PO(M^4(P))),$$

and the resulting sequence is an LLS period of M repeated twice.

Proof. By Proposition 8.23 the difference

$$LLS(\angle PO(M^6(P))) - LLS(\angle PO(M^4(P)))$$

exists and is a period for M^2. Finally by the above the resulting sequence is a period of the LLS sequence for M repeated twice by Proposition 10.14.   □

Remark 8.28. The proof of Corollary 8.27 is based on Proposition 8.23 which we prove after giving several important results of the classical Gauss Reduction theory are given. Still we prefer to discuss Corollary 8.27 in this section for convenience of the material exposition.

The situation of Corollary 8.27 dramatically changes if we take an integer matrix that is not in $GL(2,\mathbb{Z})$. Here we consider matrices with distinct irrational eigenvalues. Such matrices still have periodic LLS sequences, however they do not determine shifts of their sails. We have the following conjecture.

Conjecture 3. For any integers $m > n$ there exists a matrix $M \notin GL(2,\mathbb{Z})$ with distinct irrational eigenvalues such that, for every integer point P

$$LLS(\angle PO(M^m(P))) - LLS(\angle PO(M^n(P)))$$

does not contain a period for the sail of M.

(Experiments suggest that $\angle M(P)OM^{2n}(P)$ might not contain a single fundamental domain of the corresponding invariant cone for M.)

8.7 Exercises

Exercise 8.1. Draw a geometric continued fraction defined by the lines $x - 2y = 1$ and $3x + 4y = 3$ and calculate the LLS sequences for all the sails.

Exercise 8.2. Is the statement of Proposition 8.6 true for rational angles (R-angles, L-angles)? Find a correct analogue in these cases.

Exercise 8.3. Prove Proposition 8.9.

Exercise 8.4. Find the LLS cycles for the following matrices

(a) $\begin{pmatrix} 0 & 1 \\ 1 & 1 \end{pmatrix}$;   (b) $\begin{pmatrix} 3 & -1 \\ -2 & 1 \end{pmatrix}$;   (c) $\begin{pmatrix} 2 & 7 \\ 3 & 10 \end{pmatrix}$.

Chapter 9
Continuant Representation of $\mathrm{GL}(2,\mathbb{Z})$ Matrices

Let us now discuss two remarkable representations of $\mathrm{SL}(2,\mathbb{Z})$ and $\mathrm{GL}(2,\mathbb{Z})$ respectively. We start with a classical representation of $\mathrm{SL}(2,\mathbb{Z})$ in terms of the modular group (see Section 9.1). It turns out that $\mathrm{SL}(2,\mathbb{Z})$ can be generated by two elements. Further we discuss an alternative approach to represent elements of $\mathrm{GL}(2,\mathbb{Z})$ (which includes $\mathrm{SL}(2,\mathbb{Z})$) that is based on continuant representation of matrices. In Section 9.2 we introduce supplementary material for that approach: we set elementary matrices and write their products in terms of continuants. After that in Section 9.3 we provide matrix continuant representations mentioned above. In fact the representation exists for almost all $GL(2,\mathbb{Z})$ matrices (see Theorem 9.6). Finally in Section 9.4 we show how to express matrices of $\mathrm{SL}(2,\mathbb{Z})$ via classic generators discussed in Section 9.1 using continuants.

9.1 Generators of $\mathrm{SL}(2,\mathbb{Z})$ and the modular group

The group $\mathrm{SL}(2,\mathbb{Z})$ is very well studied due to its relations to the modular group in hyperbolic geometry. Let us just say a few words about it. Recall that the *modular group* is the group of linear fractional transformations with integer coefficients of the upper half of the complex plane. Such transformations can be written in the form

$$z \mapsto \frac{az+d}{cz+d}$$

(see e.g. in [4]) and represented by two matrices $\pm \begin{pmatrix} a & b \\ c & d \end{pmatrix}$. Denote this transformation by $\begin{bmatrix} a & b \\ c & d \end{bmatrix}$.

Here the matrix can be taken up to the projective invariance, i.e., up to the multiplication by -1. By that reason the modular group is denoted by $\mathrm{PSL}(2,\mathbb{Z})$ (see more about modular groups in [184]). One can say, that

© Springer-Verlag GmbH Germany, part of Springer Nature 2022
O. N. Karpenkov, *Geometry of Continued Fractions*,
Algorithms and Computation in Mathematics 26,
https://doi.org/10.1007/978-3-662-65277-0_9

$$PSL(2,\mathbb{Z}) = SL(2,\mathbb{Z})/\{\pm Id\}.$$

It is well known that $PSL(2,\mathbb{Z})$ is generated by two elements (see, e.g., [199]):

$$S: z \mapsto -\frac{1}{z} \quad \text{and} \quad T: z \mapsto z+1.$$

Moreover the modular group can be defined by generators and relations as follows:

$$PSL(2,\mathbb{Z}) = \langle S, T | S^2 = I, (ST)^3 = I \rangle,$$

where I is an identical transformation of the upper half plain.

In analogy to $PSL(2,\mathbb{Z})$, the group $SL(2,\mathbb{Z})$ can be generated by matrices representing the transformations S and T, namely by

$$M_S = \begin{pmatrix} 0 & -1 \\ 1 & 0 \end{pmatrix} \quad \text{and} \quad M_T = \begin{pmatrix} 1 & 1 \\ 0 & 1 \end{pmatrix}.$$

Here we have

$$SL(2,\mathbb{Z}) = \langle M_S, M_T | M_S^4 = Id, (M_S M_T)^6 = Id \rangle,$$

Remark 9.1. Our intention for this section is to study an alternative representation of $GL(2,\mathbb{Z})$ matrices in terms of products of the matrices of type

$$\begin{pmatrix} 0 & 1 \\ 1 & a \end{pmatrix}.$$

Such approach is based on geometry of numbers. We discuss it further in this chapter.

9.2 Basic properties of matrices $M_{a_1,\ldots,a_n}$

Let us set the following general notation.

Definition 9.2. Let a be a real number, denote by M_a the following matrix:

$$M_a = \begin{pmatrix} 0 & 1 \\ 1 & a \end{pmatrix}.$$

Now let $(a_1,\ldots,a_n)$ be any sequence of real numbers, we set

$$M_{a_1,\ldots,a_n} = \prod_{k=1}^{n} \begin{pmatrix} 0 & 1 \\ 1 & a_k \end{pmatrix}.$$

We continue with the following remarkable result for matrices $M_{a_1,\dots,a_n}$.

Proposition 9.3. *Let $n \geq 0$ and let $(a_1,\dots,a_n)$ be any sequence of real numbers. Then we have*

$$M_{a_1,\dots,a_n} = \begin{pmatrix} K_{n-2}(a_2,\dots,a_{n-2}) & K_{n-1}(a_2,\dots,a_n) \\ K_{n-1}(a_1,a_2,\dots,a_{n-1}) & K_n(a_1,a_2,\dots,a_n) \end{pmatrix}.$$

In addition, we have

$$\det M = (-1)^n.$$

Example 9.4. Let us consider the following simple example of a $GL(2,\mathbb{Z})$ matrix

$$M = \begin{pmatrix} 7 & 32 \\ 19 & 87 \end{pmatrix}.$$

The continuant form for M and the corresponding product matrix representation are as follows

$$M = \begin{pmatrix} K_2(-3,-2) & K_3(-3,-2,5) \\ K_3(3,-3,-2) & K_4(3,-3,-2,5) \end{pmatrix}.$$

Hence M is represented by the following even sequence:

$$(3,-3,-2,5).$$

Therefore, we have

$$M = M_{3,-3,-2,5} = M_3 \cdot M_{-3} \cdot M_{-2} \cdot M_5.$$

Note also that

$$\det M = (-1)^4 = 1.$$

Proof. The proof of the proposition is by induction in n.

Base of induction. For $n = 1$ we have

$$M_{a_1} = \begin{pmatrix} 0 & 1 \\ 1 & a_1 \end{pmatrix} = \begin{pmatrix} K_{-1}() & K_0() \\ K_0() & K_1(a_1) \end{pmatrix}.$$

For $n = 2$ we have

$$M_{a_1,a_2} = M_{a_1} M_{a_2} = \begin{pmatrix} 1 & a_2 \\ a_1 & 1+a_1a_2 \end{pmatrix} = \begin{pmatrix} K_0() & K_1(a_2) \\ K_1(a_1) & K_2(a_1,a_2) \end{pmatrix}.$$

Step of induction. We have

$$M_{a_1,\ldots,a_{n+1}} = M_{a_1,\ldots,a_n} \cdot M_{a_{n+1}} =$$

$$\begin{pmatrix} K_{n-2}(a_1,\ldots,a_{n-2}) & K_{n-1}(a_2,\ldots,a_n) \\ K_{n-1}(a_1,\ldots,a_{n-1}) & K_n(a_1,\ldots,a_n) \end{pmatrix} \cdot \begin{pmatrix} 0 & 1 \\ 1 & a_{n+1} \end{pmatrix} =$$

$$\begin{pmatrix} K_{n-1}(a_2,\ldots,a_n) & K_{n-2}(a_2,\ldots,a_{n-2}) + a_{n+1}K_{n-1}(a_2,\ldots,a_n) \\ K_n(a_1,\ldots,a_n) & K_{n-1}(a_1,\ldots,a_{n-1}) + a_{n+1}K_n(a_1,\ldots,a_n) \end{pmatrix} =$$

$$\begin{pmatrix} K_{n-1}(a_2,\ldots,a_n) & K_n(a_2,\ldots,a_{n+1}) \\ K_n(a_1,\ldots,a_n) & K_{n+1}(a_1,\ldots,a_{n+1}) \end{pmatrix}.$$

The last inequality is a direct corollary of Proposition 1.13. This concludes the proof for the induction step.

Finally, since $\det M_a = -1$ we have

$$\det M = (-1)^n.$$

$\square$

9.3 Matrices of GL(2, ℤ) in terms of continuants

Let us start this section with the following definition.

Definition 9.5. Let n be a positive integer. Consider the following set of sequences:

$$I_{-n-} = \{(a_1,\ldots,a_n) \mid \ a_1 \in \mathbb{Z}; \ a_2,\ldots,a_{n-1} \in \mathbb{Z}_+; \ a_n \in \mathbb{Z}\}.$$

Set also

$$I_{-\infty-} = \bigcup_{k=1}^{\infty} I_{-k-}.$$

We have the following statement.

Theorem 9.6. *For every matrix*

$$M = \begin{pmatrix} a & c \\ b & d \end{pmatrix}$$

in GL(2, ℤ) *excluding matrices*

$$\pm \begin{pmatrix} 0 & -1 \\ 1 & k \end{pmatrix} \quad k=1,2,3,\ldots$$

there exist a unique integer $n > 0$ and a unique sequence of numbers $(a_1,\ldots,a_n) \in I_{-n-}$ such that the following holds.
(i) If $a \neq 0$ then

$$M = \text{sgn}(a)M_{a_1,\dots,a_n} = \text{sgn}(a)\begin{pmatrix} K_{n-2}(a_2,\dots,a_{n-2}) & K_{n-1}(a_2,\dots,a_n) \\ K_{n-1}(a_1,\dots,a_{n-1}) & K_n(a_1,\dots,a_n) \end{pmatrix}.$$

Here the sequence $(a_1,\dots,a_n)$ satisfies

$$\frac{b}{a} = [a_1;a_2 : \cdots : a_{n-1}],$$

where we take the odd regular continued fraction for b/a when $\det(M) = 1$, and the even regular continued fraction for b/a when $\det(M) = -1$.

Finally, the value of a_n is the solution of the following linear equation:

$$K_{n-1}(a_2,\dots,a_{n-1},x) = c.$$

(ii) For the case $a = 0$ we have:

$$M = \begin{pmatrix} 0 & 1 \\ 1 & k \end{pmatrix} = M_k; \qquad M = \begin{pmatrix} 0 & -1 \\ -1 & -k \end{pmatrix} = -M_k.$$

Finally the matrices

$$\pm \begin{pmatrix} 0 & -1 \\ 1 & k \end{pmatrix}$$

are the only matrices that (up to multiplication by -1) do not have the continuant representations satisfying the above conditions.

Remark 9.7. Note that SL$(2,\mathbb{Z})$ matrices correspond to sequences with an even number of elements, while GL$(2,\mathbb{Z})$ with negative determinant are represented by sequences with an odd number of elements. This follows directly from Proposition 9.3.

Remark 9.8. As one can see, we have many expressions for GL$(2,\mathbb{Z})$ matrices defined up to a sign of a matrix. This is closely related to the modular group PSL$(2,\mathbb{Z})$ discussed above. Similar to PSL$(2,\mathbb{Z})$ one can define PGL$(2,\mathbb{Z})$: here we take all the matrices of GL$(2,\mathbb{Z})$ up to the same equivalence $M \sim -M$ and denote them by $[M]$.

Theorem 9.6 has the following nice corollary on the structure of the group PGL$(2,\mathbb{Z})$.

Corollary 9.9. *The mapping*

$$\{\pm M_{a_1,\dots,a_n}\} \mapsto (a_1,\dots,a_n)$$

is a bijection of

$$\text{PGL}(2,\mathbb{Z}) \setminus \left\{ \begin{bmatrix} 0 & -1 \\ 1 & k \end{bmatrix} \middle| k \in \mathbb{Z} \right\}$$

and the set of all sequences $I_{-\infty-}$.

Proof. This corollary follows directly from the formulae of Theorem 9.6. $\qquad\square$

Example 9.10. Let us start with several examples of matrices containing a zero element. On the left hand side we write a matrix M and on the right hand side we show the sequence $(a_1, \ldots, a_n)$ such that

$$M = M_{a_1, \ldots, a_n}.$$

We have the following cases

$\begin{pmatrix} 1 & k \\ 0 & 1 \end{pmatrix}$	$(0, k)$	$\begin{pmatrix} 1 & k \\ 0 & -1 \end{pmatrix}$	$(-1, 1, k-1)$
$\begin{pmatrix} 1 & 0 \\ k & 1 \end{pmatrix}$	$(k, 0)$	$\begin{pmatrix} 1 & 0 \\ k & -1 \end{pmatrix}$	$(k-1, 1, -1)$
$\begin{pmatrix} 0 & 1 \\ 1 & k \end{pmatrix}$	(k)	$\begin{pmatrix} 0 & -1 \\ 1 & k \end{pmatrix}$	Not available
$\begin{pmatrix} k & 1 \\ 1 & 0 \end{pmatrix}$	$(0, k, 0)$ $k > 0$	$\begin{pmatrix} k & -1 \\ 1 & 0 \end{pmatrix}$	$(0, k-1, 1, -1)$ $k > 1$

with the last irregular matrix in the last family for $k = 1$:

$\begin{pmatrix} 1 & -1 \\ 1 & 0 \end{pmatrix}$	$(1, -1)$

Example 9.11. Let us now discuss cases formed by different variations of vectors $(3, 4)$ and $(2, 3)$. Here we multiply one or two of the vectors by ± 1 and/or simultaneously swap the order of their coordinates or their orders.

$\begin{pmatrix} 3 & 2 \\ 4 & 3 \end{pmatrix}$	$(1, 2, 1, 0)$	$\begin{pmatrix} 3 & 2 \\ -4 & -3 \end{pmatrix}$	$(-2, 1, 1, 1, 0)$
$\begin{pmatrix} 3 & -2 \\ 4 & -3 \end{pmatrix}$	$(1, 3, -1)$	$\begin{pmatrix} 3 & -2 \\ -4 & 3 \end{pmatrix}$	$(-2, 1, 2, -1)$
$\begin{pmatrix} 4 & 3 \\ 3 & 2 \end{pmatrix}$	$(0, 1, 2, 1, 0)$	$\begin{pmatrix} 4 & 3 \\ -3 & -2 \end{pmatrix}$	$(-1, 3, 1, 0)$
$\begin{pmatrix} 4 & -3 \\ 3 & -2 \end{pmatrix}$	$(0, 1, 3, -1)$	$\begin{pmatrix} 4 & -3 \\ -3 & 2 \end{pmatrix}$	$(-1, 4, -1)$

Proof of Theorem 9.6. Item (i). Here we assume that $a \neq 0$. We start with the case of $SL(2, \mathbb{Z})$ matrices (i.e., the integer matrices with unit determinant). Recall that b/a has a unique odd regular continued fraction:

$$\frac{b}{a} = [a_1; \cdots : a_{2n-1}].$$

So we set

$$a = K_{2n-1}(a_1, \ldots, a_{2n-1}) \quad \text{and} \quad b = K_{2n-1}(a_1, \ldots, a_{2n-1})$$

keeping in mind a general expression of Remark 1.18:

$$\frac{K_{2n-1}(a_1,\ldots,a_{2n-1})}{K_{2n-2}(a_2,\ldots,a_{2n-1})} = [a_1;\cdots:a_{2n-1}].$$

The uniqueness of the odd continued fraction implies the uniqueness of the elements $(a_1,\cdots,a_{2n-1})$ defining the continuants for a and b.

Let us show how to find a_n. Here we note that once (a,b) is fixed then the integer distance from (c,d) to the line through $(0,0)$ and (a,b) is 1, and in addition (a,b) and (c,d) should form a positively oriented basis of $\mathbb{R}^2$ (since $\det M = 1$). Such points (c,d) should be an integer point of the line

$$\begin{pmatrix} K_{2n-2}(a_2,\ldots,a_{2n-1}) \\ K_{2n-1}(a_1,\ldots,a_{2n-1}) \end{pmatrix} + a_n \begin{pmatrix} K_{2n-3}(a_2,\ldots,a_{2n-2}) \\ K_{2n-2}(a_1,\ldots,a_{2n-2}) \end{pmatrix},$$

where all integer points are attained for integer values of the parameter a_n. It is clear that once we know the integer point (c,d), the value of a_n is uniquely defined (see Fig. 9.1, Left) as the solution of the equation:

$$\begin{pmatrix} c \\ d \end{pmatrix} = \begin{pmatrix} K_{2n-2}(a_2,\ldots,a_{2n-1}) \\ K_{2n-1}(a_1,\ldots,a_{2n-1}) \end{pmatrix} + x \begin{pmatrix} K_{2n-3}(a_2,\ldots,a_{2n-2}) \\ K_{2n-2}(a_1,\ldots,a_{2n-2}) \end{pmatrix},$$

which is equivalent to

$$\begin{pmatrix} c \\ d \end{pmatrix} = \begin{pmatrix} K_{2n-1}(a_2,\ldots,a_{2n-1},x) \\ K_{2n}(a_1,\ldots,a_{2n-1},x) \end{pmatrix}.$$

In particular (since $\det(M) = 1$) the last system is equivalent to the linear equation

$$K_{2n-1}(a_2,\ldots,a_{2n-1},x) = c.$$

This concludes the proof for $SL(2,\mathbb{Z})$ matrices.

The proof for $GL(2,\mathbb{Z})$ matrices with negative determinants is analogous, it is left as an exercise. This concludes the proof of Item (i).

Item (ii). The first statements of this item are straightforward. The matrix

$$\pm \begin{pmatrix} 0 & -1 \\ 1 & k \end{pmatrix}$$

does not have a continuant representation by the following reason. One gets $a = 0$ if and only if

$$K_{n-2}(a_2,\ldots,a_{n-1}) = 0$$

with non-negative values $a_2,\ldots,a_{n-1}$. This is only possible in the exceptional case when $n = 1$, i.e., for the sequences (a_1). It remains to say that

$$M_{a_1} = \begin{pmatrix} 0 & 1 \\ 1 & a_1 \end{pmatrix} \neq \begin{pmatrix} 0 & -1 \\ 1 & k \end{pmatrix}$$

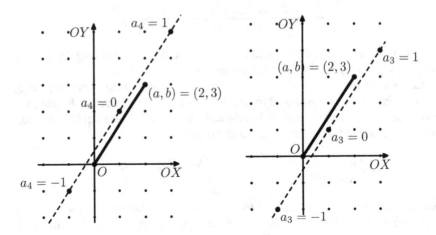

Fig. 9.1 Possible values for (c,d) with $(a,b) = (2,3)$ are on the two lines lines: one line for the case of SL(2, ℤ) matrices (Left) and one line for the case of $\det(M) = -1$ (Right).

for any values of a and k. This concludes the proof. □

Let us illustrate the proof of Theorem 9.6(i) with the following example.

Example 9.12. *Let us find all possible integers c and d such that the matrix*

$$\begin{pmatrix} 2 & c \\ 3 & d \end{pmatrix}$$

is in GL(2, ℤ).

First, let us find all SL(2, ℤ) matrices. In this case we consider the odd continued fraction for 3/2:

$$\frac{3}{2} = [1; 1 : 1].$$

Now, by Theorem 9.6 all SL(2, ℤ) matrices will be listed by the sequences

$$(1, 1, 1, a_4).$$

Then, we have

$$\begin{pmatrix} 2 & K_3(1, 1, a_4) \\ 3 & K_4(1, 1, 1, a_4) \end{pmatrix} = \begin{pmatrix} 2 & 2a_4 + 1 \\ 3 & 3a_4 + 2 \end{pmatrix},$$

where a_4 is any integer. For instance if $a_4 = -1, 0, 1$, then the vectors (c, d) are respectively as follows

$$(-1, -1), \qquad (1, 2), \qquad (3, 5)$$

(see Fig. 9.1, Left).

Now for the case of negative determinant we have:

$$\frac{3}{2} = [1;2].$$

Now, by Theorem 9.6 all $SL(2,\mathbb{Z})$ matrices will be listed by the sequences

$$(1,2,a_3).$$

Then, we have

$$\begin{pmatrix} 2 & K_2(2,a_3) \\ 3 & K_3(1,2,a_3) \end{pmatrix} = \begin{pmatrix} 2 & 2x+1 \\ 3 & 3x+1 \end{pmatrix},$$

where a_3 is any integer. For instance if $a_3 = -1,0,1$, then the vectors (c,d) are respectively as follows

$$(-1,-2), \qquad (1,1), \qquad (3,4)$$

(see Fig. 9.1, Right).

9.4 An expression of matrices in terms of M_S and M_T

Finally we would like to mention that there is a simple way to write down an expression for any $SL(2,\mathbb{Z})$ matrix in terms of generators M_S and M_T (introduced in Section 9.1 above) based on the statement of Theorem 9.6.

We start with expressions for the exceptional matrices

$$\begin{pmatrix} 0 & -1 \\ 1 & k \end{pmatrix} = M_T^k \quad \text{and} \quad -\begin{pmatrix} 0 & -1 \\ 1 & k \end{pmatrix} = S^2 \cdot M_T^k.$$

By Theorem 9.6 any other $SL(2,\mathbb{Z})$ matrix M can be represented as

$$M = M_{a_1,\dots,a_{2n}} \quad \text{or} \quad M = -M_{a_1,\dots,a_{2n}}$$

for some sequence of even number of integers as in Theorem 9.6.

Now the formula for M follows from the following two observations:

$$M_{ab} = M_S^2 \cdot (M_S \cdot M_T^{-a}) \cdot (M_S \cdot M_T^b) \quad \text{and} \quad S^2 = -1,$$

and the fact that

$$M_{a_1,\dots,a_{2n}} = \prod_{i=1}^{n} M_{a_{2i-1},a_{2i}}.$$

We have

$$M_{a_1,\dots,a_{2n}} = \prod_{i=1}^{n} \left(M_S^2 \cdot (M_S \cdot M_T^{-a_{2i-1}}) \cdot (M_S \cdot M_T^{a_{2i}}) \right) \quad \text{and}$$

$$-M_{a_1,\dots,a_{2n}} = S^2 \prod_{i=1}^{n} \left(M_S^2 \cdot (M_S \cdot M_T^{-a_{2i-1}}) \cdot (M_S \cdot M_T^{a_{2i}}) \right).$$

Remark 9.13. Note that the above observation provides the evidence to the fact that M_S and M_T generate SL(2,$\mathbb{Z}$).

9.5 Exercises

Exercise 9.1. Find the representation in terms of the continuants of the following matrices

(a) $\begin{pmatrix} 3 & 4 \\ 5 & 7 \end{pmatrix}$;   (b) $\begin{pmatrix} -3 & 2 \\ 5 & -3 \end{pmatrix}$;   (c) $\begin{pmatrix} 11 & 9 \\ -6 & -5 \end{pmatrix}$.

Exercise 9.2. Find a decomposition in M_S and M_T of the following matrices

(a) $\begin{pmatrix} 1 & 5 \\ 2 & 11 \end{pmatrix}$;   (b) $\begin{pmatrix} 7 & -13 \\ -5 & 8 \end{pmatrix}$;   (c) $\begin{pmatrix} -3 & -2 \\ -4 & -3 \end{pmatrix}$.

Exercise 9.3. Find all possible integers c and d such that the matrix

$$\begin{pmatrix} 7 & c \\ 5 & d \end{pmatrix}$$

is in GL(2,$\mathbb{Z}$).

Exercise 9.4. Write the coefficients of the matrix $M_{a_1,\dots,a_n}$, for $(a_1,\dots,a_n)$ being as follows:
(a) (1);   (b) $(1,2,3,4,5)$;   (c) $(-3,2,1,4,-2)$;   (d) $(0,2,4,2,0)$.

Exercise 9.5. Finish the proof of Theorem 9.6(i), namely study the case of GL(2,$\mathbb{Z}$) matrices with negative determinants.

Chapter 10
Semigroup of Reduced Matrices

There are several ways to construct reduced matrices, however as a rule they are closely related with each other. The reason for that might be the structure of the group. We should mention that the approach here is rather different to the classical approach for closed fields via Jordan blocks.

In this chapter we choose reduced matrices in a rather natural way such that they form a semigroup with respect to the matrix multiplication. An extra benefit of this choice is as follows: there exists a simple description of such matrices, we discuss it in Section 10.1 (see Theorem 10.7).

An important property of reduced matrices is that every integer conjugacy class of $GL(2,\mathbb{Z})$ possesses at least one reduced matrix in it, we show this in Section 10.2. In general there might be more than one reduced matrix in a conjugacy class, but as we show in Chapter 11 the number of such reduced matrices is finite.

10.1 Definition and basic properties of reduced matrices

We start in Subsection 10.1.1 with the definition of a reduced matrix. Then we compute continuant representations of all reduced matrices in Subsection 10.1.2. Further In Subsection 10.1.3 we prove a simple necessary and sufficient condition for a matrix to be reduced. Finally in Subsection 10.1.4 we specify the LLS cycles for reduced matrices (via the elements of the matrix).

10.1.1 Reduced matrices

We start with the elementary building blocks for generating all reduced matrices.

Definition 10.1. For a positive integer a we say that the matrix

© Springer-Verlag GmbH Germany, part of Springer Nature 2022
O. N. Karpenkov, *Geometry of Continued Fractions*,
Algorithms and Computation in Mathematics 26,
https://doi.org/10.1007/978-3-662-65277-0_10

$$M_a = \begin{pmatrix} 0 & 1 \\ 1 & a \end{pmatrix}$$

is the *elementary reduced matrix*.

Starting with elementary reduced matrices we generate the semigroup of reduced matrices.

Definition 10.2. We say that a matrix is *reduced* if it is a product of elementary reduced matrices.

Directly from the definition of reduced matrices we get the following remarkable property.

Proposition 10.3. *The set of all reduced matrices is a semigroup with respect to the matrix multiplication; it is freely generated by elementary reduced matrices.*

Proof. The fact that the set of all reduce matrices is a semigroup follows directly from the definition. It is freely generated by the next theorem (Theorem 10.5), where we show that the LLS cycle for $M_{a_1,\dots,a_n}$ is $(a_1,\dots,a_n)$. Therefore, different products of elementary reduced operators have different LLS cycles. Recall that LLS cycles are invariants of matrices, and hence the semigroup is free generated.   □

Remark 10.4. (**On a complete invariant of reduced matrices.**) Proposition 10.3 implies that the set of hyperbolic reduced matrices is in one-to-one correspondence with the set of finite sequences of positive integer elements. Namely, the map

$$(a_1,\dots,a_n) \mapsto M_{a_1,\dots,a_n}$$

is a bijection. Furthermore, as we show later in Theorem 10.5 the sequence

$$(a_1,\dots,a_n)$$

is the LLS cycle for $M_{a_1,\dots,a_n}$.

10.1.2 Continuant representations of reduced matrices

In the proof of Theorem 10.7 we use the following important structural theorem.

Theorem 10.5. *Let n, $a_1,\dots,a_n$ be positive integers. Consider the matrix*

$$M = M_{a_1,\dots,a_n} \left(= \prod_{i=1}^{n} M_{a_i} \right).$$

Then the following two statements hold.
(i) We have

$$M = \begin{pmatrix} K_{n-2}(a_2,\ldots,a_{n-1}) & K_{n-1}(a_2,\ldots,a_n) \\ K_{n-1}(a_1,a_2,\ldots,a_{n-1}) & K_n(a_1,a_2,\ldots,a_n) \end{pmatrix}.$$

(ii) The LLS cycle of M is

$$(a_1,a_2,\ldots,a_n).$$

Example 10.6. For the case of the above matrix

$$M = \begin{pmatrix} 3 & 7 \\ 5 & 12 \end{pmatrix}$$

we have

$$M = \begin{pmatrix} K_2(1,2) & K_3(1,2,2) \\ K_3(1,1,2) & K_4(1,1,2,2) \end{pmatrix}$$

and the LLS cycle for M is $(1,1,2,2)$.

Proof. Item (i) follows directly from Proposition 9.3.

Item (ii). Consider the sequence of integer points

$$(x_k, y_k) = M^k(1,0), \quad \text{for } k = 1,2,3,\ldots$$

By Item (i) for every k the coordinates x_k and y_k are relatively prime and

$$\frac{y_k}{x_k} = [(a_1;a_2 : \cdots : a_n)^k].$$

Therefore, by Theorem 3.1, all the points (x_k, y_k) are vertices of the principal parts of the sail the periodic continued fraction

$$\alpha = [(a_1;a_2 : \cdots : a_n)].$$

Therefore, the direction of the vector $(1,\alpha)$ is the limiting direction for the sequence of directions for the vectors (x_k, y_k), and thus we have

$$\lim_{k \to \infty} \frac{y_k}{x_k} = \alpha.$$

Hence $(1,\alpha)$ is one of the eigenvectors corresponding to the maximal eigenvalue. By construction, the LLS sequence for α is periodic with period

$$(a_1,a_2,\ldots,a_n).$$

Finally by Proposition 4.13 the sail for α starting from some element coincides with the sail for M. Therefore, since the sail for M is periodic, the sail for α has the same period, i.e.,

$$(a_1,a_2,\ldots,a_n).$$

Furthermore,

$$M^2(1,0) = M_{a_1,\ldots,a_n,a_1,\ldots,a_n}(1,0)$$
$$= \left(K_{2n-2}(a_2,\ldots,a_n,a_1,\ldots,a_{n-1}, K_{2n-1}(a_1,\ldots,a_n,a_1,\ldots,a_{n-1})\right).$$

Hence M^2 acts as a shift by n vertices on the sails associated to M. Therefore, by Definition 8.13 the LLS cycle of M is $(a_1,a_2,\ldots,a_n)$. This concludes the proof of Item (ii). □

10.1.3 A necessary and sufficient condition for a matrix to be reduced

Now we formulate an important property of reduced matrices that allows us immediately to decide whether a given matrix is reduced or not.

Theorem 10.7. *A matrix*

$$M = \begin{pmatrix} a & c \\ b & d \end{pmatrix}$$

in $GL(2,\mathbb{Z})$ *is reduced if and only if*

$$d \geq b \geq a \geq 0.$$

Example 10.8. For instance, the matrix

$$\begin{pmatrix} 3 & 7 \\ 5 & 12 \end{pmatrix}$$

is reduced. In fact, this matrix is written as the product of elementary reduced matrices as follows

$$\begin{pmatrix} 3 & 7 \\ 5 & 12 \end{pmatrix} = \begin{pmatrix} 0 & 1 \\ 1 & 1 \end{pmatrix}^2 \cdot \begin{pmatrix} 0 & 1 \\ 1 & 2 \end{pmatrix}^2.$$

Remark 10.9. In some manuscripts Theorem 10.7 is considered as the main definition of reduced matrices.

Proof of Theorem 10.7. Necessary condition. Let M be a reduced matrix. Then by Theorem 10.5 there exist positive integers $n, a_1,\ldots,a_n$ such that

$$M = \begin{pmatrix} K_{n-2}(a_2,\ldots,a_{n-1}) & K_{n-1}(a_2,\ldots,a_n) \\ K_{n-1}(a_1,a_2,\ldots,a_{n-1}) & K_n(a_1,a_2,\ldots,a_n) \end{pmatrix}.$$

Hence the condition

$$d \geq b \geq a \geq 0$$

is equivalent to the condition

$$K_n(a_1,\ldots,a_n) \geq K_{n-1}(a_1,\ldots,a_{n-1}) \geq K_{n-2}(a_2,\ldots,a_{n-1}) \geq 0. \qquad (10.1)$$

The first inequality follows directly from

$$K_n(a_1,\dots,a_n) = a_n K_{n-1}(a_1,\dots,a_{n-1}) + K_{n-2}(a_1,\dots,a_{n-2})$$
$$\geq K_{n-1}(a_1,\dots,a_{n-1}).$$

Here we have the equality if and only if $n = 1$ and $a_1 = 1$.

The second inequality of Inequality 10.1 is proven by analogy (keeping in mind symmetry of the continuant for palindromic substitutions).

Finally, by the construction, the continuants are nonnegative. This concludes the proof of the necessary condition of the theorem.

Sufficient condition. Let us now start with a matrix

$$M = \begin{pmatrix} a & c \\ b & d \end{pmatrix}$$

that satisfies the condition

$$d > b \geq a \geq 0.$$

First of all in the case where $a = 0$ we immediately have that $b = 1$. Hence we have reduced matrices.

Now let $a > 0$. In this situation we consider two different cases with respect to the sign of the determinant of the matrix M.

Case I. Let $\det M = 1$. Then consider the odd continued fraction for b/a:

$$\frac{b}{a} = [a_1; a_2 : \cdots : a_{2n-1}].$$

In particular, we have

$$a = K_{2n-2}(a_2,\dots,a_{2n-1});$$
$$b = K_{2n-1}(a_1,a_2,\dots,a_{2n-1}).$$

Since $\det M = 1$, there exists a positive integer λ, such that

$$c = \lambda K_{2n-2}(a_2,\dots,a_{2n-1}) + K_{2n-3}(a_2,\dots,a_{2n-2});$$
$$d = \lambda K_{2n-1}(a_1,a_2,\dots,a_{2n-1}) + K_{2n-2}(a_1,a_2,\dots,a_{2n-2}),$$

or, equivalently,

$$c = K_{2n-1}(a_2,\dots,a_{2n-1},\lambda);$$
$$d = K_{2n}(a_1,\dots,a_{2n-1},\lambda).$$

Hence

$$M = \begin{pmatrix} K_{2n-2}(a_2,\dots,a_{2n-1}) & K_{2n-1}(a_2,\dots,a_{2n-1},\lambda) \\ K_{2n-1}(a_1,a_2,\dots,a_{2n-1}) & K_{2n}(a_1,\dots,a_{2n-1},\lambda) \end{pmatrix}.$$

Therefore, by Theorem 10.5

$$M = M_{a_1} \cdot \ldots \cdot M_{a_{2n-1}} \cdot M_\lambda.$$

Case II. Now let $\det M = -1$. Then consider the even continued fraction for b/a:

$$\frac{b}{a} = [a_1; a_2 : \cdots : a_{2n}]$$

and similarly construct λ from the condition $\det M = -1$.

In analogy to Case I, we have

$$M = M_{a_1} \cdot \ldots \cdot M_{a_{2n}} \cdot M_\lambda.$$

We leave further details of the proof as an exercise for the reader and conclude Case II.

Finally it remains to say that for the case $d = b \geq a \geq 0$ the only possible matrix is

$$\begin{pmatrix} 0 & 1 \\ 1 & 1 \end{pmatrix}$$

which is reduced by definition. This concludes the proof of the sufficient condition of the theorem.   □

10.1.4 LLS cycles of reduced matrices

It turns out that it is very simple to write the LLS cycles for reduced matrices with given coefficients.

Corollary 10.10. *Let*

$$M = \begin{pmatrix} a & c \\ b & d \end{pmatrix}$$

be a reduced matrix. The LLS cycle for M is

$$(a_1, a_2, \ldots, a_n).$$

Here n is even for $\mathrm{SL}(2, \mathbb{Z})$ *matrices and n is odd for* $\mathrm{GL}(2, \mathbb{Z})$ *matrices with negative determinant. Then the numbers* $a_1, a_2, \ldots, a_n$ *are defined as the elements of the regular continued fraction*

$$\frac{d}{c} = [a_1; a_2 : \cdots : a_n].$$

In addition we have

$$M = M_{a_1, a_2, \ldots, a_n}.$$

Proof. By Theorem 10.5 and Remark 1.18 we have

$$\frac{d}{c} = \frac{K_n(a_1, a_2, \ldots, a_n)}{K_{n-1}(a_2, a_2, \ldots, a_n)} = [a_1; a_2 : \cdots : a_n]$$

hence

$$M = M_{a_1,a_2,\dots,a_n},$$

and the LLS cycle for M is $(a_1, a_2, \dots, a_n)$.

□

Example 10.11. Let us write a period of the LLS sequence for the matrix

$$M = \begin{pmatrix} 3 & 7 \\ 5 & 12 \end{pmatrix}.$$

First note that

$$\det(M) = 1$$

and hence we should take the even regular continued fraction. Then

$$\frac{d}{c} = \frac{12}{7} = [1; 1 : 2 : 2].$$

Therefore, one of the LLS periods for M is $(1, 1, 2, 2)$.

10.2 Existence of reduced matrices in every integer conjugacy class of $GL(2,\mathbb{Z})$

We have the following rather straightforward statement on existence of a reduced matrix integer conjugate to a given one.

Theorem 10.12. *Let M be a hyperbolic $GL(2,\mathbb{Z})$ matrix with irrational eigenvalues. Then there exists a reduced matrix M' such that M is integer conjugate either to M' or to $-M'$.*

Proof. Let $(a_1, \dots, a_n)$ be the LLS cycle of M. Set $M' = M_{a_1,\dots,a_n}$. Then by Theorem 10.5(ii) the reduced matrix M' has the same LLS cycle as M. Hence by Corollary 8.16 the matrix M is integer conjugate to either M' or to $-M'$. □

Remark 10.13. Reduced matrices are used further in the Gauss Reduction Theory. We list all reduced matrices integer conjugate to a given one later in Theorem 11.8.

We conclude this chapter with the following important result.

Proposition 10.14. *Let M be a hyperbolic $GL(2,\mathbb{Z})$ matrix with irrational eigenvalues. Then the LLS cycle of M^2 is a twice repeated sequence of integers.*

Remark 10.15. A non-trivial case here is the case of matrices that have negative eigenvalues. Such matrices have odd LLS cycles.

Proof. First of all let us study the LLS cycles of reduced matrices. Let

$$M = M_{a_1,\dots,a_n}$$

be a reduced matrix for the sequence of positive integers $(a_1, \ldots, a_n)$. Then from Definition 9.2 we have

$$M^2 = M^2_{a_1, \ldots, a_n} = M_{a_1, \ldots, a_n, a_1, \ldots, a_n}.$$

Hence the LLS cycle for M^2 is twice the LLS cycle for M.

For an arbitrary M in $GL(2, \mathbb{Z})$ with irrational real eigenvalues it is known that M is integer conjugate to $\pm M'$, where M' is a reduced matrix (see Theorem 10.12). Now recall that the LLS cycles are invariant under both integer conjugation and multiplication by -1. Therefore, the period of LLS sequence corresponding to M^2 will be twice the period of the LLS sequence for M. $\qquad \square$

10.3 Exercises

Exercise 10.1. Write the following reduced matrices as products of elementary reduced matrices:

(a) $\begin{pmatrix} 7 & 23 \\ 17 & 56 \end{pmatrix}$;   (b) $\begin{pmatrix} 5 & 13 \\ 8 & 21 \end{pmatrix}$;   (c) $\begin{pmatrix} 7 & 36 \\ 22 & 113 \end{pmatrix}$.

Chapter 11
Continued Fractions and $\mathrm{SL}(2,\mathbb{Z})$ Conjugacy Classes. Elements of Gauss's Reduction Theory

In this chapter we study the structure of the conjugacy classes of $\mathrm{GL}(2,\mathbb{Z})$. Recall that $\mathrm{GL}(2,\mathbb{Z})$ is the group of all invertible matrices with integer coefficients. The group $\mathrm{GL}(2,\mathbb{Z})$ has another commonly used notation: $\mathrm{SL}_2^{\pm}\mathbb{Z}$, indicating that all matrices of the group has determinants equal either to 1 or to -1.

We say that the matrices A and B from $\mathrm{GL}(2,\mathbb{Z})$ are *integer conjugate* if there exists an $\mathrm{GL}(2,\mathbb{Z})$ matrix C such that $B = CAC^{-1}$.

Notice that in the case of algebraically closed fields (say in $\mathbb{C}$), every matrix is conjugate to its Jordan normal form. The situation with $\mathrm{GL}(n,\mathbb{Z})$ is much more complicated, since $\mathbb{Z}$ is not a field. A description of integer conjugacy classes in the two-dimensional case is the subject of Gauss's reduction theory, where conjugacy classes are classified by periods of certain periodic continued fractions (for additional information we refer to [83, 143, 113, 145]). We present first steps in the study of multidimensional Gauss reduction theory in Chapter 25, in the second part of this book.

In this chapter we discuss the main elements of classical Gauss reduction theory based on the theory of geometric continued fractions studied in previous chapters (see also [105]). In particular, we discuss some statistical questions and formulate several open problems.

11.1 Conjugacy classes of $\mathrm{GL}(2,\mathbb{Z})$ in general

It is natural to split $\mathrm{GL}(2,\mathbb{Z})$ into the following three cases.

Elliptic case (complex spectra): Consider $\mathrm{GL}(2,\mathbb{Z})$ matrices whose characteristic polynomials have a pair of complex conjugate roots. There are exactly three integer conjugacy classes of such matrices (see Theorem 11.2), represented by

$$\begin{pmatrix} 1 & 1 \\ -1 & 0 \end{pmatrix}, \quad \begin{pmatrix} 0 & 1 \\ -1 & 0 \end{pmatrix}, \quad \text{and} \quad \begin{pmatrix} 0 & 1 \\ -1 & -1 \end{pmatrix}.$$

© Springer-Verlag GmbH Germany, part of Springer Nature 2022
O. N. Karpenkov, *Geometry of Continued Fractions*,
Algorithms and Computation in Mathematics 26,
https://doi.org/10.1007/978-3-662-65277-0_11

Parabolic case (rational spectra): Let us now study matrices whose characteristic polynomials have rational roots. (This contains the degenerate case of two coinciding roots as in this case that would be a double root of a quadratic polynomial with integer coefficients, which actually equal to 1). Such matrices have eigenvalues equal to ± 1, each of these matrices is integer conjugate to exactly one matrix of the following families:

$$\begin{pmatrix} 1 & m \\ 0 & 1 \end{pmatrix} \quad \text{for} \quad m \geq 0;$$

$$\begin{pmatrix} -1 & -m \\ 0 & -1 \end{pmatrix} \quad \text{for} \quad m \geq 0;$$

$$\begin{pmatrix} 1 & 0 \\ 0 & -1 \end{pmatrix} \quad \text{and} \quad \begin{pmatrix} 1 & 1 \\ 0 & -1 \end{pmatrix}.$$

See Theorem 11.6 below (see also in [16]).

Hyperbolic case (real irrational spectra): The case of hyperbolic matrices with real distinct irrational eigenvalues is the most complicated. The conjugacy classes of matrices can be represented by reduced matrices introduced in Chapter 10. LLS-cycles here will be complete invariants (up to the multiplication by ± 1). However there can be more than one reduced matrix in a conjugacy class; the number of such matrices is finite and it is equivalent to the minimal period of the LLS sequence for the corresponding to conjugacy class of matrices (see Theorem 11.8). For additional information on Gauss Reduction Theory we refer to [83, 143, 113, 145] (including questions of matrix similarity in [38]).

11.2 Elliptic case

The matrices that we study in this section are rather important as all of them are of finite order. In some sense they are similar to finite order rotation matrices of Euclidean geometry. Invariance under the action of such matrices implies symmetry properties for the objects. Despite the Euclidean case, the orders of symmetries can take only the values 1, 2, 3, 4, and 6. This in particular implies that there are no integer regular polygons of orders 5 and $n \geq 7$ (see Proposition 2.26). To show this we prove the following proposition.

Proposition 11.1. *Let A be a* $GL(2, \mathbb{Z})$*-matrix with complex conjugate eigenvalues, then either*

$$A^3 = \text{Id}, \quad or \quad A^4 = \text{Id}, \quad or \quad A^6 = \text{Id}.$$

Proof. First of all note that the characteristic polynomial of A in variable t is

$$t^2 - \text{tr}(A)t + \det(A)$$

(here $\text{tr}(A)$ denotes the trace of A) and its discriminant is

$$\mathrm{tr}^2(A) - 4\det(A).$$

If $\det A = -1$, then the discriminant is positive, and the eigenvalues are real (we do not consider such case). Therefore, $\det A = 1$, and the discriminant is

$$\mathrm{tr}^2(A) - 4.$$

Since all the elements of A are integers, there are only three values of the trace that secure negative values of the discriminant: $-1, 0, 1$. The corresponding characteristic polynomials are respectively factors of the polynomials

$$x^3 - 1, \qquad x^4 - 1, \quad \text{and} \quad x^6 - 1.$$

Therefore, all the eigenvalues are the corresponding roots of the unity. Since matrix A is real and its eigenvalues are complex conjugate, the eigenvalues are distinct to each other. This implies that matrix A is diagonalizable in some complex basis, and, therefore, either A^3, or A^4, or A^6 are identity matrices.

□

Let us prove the classification theorem now.

Theorem 11.2. *Any* $\mathrm{GL}(2,\mathbb{Z})$*-matrix with complex conjugate eigenvalues is integer conjugate to precisely one of the following three matrices:*

$$\begin{pmatrix} 0 & 1 \\ 1 & -1 \end{pmatrix}, \qquad \begin{pmatrix} 0 & 1 \\ 1 & 0 \end{pmatrix}, \quad \text{and} \quad \begin{pmatrix} 0 & 1 \\ 1 & 1 \end{pmatrix}.$$

Proof. Consider a $\mathrm{GL}(2,\mathbb{Z})$-matrix A whose eigenvalues are complex conjugate numbers. Then the level sets of the MD-characteristic for A are ellipses.

Consider an integer point v with minimal possible value of the MD-characteristic. This point will be on the ellipse that does not contain integer points in its interior except for the origin. By Proposition 7.11 $A(v)$ is also on this ellipse, and hence the triangle with vertices O, v, and $A(v)$ is empty.

Hence the vectors v and $A(v)$ generate $\mathbb{Z}^2$. Then in the basis $(v, A(v))$ the matrix A will be of the form

$$\begin{pmatrix} 0 & m \\ 1 & n \end{pmatrix}.$$

Firstly, since $\det A = 1$, we have $m = -1$.
Secondly, since $\mathrm{tr} A \in \{-1, 0, 1\}$, we have $n \in \{-1, 0, 1\}$.
Therefore, A is integer conjugate to precisely one of the following three matrices:

$$\begin{pmatrix} 0 & 1 \\ 1 & -1 \end{pmatrix}, \qquad \begin{pmatrix} 0 & 1 \\ 1 & 0 \end{pmatrix}, \quad \text{and} \quad \begin{pmatrix} 0 & 1 \\ 1 & 1 \end{pmatrix}.$$

This concludes the proof.

□

Remark 11.3. Assume we are given a $GL(2,\mathbb{Z})$ matrix M with complex eigenvalues. Then it is very simple to determine which of one of the three above matrices is integer conjugate to M. We should compute the trace of M. It could be either -1, or 0, or 1 which will completely determine the integer conjugacy class of the matrix.

11.3 Parabolic case

In this subsection we study conjugacy classes of matrices that have a rational eigenvalue. Let us start with the following preliminary proposition.

Proposition 11.4. *Let M be a $GL(2,\mathbb{Z})$ matrix with a rational eigenvalue. Then the following hold.*
(i) All the eigenspaces of M are integer, namely, they contain integer points.
(ii) There exists an integer m such that the matrix M is integer conjugate to either

$$\begin{pmatrix} 1 & m \\ 0 & 1 \end{pmatrix} \quad or \quad \begin{pmatrix} 1 & m \\ 0 & -1 \end{pmatrix} \quad or \quad \begin{pmatrix} -1 & m \\ 0 & -1 \end{pmatrix}.$$

Remark 11.5. In particular, if a $GL(2,\mathbb{Z})$-matrix M has a rational eigenvalue, then both its eigenvalues are in the set $\{-1,1\}$.

Proof. Item (i). This follows directly from the standard procedure of finding eigenvectors. We leave this as an easy exercise on linear algebra.

Item (ii). Assume M has a rational eigenvector. If M is a scalar multiplication matrix, then the fact that $|\det M| = 1$ implies that M is $\pm\mathrm{Id}$. In the case where M is not a scalar multiplication matrix, there exists a one-dimensional eigenspace corresponding to the rational value, that is integer by Item (i). Let v be an elementary integer vector (i.e., with relatively prime coordinates) of this eigenspace. Now consider any integer basis of the lattice $\mathbb{Z}^2$, that contains v as the first element of the basis. In this basis the matrix M has a form

$$\begin{pmatrix} l & m \\ 0 & n \end{pmatrix}.$$

Since $M \in GL(2,\mathbb{Z})$, we have

$$l = \pm 1 \quad \text{and} \quad n = \pm 1.$$

Therefore, the eigenvalues of M are either 1 or -1. In the case where M has an eigenvalue 1, from the above we have the following: M is integer conjugate to either

$$\begin{pmatrix} 1 & m \\ 0 & 1 \end{pmatrix} \quad \text{or} \quad \begin{pmatrix} 1 & m \\ 0 & -1 \end{pmatrix}$$

Finally if both of the eigenvalues equal -1, then M is integer conjugate to

$$\begin{pmatrix} -1 & m \\ 0 & -1 \end{pmatrix}.$$

This concludes the proof. □

The classification theorem in the parabolic case of rational spectra matrices is as follows.

Theorem 11.6. *Let M be a* $GL(2,\mathbb{Z})$ *matrix with a rational eigenvalue. Then M is integer conjugate to a precisely one matrix of the following list:*

- $SL(2,\mathbb{Z})$ *matrices* $\begin{pmatrix} 1 & m \\ 0 & 1 \end{pmatrix}$, *where* $m \in \mathbb{Z}_{\geq 0}$;

- $SL(2,\mathbb{Z})$ *matrices* $\begin{pmatrix} -1 & -m \\ 0 & -1 \end{pmatrix}$, *where* $m \in \mathbb{Z}_{\geq 0}$;

- *finally* $GL(2,\mathbb{Z}) \setminus SL(2,\mathbb{Z})$ *is represented by the following two matrices*

$$\begin{pmatrix} 1 & 0 \\ 0 & -1 \end{pmatrix} \quad \text{and} \quad \begin{pmatrix} 1 & 1 \\ 0 & -1 \end{pmatrix}.$$

First of all we prove the following lemma.

Lemma 11.7. *Two matrices*

$$M_1 = \begin{pmatrix} 1 & m_1 \\ 0 & 1 \end{pmatrix} \quad \text{and} \quad M_2 = \begin{pmatrix} 1 & m_2 \\ 0 & 1 \end{pmatrix}$$

are integer conjugate if and only if $m_1 = \pm m_2$.

The proof of this lemma is a nice exercise in linear algebra.

Proof. Recall that M_1 and M_2 are integer conjugate if and only if there exists a $GL(2,\mathbb{Z})$ matrix A such that

$$M_1 = A M_2 A^{-1}$$

or, equivalently,

$$M_1 A = A M_2.$$

Assume that

$$A = \begin{pmatrix} a & c \\ b & d \end{pmatrix}.$$

Then we have

$$\begin{pmatrix} 1 & m_1 \\ 0 & 1 \end{pmatrix} \cdot \begin{pmatrix} a & c \\ b & d \end{pmatrix} = \begin{pmatrix} a+m_1b & c+m_1d \\ b & d \end{pmatrix};$$

$$\begin{pmatrix} a & c \\ b & d \end{pmatrix} \cdot \begin{pmatrix} 1 & m_2 \\ 0 & 1 \end{pmatrix} = \begin{pmatrix} a & m_2a+c \\ b & m_2b+d \end{pmatrix}.$$

If the case where the above two matrices coincide we have

$$a = a + m_1 b$$

(here we compare the top diagonal element of the matrix). Therefore,

$$m_1 b = 0.$$

If $m_1 = 0$, then M_1 is the identity matrix and it is not (integer) conjugate to any other matrix. If $m_1 \neq 0$ then $b = 0$, and hence

$$a = \pm 1 \qquad \text{and} \qquad d = \pm 1.$$

Now let us compare the top right elements of the matrices. We have

$$c + m_1 d = m_2 a + c.$$

The last is equivalent to

$$m_1 d = m_2 a.$$

Since $a, d = \pm 1$ we have

$$m_1 = \pm m_2.$$

So if the absolute values of m_1 and m_2 are distinct the matrices M_1 and M_2 are not integer conjugate.

Finally it remains to show that if $m_1 = -m_2$ then M_1 is integer conjugate to M_2. Indeed, we have

$$\begin{pmatrix} 1 & -m \\ 0 & 1 \end{pmatrix} = \begin{pmatrix} 1 & 0 \\ 0 & -1 \end{pmatrix} \cdot \begin{pmatrix} 1 & m \\ 0 & 1 \end{pmatrix} \cdot \begin{pmatrix} 1 & 0 \\ 0 & -1 \end{pmatrix}.$$

This concludes the proof for the case of two unit eigenvalues.   □

Proof of Theorem 11.6. Let us study first the case when both eigenvalues of M equal 1. Then by Proposition 11.4 the matrix M is integer conjugate to

$$\begin{pmatrix} 1 & m \\ 0 & 1 \end{pmatrix}.$$

and hence we get the matrices of the first item in Theorem 11.6 by Lemma 11.6.

The case of two eigenvalues equal to -1 is similar to the above one. Here we have a similar classification after multiplying matrices by -1.

Finally let one of the eigenvalues of M equal 1 and the other equal -1. By Proposition 11.4 the matrix M is equivalent to

$$\begin{pmatrix} 1 & m \\ 0 & -1 \end{pmatrix}$$

for some integer m. The last matrix has two eigenvectors:

$$v = (1, 0) \qquad \text{and} \qquad w = (-n, 2).$$

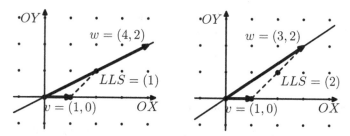

Fig. 11.1 Two options for the case of distinct rational eigenvalues: $LLS = (1)$ (Left) and $LLS = (2)$ (Right).

Now two matrices with rational eigenvectors are equivalent if and only if the LLS sequences of the angles with edges containing eigenvectors coincide. In our case we have two different cases here:

- the LLS sequence is equal to (1) when n is even;
- the LLS sequence is equal to (2) when n is odd.

(See Fig. 11.1.)

The integer conjugacy classes of these two cases of matrices are represented by

$$\begin{pmatrix} 1 & 0 \\ 0 & -1 \end{pmatrix} \quad \text{and} \quad \begin{pmatrix} 1 & 1 \\ 0 & -1 \end{pmatrix}.$$

This concludes the proof for the last case of eigenvalues of different signs.   □

11.4 Hyperbolic case

In this section we study conjugacy classes of $GL(2, \mathbb{Z})$-matrices with real irrational eigenvalues. In particular we show that LLS cycles are complete invariants of the conjugacy classes and list all reduced matrices integer conjugate to a given one. As an important consequence we get the statement on algebraicity of periodic sails.

11.4.1 The set of reduced matrices integer conjugate to a given one

The following structural theorem provides description of conjugacy classes in $GL(2, \mathbb{Z})$.

Theorem 11.8. *Let M be a hyperbolic $GL(2, \mathbb{Z})$ matrix with irrational eigenvalues and let*

$$(a_1, \ldots, a_n)$$

be its LLS cycle. Finally let m be the minimal length of the period of the LLS cycle.
Then the list of all reduced matrices integer conjugate simultaneously either to M
or to −M consists of m matrices of the form

$$M_{a_{1+k},\dots,a_{n+k}} \quad \text{for } k = 1, \dots, m.$$

(Here we define indices cyclically modulo n.)

Proof. We know that two matrices have the same LLS sequences if and only if
their unions of eigenlines are integer congruent to each other. Hence M could be
conjugate only to reduced matrices commuting with

$$M_{a_{1+k},\dots,a_{n+k}} \quad \text{for } k = 1, \dots, m.$$

(These are the only matrices that have such LLS sequences.) Such matrices are some
powers of these matrices.

Finally the shift of the LLS sequence of M^2 by n vertices uniquely determines
the reduced matrices. $\qquad\qquad\Box$

11.4.2 Complete invariant of integer conjugacy classes

Theorem 11.8 implies the following fundamental result.

Corollary 11.9. (On complete invariant of integer conjugacy classes.) *LLS cycles*
are complete invariants of conjugacy classes of $\mathrm{PGL}(2,\mathbb{Z})$ *matrices with distinct*
positive real eigenvalues (i.e., up to the multiplication by -1*). Namely, we have the*
following two statements:
(i) If matrices M_1 and M_2 have the same LLS cycles, then M_1 is integer conjugate
to either M_2 or $-M_2$.
(ii) Any arbitrary cyclic sequence is realised as the LLS cycle for some integer
conjugacy class of hyperbolic $\mathrm{GL}(2,\mathbb{Z})$ *matrices with irrational eigenvalues.* $\quad\Box$

11.4.3 Algebraicity of matrices with periodic LLS sequences

We conclude this section with a general statement on angles with periodic LLS
sequences.

Corollary 11.10. *A sail with a periodic LLS sequence is algebraic (i.e., a sail of*
some hyperbolic matrix in $\mathrm{GL}(2,\mathbb{Z})$*).*

Proof. From Corollary 11.9 we have reduced matrices for all finite sequences as periods. Then in Proposition 8.9 we showed that sails with equivalent LLS sequences
are either integer congruent or dual. Therefore, any sail with a periodic LLS sequence is algebraic. $\qquad\qquad\Box$

11.5 Computation of all reduced matrices integer conjugate to a given one

11.5.1 Explicit computation via LLS cycles

Remark 11.11. Theorem 11.8 provides technique to compute all reduced matrices integer conjugate to a given matrix M. Here one could use the algorithm to construct the LLS cycles for matrices (see Sections 8.6). After the LLS cycle for a given matrix is found, say it is

$$(a_1, \ldots, a_n),$$

and let m be its minimal period, we can then write down the reduced matrices in accordance with Theorem 10.5. All the reduced matrices conjugate to M will be of the form

$$\begin{pmatrix} K_{n-2}(a_{k+2}, \ldots, a_{k+n-1}) & K_{n-1}(a_{k+2}, \ldots, a_{k+n}) \\ K_{n-1}(a_{k+1}, a_{k+2}, \ldots, a_{k+n-1}) & K_n(a_{k+1}, a_{k+2}, \ldots, a_{k+n}) \end{pmatrix},$$

here $k = 0, \ldots, m-1$.

We continue with the following example.

Example 11.12. Consider

$$M = \begin{pmatrix} 1519 & 1164 \\ -1964 & -1505 \end{pmatrix}.$$

The LLS cycle of M is $(1, 2, 1, 2)$. (Here we omit the computation of the LLS cycle for M, they are done following the techniques discussed in Section 8.6. The computations are left as an exercise for the reader.) Hence there are exactly two reduced matrices represented by the sequences $(1, 2, 1, 2)$ and $(2, 1, 2, 1)$. For the case of $(1, 2, 1, 2)$ we have

$$\begin{pmatrix} K_2(2, 1) & K_3(2, 1, 2) \\ K_3(1, 2, 1) & K_4(1, 2, 1, 2) \end{pmatrix} = \begin{pmatrix} 3 & 8 \\ 4 & 11 \end{pmatrix},$$

and for the case of $(2, 1, 2, 1)$ we have

$$\begin{pmatrix} K_2(1, 2) & K_3(1, 2, 1) \\ K_3(2, 1, 2) & K_4(2, 1, 2, 1) \end{pmatrix} = \begin{pmatrix} 3 & 4 \\ 8 & 11 \end{pmatrix}.$$

So the set of reduced matrices integer conjugate to M is

$$\left\{ \begin{pmatrix} 3 & 8 \\ 4 & 11 \end{pmatrix}, \begin{pmatrix} 3 & 4 \\ 8 & 11 \end{pmatrix} \right\}.$$

11.5.2 Algorithmic computation: Gauss Reduction theory

In this subsection we are aiming to construct a reduced matrix integer conjugate to a given one using the algorithm similar to the Euclidean algorithm. Such algorithms are sometimes treated as Gauss reduction algorithms.

For the remainder of this section we denote by $[[p,r][q,s]]$ the matrix

$$\begin{pmatrix} p & r \\ q & s \end{pmatrix}.$$

Algorithm to construct reduced matrices.

Input data. We are given an SL$(2,\mathbb{Z})$ matrix $M = [[p,r][q,s]]$. In addition, we suppose that the characteristic polynomial of the matrix is irreducible over $\mathbb{Q}$ (or, equivalently, that it does not have ± 1 as roots) and has two positive real roots.

Goal of the algorithm. To construct one of the periods of the LLS sequence for M.

Step 1. If $q < 0$, then we multiply the matrix $[[p,r][q,s]]$ by $-$Id. Go to Step 2.

Step 2. We have $q \geq 0$. After conjugation of the matrix $[[p,r][q,s]]$ by the matrix $[[1,-\lfloor p/q \rfloor][0,1]]$ we get the matrix $[[p',r'][q',s']]$, where $0 \leq p' \leq q'$. Go to Step 3.1.

Step 3.1. Suppose that $q' = 1$. Then $p' = 0$, $r' = -1$. In addition we have $|s'| > 2$, otherwise, the matrix has either complex roots or rational roots. The algorithm stops, and the output of the algorithm is the matrix $[[0,-1][1,s']]$.

Step 3.2.1. Suppose that $q' > 1$ and $s' > q'$. Then the algorithm stops, and the output of the algorithm is the matrix $[[p',r'][q',s']]$.

Step 3.2.2. Suppose that $q' > 1$ and $s' < -q'$. Conjugate the matrix $[[p',r'][q',s']]$ by the matrix $[[-1,1][0,1]]$ and multiply by $-$Id. As a result we have the matrix $[[p'',r''][q'',s'']]$ with $q'' = q'$, $p'' = q' - p'$, and $s'' = -q' - s' > 0$. Go to Step 3.2.1 or to Step 3.2.3 depending on s'' and q''.

Step 3.2.3. Suppose that $q' > 1$ and $|s'| \leq |q'|$. Notice that the absolute values of q' and s' do not coincide, since the matrix has unit determinant. Hence $|s| < |q|$. Then we have

$$|r'| = \left| \frac{p's' - 1}{q'} \right| \leq \frac{(q'-1)^2 + 1}{q'} \leq q' - 1.$$

Go to Step 1 with the matrix $[[s',q'][r',p']]$ with $|r'| < |q'|$. This is the matrix obtained from $[[p',r'][q',s']]$ by conjugation with $[[0,-1][-1,0]]$.

Output. The reduced integer matrix that is conjugate to $\pm M$. The sign is defined by the sign of the trace of the original matrix (i.e., of $p + s$).

Remark 11.13. During the algorithm we only use conjugations with GL$(2,\mathbb{Z})$ matrices and multiply by $-$Id. So in the output one should expect a reduced integer matrix that is conjugate either to M or to $-M$. Nevertheless, since M, as well as any reduced matrix, has all positive eigenvalues, the resulting matrix is conjugate exactly to $\pm M$.

Remark 11.14. The algorithm works in a finite number of steps, since after each iteration the integer absolute value of q is decreases.

11.6 Statistical properties of reduced $SL(2,\mathbb{Z})$ matrices

Without loss of generality in this section we restrict to the case of $SL(2,\mathbb{Z})$ matrices. Note that there are many different settings to study statistical properties of conjugacy classes of $SL(2,\mathbb{Z})$ matrices. For instance in [40] the conjugacy classes of elements of $SL(2,\mathbb{Z})$ with given trace are counted. Here we approach the question via complexity of reduced operators in the conjugacy classes.

11.6.1 Complexity of reduced matrices

As we have already seen, in the majority of conjugacy classes there is more than one reduced matrix. Then it is natural to introduce some additional notion of complexity to reduce the number of the reduced matrices. Let us introduce one important notion of complexity, whose extended definition we will use later in Chapter 25 in the multidimensional case.

Definition 11.15. Let $M = [[p,r][q,s]]$ be a reduced matrix. A ς-*complexity* of M is the number q. We denote it by $\varsigma(M)$.

A geometric interpretation of $\varsigma(M)$ is as the integer area of the triangle with vertices $O(0,0)$, $v = (1,0)$, and $M(v)$.

Problem 4. Find a reduced matrix with minimal ς-complexity from a given conjugacy class.

Remark. The minimal ς-complexity coincides with the minimal value of $\mathrm{lsin}\,POQ$, where P is an arbitrary integer point distinct from O, and $Q = A(P)$.

If the LLS cycle of M is $(a_1,\ldots,a_{2n})$, then the above problem is equivalent to finding the minimal numerator among the numerators of the rational numbers:

$$[a_1;\cdots:a_{2n-1}], \quad [a_2;\cdots:a_{2n}], \quad [a_3;\cdots:a_{2n}:a_1], \quad \ldots \quad ,[a_{2n};a_1:\cdots:a_{2n-2}].$$

Example 11.16. For instance, if the LLS cycle of M is (a,b), where $a < b$, then the minimum for the numerators is a.

Example 11.17. Let the minimal period of the LLS sequence consists of three elements: (a,b,c,a,b,c). The numerator of the fraction $[a;b:c:a:b]$ is minimal if and only if $c \geq a,b$.

The following example shows that it is not correct to skip the maximal entry of the sequence.

Example 11.18. Consider the LLS cycle $(1,4,5,4,1,4)$. The minimum of the numerators is attained at the fraction $[1;4:5:4:1]$, and not at the fraction $[4;1:4:1:4]$.

11.6.2 Frequencies of reduced matrices

We start with a proper probabilistic space. Let $P = (a_1, a_2, \ldots, a_{2n-1}, a_{2n})$ be an arbitrary sequence of positive integer elements. Denote by $\#_N(P)$ the number of matrices M satisfying the following:

(i). The absolute value of every entry of M does not exceed N.

(ii). The sequence P is the LLS cycle of M.

(iii). The algorithm of the previous subsection with M as an input produces the reduced matrix $[[p,r][q,s]]$ with $(p,q) = (0,1)$ for the case of $P = (1,a_2)$, and

$$q/p = [a_1; a_2 : \cdots : a_{2n-1}]$$

for all the remaining cases.

Let us formulate several pertinent open questions.

Problem 5. (i) Which reduced matrix with a given trace (or with a given LLS sequence) is the most frequent as N tends to infinity?

(ii) What is the relative probabilities of two sequences representing the same LLS cycle as N tends to infinity?

(iii) Is it true that the reduced matrix with minimal ς-complexity is the most frequent among the reduced matrices of a given LLS cycle?

In Table 11.6.2 we show some values of $\#_{25000}(P)$ for matrices with absolute value of the trace less than 11.

Observe that $\mathrm{SL}(2,\mathbb{Z})$ matrices corresponding to the sequence $(1,2)$ are more frequent than those corresponding to $(1,1)$. This is due to the fact that matrices with LLS cycles $(1,1)$ have integer congruent dual sails. One may enumerate the matrices with multiplicities equivalent to the number of integer congruent sails and obtain

$$4\#_{25000}(1,1) > 2\#_{25000}(1,2) + 2\#_{25000}(2,1).$$

We conclude with the following question. Denote by $\mathrm{GK}(P)$ the frequency of a sequence $P = (a_1, a_2, \ldots, a_{2n-1})$ in the sense of Gauss—Kuzmin:

$$\mathrm{GK}(P) = \frac{1}{\ln(2)} \ln\left(\frac{(\alpha_1 + 1)\alpha_2}{\alpha_1(\alpha_2 + 1)} \right),$$

where

Absolute value of the trace	Notation for classes of equivalent matrices	Period P	Matrix $[[p,r][q,s]]$	Value of $\#_{25000}(P)$
3	L_3	$(1,1)$	$[[0,1][-1,3]]$	663160
4	L_4	$(1,2)$	$[[0,1][-1,4]]$	834328
		$(2,1)$	$[[1,2][1,3]]$	304776
5	L_5	$(1,3)$	$[[0,1][-1,5]]$	818200
		$(3,1)$	$[[1,3][1,4]]$	194528
6	$L_{6,1}$	$(1,4)$	$[[0,1][-1,6]]$	777128
		$(4,1)$	$[[1,4][1,5]]$	141784
	$L_{6,2}$	$(2,2)$	$[[1,2][2,5]]$	446432
7	$L_{7,1}$	$(1,5)$	$[[0,1][-1,7]]$	734904
		$(5,1)$	$[[1,5][1,6]]$	110848
	$L_{7,2}$	$(1,1,1,1)$	$[[2,3][3,5]]$	201744
8	$L_{8,1}$	$(1,6)$	$[[0,1][-1,8]]$	695560
		$(6,1)$	$[[1,6][1,7]]$	90688
	$L_{8,2}$	$(2,3)$	$[[1,2][3,7]]$	435472
		$(3,2)$	$[[1,3][2,7]]$	310872
9	L_9	$(1,7)$	$[[0,1][-1,9]]$	660984
		$(7,1)$	$[[1,7][1,8]]$	76552
10	$L_{10,1}$	$(1,8)$	$[[0,1][-1,10]]$	630592
		$(8,1)$	$[[1,8][1,9]]$	66064
	$L_{10,2}$	$(2,4)$	$[[1,2][4,9]]$	408216
		$(4,2)$	$[[1,4][2,9]]$	239712
	$L_{10,3}$	$(1,1,1,2)$	$[[2,3][5,8]]$	260872
		$(2,1,1,1)$	$[[2,5][3,8]]$	114084
		$(1,2,1,1)$	$[[3,4][5,7]]$	149832
		$(1,1,2,1)$	$[[3,5][4,7]]$	114084

Table 11.1 Values of $\#_{25000}(P)$ for matrices with small absolute value of the trace.

$$\alpha_1 = [a_1; a_2 : \cdots : a_{2n-2} : a_{2n-1}] \quad \text{and} \quad \alpha_2 = [a_1; a_2 : \cdots : a_{2n-2} : a_{2n-1}+1].$$

Problem 6. Let

$$P_1 = (a_1, a_2, \ldots a_{2n-1}, a_{2n}), \quad P'_1 = (a_1, a_2, \ldots a_{2n-1}),$$
$$P_2 = (a_2, a_3, \ldots a_{2n}, a_1), \quad P'_2 = (a_2, a_3, \ldots a_{2n}).$$

Is the following true:

$$\lim_{n \to \infty} \frac{\#_n(P_1)}{\#_n(P_2)} = \frac{\mathrm{GK}(P'_1)}{\mathrm{GK}(P'_2)} ?$$

For further references to Gauss—Kuzmin statistics, see Section 13.

11.7 Exercises

Exercise 11.1. Is it true that two $\mathrm{SL}(2, \mathbb{R})$ matrices are integer conjugate if and only if their geometric continued fractions are integer congruent?

Exercise 11.2. Consider a hyperbolic $SL(2,\mathbb{Z})$ matrix. Let its eigen lines be $y = \alpha_{1,2}x$. Then α_1 and α_2 are conjugate quadratic irrationalities.

Exercise 11.3. Let A be a hyperbolic (2×2) matrix with irrational eigenvalues. Find all the matrices commuting with A.

Exercise 11.4. Find all reduced matrices for the matrix

$$\begin{pmatrix} 103 & 69 \\ 100 & 67 \end{pmatrix}.$$

Exercise 11.5. Prove Proposition 11.4(i).

Chapter 12
Lagrange's Theorem

The aim of this chapter is to study questions related to the periodicity of geometric and regular continued fractions. The main object here is to prove Lagrange's theorem stating that every quadratic irrationality has a periodic continued fraction, conversely that every periodic continued fraction is a quadratic irrationality. One of the ingredients to the proof of Lagrange theorem is the classical theorem on integer solutions of Pell's equation

$$m^2 - dn^2 = 1.$$

So, there is a strong relation between periodic fractions and quadratic irrationalities. It is natural to ask what happens in cases of cubic, quartic, etc., irrationalities? How can one generalize Lagrange's theorem to the multidimensional case? We give a partial answer to such questions in Chapter 22.

We start in Section 12.1 with the study of so-called *Dirichlet groups*, which are the subgroups of $GL(2, \mathbb{Z})$ preserving certain pairs of lines. These groups are closely related to the periodicity of sails. The structure of a Dirichlet group is induced by the structure of the group of units in orders (we will discuss this later in more detail for the multidimensional case in Chapter 21); here we restrict ourselves to the simplest two-dimensional case. In Section 12.2 we take a break and show how to take nth roots of matrices using Gauss's reduction theory. In Section 12.3 we study the solutions of Pell's equation. And finally, in Section 12.4 we prove Lagrange's theorem.

12.1 The Dirichlet group

Let A be a $GL(2, \mathbb{Z})$ matrix with two distinct eigenvalues. Denote the set of all $GL(2, \mathbb{Z})$ matrices commuting with A by $\Xi(A)$, i.e.,

$$\Xi(A) = \{B \in GL(2, \mathbb{Z}) \mid AB = BA\}.$$

© Springer-Verlag GmbH Germany, part of Springer Nature 2022
O. N. Karpenkov, *Geometry of Continued Fractions*,
Algorithms and Computation in Mathematics 26,
https://doi.org/10.1007/978-3-662-65277-0_12

The set $\varXi(A)$ is a group under the operation of matrix multiplication. We call it the *Dirichlet group* in dimension 2. We study the multidimensional case later, in Chapter 21.

In case of A with only real eigenvalues, we are interested in the subgroup of the Dirichlet group $\varXi(A)$ containing all matrices with positive eigenvalues. This group is called the *positive Dirichlet group* and denoted by $\varXi_+(A)$.

Definition 12.1. We say that two Dirichlet groups (or positive Dirichlet groups) $\varXi_1$ and $\varXi_2$ are integer conjugate if there exists $B \in \mathrm{GL}(2,\mathbb{Z})$ such that

$$A \to B^{-1}AB$$

is an isomorphism of $\varXi_1$ and $\varXi_2$.

Proposition 12.2. *Two Dirichlet groups (or positive Dirichlet groups) $\varXi_1$ and $\varXi_2$ are integer congruent if there exist $A \in \varXi_1$ with distinct eigenvalues and $B \in \mathrm{GL}(2,\mathbb{Z})$ such that $B^{-1}AB \in \varXi_2$.* $\qquad\qquad\square$

Let us study the structure of Dirichlet groups for the matrices of $\mathrm{GL}(2,\mathbb{Z})$. There are three essentially different cases here (this is similar to the complex, rational and real irrational cases of Section 11). The first case is for elliptic matrices. In the second and the third cases we have hyperbolic matrices, i.e., matrices all of whose eigenvalues are real. In the second case we consider matrices with rational eigenvalues (parabolic case): the eigenvalues of such $\mathrm{GL}(2,\mathbb{Z})$ matrices are ± 1. In the third case we study matrices with irrational eigenvalues (hyperbolic case): the eigenvalues of such matrices are quadratic irrationalities (i.e., $\frac{a \pm b\sqrt{c}}{d}$, where a,b,d are integers and $c > 1$ is a square-free integer).

Remark 12.3. Since conjugate matrices have isomorphic Dirichlet groups, it is enough to study only one representative of each conjugacy class of matrices in $\mathrm{GL}(2,\mathbb{Z})$.

Recall also that the Dirichlet groups are defined only for matrices with distinct eigenvalues, so we will consider only matrices with distinct eigenvalues.

Elliptic case. There are exactly two Dirichlet groups of such type up to conjugation. The representative matrices of these groups can be chosen from the group $\mathrm{SL}(2,\mathbb{Z})$. We have already seen them in Chapter 11:

$$\begin{pmatrix} 0 & 1 \\ -1 & 0 \end{pmatrix}; \qquad \begin{pmatrix} 1 & 1 \\ -1 & 0 \end{pmatrix}.$$

The first matrix is the matrix of rotation by the angle $\pi/2$. Only its powers commute with it, and hence the Dirichlet group is isomorphic to $\mathbb{Z}/4\mathbb{Z}$. The second matrix represents the 6-symmetry of a lattice. The related Dirichlet group is isomorphic to $\mathbb{Z}/6\mathbb{Z}$.

Real spectra case I: rational eigenvalues (parabolic matrices). Here we have two conjugate classes of matrices represented by

$$A_1 = \begin{pmatrix} 1 & 0 \\ 0 & -1 \end{pmatrix}; \qquad A_2 = \begin{pmatrix} 1 & -1 \\ 0 & -1 \end{pmatrix}.$$

The Dirichlet group for $\Xi(A_i)$ is isomorphic to $\mathbb{Z}/2\mathbb{Z} \oplus \mathbb{Z}/2\mathbb{Z}$, and it is generated by $\pm A_i$ (for $i = 1, 2$ respectively).

Real spectra case II: irrational eigenvalues (hyperbolic matrices). This is the largest set of irreducible matrices whose conjugacy classes are described via Gauss's reduction theory. We start with a $\mathrm{GL}(2, \mathbb{Z})$ matrix A with distinct irrational eigenvalues (as we have already mentioned, the eigenvalues are conjugate quadratic irrationalities). First, note that A^2 is in $\mathrm{SL}(2, \mathbb{Z})$ and has distinct eigenvalues. Hence without loss of generality we restrict ourselves to $\mathrm{SL}(2, \mathbb{Z})$ matrices. Such matrices have periodic geometric continued fractions with periodic LLS sequences. Due to Gauss's reduction theory and Proposition 8.9 there exists a one-to-one correspondence between integer conjugacy classes of Dirichlet groups and minimal periods of LLS sequences.

Theorem 12.4. *Consider an irreducible hyperbolic matrix $A \in \mathrm{SL}(2, \mathbb{Z})$. Then its Dirichlet group $\Xi(A)$ is isomorphic to $\mathbb{Z} \oplus \mathbb{Z}/2\mathbb{Z}$.*

Proof. The proof is straightforward. Any matrix commuting with A has the same eigenlines, so it either acts as a shift on the sail, or exchanges the sails. The group of shifts is homeomorphic to $\mathbb{Z}$. Suppose it is generated by $\hat{A}$. If the period is even, then the dual sails are not congruent, and the group is $\mathbb{Z} \oplus \mathbb{Z}/2\mathbb{Z}$ generated by $\hat{A}$ and $-\mathrm{Id}$. If the period of the LLS sequence of the sail is odd, then one can extract a square root of $\hat{A}$ in $\mathrm{SL}(2, \mathbb{Z})$. This square root will exchange dual sails, which are congruent in this case. The Dirichlet group is again $\mathbb{Z} \oplus \mathbb{Z}/2\mathbb{Z}$, and it is generated by $\hat{A}^{1/2}$ and $-\mathrm{Id}$. $\qquad\square$

Remark 12.5. Let us explain how to find generators of a two-dimensional Dirichlet group $\Xi(A)$ for a hyperbolic matrix with irrational eigenvalues. Take A^2. This matrix has all positive eigenvalues, and therefore, it preserves every sail. Construct a sail of A^2. The matrix A^2 defines a shift of the LLS sequence and hence one of its periods. Let this period be n times the minimal period. Then the second generator of $\Xi(A)$ is $(A^2)^{1/n}$ (which is easily constructed using a Jordan basis for A in which the matrix is diagonal).

Corollary 12.6. *Consider an irreducible hyperbolic matrix $A \in \mathrm{SL}(2, \mathbb{Z})$. Then its positive Dirichlet group $\Xi_+(A)$ is homeomorphic to $\mathbb{Z}$.* $\qquad\square$

Example 12.7. Consider the reduced matrix

$$A = \begin{pmatrix} 11 & 30 \\ 15 & 41 \end{pmatrix} = \begin{pmatrix} 11 & 8+2 \cdot 11 \\ 15 & 11+2 \cdot 15 \end{pmatrix}.$$

The period of the LLS sequence related to this continued fraction is $(1, 2, 1, 2, 1, 2)$. So there exists a cube root of A in $\mathrm{SL}(2, \mathbb{Z})$. This cube root is also reduced and has period of LLS sequence $(1, 2)$. Hence by Theorem 11.8 it is

$$\begin{pmatrix} 1 & 0+2\cdot1 \\ 1 & 1+2\cdot1 \end{pmatrix} = \begin{pmatrix} 1 & 2 \\ 1 & 3 \end{pmatrix}.$$

Denote this matrix by $\hat{A}$. The group $\varXi(A)$ is generated by $\hat{A}$ and $-\mathrm{Id}$.

Test for two Hyperbolic Dirichlet groups to be integer conjugate. Consider two $GL(2,\mathbb{Z})$ matrices A and B with real irrational eigenvalues. We would like to check if they generate integer conjugate Dirichlet groups (i.e., that all elements of one group are integer conjugate to the elements of the other). This is similar to checking if the arrangements of eigenlines generated by matrices A and B are integer congruent to each other.

Equivalently, we can write as follows. Consider two $GL(2,\mathbb{Z})$ matrices A and B with real irrational eigenvalues. Do there exist integers n, m such that A^n is integer conjugate to B^m.

In order to answer such questions one should compute periods of LLS sequences for A and B and find minimal periods for them (e.g., this could be done using the result of Theorem 8.27). These periods are the same up to a cyclic shift and possibly a change of the order of all the elements if and only if the corresponding cones are integer congruent to each other.

12.2 Construction of the integer nth root of a $GL(2,\mathbb{Z})$ matrix

In this section we study how to check whether a certain matrix has roots that are also integer matrices. We start with $SL(2,\mathbb{Z})$ matrices and further say a few words about $GL(2,\mathbb{Z})$ case.

Algorithm to construct an integer nth root of a $GL(2,\mathbb{Z})$ matrix.

Data. Suppose that we are given an $SL(2,\mathbb{Z})$ matrix $A = [[p,r][q,s]]$ with positive discriminant (this matrix has real spectrum with irrational eigenvalues) and a positive integer n.

Algorithm goals. To verify whether there exists and if so to find a matrix $B \in GL(2,\mathbb{Z})$ such that $A = B^n$.

Step 1. First we find one of the reduced matrices $A_2 = [[p_2,r_2][q_2,s_2]]$ that is congruent to A and remember the congruence matrix C (i.e., $A_2 = CAC^{-1}$).

Step 2. By Theorem 11.8 we have the formulas to write the LLS periods for A_2. There exists $A_3 = (A_2)^{1/n}$ that belongs to $GL(2,\mathbb{Z})$ if and only if the constructed LLS periods have subperiods whose lengths are $1/n$ that of A_2.

Step 3. Suppose that the period repeats the sequence $(a_1,\dots,a_k)$ n times. Then the matrix A_3 is constructed by the formula $A_3 = [[p_3,r_3][q_3,s_3]]$, where p_3 and q_3 are relatively prime positive numbers defined by

$$\frac{q_3}{p_3} = [a_1; \cdots : a_{k-1}].$$

The remaining two coefficients are

$$r_3 = \frac{q_3 r_2}{q_2}, \qquad s_3 = \frac{q_2 p_3 + (s_2 - p_2) q_3}{q_2}.$$

Since $q_2 > 0$, all the coefficients are well defined.

Output. We have verified that an nth root exists. If it exists, the resulting matrix B of this root is as follows:

$$B = C^{-1} A_3 C.$$

Proposition 12.8. *The matrix A_3 constructed in Step 3 is an nth root of the matrix A_2.*

Proof. There exists a matrix $M \in \Xi(A_2)$ mapping $(1,0)$ to (a_3, b_3). This matrix corresponds to the sail integer self-congruence related to the shift for the period $(a_1, \ldots, a_k)$ if k is even. If k is odd, we have an integer congruence of dual sails. In both cases the matrix M is the nth root of A_2 contained in $\Xi(A_2)$.

Since $M(1,0) = (a_3, b_3)$, the first row of M coincides with the first row of A_3. The coefficients in the second rows of A_3 and M are uniquely defined by the fact that $A_2 M = M A_2$ and $A_3 M = M A_3$. The formulas are obtained from these matrix equations (which are four equations in the elements). Hence $A_3 = M \in \Xi(A_2) \subset \mathrm{GL}(2, \mathbb{Z})$, and it is constructed by the formulas in the algorithm. $\square$

Remark 12.9. Now let $A \in \mathrm{GL}(2, \mathbb{Z})$ and $\det(A) = -1$. It is clear that A does not have integer even roots. Suppose n is odd. Then to find the nth root one should check the $(2n)$th root for the matrix A^2 in $\mathrm{SL}(2, \mathbb{Z})$. The matrix $(A^2)^{1/2n}$ (if it is in $\mathrm{GL}(2, \mathbb{Z})$) is one of the nth roots of A.

12.3 Pell's equation

Before studying Lagrange's theorem, we focus on the classical Pell's equation, which we will use in the proof of Lagrange theorem. *Pell's equation* is a Diophantine equation of the form

$$m^2 - dn^2 = 1,$$

where n and m are integer variables and d is not the square of an integer. Pell's equation has trivial solutions $(\pm 1, 0)$. The problem of finding nontrivial solutions of Pell's equation was posed by P. Fermat in 1657. The solution of Pell's equation was mistakenly attributed to J. Pell. In fact, W. Brouncker found a method to solve this equation. The first to publish a strict proof was J.-L. Lagrange, in 1768 (see [140]).

Theorem 12.10. *Let d be a nonsquare integer. Then Pell's equation*

$$m^2 - dn^2 = 1$$

has positive solutions. In addition, there exists a positive solution (m_1, n_1) *such that the set of all positive solutions coincides with the set of pairs* (m_k, n_k) *with positive integer parameter k defined as follows:*

$$\binom{m_k}{n_k} = \begin{pmatrix} m_1 & dn_1 \\ n_1 & m_1 \end{pmatrix}^k \binom{1}{0}.$$

Proof. **Existence of a nonnegative solution.** By Theorem 1.22 there are infinitely many integer solutions of the inequality

$$\left| \sqrt{d} - \frac{p}{q} \right| < \frac{1}{\sqrt{5}q^2}.$$

Letting (p, q) be any of them, then

$$|q^2 d - p^2| = |q\sqrt{d} - p||q\sqrt{d} + p| = \left| \sqrt{d} - \frac{p}{q} \right| \left| 2\sqrt{d} + \left(\sqrt{d} - \frac{p}{q} \right) \right| q^2 < \frac{2\sqrt{d} + 1}{\sqrt{5}}.$$

Therefore, there exists an integer c such that the equation

$$q^2 d - p^2 = c$$

has infinitely many integer solutions. Choose among these solutions (m_1, n_1) and (m_2, n_2) such that

$$m_1 \equiv m_2 \pmod{c} \qquad \text{and} \qquad n_1 \equiv n_2 \pmod{c}.$$

(This is possible, since the number of distinct remainder pairs is finite.)
    Now take

$$\hat{m} = \frac{m_1 m_2 - dn_1 n_2}{c} \qquad \text{and} \qquad \hat{n} = \frac{m_2 n_1 - dm_1 n_2}{c}.$$

Notice that

$$\hat{m} \equiv m_1^2 - dn_1^2 = c \equiv 0 \pmod{c} \qquad \text{and} \qquad \hat{n} \equiv m_1 n_1 - m_1 n_1 = 0 \pmod{c}.$$

Hence the point $(\hat{m}, \hat{n})$ is integer. Now let us consider

$$\hat{m}^2 - d\hat{n}^2 = (\hat{m} - \sqrt{d}\hat{n})(\hat{m} + \sqrt{d}\hat{n}) = \frac{m_1 - \sqrt{d}n_1}{m_2 - \sqrt{d}n_2} \cdot \frac{m_1 + \sqrt{d}n_1}{m_2 + \sqrt{d}n_2} = \frac{m_1^2 - dn_1^2}{m_2^2 - dn_2^2} = \frac{c}{c} = 1.$$

(The second equality is established by direct calculations.) This concludes the proof of the existence of nonnegative solutions.

Structure of the set of solutions. Let (m,n) be a positive solution of Pell's equation. Then

$$\binom{m}{n} = A_{m,n} \binom{1}{0}, \quad \text{where} \quad A_{m,n} = \begin{pmatrix} m & dn \\ n & m \end{pmatrix}.$$

Notice that $A_{m,n}$ is an integer matrix with characteristic polynomial $t^2 - 2mt + 1$ in variable t. Hence, $A_{m,n}$ is in $SL(n,\mathbb{Z})$, and all its eigenvalues are distinct positive real numbers.

Consider an angle $\alpha_{\sqrt{d}}$ with vertex at the origin and edges $y = \pm x/\sqrt{d}$ for $x > 0$. Let us show that $A_{m,n}$ preserves $\alpha_{\sqrt{d}}$.

The union of lines containing $\alpha_{\sqrt{d}}$ is defined by the equation

$$x^2 - dy^2 = 0.$$

Let us find the image of this union with respect to the action of $A_{m,n}$. The equation for the image is

$$(mx + dny)^2 - d(nx + my)^2 = 0,$$

which is equivalent (after simplification) to the initial equation

$$x^2 - dy^2 = 0.$$

Hence $A_{m,n}$ preserves the union of two lines. There are two possible situations here: either these lines are invariant lines of $A_{m,n}$, or $A_{m,n}$ interchanges the lines. Since points with positive coordinates are taken to points with positive coordinates, the lines are eigenlines. Finally, since the eigenvalues are positive real numbers, the rays defining the angle $\alpha_{\sqrt{d}}$ are preserved.

So the matrix $A_{m,n}$ is in the positive Dirichlet group Ξ_+ preserving the angle $\alpha_{\sqrt{d}}$. It is easy to see that the point $(1,0)$ is on the sail of $\alpha_{\sqrt{d}}$ (it is the only point with $x \le 1$ in the angle).

We have shown that every positive solution of Pell's equation is defined by a shift of a sail by a matrix of the positive Dirichlet group Ξ_+. From the other point of view, every matrix of the Dirichlet group preserves the value $x^2 - dy^2$ (which is 1 in our case). Recall that Ξ_+ is isomorphic to $\mathbb{Z}$. Letting A generate this group (we choose A such that it takes $(1,0)$ to the half-plane $y > 0$), then all the solutions with positive x-coordinate are of the form $A^k(1,0)$. We choose only a positive parameter k, in order to have positive y-coordinates of the solutions. Finally, denoting by (m_1, n_1) the point $A(1,0)$, then the matrix

$$\begin{pmatrix} m_1 & dn_1 \\ n_1 & m_1 \end{pmatrix}$$

is in Ξ_+ and defines the same shift of the sail (as A), hence equaling A. This concludes the proof of the second statement of the theorem. □

12.4 Periodic continued fractions and quadratic irrationalities

A continued fraction $[a_0; a_1 : \cdots]$ is called *periodic* if there exist positive integers k_0 and h such that for every $k > k_0$,

$$a_{k+h} = a_k.$$

We denote it by $[a_0; a_1 : \cdots : a_{k_0} : (a_{k_0+1} : \cdots : a_{k_0+h})]$.

Theorem 12.11. (Lagrange.) *Every periodic regular continued fraction is a quadratic irrationality (i.e., $\dfrac{a+b\sqrt{c}}{d}$ for some integers a, b, c, and d, where $b \neq 0$, $c > 1$, $d > 0$, and c is square-free). The converse is also true: every quadratic irrationality has a periodic regular continued fraction.*

Lemma 12.12. *For every quadratic irrationality ξ there exists an* $\mathrm{SL}(2, \mathbb{Z})$ *matrix such that one of its eigenvectors is* $(1, \xi)$.

Proof. Let ξ be a root of the equation $c_2 x^2 + c_1 x + c_0 = 0$ with integer coefficients c_0, c_1, c_2. Consider an arbitrary matrix

$$A = \begin{pmatrix} a_{11} & a_{12} \\ a_{21} & a_{22} \end{pmatrix}.$$

Its eigenvectors are

$$\left(1, \frac{a_{22} - a_{11} \pm \sqrt{(a_{22} - a_{11})^2 + 4a_{12}a_{21}}}{2a_{12}} \right).$$

Notice that ξ and its conjugate root are of the form

$$\frac{-c_1 \pm \sqrt{c_1^2 - 4c_0 c_2}}{2c_2}.$$

So the matrix A has a root $(1, \xi)$ if the following system is satisfied:

$$\begin{cases} nc_0 = -a_{21}, \\ nc_1 = a_{11} - a_{22}, \\ nc_2 = a_{12}, \end{cases}$$

for some $n \neq 0$.

Let us find a matrix of $\mathrm{SL}(2, \mathbb{Z})$ satisfying this system for some integer n. Since n is an integer, the coefficients a_{12} and a_{21} are integers as well. Since $\det A = 1$, we have

$$\det A = a_{11}a_{22} - a_{12}a_{21} = a_{11}(na_{11} - c_1) + n^2 c_0 c_2 = 1.$$

Therefore,

$$a_{11} = \frac{nc_1 \pm \sqrt{n^2(c_1^2 - 4c_0c_2) - 4}}{2}.$$

The coefficient a_{11} is an integer if and only if there exists an integer m satisfying

$$m^2 = n^2(c_1^2 - 4c_0c_2) - 4.$$

Set $D = c_1^2 - 4c_0c_2$, $m' = m/2$, and $n' = n/2$ and rewrite the equation

$$m'^2 - Dn'^2 = 1.$$

We end up with Pell's equation. Since ξ is irrational, the discriminant $D = c_1^2 - 4c_0c_2$ is not the square of some integer (since ξ is real, $D \geq 0$). Hence, by Theorem 12.10, it has an integer solution (m_0', n_0') with $n_0' \neq 0$. Hence the matrix

$$\begin{pmatrix} m_0' - n_0'c_1 & 2n_0'c_2 \\ -2n_0'c_0 & m_0' + n_0'c_1 \end{pmatrix}$$

has integer elements and unit determinant. Therefore, it is in $\mathrm{SL}(2, \mathbb{Z})$. □

Proof of Lagrange's theorem. First, let us show that *every periodic continued fraction represents a quadratic irrationality.* Since the continued fraction is infinite, the corresponding number is irrational.

Suppose that the periodic continued fraction for ξ does not have a preperiod, i.e.,

$$\xi = [(a_0; a_1 : \cdots : a_n)].$$

Then

$$\xi = [a_0; a_1 : \cdots : a_n : \xi] = \frac{p_n\xi + p_{n-1}}{q_n\xi + q_{n-1}}.$$

(The last equality holds by Theorem 1.13.) Notice that the denominator $q_{n-1}\xi + q_{n-2}$ is nonzero, since ξ is irrational. Therefore ξ satisfies

$$q_{n-1}\xi^2 + (q_{n-1} - p_{n-1})\xi - p_{n-2} = 0.$$

Hence ξ is a quadratic irrationality.

Suppose now that

$$\xi = [a_0; a_1 : \cdots : a_n : (a_{n+1} : a_{n+2} : \cdots : a_{n+m})].$$

Set

$$\hat{\xi} = [(a_{n+1}; a_{n+2} : \cdots : a_{n+m})].$$

Then by Theorem 1.13 we have

$$\xi = [a_0; a_1 : \cdots : a_n : \hat{\xi}] = \frac{p_n\hat{\xi} + p_{n-1}}{q_n\hat{\xi} + q_{n-1}}.$$

We have already shown that $\hat{\xi}$ is a quadratic irrationality. Hence ξ is a quadratic irrationality as well.

Secondly, we prove that *every quadratic irrationality has a periodic continued fraction*. Let $\xi > 1$ be a quadratic irrationality. By Lemma 12.12 there exists an $SL(2,\mathbb{Z})$ matrix A with an eigenvector $(1,\xi)$. By Theorem 11.9 the geometric continued fraction of A has periodic *LLS* sequence. The *LLS* sequence for the angle α generated by the two vectors $(1,0)$ and $(1,\xi)$ is one-side infinite, and by Proposition 4.13, from some element it coincides (up to a shift) with the *LLS* sequence for the matrix A. Hence, the *LLS* sequence for α is periodic from some point on. Hence from Corollary 3.7 and Theorem 3.8, the continued fraction for ξ is periodic.

Suppose now that $\xi < 1$. The number $\hat{\xi} = \xi - \lfloor \xi \rfloor + 1$ is quadratic and greater than 1, and hence it is periodic by the above. The continued fractions for ξ and $\hat{\xi}$ are distinct only in the first element. Hence the continued fraction for ξ is periodic as well.   □

12.5 Exercises

Exercise 12.1. Prove Proposition 12.2.

Exercise 12.2. Prove that the two Dirichlet groups of hyperbolic matrices with rational eigenvalues

$$\begin{pmatrix} 1 & 0 \\ 0 & -1 \end{pmatrix}; \qquad \begin{pmatrix} 1 & -1 \\ 0 & -1 \end{pmatrix}$$

are not integer congruent.

Exercise 12.3. Let A be a (2×2) matrix with distinct eigenvalues. Then every matrix commuting with A has the same eigenlines.

Exercise 12.4. Do the following matrices have a cube root in $GL(2,\mathbb{Z})$:

$$\begin{pmatrix} 25 & 32 \\ 32 & 41 \end{pmatrix}; \qquad \begin{pmatrix} 89 & 144 \\ 144 & 233 \end{pmatrix}?$$

Exercise 12.5. Solve explicitly Pell's equation

$$m^2 - dn^2 = 1$$

for $d = 2, 3, 5$.

Exercise 12.6. Find an $SL(2,\mathbb{Z})$ matrix with eigenvector $(1, \sqrt{11})$.

Chapter 13
Gauss—Kuzmin Statistics

It turns out that for almost all real numbers the frequency of a positive integer k in a continued fraction is equal to

$$\frac{1}{\ln 2} \ln \left(1 + \frac{1}{k(k+2)} \right),$$

i.e., for a general real x we have 42% of 1, 17% of 2, 9% of 3, etc. This distribution is traditionally called the *Gauss—Kuzmin* distribution. The statistics of the elements in continued fractions first appeared in the letters of K.F. Gauss to P.S. Laplace at the beginning of the nineteenth century (see [70], and also in [14]). The first proof with additional estimates was developed by R.O. Kuzmin in [132] in 1928 (see also [133]), and a little later, another proof with new estimates was given by P. Lévy in [142]. Further investigations in this directions were made by E. Wirsing in [224].

In this chapter we describe two strategies to study distributions of elements in continued fractions. A classical approach to the Gauss—Kuzmin distribution is based on the ergodicity of the Gauss map. The second approach is related to the geometry of continued fractions and its projective invariance. It is interesting to note that the frequencies of elements has an unexpected interpretation in terms of cross-ratios (see Remark 13.31). Unfortunately, the classical approach does not have a generalization to the case of multidimensional sails, since it is not clear what map is the multidimensional analogue of the Gauss map. We avoid this problem by using a geometric approach to define and investigate multidimensional statistical questions for multidimensional sails. We describe the multidimensional case in Chapter 23.

In the first five sections of this chapter we discuss the classical ergodic approach to the Gauss—Kuzmin distribution. In Section 13.1 we give some basic notions and definitions of ergodic theory. Further, in Sections 13.2 and 13.3, we present a measure related to continued fractions and the Gauss map. We prove the pointwise Gauss—Kuzmin theorem and formulate the original Gauss—Kuzmin theorem in Sections 13.4 and 13.5 respectively.

© Springer-Verlag GmbH Germany, part of Springer Nature 2022
O. N. Karpenkov, *Geometry of Continued Fractions*,
Algorithms and Computation in Mathematics 26,
https://doi.org/10.1007/978-3-662-65277-0_13

In the last five sections we study the statistic of edges of geometric one-dimensional continued fractions. After a brief discussion of cross-ratios (Section 13.6) we define a structure of a smooth manifold on the set of geometric continued fractions CF_1 in Section 13.7. Further, in Section 13.8, we define the Möbius measure on CF_1 which is invariant under the group $\mathrm{PGL}(\mathbb{R}, 2)$ acting on CF_1; we write the Möbius form explicitly in Section 13.9. Finally, in Section 13.10, we define related frequencies of edges of continued fractions and show that they coincide with the Gauss—Kuzmin statistics of the elements of continued fractions.

13.1 Some information from ergodic theory

Let X be a set, Σ a σ-algebra on X, and μ a measure on the elements of Σ. The collection (X, Σ, μ) is called a *measure space*. If $\mu(X) = 1$, the measure space is called a *probability measure space*.

Given a transformation T of a set X to itself, for every μ-integrable function f on X one can define the *time average* for f at the point x to be

$$\overline{f}(x) = \lim_{n \to \infty} \frac{1}{n} \sum_{k=0}^{n-1} f(T^k x).$$

The *space average* I_f is

$$I_f = \frac{1}{\mu(X)} \int f d\mu.$$

The space average always exists. The time average does not exist for all x. Nevertheless, in the case in which we are interested, it exists for almost all x. We formulate the related theorem after one important definition.

Definition 13.1. Let (X, Σ, μ) be a measure space. A transformation $T : X \to X$ is *measure-preserving* if it is measurable and

$$\mu(T^{-1}(A)) = \mu(A)$$

for every set A of Σ.

For measure-preserving transformations we have the following theorem.

Theorem 13.2. (Birkhoff's Pointwise Ergodic Theorem.) *Consider an arbitrary measure space (X, Σ, μ) and a measure preserving transformation T on X. Let f be a μ-integrable function on X. Then the time average converges almost everywhere to an invariant function $\overline{f}$.*

Definition 13.3. Consider a probability measure space (X, Σ, μ). A measure-preserving transformation T on X is *ergodic* if for every $X' \in \Sigma$ satisfying $T^{-1}(X') = X'$, either $\mu(X') = 0$ or $\mu(X') = 1$.

Theorem 13.4. (Birkhoff—Khinchin's Ergodic Theorem.) *Consider a probability measure space* (X, Σ, μ) *and a measure preserving transformation* T. *Suppose that* T *is ergodic. Then the values of the time average function are equal to the space average (i.e.,* $\overline{f}(x) = I_f$*) almost everywhere.*   □

13.2 The measure space related to continued fractions

In this section we define a measure space that is closely related to distributions of the elements of continued fractions. For this measure we formulate a statement on the density of points for measurable subsets, which we use in the essential way in the proofs below.

13.2.1 Definition of the measure space related to continued fractions

Consider the measure space of the segment $I = \{x \mid 0 \le x < 1\}$ with the Borel σ-algebra Σ and the measure $\hat{\mu}$ defined on a measurable set S as

$$\hat{\mu}(S) = \frac{1}{\ln 2} \int\limits_{S} \frac{dx}{1+x}.$$

The coefficient $1/\ln 2$ is taken such that the measure of the segment I equals 1.

13.2.2 Theorems on density points of measurable subsets

We start with a classical theorem on Lebesgue measure space. Denote by $B(x, \varepsilon)$ the standard ball of radius ε centered at x.

Theorem 13.5. (Lebesgue density.) *Let* λ *be the n-dimensional Lebesgue measure on* $\mathbb{R}^n$. *If* $A \subset \mathbb{R}^n$ *is a Borel measurable set, then almost every point* $x \in A$ *is a Lebesgue density point:*

$$\lim_{\varepsilon \to 0} \frac{\lambda \left(A \cap B(x, \varepsilon) \right)}{\lambda \left(B(x, \varepsilon) \right)} = 1.$$

□

Here "almost every point" means "except for a subset of zero measure".

The measure $\hat{\mu}$ is equivalent to the one-dimensional Lebesgue measure λ on the segment $[0, 1]$ (for more information on measure theory, see [148]). Hence we have a similar statement in the case of the measure space $(X, \Sigma, \hat{\mu})$.

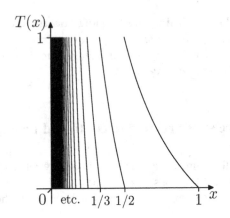

Fig. 13.1 The Gauss map.

Corollary 13.6. (*$\hat{\mu}$-density.*) *Let $X = [0,1]$ and let $\hat{\mu}$ be as above. If $A \subset X$ is a $\hat{\mu}$-measurable set with positive measure $\hat{\mu}(A)$, then almost every point in A satisfies*

$$\lim_{\varepsilon \to 0} \frac{\hat{\mu}\left(A \cap B(x,\varepsilon)\right)}{\hat{\mu}\left(B(x,\varepsilon)\right)} = 1.$$

□

13.3 On the Gauss map

Let us introduce a transformation whose ergodic properties will form the basis for the proof of the Gauss—Kuzmin theorem.

13.3.1 The Gauss map and corresponding invariant measure

We consider the measure space $(X, \Sigma, \hat{\mu})$ defined in the previous section. Define the *Gauss map* T of a segment $[0,1]$ to itself as follows:

$$T(x) = \{1/x\},$$

where $\{r\}$ denotes the fractional part $r - \lfloor r \rfloor$ (see Fig. 13.7).

Proposition 13.7. *The Gauss map T is measure-preserving for the measure space $(X, \Sigma, \hat{\mu})$.*

We start with the following lemma.

Lemma 13.8. *Let* $x = [0; a_1 : a_2 : \cdots]$. *Then*

$$T^{-1}(x) = \{[0; k : a_1 : a_2 : \cdots] | k \in \mathbb{Z}_+\} = \left\{ \frac{1}{x+k} \middle| k \in \mathbb{Z}_+ \right\}.$$

Proof. The first equality follows directly from the fact that

$$T([0; b_1 : b_2 : \cdots]) = [0; b_2 : b_3 \ldots]$$

and the fact that every real number has a unique regular continued fraction expansion with the last element not equal to 1.

The second equality is straightforward. □

Proof of Proposition 13.7. Consider a measurable set S. From Lemma 13.8 it follows that

$$\hat{\mu}(T^{-1}(S)) = \frac{1}{\ln 2} \int_{T^{-1}(S)} \frac{dx}{1+x} = \frac{1}{\ln 2} \sum_{k=1}^{\infty} \left(\int_{T^{-1}(S) \cap [\frac{1}{k+1}, \frac{1}{k}]} \frac{dx}{1+x} \right).$$

Notice that on each open (i.e., without the boundary points) segment $]1/k, 1/(k+1)[$ the operator T is in one-to-one correspondence with the open segment $]0, 1[$. Let us denote the inverse function to T on the segment $]1/k, 1/(k+1)[$ by $T_{(k)}^{-1}$. Therefore,

$$T\left(T^{-1}(S) \cap \left[\frac{1}{k+1}, \frac{1}{k}\right]\right) = T(T_{(k)}^{-1}(S)) = S,$$

and we can apply the rule of differentiation of a composite function. Recall that from Lemma 13.8 we know, that

$$T_{(k)}^{-1}(x) = \frac{1}{x+k}.$$

Then we have

$$\int_{T^{-1}(S) \cap [\frac{1}{k+1}, \frac{1}{k}]} \frac{dx}{1+x} = \int_{T_{(k)}^{-1}(S)} \frac{dx}{1+x} = \int_{T(T_{(k)}^{-1}(S))} \frac{dT_{(k)}^{-1}(x)}{1+T_{(k)}^{-1}(x)} = \int_S \frac{-d\left(\frac{1}{x+k}\right)}{1+\frac{1}{x+k}}$$

$$= \int_S \frac{dx}{(x+k)(x+k+1)}$$

(the negative sign is taken, since the map $T_{(k)}^{-1} : x \to \frac{1}{x+k}$ changes the orientation). So we have

$$\hat{\mu}(T^{-1}(S)) = \frac{1}{\ln 2} \sum_{k=1}^{\infty} \int_S \frac{dx}{(x+k)(x+k+1)}.$$

Since the integrated functions are nonnegative, we can change the order of the summation and the integration operations. We get

$$\hat{\mu}(T^{-1}(S)) = \frac{1}{\ln 2} \int\limits_{S} \left(\sum_{k=1}^{\infty} \frac{1}{(x+k)(x+k+1)} \right) dx = \frac{1}{\ln 2} \int\limits_{S} \left(\sum_{k=1}^{\infty} \left(\frac{1}{x+k} - \frac{1}{x+k+1} \right) \right) dx$$

$$= \frac{1}{\ln 2} \int\limits_{S} \frac{dx}{x+1} = \hat{\mu}(S).$$

So for every measurable set S we have

$$\hat{\mu}(T^{-1}(S)) = \hat{\mu}(S).$$

Therefore, the Gauss map T preserves the measure $\hat{\mu}$. $\square$

Remark 13.9. **(On the Euler—Mascheroni constant.)** By definition, the *Euler—Mascheroni constant* (traditionally denoted by γ) is the following infinite sum

$$\gamma = \lim_{n \to \infty} \left(\sum_{k=1}^{n} \frac{1}{k} - \ln n \right).$$

It was first studied by L. Euler in 1734. It is not known whether γ is irrational. It turns out that the Euler—Mascheroni constant can be expressed as an integral of the Gauss map with respect to Lebesgue measure:

$$\gamma = 1 - \int\limits_{0}^{1} T(x)dx.$$

13.3.2 An example of an invariant set for the Gauss map

Let us consider one example of a measurable set that is invariant under the Gauss map.

Denote by Ψ the set of all irrational numbers in the segment $[0,1]$ whose continued fractions contain only finitely many 1's. It is clear that

$$T^{-1}(\Psi) = \Psi,$$

since the operation T^{-1} shifts elements of continued fractions by one and inserts the first element.

Proposition 13.10. *The set Ψ is measurable (i.e., $\Psi \in \Sigma$).*

Proof. Denote by Υ_n the set of all irrational numbers that contain the element 1 exactly at place n. Notice that

$$\Upsilon_1 = [1/2, 1],$$

and therefore, it is measurable. Hence for every n the set

$$\Upsilon_{n+1} = T^n(\Upsilon_1)$$

is measurable.

Denote by Ψ_0 the set of all irrational numbers that do not contain an element '1'. Since

$$\Psi_0 = X \setminus \bigcup_{n=1}^{\infty} \Upsilon_n,$$

the set Ψ_0 is also measurable. Then

$$T^{-n}(\Psi_0)$$

is measurable for any positive integer n. Hence

$$\Psi = \bigcup_{n=1}^{\infty} T^{-n}(\Psi_0)$$

is measurable. □

We will prove later that the Gauss map is ergodic, and therefore, Ψ is either of zero measure or full measure in X.

13.3.3 Ergodicity of the Gauss map

In this subsection we prove the ergodicity of the Gauss map.

Proposition 13.11. *The Gauss map is ergodic.*

Before proving Proposition 13.11 we introduce some supplementary notation and prove two lemmas.

For a sequence of positive integers $(a_1, \ldots, a_n)$ denote by $I_{(a_1, \ldots, a_n)}$ the segment with endpoints $[0; a_1 : \cdots : a_{n-1} : a_n]$ and $[0; a_1 : \cdots : a_{n-1} : a_n + 1]$. It is clear that the map

$$T^n : I_{(a_1, \ldots, a_n)} \to [0, 1]$$

is one-to-one on the segment $I_{(a_1, \ldots, a_n)}$, and the inverse to T^n is

$$T^{-1}_{(a_1, \ldots, a_n)} : x \to [0; a_1 : \cdots : a_n : 1/x].$$

In terms of k-convergents $p_k/q_k = [0; a_1 : \cdots : a_k]$, the expression for $T^{-1}_{(a_1, \ldots, a_n)}(x)$ is as follows (see Proposition 1.13):

$$T^{-1}_{(a_1, \ldots, a_n)}(x) = \frac{p_n/x + p_{n-1}}{q_n/x + q_{n-1}} = \frac{p_n + p_{n-1}x}{q_n + q_{n-1}x}.$$

Lemma 13.12. *The measure of a segment $I_{(a_1, \ldots, a_n)}$ satisfies the following inequality:*

$$\hat{\mu}(I_{(a_1,\ldots,a_n)}) < \frac{1}{\ln 2(q_n+q_{n-1})(p_n+q_n)}.$$

Proof. We have

$$\hat{\mu}(I_{(a_1,\ldots,a_n)}) = \frac{1}{\ln 2} \int_{I_{(a_1,\ldots,a_n)}} \frac{dx}{1+x} = \frac{1}{\ln 2} \left| \int_{[0;a_1:\cdots:a_n]}^{[0;a_1:\cdots:a_n:1]} \frac{dx}{1+x} \right|$$

$$= \frac{1}{\ln 2} \left| \ln \left(\frac{1+\frac{p_n+p_{n-1}}{q_n+q_{n-1}}}{1+\frac{p_n}{q_n}} \right) \right| = \frac{1}{\ln 2} \left| \ln \left(1 + \frac{1}{(q_n+q_{n-1})(p_n+q_n)} \right) \right|$$

$$< \frac{1}{\ln 2(q_n+q_{n-1})(p_n+q_n)}.$$

The last inequality follows from the concavity of the natural logarithm function.   □

Lemma 13.13. *For any invariant set* S *of positive measure and any interval* $I_{(a_1,\ldots,a_n)}$,

$$\hat{\mu}(S \cap I_{(a_1,\ldots,a_n)}) \geq \frac{\ln 2}{2} \hat{\mu}(S)\hat{\mu}(I_{(a_1,\ldots,a_n)}).$$

Proof. Since the map T is surjective, we also have

$$T^n\left(S \cap I_{(a_1,\ldots,a_n)}\right) = S.$$

We have

$$\frac{1}{\ln 2} \int_{S \cap I_{(a_1,\ldots,a_n)}} \frac{dx}{1+x} = \frac{1}{\ln 2} \int_S \frac{d\left(\frac{p_n+p_{n-1}x}{q_n+q_{n-1}x}\right)}{1+\frac{p_n+p_{n-1}x}{q_n+q_{n-1}x}}$$

$$= \frac{1}{\ln 2} \int_S \frac{dx}{(q_n+q_{n-1}x)(q_n+q_{n-1}x+p_n+p_{n-1}x)}$$

$$\geq \frac{1}{\ln 2 \cdot q_n(q_n+q_{n-1}+p_n+p_{n-1})} \int_S \frac{dx}{1+x}$$

$$= \frac{1}{q_n(q_n+q_{n-1}+p_n+p_{n-1})} \hat{\mu}(S)$$

$$\geq \frac{1}{(q_n+q_{n-1})(2p_n+2q_n)} \hat{\mu}(S) \geq \frac{\ln 2}{2} \hat{\mu}(S)\hat{\mu}(I_{(a_1,\ldots,a_n)}).$$

The last inequality follows from Lemma 13.12.   □

Proof of Proposition 13.11. Let S be a measurable subset of the unit interval such that $T^{-1}(S) = S$. Suppose also $\hat{\mu}(S) > 0$.

For any irrational number $y = [0; a_1 : a_2 : \cdots]$, from Lemma 13.13 we have

$$\frac{\hat{\mu}\left((X\setminus S)\cap B(y,\hat{\mu}(I_{(a_1,\ldots,a_n)}))\right)}{\hat{\mu}\left(B(y,\hat{\mu}(I_{(a_1,\ldots,a_n)}))\right)} \leq 1 - \frac{\ln 2}{2}\frac{\hat{\mu}(S)\hat{\mu}(I_{(a_1,\ldots,a_n)})}{\hat{\mu}\left(B(y,\hat{\mu}(I_{(a_1,\ldots,a_n)}))\right)} = 1 - \frac{\ln 2}{4}\hat{\mu}(S).$$

Hence y is not a $\hat{\mu}$-density point of $X\setminus S$. Therefore, by Corollary 13.6 almost every point of $[0,1]\setminus\mathbb{Q}$ is not in $X\setminus S$, and therefore, it is in S. Hence

$$\hat{\mu}(S) \geq \hat{\mu}([0,1]\setminus\mathbb{Q}) = 1$$

Hence $\hat{\mu}(S) = 1$, concluding the proof of ergodicity of T. $\qquad\square$

13.4 Pointwise Gauss—Kuzmin theorem

Consider x in the segment $[0,1]$. Let the regular continued fraction for x be $[0; a_1 : \cdots : a_n]$ (odd or infinite). For a positive integer k, set

$$\hat{P}_{n,k}(x) = \frac{\#(k,n)}{n},$$

where $\#(k,n)$ is the number of integer elements a_i equal to k for $i = 1,\ldots,n$. Define

$$\hat{P}_k(x) = \lim_{n\to\infty}\hat{P}_{n,k}(x).$$

Theorem 13.14. *For every positive integer k and almost every x (i.e., in the complement of a set of zero measure) the following holds:*

$$\hat{P}_k(x) = \frac{1}{\ln 2}\ln\left(1 + \frac{1}{k(k+2)}\right).$$

We consider this theorem a *pointwise Gauss—Kuzmin theorem*. To prove this pointwise Gauss—Kuzmin theorem we use Birkhoff's ergodic theorems.

Proof. Consider a subset $S\subset I$. Let χ_S be the *characteristic function* of S, i.e.,

$$\chi_S(x) = \begin{cases} 1, & \text{if } x \in S, \\ 0, & \text{otherwise.} \end{cases}$$

Then

$$\hat{P}_{n,k}(x) = \frac{1}{n}\sum_{s=0}^{n-1}\chi_{]\frac{1}{k+1},\frac{1}{k}]}(T^s x).$$

Hence, by Birkhoff's pointwise ergodic theorem, the limit $\hat{P}_k(x)$ exists almost everywhere. Since the transformation T is ergodic, we apply the Birkhoff—Khinchin ergodic theorem and get

$$\hat{P}_k(x) = \int_0^1 \chi_{]\frac{1}{k+1},\frac{1}{k}]} d\hat{\mu} = \frac{1}{\ln 2} \int_{1/(k+1)}^{1/k} \frac{dx}{1+x} = \frac{1}{\ln 2} \ln\left(1 + \frac{1}{k(k+2)}\right).$$

$\square$

13.5 Original Gauss—Kuzmin theorem

Let α be some irrational number between zero and one, and let $[0; a_1 : a_2 : a_3 : \cdots]$ be its regular continued fraction.

Let $m_n(x)$ denote the Lebesgue measure of the set of real numbers α contained in the segment $[0,1]$ such that $T^n(\alpha) < x$ (here T is the Gauss map). In his letters to P.S. Laplace C.F. Gauss formulated without proofs the following theorem.

Theorem 13.15. Gauss—Kuzmin. *For $0 \leq x \leq 1$ the following holds:*

$$\lim_{n \to \infty} m_n(x) = \frac{\ln(1+x)}{\ln 2}.$$

$\square$

This theorem is technically complicated. For the proof we refer to the original manuscripts of R.O. Kuzmin [132] and [133] (see also A.Ya. Khinchin [116]).

Denote by $P_n(k)$, for an arbitrary integer $k > 0$, the measure of the set of all real numbers α of the segment $[0,1]$ such that each of them has the number k at the nth position. The limit $\lim_{n \to \infty} P_n(k)$ is called the *frequency of k* for regular continued fractions and is denoted by $P(k)$.

Corollary 13.16. *For every positive integer k, the following holds:*

$$P(k) = \frac{1}{\ln 2} \ln\left(1 + \frac{1}{k(k+2)}\right).$$

Proof. Notice that $P_n(k) = m_n(\frac{1}{k}) - m_n(\frac{1}{k+1})$. Now the statement of the corollary follows from the Gauss—Kuzmin theorem. $\square$

13.6 Cross-ratio in projective geometry

In this section we switch to the multidimensional case for a while in order to give some definitions that are similar to the one-dimensional case (we will use these later, in Chapter 23).

13.6.1 Projective linear group

The *projective linear group* (or the *group of projective transformations*) is the quotient group

$$PGL(\mathbb{R},n) = GL(\mathbb{R},n)/Z(\mathbb{R},n),$$

where $Z(\mathbb{R},n)$ is the one-dimensional subgroup of all nonzero scalar transformations of $\mathbb{R}^n$. The group $PGL(\mathbb{R},n)$ acts on the equivalence classes of vectors in $\mathbb{R}^n$ with respect to $Z(\mathbb{R},n)$. We have

$$\mathbb{R}^n/Z(\mathbb{R},n) = \mathbb{R}P^{n-1}.$$

Consider the affine part $\mathbb{R}^{n-1} \subset \mathbb{R}P^{n-1}$. The stabilizer for the affine part is exactly the group $Aff(\mathbb{R},n-1)$.

13.6.2 Cross-ratio, infinitesimal cross-ratio

Consider an arbitrary line in $\mathbb{R}^{n-1}$ with a Euclidean coordinate on it.

Definition 13.17. Consider a 4-tuple of points on a line with coordinates z_1, z_2, z_3, and z_4. The value

$$\frac{(z_1 - z_3)(z_2 - z_4)}{(z_2 - z_3)(z_1 - z_4)}$$

is called the *cross-ratio* of the 4-tuple.

It is clear that the cross-ratio does not depend on the choice of the Euclidean coordinate on the line, and therefore, it a function on the space of ordered 4-tuples of distinct points in a line.

As we have already noted above, the space $\mathbb{R}^{n-1}$ can be considered an affine chart $\mathbb{R}^{n-1} \subset \mathbb{R}P^{n-1}$. Hence the action of $PGL(\mathbb{R},n)$ is well defined on the closure of $\mathbb{R}^{n-1}$ (which is actually $\mathbb{R}P^{n-1}$). The projective transformations take planes to planes and, in particular, lines to lines. So it is natural to ask what happens to the cross-ratios of four points on a line.

Proposition 13.18. *The cross-ratio of four points is an invariant of projective transformations of $\mathbb{R}^n$.* □

We are interested in the *infinitesimal cross-ratio*, which is the following 2-form:

$$\frac{dxdy}{(x-y)^2}.$$

Notice that in the denominator we have

$$(x-y)^2 = \lim_{\varepsilon \to 0} (x - (y + \varepsilon dy)) \cdot ((x + \varepsilon dx) - y).$$

Corollary 13.19. *The infinitesimal cross-ratio is an invariant of projective transformations of* $\mathbb{R}^{n-1}$.

Proof. The density of the infinitesimal cross-ratio coincides with the asymptotic coefficient at ε^2 of the cross-ratios of 4-tuples of points:

$$x, \quad y, \quad x+\varepsilon dx, \quad y+\varepsilon dy,$$

as ε tends to 0. Therefore, the infinitesimal cross-ratio is a projective invariant.   □

13.7 Smooth manifold of geometric continued fractions

Denote the set of all geometric continued fractions by CF_1. Consider an arbitrary element of CF_1. It is a continued fraction defined by an (unordered) pair of nonparallel lines (ℓ_1, ℓ_2) passing through integer points.

Denote the sets of all ordered collections of two independent and dependent straight lines by FCF_1 and Δ_1 respectively. We say that FCF_1 is a space of *geometric framed continued fractions*. We have

$$FCF_1 = \left(\mathbb{R}P^1 \times \mathbb{R}P^1\right) \setminus \Delta_1 = T^2 \setminus \Delta_1 \quad \text{and} \quad CF_1 = FCF_1/(\mathbb{Z}/2\mathbb{Z}),$$

where $\mathbb{Z}/2\mathbb{Z}$ is the group transposing the lines in geometric continued fractions. Note that FCF_1 is a 2-fold covering of CF_1. We call the map of "forgetting" the order in the ordered collections the *natural projection* of the manifold FCF_1 to the manifold CF_1 and denote it by p (i.e., $p: FCF_1 \to CF_1$).

Notice that FCF_1 is homeomorphic to the annulus and CF_1 is homeomorphic to the Möbius band.

13.8 Möbius measure on the manifolds of continued fractions

The group $\mathrm{PGL}(2, \mathbb{R})$ of transformations of $\mathbb{R}P^1$ takes the set of all straight lines passing through the origin in the plane into itself. Hence, $\mathrm{PGL}(2, \mathbb{R})$ naturally acts on CF_1 and FCF_1. It is clear that the action of $\mathrm{PGL}(2, \mathbb{R})$ is transitive, i.e., it takes any (framed) continued fraction to any other. Notice that a stabilizer of any geometric continued fraction is one-dimensional.

Definition 13.20. A form on the manifold CF_1 (respectively FCF_1) is said to be a *Möbius form* if it is invariant under the action of $\mathrm{PGL}(2, \mathbb{R})$.

Remark 13.21. The name for the invariant forms comes from theory of energies of knots and graphs in low dimensional topology, where these forms are used as densities for Möbius energies that are invariant under the group of Möbius transformations in $\mathbb{R}^3$ (we refer the interested reader to the book by J. O'Hara [163]).

Proposition 13.22. *All Möbius forms of the manifolds CF_1 and FCF_1 are proportional.*

Proof. Transitivity of the action of $PGL(2,\mathbb{R})$ implies that all Möbius forms of the manifolds CF_1 and FCF_1 are proportional. □

Let ω be some volume form of the manifold M. Denote by μ_ω a measure of the manifold M that for every open measurable set S contained in the same piecewise connected component of M is defined by the equality

$$\mu_\omega(S) = \left| \int_S \omega \right|.$$

Definition 13.23. A measure μ of the manifold CF_1 (FCF_1) is said to be a *Möbius measure* if there exists a Möbius form ω of CF_1 (FCF_1) such that $\mu = \mu_\omega$.

From Proposition 13.22 we have the following.

Corollary 13.24. *Any two Möbius measures are proportional.* □

Remark 13.25. The projection p takes the Möbius measures of the manifold FCF_1 to the Möbius measures of the manifold CF_1. This establishes an isomorphism between the spaces of Möbius measures for CF_1 and FCF_1. Since the manifold of framed continued fractions possesses simpler chart systems, all formulas of the work are given for the case of the framed continued fraction manifold. To calculate a measure of some set F of the unframed continued fraction manifold, one should: take $p^{-1}(F)$; calculate the Möbius measure of the obtained set of the manifold of framed continued fractions, and divide the result by 2.

13.9 Explicit formulas for the Möbius form

Let us write down Möbius forms of the framed one-dimensional continued fraction manifold FCF_1 explicitly in special charts.

Consider a vector space $\mathbb{R}^2$ equipped with standard metrics on it. Letting l be an arbitrary straight line in $\mathbb{R}^2$ that does not pass through the origin, choose some Euclidean coordinates $O_l X_l$ on it. Denote by $FCF_{1,l}$ a chart of the manifold FCF_1 that consists of all ordered pairs of straight lines both intersecting l. Let us associate to any point of $FCF_{1,l}$ (i.e. to a collection of two straight lines) coordinates (x_l, y_l), where x_l and y_l are the coordinates on l for the intersections of l with the first and the second straight lines of the collection respectively. Denote by $|\bar{v}|_l$ the Euclidean length of a vector $\bar{v}$ in the coordinates $O_l X_l Y_l$ of the chart $FCF_{1,l}$. Note that the chart $FCF_{1,l}$ is the space $\mathbb{R} \times \mathbb{R}$ minus its diagonal.

Consider the following form in the chart $FCF_{1,l}$:

$$\omega_l(x_l, y_l) = \frac{dx_l \wedge dy_l}{|x_l - y_l|_l^2}.$$

Proposition 13.26. *The measure μ_{ω_l} coincides with the restriction of some Möbius measures to $FCF_{1,l}$.*

Proof. Notice that the form $\omega_l(x_l, y_l)$ coincides with the infinitesimal cross-ratio on the line l. Hence it is invariant under projective transformations of l (on an everywhere dense subset) in the chart $FCF_{1,l}$. Therefore, the measure μ_{ω_l} coincides with the restriction of some Möbius measures to $FCF_{1,l}$.   □

Corollary 13.27. *A restriction of an arbitrary Möbius measure to the chart $FCF_{1,l}$ is proportional to μ_{ω_l}.*

Proof. The statement follows from the proportionality of any two Möbius measures.   □

Consider now the manifold FCF_1 as a set of ordered pairs of distinct points on a circle $\mathbb{R}/\pi\mathbb{Z}$ (this circle is a one-dimensional projective space obtained from the unit circle by identifying antipodal points). The doubled angular coordinate φ of the circle $\mathbb{R}/\pi\mathbb{Z}$ induced by the coordinate x of the straight line $\mathbb{R}$ naturally defines the coordinates (φ_1, φ_2) of the manifold FCF_1.

Proposition 13.28. *The form $\omega_l(x_l, y_l)$ is extendible to some form ω_1 of FCF_1. In coordinates (φ_1, φ_2), the form ω_1 can be written as follows:*

$$\omega_1 = \frac{1}{4}\cot^2\left(\frac{\varphi_1 - \varphi_2}{2}\right)d\varphi_1 \wedge d\varphi_2.$$

The proof of Proposition 13.28 is left as an exercise for the reader.

13.10 Relative frequencies of edges of one-dimensional continued fractions

Without loss of generality, in this section we consider only the Möbius form ω_1 of Proposition 13.28. Denote the natural projection of the form μ_{ω_1} to the manifold of one-dimensional continued fractions CF_1 by μ_1.

Consider an arbitrary segment F with vertices at integer points. Denote by $CF_1(F)$ the set of continued fractions that contain the segment F as an edge.

Definition 13.29. The quantity $\mu_1(CF_1(F))$ is called the *relative frequency* of the edge F.

Note that the relative frequencies of edges of the same integer-linear type are equivalent. Every edge of a one-dimensional continued fraction is at unit integer distance from the origin. Thus, the integer-linear type of a segment is defined by its integer length (the number of inner integer points plus unity). Denote the relative frequency of the edge of integer length k by $\mu_1("k")$.

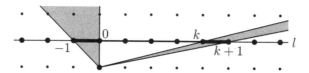

Fig. 13.2 Rays defining a continued fraction should lie in the domain shaded in gray.

Proposition 13.30. *For every positive integer k, the following holds:*

$$\mu_1("k") = \ln\left(1 + \frac{1}{k(k+2)}\right).$$

Proof. Consider a particular representative of an integer-linear type of a length-k segment: the segment with vertices $(0,1)$ and $(k,1)$. The one-dimensional continued fraction contains the segment as an edge if and only if one of the straight lines defining the fraction intersects the interval with vertices $(-1,1)$ and $(0,1)$ while the other straight line intersects the interval with vertices $(k,1)$ and $(k+1,1)$ (see Fig. 13.2).

For the straight line l defined by the equation $y = 1$, we calculate the Möbius measure of the Cartesian product of the described pair of intervals. By the last section it follows that this quantity coincides with the relative frequency $\mu_1("k")$. So,

$$\mu_1("k") = \int_{-1}^{0}\int_{k}^{k+1} \frac{dx_l dy_l}{(x_l - y_l)^2} = \int_{k}^{k+1}\left(\frac{1}{y_l} - \frac{1}{y_l + 1}\right) dy_l$$

$$= \ln\left(\frac{(k+1)(k+1)}{k(k+2)}\right) = \ln\left(1 + \frac{1}{k(k+2)}\right).$$

This proves the proposition. □

Remark 13.31. Note that the argument of the logarithm $\frac{(k+1)(k+1)}{k(k+2)}$ is the cross-ratio of points $(-1,1)$, $(0,1)$, $(k,1)$, and $(k+1,1)$.

Corollary 13.32. *The relative frequency* $\mu_1("k")$*, up to the factor*

$$\ln 2 = \int_{-1}^{0}\int_{1}^{+\infty} \frac{dx_l dy_l}{(x_l - y_l)^2},$$

coincides with the Gauss—Kuzmin frequency $P(k)$. □

13.11 Exercises

Exercise 13.1. (a) Prove that the measure

$$\mu(S) = \frac{1}{\ln 2} \int\limits_S \frac{dx}{1+x}$$

is a probability measure on the segment $[0,1]$, i.e., $\mu([0,1]) = 1$.
(b) Find $\mu([a,b])$ for $0 \le a < b \le 1$, where μ is as above.

Exercise 13.2. Ergodicity of the doubling map. Consider the space (S^1, Σ, λ), where X is the unit circle, Σ is the Borel σ-algebra, and λ is the Lebesgue measure. Consider the doubling map $T : S^1 \to S^1$ such that

$$T(\varphi) = 2\varphi.$$

Prove that T is measure-preserving and ergodic.

Exercise 13.3. Define the frequencies of subsequences in continued fractions. What is the frequency of the sequence $(1,2,3)$?

Exercise 13.4. Prove the $\hat{\mu}$-density theorem from the Lebesgue density theorem.

Exercise 13.5. Recall that Ψ_0 is the subset irrational numbers in $[0,1]$ whose continued fractions do not contain 1 as an element. Prove by elementary means (without using ergodic theorems) that

$$\hat{\mu}(\Psi_0) = 0.$$

Exercise 13.6. Prove the projective invariance of the cross-ratio.

Exercise 13.7. Prove that any two triples of points on a line are projectively equivalent. Find a criterion for two 4-tuples of points on a line to be projectively equivalent.

Exercise 13.8. Prove that
(a) $\mathbb{R}P^1$ is homeomorphic to a circle;
(b) FCF_1 is homeomorphic to an annulus;
(c) CF_1 is homeomorphic to the Möbius band.

Exercise 13.9. Prove Proposition 13.28.

Chapter 14
Geometric Aspects of Approximation

The approximation properties of continued fractions have attracted researchers for centuries. There are many different directions of investigation in this important subject (the study of best approximations, badly approximable numbers, etc.). In this chapter we consider two geometric questions of approximations by continued fractions. In Section 14.1 we prove two classical results on best approximations of real numbers by rational numbers. Further, in Section 14.2, we describe a rather new branch of generalized Diophantine approximations concerning arrangements of two lines in the plane passing through the origin. In this chapter we use some material related to basic properties of continued fractions of Chapter 1, to geometry of numbers of Chapter 3, and to Markov—Davenport forms of Chapter 11.

14.1 Two types of best approximations of rational numbers

In this section we study the *problem of best approximations of real numbers* by rational numbers. We prove here a classical result that all best approximations p/q are represented by the points (p,q) of the corresponding sails. Conversely, the point (p,q) of the sails is a best approximation if the denominator q satisfies the so-called half-rule. Further, we discuss strong best approximations, which are exactly convergents of the corresponding continued fractions.

14.1.1 Best Diophantine approximations

We say that a rational number p/q (where $q > 0$) is a *best approximation* of a real number α if for any other fraction p'/q' with $0 < q' < q$ we get

$$\left| \alpha - \frac{p'}{q'} \right| > \left| \alpha - \frac{p}{q} \right|.$$

© Springer-Verlag GmbH Germany, part of Springer Nature 2022
O. N. Karpenkov, *Geometry of Continued Fractions*,
Algorithms and Computation in Mathematics 26,
https://doi.org/10.1007/978-3-662-65277-0_14

Remark 14.1. By the definition of best approximations, every integer is a best approximation of every real number α. To be more consistent we should say that the best approximations with integer denominator are $\lfloor \alpha \rfloor$ and $\lfloor \alpha \rfloor + 1$.

Remark 14.2. From the definition of best approximations it is not immediate that there are no two best approximations with the same denominator (greater than 1). This is a direct corollary of Theorem 14.3 below.

Theorem 14.3. *Let α be a positive real number with regular continued fraction $[a_0; a_1 : \cdots]$. Denote by p_k/q_k its k-convergent and by r_k the remainder part $[a_k; a_{k+1} : \cdots]$. A rational number p/q (with $q > 1$) is a best approximation of α if and only if there exist $k \geq 0$ and an integer x such that*

$$p/q = [a_0; a_1 : \cdots : a_k : x],$$

where x satisfies either

$$\left\lfloor \frac{a_{k+1}}{2} \right\rfloor < x \leq a_{k+1}$$

or alternatively $x = a_{k+1}/2$ in the specific case in which a_{k+1} is divisible by 2 and α satisfies the inequality

$$r_{k+1} < \frac{q_{k+1}}{q_k}$$

(which is sometimes called the half-rule condition).

Remark 14.4. The condition

$$r_{k+1} < \frac{q_{k+1}}{q_k}$$

is equivalent to the condition

$$[a_{k+1}; a_{k+2} : \cdots] < [a_{k+1}; a_k : \cdots : a_0].$$

Before proving this theorem, we need to prove several preparatory statements.

Proposition 14.5. *Consider an arbitrary nonzero real number α. Then all its convergents are best approximations for α.*

Proof. Let p_k/q_k be the k-convergent to α (we assume that the integers p_k and q_k are relatively prime). We prove the proposition by induction on k.

Base of induction. If the denominator q_k equals 1, then we have either $\lfloor \alpha \rfloor$ for $k = 0$ or sometimes $\lfloor \alpha \rfloor + 1$ for $k = 1$ when $a_1 = 1$. As we have agreed in Remark 14.1, these two numbers are best approximations.

Step of induction. Suppose that the $(k-1)$-convergent p_{k-1}/q_{k-1} is a best approximation. Let us prove that the k-convergent p_k/q_k is a best approximation (we suppose also that $q_k > 1$). Set

$$A = (p_{k-1}, q_{k-1}) \quad \text{and} \quad B = (p_k, q_k).$$

From Theorem 3.1 it follows that A and B are vertices of two different sails for two angles defined by the ray $y = \alpha x$ together with two positive coordinate rays.

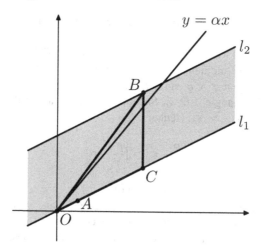

Fig. 14.1 There is no integer points in the strip between the lines l_1 and l_2.

Denote by l_1 the line passing through the points O and A, and by l_2 the line passing through the points B and parallel to l_1. Draw the line through B parallel to the line $x = 0$ and denote the intersection of this line with the ray OA by C (see Fig. 14.1).

From Proposition 1.15 it follows that the triangle $\triangle AOB$ is empty. Hence, there are no integer points in the strip between the lines l_1 and l_2 (which is gray in Fig. 14.1). Thus every integer point of the triangle $\triangle BOC$ is either in the line OC or coincides with B. Therefore, every integer point (p,q) with $q_{k-1} \le q < q_k$ satisfies

$$\left| \frac{p}{q} - \alpha \right| > \min \left(\left| \frac{p_{k-1}}{q_{k-1}} - \alpha \right|, \left| \frac{p_k}{q_k} - \alpha \right| \right) = \left| \frac{p_k}{q_k} - \alpha \right|$$

(the last equality follows from Proposition 1.21). By induction, p_{k-1}/q_{k-1} is a best approximation. Hence if $q < q_{k-1}$, then

$$\left| \frac{p}{q} - \alpha \right| > \left| \frac{p_{k-1}}{q_{k-1}} - \alpha \right| > \left| \frac{p_k}{q_k} - \alpha \right|$$

(the second inequality follows from Proposition 1.21). Therefore, the k-convergent p_k/q_k is a best approximation. The induction step is complete.   □

Lemma 14.6. *Consider a nonzero real number $\alpha > 0$ with regular continued fraction $[a_0; a_1 : \cdots]$. Let p_i/q_i be its i-convergent for $i = 0, 1, 2, \ldots$. Then for every $k \ge 1$ every best approximation p/q of α satisfying*

$$q_{k-1} < q \le q_k$$

lies on the line passing through the point (p_k, q_k) and parallel to the vector (p_{k-1}, q_{k-1}) (it is the line l_2 in Fig. 14.1).

Proof. Let $A = (p_{k-1}, q_{k-1})$ and $B = (p_k, q_k)$. Suppose that the lines l_1 and l_2 are as above (see Fig. 14.1).

Consider $E(p,q)$. If $\mathrm{ld}(E, OA) > 1$, then the triangle $\triangle EOA$ contains at least one integer point $E'(p', q')$ incident with $\triangle EOA$ distinct from E and not contained in AO. First, it is clear that $q' < q$, since E has the greatest second coordinate among all the points of the triangle $\triangle OAE$. Secondly, E and E' are in the same half-plane with respect to the line $y = \alpha x$. Hence,

$$\left| \frac{p}{q} - \alpha \right| > \left| \frac{p'}{q'} - \alpha \right|.$$

Therefore, p/q is not a best approximation of α.

Suppose now that the point E is in the same half-plane with the point A with respect to the line $y = \alpha x$. Then the angle between OA and $y = \alpha x$ is not greater than the angle between OE and $y = \alpha x$. Therefore, E is not a best approximation.

So if E is a best approximation satisfying the conditions of the lemma, then $\mathrm{ld}(E, OA) = 1$ and E is in the same half-plane with the point B. Therefore, E is on the line l_2. □

Corollary 14.7. *All best approximations of* $\alpha = [a_0; a_1 : \cdots]$ *satisfying*

$$q_k < q \leq q_{k+1}$$

are of the form $[a_0; \cdots : a_k : m]$, *where* m *is a positive integer not greater than* a_k.

Proof. From Lemma 14.6 we know that all best approximations satisfying the condition of the corollary are on a line l_2. All integer points of this line are of the form $B + mOA$ for an integer m, or in other words,

$$(p_k, q_k) + m(p_{k-1}, q_{k-1}).$$

From Proposition 1.13 it follows that

$$\frac{p_k + m p_{k-1}}{q_k + m q_{k-1}} = [a_0; \cdots : a_k : m].$$

For $m \leq 0$ the point $B + mOA$ is not in the first octant. For $m > a_k$ the second coordinate of the point $B + mOA$ is greater than q_k. Therefore, we have $0 < m \leq a_k$. □

Proof of Theorem 14.3. By Corollary 14.7 we know that all best approximations are of the form

$$[a_0; \cdots : a_k : x],$$

where $0 < x \leq a_{k+1}$. By Proposition 14.5 we get that all the convergents (i.e., the case $x = a_{k+1}$) are best approximations. It is easy to see that

$$[a_0; \cdots : a_k : x] > [a_0; \cdots : a_k : x+1]$$

for $x = 1, 2, \ldots, a_k - 2$. Therefore, if $[a_0; \cdots : a_k : x]$ is a best approximation, then $[a_0; \cdots : a_k : x+1]$ is also a best approximation.

Now it remains only to find the smallest number x in the set $\{1, 2, \ldots, a_k - 2\}$ such that the number $[a_0; \cdots : a_k : x]$ is a best approximation. By Proposition 14.5 we know that the previous best approximation is exactly p_k/q_k. So we should check when the following inequality holds:

$$\bigl|[a_0; \cdots : a_k : x] - \alpha\bigr| < \bigl|[a_0; \cdots : a_{k-1} : a_k] - \alpha\bigr|. \tag{14.1}$$

Recall that

$$\alpha = \frac{p_k r_{k+1} + p_{k-1}}{q_k r_{k+1} + q_{k-1}};$$

$$[a_0; \cdots : a_k] = \frac{p_k}{q_k};$$

$$[a_0; \cdots : a_k : x] = \frac{p_k x + p_{k-1}}{q_k x + q_{k-1}}.$$

Therefore,

$$\bigl|[a_0; \cdots : a_{k-1} : a_k] - \alpha\bigr| = \frac{|p_k q_{k-1} - q_k p_{k-1}|}{q_k(q_k r_{k+1} + q_{k-1})} = \frac{1}{q_k(q_k r_{k+1} + q_{k-1})}$$

and

$$\bigl|[a_0; \cdots : a_k : x] - \alpha\bigr| = \frac{|p_k q_{k-1} - q_k p_{k-1}|(r_{k+1} - x)}{(q_k x + q_{k-1})(q_k r_{k+1} + q_{k-1})} = \frac{r_{k+1} - x}{(q_k x + q_{k-1})(q_k r_{k+1} + q_{k-1})}.$$

Here $|p_k q_{k-1} - q_k p_{k-1}| = 1$ holds by Proposition 1.15. Notice that we do not consider the case $x \geq r_{k+1}$, since then the denominator of the corresponding fraction is greater than q_{k+1}. Hence inequality (14.1) is equivalent to the inequality

$$\frac{r_{k+1} - x}{(q_k x + q_{k-1})} < \frac{1}{q_k},$$

which is equivalent to the following one:

$$x > \frac{r_{k+1} - \frac{q_{k-1}}{q_k}}{2}.$$

Suppose that a_{k+1} is odd. Then

$$\left\lfloor \frac{a_{k+1}}{2} \right\rfloor < \frac{r_{k+1} - \frac{q_{k-1}}{q_k}}{2} < \left\lfloor \frac{a_{k+1}}{2} \right\rfloor + 1.$$

Therefore, for an integer x, inequality (14.1) is equivalent to

$$x > \frac{a_{k+1}}{2}.$$

Now suppose that a_{k+1} is even. Then

$$\left\lfloor \frac{a_{k+1}}{2} - 1 \right\rfloor < \frac{r_{k+1} - \frac{q_{k-1}}{q_k}}{2} < \left\lfloor \frac{a_{k+1}}{2} \right\rfloor + 1.$$

Therefore, for an integer x, inequality (14.1) is satisfied for

$$x > \frac{a_{k+1}}{2}.$$

Finally, it remains to study the case of $x = a_{k+1}/2$. We should check whether the following inequality holds:

$$a_{k+1}/2 > \frac{r_{k+1} - \frac{q_{k-1}}{q_k}}{2}.$$

It is equivalent to

$$a_{k+1} > r_{k+1} - \frac{q_{k-1}}{q_k},$$

and hence

$$\frac{a_{k+1}q_k - q_{k-1}}{q_k} > r_{k+1}.$$

According to Proposition 1.13, the numerator of the left part equals to q_{k+1}. So for $x = a_{k+1}/2$, inequality (14.1) is equivalent to

$$r_{k+1} < \frac{q_{k+1}}{q_k}.$$

We have described all solutions of inequality (14.1). This concludes the proof of Theorem 14.3.   □

14.1.2 Strong best Diophantine approximations

We say that a rational number p/q (where $q > 0$) is a *strong best approximation* of a real number α if for every fraction p'/q' with $0 < q < q'$ we get

$$|q'\alpha - p'| > |q\alpha - p|.$$

In this subsection we show that all strong best approximations are convergents of the corresponding number. In particular, every strong best approximation is a best approximation.

Proposition 14.8. *Every strong best approximation of a real number α is a best approximation of α.*

Proof. Consider a strong best approximation p/q of α. Then for every rational number p'/q' with $0 < q < q'$ we get

$$|q'\alpha - p'| > |q\alpha - p|.$$

Since $0 < q' < q$, we get

$$\frac{1}{q'}|q'\alpha - p'| > \frac{1}{q}|q\alpha - p|,$$

which is equivalent to

$$\left|\alpha - \frac{p'}{q'}\right| > \left|\alpha - \frac{p}{q}\right|.$$

Hence p/q is a best approximation. □

Theorem 14.9. *The set of all strong best approximations of a real number α coincides with the set of all convergents of α.*

Remark 14.10. In the case of rational $\alpha = [a_0; a_1 : \cdots : a_n]$ (where $a_n \neq 1$) the number $p'/q' = [a_0; a_1 : \cdots : a_n - 1]$ is not a strong best approximation of α. In this case there is a unique number $p_{n-1}/q_{n-1} = [a_0; a_1 : \cdots : a_{n-1}]$ with smaller denominator that is as good as p'/q':

$$|q'\alpha - p'| = |q_{n-1}\alpha - p_{n-1}|.$$

Proof. Let p_i/q_i denote the i-convergent of $\alpha = [a_0; a_1 : a_2 : \cdots]$.

From Proposition 14.8 it follows that every strong best approximation is a best approximation. By Theorem 14.3 every best approximation is on a sail of the corresponding angles. Therefore, by Theorem 3.1 and Proposition 1.13, every best approximation can be written for some $k \geq 1$ as

$$\frac{hp_k + p_{k-1}}{hq_k + q_{k-1}}, \tag{14.2}$$

where h varies from 0 to a_{k+1}. For $h = 0$ and $h = a_{k+1}$ we get the k-convergent and the $(k+1)$-convergent respectively. If $0 < h < a_{k+1}$, then on the one hand,

$$\begin{aligned}
&|(hq_k + q_{k-1})\alpha - (hp_k + p_{k-1})| \\
=\ &|((h+1)q_k + q_{k-1})\alpha - ((h+1)p_k + p_{k-1}) - (q_k\alpha - p_k)| \\
=\ &|((h+1)q_k + q_{k-1})\alpha - ((h+1)p_k + p_{k-1})| + |(q_k\alpha - p_k)| \\
>\ &|q_k\alpha - p_k|;
\end{aligned}$$

but on the other hand, $hq_k + q_{k-1} > q_k$. Hence for $0 < h < a_{k+1}$, the rational number of expression (14.2) is not a strong best approximation. Therefore, every strong best approximation of α is a convergent of α.

For the case of convergents we have

$$\begin{aligned}
|q_{k+1}\alpha - p_{k+1}| &= |(q_{k+1} + q_k)\alpha - (p_{k+1} + p_k) - (q_k\alpha - p_k)| \\
&< |q_k\alpha - p_k| - |(q_{k+1} + q_k)\alpha - (p_{k+1} + p_k)| < |q_k\alpha - p_k|.
\end{aligned}$$

The first inequality holds since the integer points (p_{k+1}, q_{k+1}) and $(p_{k+1} + p_k, q_{k+1} + q_k)$ are in different half-planes with respect to the line $y = \alpha x$. Therefore, all the convergents are strong best approximations.   □

Example 14.11. Consider the example of $\alpha = \frac{47}{21} = [2; 4 : 5]$. All best approximations and strong best approximations are shown in the following table.

Approximations	[2]	[2;3]	[2;4]	[2;4:3]	[2;4:4]	[2;4:5]
Best	+	+	+	+	+	+
Strong Best	+	-	+	-	-	+

14.2 Rational approximations of arrangements of two lines

In the previous section we studied the classical problem of an approximation of real numbers by rational numbers. In geometric language this means that we approximate some real ray with vertex at the origin by integer rays with vertices at the origin. Now we say a few words about the approximation theory of arrangements of two lines intersecting at the origin. On the one hand, the situation here reminds us of the situation of simultaneous approximation, while on the other hand, algebraic aspects of arrangement approximation are not visible for simultaneous approximation. We briefly touch on the multidimensional case in Chapter 26. (For extra details we refer to [109].)

In Section 14.2.1 we associate to any arrangement of two lines a *Markov—Davenport form*. (Actually, we have already seen these formlys while studying the Markov spectrum in Chapter 7.) Further, in Section 14.2.2, we say a few words about a natural class of integer approximations for arrangements and introduce the notion of sizes for integer approximations. We define the distance function (discrepancy) between two arrangements in Section 14.2.3.

In Section 14.2.4 we prove a general estimate for the discrepancy of best approximations for the arrangements defined by pairs of continued fractions with bounded elements: it turns out that in this case, the discrepancy is proportional to N^{-2}, where N is the size of the corresponding best approximation. Further, we give an example of an arrangement one of whose continued fractions has a rapid growing sequence of elements. In this case the discrepancy is asymptotically not better than $N^{-1-\varepsilon}$. We study best approximations of arrangements whose lines are eigenlines of certain matrices in $\mathrm{SL}(2, \mathbb{Z})$ in Section 14.2.5. Finally, in Section 14.2.6 we show how to simplify the calculation of best approximations in practice.

14.2.1 Regular angles and related Markov—Davenport forms

In this section we study arrangements of two distinct lines passing through the origin.

We start with the notion of Markov—Davenport forms.

Definition 14.12. Consider an arbitrary quadratic form

$$f(x,y) = ax^2 + bxy + cy^2$$

with real coefficients and positive discriminant $\Delta(f) = b^2 - 4ac$. Then the form

$$\frac{f(x,y)}{\sqrt{\Delta(f)}}$$

is called the *Markov—Davenport form*.

Consider an arbitrary arrangement R of lines l_1 and l_2. Let the linear forms $a_1x + b_1y$ and $a_2x + b_2y$ attain zero values at all vectors of l_1 and l_2 respectively. There is a unique, up to a sign, Markov—Davenport form associated to R whose zero set coincides with the union $l_1 \cup l_2$. It is written as

$$\frac{(a_1x + b_1y)(a_2x + b_2y)}{a_1b_2 - a_2b_1}.$$

We say that this form is *associated* to the arrangement R and denote it by Φ_R.

Example 14.13. Let us write the Markov—Davenport form for the arrangement of eigenlines of the Fibonacci operator

$$\begin{pmatrix} 1 & 1 \\ 1 & 0 \end{pmatrix}.$$

The Fibonacci operator has two eigenlines,

$$y = -\theta x \quad \text{and} \quad y = \theta^{-1}x,$$

where θ is the golden ratio, i.e., $\theta = \frac{1+\sqrt{5}}{2}$. The Markov—Davenport form of the Fibonacci operator is

$$\frac{(y + \theta x)(y - \theta^{-1}x)}{-\theta - \theta^{-1}} = \frac{1}{\sqrt{5}}(x^2 - xy - y^2).$$

14.2.2 Integer arrangements and their sizes

To construct an approximation theory for arrangements one should choose the set of all approximations and define the quality of each of the approximations of this subset. We consider the set of all integer arrangements as the set of approximations: an arrangement of two lines passing through the origin is called *integer* if both its lines contain integer points distinct from the origin. Let us define the quality of the approximation (which we call the *size*).

For a vector $v = (a,b)$ denote by $|v|$ the norm $\max(|a|,|b|)$.

Definition 14.14. Consider an integer arrangement R of lines l_1 and l_2 passing through the origin. Let v_1 and v_2 be two integer vectors of integer length one contained in l_1 and l_2 respectively. The *size* of R is the real number

$$\max(|v_1|,|v_2|),$$

which we denote by $v(R)$.

Remark 14.15. For a generic irrational arrangement the best approximation of any size greater than 1 is unique. Nevertheless, for particular cases there could be more than one best approximation of the same size.

14.2.3 Discrepancy functional and approximation model

Let us now define a natural distance between two arrangements in the plane passing through the origin.

Definition 14.16. Let R_1, R_2 be two arrangements of pairs of lines in the plane. Consider two symmetric bilinear forms

$$\Phi_{R_1}(v) + \Phi_{R_2}(v) \quad \text{and} \quad \Phi_{R_1}(v) - \Phi_{R_2}(v).$$

Take the maximal absolute values of the coefficients of these forms. Suppose that they equal c_1 and c_2. The *discrepancy* between R_1 and R_2 is $\min(c_1,c_2)$. We denote it by $\rho(R_1,R_2)$.

Remark 14.17. In case of arrangements R_1 and R_2 of ordered lines one should consider the maximal absolute values of the coefficients only of one form

$$\Phi_{R_1}(v) - \Phi_{R_2}(v).$$

The problem of *best approximations* of an arbitrary arrangement by integer arrangements is the following.

Problem of best approximations of arrangements. *For a given arrangement R and an integer $N > 0$, find an integer arrangement R_N such that $v(R_N) \leq N$ and*

$$\rho(R,R_N) = \min\{\rho(R,R') \mid R \text{ is integer}, v(R') \leq N\}.$$

Remark 14.18. The space of all arrangements of two lines passing through the origin is two-dimensional. Nevertheless, the complexity of approximation by integer arrangements is more complicated in comparison with simultaneous approximation of a line in $\mathbb{R}^3$ (whose configuration space is also two-dimensional). This happens becouse for the approximation of the arrangement of two lines one needs two pairs of relatively prime integers, while for simultaneous approximation of a vector in $\mathbb{R}^3$, a triple of relatively prime integers suffices. Nevertheless, geometric properties of continued fractions essentially simplify the approximations of arrangements. This is relevant especially for so-called *algebraic arrangements*, whose lines are eigenlines of some $SL(2,\mathbb{Z})$ matrix. In this case all the approximations are constructed from some periodic sets that are easy to construct.

14.2.4 Lagrange estimates for the case of continued fractions with bounded elements

In this subsection we prove an analogue of Lagrange's theorem on the approximation rate of arrangements of two lines $y = \alpha_1 x$ and $y = \alpha_2 x$ whose coefficients α_1 and α_2 have continued fractions with bounded elements. In particular, this includes all algebraic arrangements (as in Remark 14.18). For the convenience of the reader we do not consider the degenerate case in which one of the lines is $x = 0$, which is not so important, since it is equivalent to the approximation of a vector in the plane.

Theorem 14.19. *Consider an arrangement R of the lines $y = \alpha_1 x$ and $y = \alpha_2 x$. Let α_1 and α_2 be real numbers having infinite continued fractions with bounded elements. Then there exist constants $C_1, C_2 > 0$ such that for every integer $N > 0$, the best approximation R_N satisfies*

$$\frac{C_1}{N^2} < \rho(R,R_N) < \frac{C_2}{N^2}.$$

This theorem is a reformulation of one of the results of [109]. We begin the proof with the following two technical lemmas.

Let R_{δ_1,δ_2} denote the arrangement of lines $y = (\alpha_i + \delta_i)x$ for $i = 1,2$.

Lemma 14.20. *Consider a real number ε satisfying*

$$0 < \varepsilon < 1/|\alpha_1 - \alpha_2|.$$

Supposing that $\rho(R,R_{\delta_1,\delta_2}) < \varepsilon$, then

$$|\delta_1| < \frac{(1+|\alpha_1|)(\alpha_1 - \alpha_2)^2}{|\alpha_2|(1-\varepsilon|\alpha_1-\alpha_2|)}\varepsilon \quad \text{and} \quad |\delta_2| < \frac{(1+|\alpha_2|)(\alpha_1 - \alpha_2)^2}{|\alpha_1|(1-\varepsilon|\alpha_1-\alpha_2|)}\varepsilon.$$

Proof. The Markov—Davenport form of R_{δ_1,δ_2} is

$$\Phi_{R_{\delta_1,\delta_2}}(x,y) = \frac{\left(y-(\alpha_1+\delta_1)x\right)\left(y-(\alpha_2+\delta_2)x\right)}{(\alpha_2+\delta_2)-(\alpha_1+\delta_1)}.$$

Then the form $\Phi_R - \Phi_{R_{\delta_1,\delta_2}}$ is as follows:

$$\begin{aligned}
&\frac{\alpha_1^2\delta_2 - \alpha_2^2\delta_1 + \alpha_1\delta_1\delta_2 - \alpha_2\delta_1\delta_2}{(\alpha_1-\alpha_2)(\alpha_1-\alpha_2+\delta_1-\delta_2)}x^2 \\
&+ \frac{\alpha_2\delta_1 - \alpha_1\delta_2}{(\alpha_1-\alpha_2)(\alpha_1-\alpha_2+\delta_1-\delta_2)}xy \\
&+ \frac{\delta_2 - \delta_1}{(\alpha_1-\alpha_2)(\alpha_1-\alpha_2+\delta_1-\delta_2)}y^2.
\end{aligned} \tag{14.3}$$

Consider the absolute values of the coefficients at y^2 and at xy for the form $\Phi_R - \Phi_{R_{\delta_1,\delta_2}}$. By the conditions of the lemma these coefficients are less then ε. We have

$$\left|\frac{\delta_2 - \delta_1}{(\alpha_1-\alpha_2)(\alpha_1-\alpha_2+\delta_1-\delta_2)}\right| < \varepsilon; \tag{14.4}$$

$$\left|\frac{\alpha_2\delta_1 - \alpha_1\delta_2}{(\alpha_1-\alpha_2)(\alpha_1-\alpha_2+\delta_1-\delta_2)}\right| < \varepsilon. \tag{14.5}$$

Inequality (14.4) implies

$$|\delta_1 - \delta_2| < \frac{(\alpha_1-\alpha_2)^2}{1-\varepsilon|\alpha_1-\alpha_2|}\varepsilon.$$

From Inequality (14.5) we have

$$|\delta_1| < \frac{|(\alpha_1-\alpha_2)(\alpha_1-\alpha_2+\delta_1-\delta_2)|\varepsilon + |\alpha_1(\delta_1-\delta_2)|}{|\alpha_2|},$$

and hence we have

$$\begin{aligned}
|\delta_1| &< \frac{|\alpha_1-\alpha_2|(|\alpha_1-\alpha_2| + \frac{(\alpha_1-\alpha_2)^2}{1-\varepsilon|\alpha_1-\alpha_2|}\varepsilon)\varepsilon + |\alpha_1|\frac{(\alpha_1-\alpha_2)^2}{1-\varepsilon|\alpha_1-\alpha_2|}\varepsilon}{|\alpha_2|} \\
&< \frac{(1+|\alpha_1|)(\alpha_1-\alpha_2)^2}{|\alpha_2|(1-\varepsilon|\alpha_1-\alpha_2|)}\varepsilon.
\end{aligned}$$

The inequality for δ_2 is obtained in a similar way. $\qquad\square$

Lemma 14.21. *Let ε be a positive real number. Suppose that real numbers δ_1 and δ_2 satisfy $|\delta_1| < \varepsilon$ and $|\delta_2| < \varepsilon$. Then the following estimate holds:*

$$\rho(R, R_{\delta_1, \delta_2}) < \frac{\max\left(2, 2(|\alpha_1| + |\alpha_2|), \alpha_1^2 + \alpha_2^2 + |\alpha_1 - \alpha_2|\varepsilon\right)}{(|\alpha_1 - \alpha_2|)(|\alpha_1 - \alpha_2| + 2\varepsilon)}\varepsilon.$$

Proof. The statement of the lemma follows directly from the estimate for $R - R_{\delta_1, \delta_2}$, shown in (14.3). For the coefficient at x^2 we have

$$\left|\frac{\alpha_1^2 \delta_2 - \alpha_2^2 \delta_1 + \alpha_1 \delta_1 \delta_2 - \alpha_2 \delta_1 \delta_2}{(\alpha_1 - \alpha_2)(\alpha_1 - \alpha_2 + \delta_1 - \delta_2)}\right| < \frac{\alpha_1^2 + \alpha_2^2 + |\alpha_1 - \alpha_2|\varepsilon}{(|\alpha_1 - \alpha_2|)(|\alpha_1 - \alpha_2| + 2\varepsilon)}\varepsilon.$$

For the coefficient at xy we have

$$\frac{\alpha_2 \delta_1 - \alpha_1 \delta_2}{(\alpha_1 - \alpha_2)(\alpha_1 - \alpha_2 + \delta_1 - \delta_2)}xy < \frac{2(|\alpha_1| + |\alpha_2|)}{(|\alpha_1 - \alpha_2|)(|\alpha_1 - \alpha_2| + 2\varepsilon)}\varepsilon.$$

For the coefficient at y^2 we have

$$\frac{\delta_2 - \delta_1}{(\alpha_1 - \alpha_2)(\alpha_1 - \alpha_2 + \delta_1 - \delta_2)}y^2 < \frac{2}{(|\alpha_1 - \alpha_2|)(|\alpha_1 - \alpha_2| + 2\varepsilon)}\varepsilon.$$

Therefore, the statement of the lemma holds by the definition of discrepancy. $\square$

Proof of Theorem 14.19. Let us start with the proof of the inequality

$$\frac{C_1}{N^2} < \rho(R, R_N).$$

Let $\alpha_1 = [a_0; a_1 : \cdots]$ and $p_i/q_i = [a_0; a_1 : \cdots : a_i]$ for $i = 0, 1, \ldots$. Without loss of generality we assume that $N > a_0$. Assume that k is the greatest integer for which $p_k \leq N$ and $q_k \leq N$. Then we have the following estimates:

$$\min\left(\left|\alpha_1 - \frac{p}{q}\right| \middle| |p| \leq N, |q| \leq N\right) \geq \left|\alpha_1 - \frac{p_{k+1}}{q_{k+1}}\right| \geq \frac{1}{q_{k+1}(q_{k+1} + q_{k+2})}$$

$$\geq \frac{1}{(a_{k+1} + 1)q_k\left((a_{k+1} + 1)q_k + (a_{k+1} + 1)(a_{k+2} + 1)q_k\right)}$$

$$\geq \frac{1}{(a_{k+1} + 1)^2(a_{k+2} + 2)} \cdot \frac{1}{N^2}.$$

The second inequality follows from Exercise 1.3 on page 20 (we leave the proof for the reader). For the third inequality we use the following inequality twice:

$$q_{i+1} = a_{i+1}q_i + q_{i-1} < (a_i + 1)q_i, \tag{14.6}$$

which follows directly from Proposition 1.13.

The same calculations are valid for α_2. After that, the constant C_1 is obtained by Lemma 14.20.

Now we prove the inequality

$$\rho(R, R_N) > \frac{C_2}{N^2}.$$

Again we assume that k is the greatest integer for which $|p_k| \leq N$ and $q_k \leq N$.
For α_1 we have the following:

$$\left| \alpha_1 - \frac{p_k}{q_k} \right| < \left| \frac{p_{k+1}}{q_{k+1}} - \frac{p_k}{q_k} \right| = \frac{1}{q_k q_{k+1}} < \frac{a_{k+1} + 1}{q_{k+1}^2} < \frac{(a_{k+1} + 1)(|\alpha_1| + 1)^2}{N^2}.$$

The first inequality follows from the fact that if the k-convergent is greater than α, then the $(k+1)$-convergent is less then α, and conversely. Proposition 1.15 implies the second equality. The third inequality follows from inequality (14.6). In the last inequality we use the assumption that either $q_{k+1} > N$ or $|p_{k+1}| > N$, in which case

$$q_{k+1} > \frac{|p_{k+1}|}{|\alpha_1| + 1} > \frac{N}{|\alpha_1| + 1}$$

(here the first inequality holds since $|\alpha - p_{k+1}/q_{k+1}| < 1$). From conditions of the theorem the set of all elements a_i of the continued fraction for α is bounded. Therefore, there exists a constant $C'_{2,1}$ such that for every N there exists an approximation of α_1 of magnitude smaller than $C'_{2,1}/N^2$.

A similar estimate holds for α_2. There exists a constant $C'_{2,2}$ such that for every N there exists an approximation of α_1 of magnitude smaller than $C'_{2,2}/N^2$. Therefore, we can apply Lemma 14.21 in order to obtain the constant C_2.   □

Remark 14.22. Now we say a few words about the case of unbounded elements of continued fractions. Let ε be a small positive real number. If the elements of a continued fraction (say for α_1) grow fast enough, then there exists a sequence N_i for which the approximations R_{N_i} are of magnitude $\frac{C}{(N_i)^{1+\varepsilon}}$. Let us illustrate this by the following example.

Example 14.23. Let M be an arbitrary positive integer. Consider an arrangement R of the lines $y = 0$ and $y = \alpha x$, where the continued fraction $[a_0; a_1 : \cdots]$ for α is defined inductively:

— $a_0 = 1$;
— let $a_0, \ldots, a_k$ be defined and $[a_0; \cdots : a_k] = \frac{p_k}{q_k}$; then $a_{k+1} = (q_k)^{M-1}$.

Let $N_k = \lfloor \frac{q_{k+1}}{2} \rfloor + 1$ for $k = 1, 2, \ldots$. Then there exists a positive constant C such that for every integer i we have

$$\rho(R, R_{N_i}) \geq \frac{C}{N_i^{1+1/M}}.$$

Proof. For every i we have

$$q_{i+1} \geq a_i q_i = q_i^{M-1} q_i = q_i^M.$$

Therefore, the magnitude of best approximations for k-convergents are as follows:

$$\left| \alpha_1 - \frac{p_k}{q_k} \right| \geq \frac{1}{q_k(q_{k+1} + q_k)} \geq \frac{1}{q_{k+1}^{1/M}(2q_{k+1})} = \frac{1}{2q_{k+1}^{1+1/M}} < \frac{1}{2^{2+1/M}N_{k+1}^{1+1/M}}.$$

From Theorem 14.3 it follows that there is no best approximation whose denominators q satisfies

$$q_k < q \leq N_k.$$

Therefore, $R_{N_k} = R_{\frac{p_k}{q_k}}$. Hence

$$\rho(R, R_{N_k}) = \rho\left(R, R_{\frac{p_k}{q_k}}\right) \geq \frac{C}{N_k^{1+1/M}}.$$

The last inequality follows from Lemma 14.20. This completes the proof. □

We suspect the existence of a *badly approximable* arrangement R and a constant C such that there are only finitely many solutions N of the inequality

$$\rho(R, R_N) \leq \frac{C}{N},$$

as in the case of simultaneous approximation of vectors in $\mathbb{R}^3$ (see, e.g., in [138]).

14.2.5 Periodic sails and best approximations in the algebraic case

14.2.5.1 Geometric w-continued fractions

Let us prove a relation between classical geometry of numbers and best simultaneous approximations.

Definition 14.24. Define inductively the *w-sails* for an arbitrary angle C with vertex at the origin.
— let the 1-*sail* be the sail of C.
— suppose that all *w-sails* for $w < w_0$ are defined. Then let the w_0-*sail* be

$$\partial \left(\text{conv}\left((C \cap \mathbb{Z}^2 \setminus \{O\}) \setminus \bigcup_{w=1}^{w_0-1} w\text{-sail} \right) \right),$$

where $\text{conv}(M)$ denotes the convex hull of M.

Consider an arbitrary arrangement R of two lines l_1 and l_2. The union of all four *w-sails* for the cones defined by the lines l_1 and l_2 is the *geometric continued w-fraction* of R.

In Fig. 14.2 we show the geometric continued 1-, 2-, and 3-fractions for a particular arrangement of two lines. It is interesting to notice that the geometric continued 3-fraction in the figure is homothetic to the geometric continued 2-fraction and to

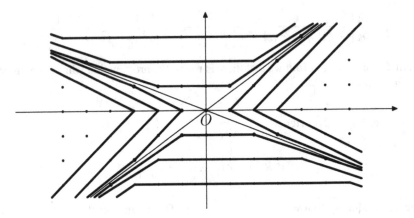

Fig. 14.2 The arrangement of two lines and its geometric continued 1-, 2-, and 3-fractions.

the geometric continued fraction (which is a 1-fraction). It turns out that it is a general fact that *the w-continued fraction of any general arrangement R (i.e., whose lines do not contain integer points except the origin) is homothetic to the geometric continued fraction of R; the coefficient of homothety is w.* We leave this statement as an exercise for the reader.

14.2.5.2 Algebraic arrangements

An arrangement of two lines l_1, l_2 is said to be *algebraic* if there exists an operator $A \in \mathrm{GL}(2, \mathbb{Z})$ with irreducible, over $\mathbb{Q}$, characteristic polynomial having two real roots whose eigenlines are l_1 and l_2. In this case the operator A acts on a w-geometric continued fraction (for any w) as a transitive shift.

Definition 14.25. The minimal absolute value of the form Φ_R on the integer lattice minus the origin is called the *Markov minimum* of Φ_R. We denote it by α_R.

Recall that Markov minima are studied in the Markov spectrum theory discussed in Chapter 7.

Lemma 14.26. *Let v be an integer point in the w-continued fraction of an algebraic arrangement R. Then*

$$|\Phi_R(v)| \geq w\alpha_R.$$

Proof. The statement follows directly from the fact that the w-continued fraction of R is homothetic to the continued fraction of R.   □

Theorem 14.27. *Let R be an algebraic arrangement. Then there exists $C > 0$ such that for every $N \in \mathbb{Z}_+$ the following holds: each of the two lines of the best approximation arrangement R_N contains an integer point of some w-sail for $w < C$.*

Proof. Let R be an arrangement of lines $y = \alpha_1 x$ and $y = \alpha_2 x$. Let also

$$v_{1,N}(x_{1,N}, y_{1,N}) \quad \text{and} \quad v_{2,N}(x_{2,N}, y_{2,N})$$

be two integer vectors of unit integer length in the lines defining R_N.

Since R is algebraic, the continued fractions for α_1 and α_2 are periodic and therefore uniformly bounded. Now we can apply Theorem 14.19 to write the estimate

$$\rho(R, R_N) < \frac{C_2}{N^2}.$$

Hence by Lemma 14.20 we have

$$\left| \frac{x_{1,N}}{y_{1,N}} - \alpha_1 \right| < \frac{\tilde{C}_2}{N^2}.$$

Notice that

$$|\Phi_R(x_{1,N}, y_{1,N})| = \left| \frac{(x_{1,N} - \alpha_1 y_{1,N})(x_{1,N} - \alpha_2 y_{1,N})}{\alpha_1 - \alpha_2} \right|$$
$$= \left| \frac{x_{1,N}}{y_{1,N}} - \alpha_1 \right| \cdot \left| \frac{x_{1,N} - \alpha_2 y_{1,N}}{\alpha_1 - \alpha_2} y_{1,N} \right|.$$

The first term above is bounded by $\tilde{C}/N^2$ for some constant $\tilde{C}$ that does not depend on N. Hence,

$$|\Phi_R(x_{1,N}, y_{1,N})| \leq \tilde{C} \left| \frac{y_{1,N}^2}{N^2} \cdot \frac{\frac{x_{1,N}}{y_{1,N}} - \alpha_2}{\alpha_1 - \alpha_2} \right| \leq \tilde{C} \left| \frac{\frac{x_{1,N}}{y_{1,N}} - \alpha_2}{\alpha_1 - \alpha_2} \right|.$$

Finally, the last expression is uniformly bounded in N. The same property holds for the values $|\Phi_R(v_{2,N})|$.

Therefore, by Lemma 14.26 the vectors of $v_{1,N}$ and $v_{2,N}$ for all N are in the union of a finite number of w-continued fractions for R. ☐

Conjecture 7. We conjecture that for all algebraic arrangements for almost all N, the vectors $v_{1,N}$ and $v_{2,N}$ defining R_N are in a 1-geometric continued fraction.

14.2.6 Finding best approximations of line arrangements

In this subsection we show a general technique for calculating best approximations for an arbitrary arrangement of lines R with eigenspaces $y = \alpha_1 x$ and $y = \alpha_2 x$ for distinct real numbers α_1 and α_2.

Proposition 14.28. *Let m and n be two integers. Supposing that $|\alpha_1 - \frac{p}{q}| < \delta$ (or $|\alpha_2 - \frac{p}{q}| < \delta$, respectively), then the following holds:*

$$\left|\alpha_1 - \frac{p}{q}\right| > \frac{|\alpha_1 - \alpha_2|}{|\alpha_1 - \alpha_2| + \delta}\frac{|\Phi_R(p,q)|}{q^2}$$

$$\left(\left|\alpha_2 - \frac{p}{q}\right| > \frac{|\alpha_1 - \alpha_2|}{|\alpha_1 - \alpha_2| + \delta}\frac{|\Phi_R(p,q)|}{q^2}\right).$$

Proof. We have

$$\left|\alpha_1 - \frac{p}{q}\right| = \frac{1}{q}\left|p - \alpha_1 q\right| = \frac{1}{q}\frac{|p - \alpha_1 q|(p - \alpha_2 q)}{p - \alpha_2 q}$$

$$= \frac{|\Phi_R(p,q)|}{q^2}\frac{|\alpha_1 - \alpha_2|}{|\alpha_1 - \alpha_2 + (\frac{p}{q} - \alpha_1)|} > \frac{|\alpha_1 - \alpha_2|}{|\alpha_1 - \alpha_2| + \delta}\frac{|\Phi_R(p,q)|}{q^2}.$$

The same holds for the case of the approximations of α_2.   □

Procedure of best approximation calculation.

Input data. We start with an arrangement R of two lines $y = \alpha_1 x$ and $y = \alpha_2 x$ and a positive integer N. The task is to construct the best approximation R_N.

Step 1. Find p_i/q_i, which is the i-convergent to α_1 with maximal possible denominator satisfying

$$1 \le p_i \le N \quad \text{and} \quad 1 \le q_i \le N.$$

Also find p'_j/q'_j, the convergent satisfying the same property for α_2.

Step 2. Consider the arrangement $\bar{R}$ of lines

$$y = \frac{p_i}{q_i}x \quad \text{and} \quad y = \frac{p'_j}{q'_j}x.$$

We know that

$$\left|\alpha - \frac{p_i}{q_i}\right| \le \frac{1}{q_i q_{i+1}} \quad \text{and} \quad \left|\alpha - \frac{p'_j}{q'_j}\right| \le \frac{1}{q'_j q'_{j+1}}.$$

By Lemma 14.21 (for $N > 1$) we have

$$\rho(R, \bar{R}) < C \max\left(\frac{1}{q_i q_{i+1}}, \frac{1}{q'_j q'_{j+1}}\right),$$

where

$$C = \frac{\max\left(2, 2(|\alpha_1| + |\alpha_2|), \alpha_1^2 + \alpha_2^2 + |\alpha_1 - \alpha_2|\right)}{(|\alpha_1 - \alpha_2|)(|\alpha_1 - \alpha_2|)}.$$

Step 3. From the definition of best approximations and the result of Step 1 we have

$$\rho(R, R_N) \le \rho(R, \bar{R}) < C \max\left(\frac{1}{q_i q_{i+1}}, \frac{1}{q'_j q'_{j+1}}\right).$$

In most cases (except for a small denominator N) this estimate satisfies

$$\rho(R, R_N) < \frac{1}{2|\alpha_1 - \alpha_2|}.$$

Choose δ_1 and δ_2 such that the lines of the best approximation R_N are $y = (\alpha_1 + \delta_1)x$ and $y = (\alpha_2 + \delta_2)x$. In the notation of Lemma 14.20 this means that $R_N = R_{\delta_1, \delta_2}$. Now we apply Lemma 14.20 to write the estimate for δ_1 and δ_2:

$$|\delta_1| < \frac{2(1 + |\alpha_1|)(\alpha_1 - \alpha_2)^2}{|\alpha_2|} \cdot C \max\left(\frac{1}{q_i q_{i+1}}, \frac{1}{q'_j q'_{j+1}}\right);$$

$$|\delta_2| < \frac{2(1 + |\alpha_2|)(\alpha_1 - \alpha_2)^2}{|\alpha_1|} \cdot C \max\left(\frac{1}{q_i q_{i+1}}, \frac{1}{q'_j q'_{j+1}}\right),$$

where C is as in Step 2.

Step 4. Suppose now that the lines of the best approximation R_N contains integer vectors (n_1, m_1) and (n_2, m_2) of unit integer length. By Proposition 14.28 we write the following estimates on the values of the Markov—Davenport form Φ_R at these vectors:

$$\frac{\Phi_R(m_1, n_1)}{n_1^2} < \frac{|\alpha_1 - \alpha_2| + \delta_1}{|\alpha_1 - \alpha_2|} \left| \alpha_1 - \frac{m_1}{n_1} \right| < \frac{|\alpha_1 - \alpha_2| + \delta_1}{|\alpha_1 - \alpha_2|} \delta_1;$$

$$\frac{\Phi_R(m_2, n_2)}{n_2^2} < \frac{|\alpha_1 - \alpha_2| + \delta_2}{|\alpha_1 - \alpha_2|} \left| \alpha_1 - \frac{m_2}{n_2} \right| < \frac{|\alpha_1 - \alpha_2| + \delta_2}{|\alpha_1 - \alpha_2|} \delta_2.$$

Step 5. This is the last step. Here we should consider all integer arrangements that contains integer points (m'_1, n'_1) and (m'_2, n'_2) satisfying the inequalities obtained in Step 4 for $\Phi_R(m_i, n_i)$ for $i = 1, 2$. Then one should choose the arrangement with the smallest discrepancy, which is R_N.

Output. In the output we have the best approximation R_N.

Remark 14.29. The above procedure of best approximation calculation seriously decreases the number of integer points in the square $N \times N$ that should be tested to find R_N. This is of a great importance especially for the case of algebraic arrangements. The procedure gives an explicit estimate on the number of w-sails whose points should be checked to construct best approximations. Since all w-sails are periodic, all its points are contained in the finite number of orbits with respect to the corresponding Dirichlet group action.

Example 14.30. Input data. Consider an arrangement of eigenlines of the Fibonacci operator:

$$\begin{pmatrix} 0 & 1 \\ 1 & 1 \end{pmatrix}.$$

Denote by F_n the nth Fibonacci number.
  Consider any integer $N \geq 100$.

Step 1. Consider a positive integer k such that $F_k \le N < F_{k+1}$ and choose an approximation $\bar{R}$ with eigenspaces $F_{k-1}y - F_k x = 0$ and $F_k y + F_{k-1}x = 0$. Then

$$\left| \alpha_1 - \frac{F_k}{F_{k-1}} \right| \le \frac{1}{F_{k-1}F_k}, \quad \left| \alpha_1 + \frac{F_{k-1}}{F_k} \right| \le \frac{1}{F_k F_{k+1}}.$$

Step 2. Since $1/(F_{k-1}F_k) < 1/(55 \cdot 89)$, we have

$$\rho(R,R_N) < \frac{\max\left(2, 2\sqrt{5}, 3 + \sqrt{5}/4895\right)}{5 + \frac{2\sqrt{5}}{4895}} \frac{1}{F_{k-1}F_k} < \frac{2\sqrt{5}}{5 + \frac{2\sqrt{5}}{4895}} \frac{(89/55)^3}{N^2} < \frac{3.79}{N^2}.$$

Step 3. Now, following Lemma 14.20 we write the estimates on δ_1 and δ_2 of the best approximation $R_N = R_{\delta_1, \delta_2}$:

$$|\delta_1| < \frac{80.35}{N^2} \quad \text{and} \quad |\delta_2| < \frac{18.97}{N^2}.$$

Step 4. Suppose that the best approximation R_N is defined by the two integer vectors (m_1, n_1) and (m_2, n_2) of unit integer length. Then we have the following two estimates:

$$\frac{|\Phi_R(m_1, n_1)|}{n_1^2} < \frac{80.65}{N^2}, \quad \frac{|\Phi_R(m_2, n_2)|}{n_2^2} < \frac{18.99}{N^2}.$$

Step 5. and Output. We have completed the computations for $N = 10^6$, and the answer in this case is the matrix with eigenspaces $F_{29}y - F_{30}x = 0$ and $F_{30}y + F_{29}x = 0$. We conjecture that for the Fibonacci matrix all the best approximations are the arrangements of lines $F_{k-1}y - F_k x = 0$ and $F_k y + F_{k-1}x = 0$ for $k = 0, 1, \ldots$.

14.3 Exercises

Exercise 14.1. Find all best approximations and all strong best approximations for the numbers $433/186$ and $575/247$.

Exercise 14.2. Let $\triangle ABC$ be an integer empty triangle. Consider a strip bounded by two lines parallel to AB and passing through the points A and C respectively. Prove that all integer points on this strip are at its boundary lines.

Exercise 14.3. Consider a positive real number α with a regular continued fraction $[a_0; a_1 : \cdots]$. Let x be an integer satisfying $1 \le x \le a_{k+1} - 1$. Prove that

$$|[a_0; a_1 : \cdots : a_k : x] - \alpha| > |[a_0; a_1 : \cdots : a_k : x + 1] - \alpha|.$$

Exercise 14.4. Let $\alpha = [a_0; a_1 : \cdots]$ be a positive real number. Denote by p_k/q_k its k-convergent and by r_k the remainder part $[a_k; a_{k+1} : \cdots]$. Then the condition

$$r_{k+1} < \frac{q_{k+1}}{q_k}$$

is equivalent to the condition

$$[a_{k+1}; a_{k+2} : \cdots] < [a_{k+1}; a_k : \cdots : a_0].$$

Exercise 14.5. Write Markov—Davenport forms for the arrangements R_1 and R_2 of eigenlines of the operators

$$\begin{pmatrix} 3 & 5 \\ 2 & 3 \end{pmatrix} \quad \text{and} \quad \begin{pmatrix} 7 & -2 \\ -3 & 1 \end{pmatrix}$$

respectively. Write the discrepancy between R_1 and R_2.

Exercise 14.6. Consider an arbitrary angle C with vertex at the origin whose edges do not contain other integer points. Prove that the w-sail of C is homothetic to the 1-sail of C and find the coefficient of homothety. What happens for the angles having integer points on their edges?

Exercise 14.7. Let R be an algebraic arrangement of two lines. Then the following statements hold.
(a). The Markov—Davenport form Φ_R attains only a finite number of values at the integer points of the corresponding continued fraction. Denote this set by S.
(b). The form Φ_R attains the value x at some vertex of the w-sail if and only if $x/w \in S$.

Exercise 14.8. Find all best approximations of sizes $1, 2, 3, 4, 5$ for the arrangement of the eigenlines of the Fibonacci operator.

Chapter 15
Geometry of Continued Fractions with Real Elements and Kepler's Second Law

In the beginning of this book we discussed the geometric interpretation of regular continued fractions in terms of LLS sequences of sails. Is there a natural extension of this interpretation to the case of continued fractions with arbitrary elements? The aim of this chapter is to answer this question.

In Sections 15.1 and 15.2 we introduce a geometric interpretation of odd or infinite continued fractions with arbitrary elements in terms of broken lines in the plane having a selected point (say the origin). Further, in Section 15.3 we consider differentiable curves as infinitesimal broken lines to define analogues of continued fractions for curves. The resulting analogues possess an interesting interpretation in terms of a motion of a body according to Kepler's second law.

15.1 Continued fractions with integer coefficients

In this section we give a formal extension of the notion of the LLS sequence for sails to the case of certain integer broken lines. Let us specify the class of broken lines we are dealing with.

A broken line is called *integer* if all its vertices are integer points. We restrict ourselves to the case of broken lines all of whose triples of consecutive vertices are not in a line.

Definition 15.1. Consider an integer broken line L in the complement to an integer point V. We say that L is *at unit integer distance from V* (or *V-broken line*, for short) if all edges of L are at unit integer distance from V.

Remark. Without loss of generality, we mostly use only O-broken lines, where O is the origin.

For arbitrary integer points A, B, and C, we consider the function of orientation sgn, namely

© Springer-Verlag GmbH Germany, part of Springer Nature 2022
O. N. Karpenkov, *Geometry of Continued Fractions*,
Algorithms and Computation in Mathematics 26,
https://doi.org/10.1007/978-3-662-65277-0_15

$$\text{sgn}(ABC) = \begin{cases} 1, & \text{if the pair of vectors } (BA, BC) \text{ defines the positive orientation,} \\ 0, & \text{if the points } A, B, \text{ and } C \text{ are in a straight line,} \\ -1, & \text{if } BA, BC \text{ defines the negative orientation.} \end{cases}$$

Let us generalize the notion of LLS sequences for sails to an arbitrary O-broken line.

Definition 15.2. Consider an O-broken line $A_0 A_1 \ldots A_n$. Put by definition

$$a_0 = \text{sgn}(A_0 V A_1) \, l\ell(A_0 A_1),$$
$$a_1 = \text{sgn}(A_0 V A_1) \, \text{sgn}(A_1 V A_2) \, \text{sgn}(A_0 A_1 A_2) \, l\sin(\angle A_0 A_1 A_2),$$
$$a_2 = \text{sgn}(A_1 V A_2) \, l\ell(A_1 A_2),$$
$$\ldots$$
$$a_{2n-3} = \text{sgn}(A_{n-2} V A_{n-1}) \, \text{sgn}(A_{n-1} V A_n) \, \text{sgn}(A_{n-2} A_{n-1} A_n) \, l\sin(\angle A_{n-2} A_{n-1} A_n),$$
$$a_{2n-2} = \text{sgn}(A_{n-1} V A_n) \, l\ell(A_{n-1} A_n).$$

The sequence $(a_0, \ldots, a_{2n-2})$ is called a *lattice signed length-sine* sequence for the the broken line $A_0 A_1 \ldots A_n$, or *LSLS sequence*, for short.
The *continued fraction for the O-broken line* $A_0 A_1 \ldots A_n$ is the element $[a_0; a_1 : \cdots : a_{2n-2}]$ of $\overline{\mathbb{R}}$.

Notice that the LSLS sequence for the sail of some angle coincides with the LLS sequence of this sail. So we consider LSLS sequences as a natural combinatoric—geometric generalization of LLS sequences. Note also that an O-broken line is uniquely defined by its LSLS sequence, the point A_0, and the direction of the vector $A_0 A_1$.

In Fig. 15.1 we show how to get signs of elements of the LSLS sequence from the local geometry of a broken line. As an example we consider the O-broken line with four vertices in Fig. 15.2.

In the next theorem we state that the LSLS sequence is the complete invariant for the action of the group of *proper integer congruences* (i.e., integer congruences preserving the orientation of the plane).

Theorem 15.3. (*i*) *Two O-broken lines are proper integer congruent if and only if their LSLS sequences coincide.*
(*ii*) *Every finite sequence of nonzero integers is realizable as an LSLS sequence for some O-broken line.*   □

We skip the proof of this theorem, since later, in the next section, we introduce a more general construction that generalize LSLS sequences for O-broken lines. For further information on O-broken lines and their LSLS sequences we refer to [102] and [104].

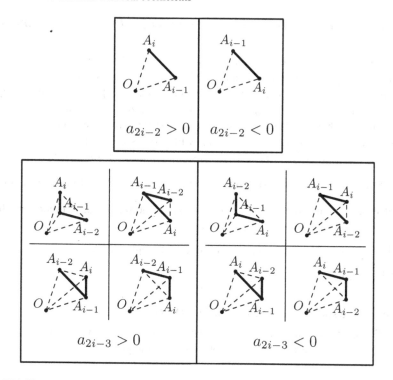

Fig. 15.1 How to identify signs of elements in LSLS sequences.

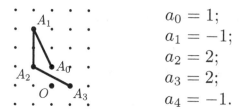

$$a_0 = 1;$$
$$a_1 = -1;$$
$$a_2 = 2;$$
$$a_3 = 2;$$
$$a_4 = -1.$$

Fig. 15.2 An O-broken line and the corresponding LSLS sequence.

15.2 Continued fractions with real coefficients

Let us describe a recent generalization of lattice properties of regular continued fractions to the case of arbitrary continued fractions (see also [106]).

Further in this chapter we work in the plane with a marked point O (say the origin). For two vectors v and w in the plane we denote the oriented volume of the parallelogram spanned by the vectors v and w by $\det(v, w)$.

15.2.1 Broken lines related to sequences of arbitrary real numbers

In Chapter 3 (see page 41) we gave a geometric algorithm to construct sails for real numbers given the regular continued fractions of the corresponding numbers. The algorithm has a sequence of positive integers $(a_0, a_1, \ldots)$ as input and its output is the sails for the number $[a_0, a_1, \ldots]$. In this subsection we slightly modify this algorithm such that it will construct a broken line starting from an arbitrary sequence of nonzero real numbers. This algorithm will give us a base to define continued fractions related to broken lines (we do this later, in the next subsection).

Algorithm to construct broken lines related to sequences of real numbers.

Input data. Given a sequence $(a_0, a_1, \ldots, a_{2n})$ of nonzero real numbers; the first vertex A_0; and the direction v of the first edge.

Goal of the algorithm. To construct a broken line $A_0 A_1 \ldots A_n$.

Step 1. Assign

$$A_1 = A_0 + \lambda v,$$

where λ is the solution of the equation $\det(OA_0, OA_1) = a_0$.

Inductive Step k. From previous steps we have the vertices $A_0, \ldots, A_k$, for $k \geq 1$. Now let us find the vertex A_{k+1}. First, we consider the point

$$P = A_k + \frac{1}{a_{2k-2}} A_{k-1} A_k.$$

Notice that the point P is on the line $A_{k-1}A_k$ and the area of $\triangle OA_kP$ equals $1/2$. Secondly, we put

$$Q = P + a_{2k-1} OA_k.$$

Thirdly, we define the point A_{k+1}(see Fig. 15.3):

$$A_{k+1} = A_k + a_{2k} A_k Q.$$

This concludes Step k.

Output. The broken line $A_0 \ldots A_n$.

Remark 15.4. The constructed broken line in some sense generalizes geometric construction from the case of sails related to regular continued fractions (see Theorem 3.1 above) to the case of broken lines related to arbitrary continued fractions (see Theorem 15.10 and Corollary 15.12 below).

Let us give a geometric meaning of $a_0, a_1, \ldots$ in terms of oriented volumes for certain vectors defined by the above broken line.

Proposition 15.5. *Let $(a_0, a_1, \ldots, a_{2n})$ be a sequence of arbitrary nonzero elements. Consider any broken line $A_0 \ldots A_n$ constructed by this sequence as above. Then the following holds:*

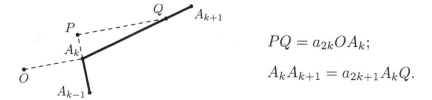

$$PQ = a_{2k}OA_k;$$

$$A_kA_{k+1} = a_{2k+1}A_kQ.$$

Fig. 15.3 Construction of the point A_{k+1}.

$$a_{2k} = \det(OA_k, OA_{k+1}), \quad k = 0, \ldots, n,$$

$$a_{2k-1} = \frac{\det(A_kA_{k-1}, A_kA_{k+1})}{a_{2k-2}a_{2k}}, \quad k = 1, \ldots, n.$$

Proof. Let us prove this proposition by induction on k.

Base of induction. Directly from the first step of the construction we get

$$\det(OA_0, OA_1) = a_0.$$

Inductive Step. Supposing that the statement holds for $k-1$, let us prove it for k.
We start with a_{2k}:

$$\det(OA_k, OA_{k+1}) = a_{2k}\det(OA_k, OQ) = a_{2k}\det(OA_k, OP) = \frac{a_{2k}\det(OA_{k-1}, OA_k)}{a_{2k-2}}$$

$$= a_{2k}.$$

The last equality follows from the induction assumption.
Now we check the expression for a_{2k-1}.

$$\frac{\det(A_kA_{k-1}, A_kA_{k+1})}{a_{2k-2}a_{2k}} = \frac{\det(A_kA_{k-1}, A_kQ)}{a_{2k-2}} = \det(PA_k, PQ) = a_{2k-1}\det(OA_k, OP)$$

$$= \frac{a_{2k-1}}{a_{2k-2}}\det(OA_{k-1}, OA_k) = a_{2k-1}.$$

The inductive step is complete. This concludes the proof. □

Example 15.6. Let us consider the first interesting case of broken lines with three
vertices. Without loss of generality we fix $A_0 = (1,0)$ and the direction $v = (0,1)$.
Letting the sequence be (a_0, a_1, a_2), we have

$$A_1 = (1, a_0).$$

The corresponding points P and Q are as follows:

$$P = (1, 1+a_0), \qquad Q = (1+a_1, 1+a_0+a_0a_1).$$

Finally, we have

$$A_2 = (1 + a_1 a_2, a_0 + a_2 + a_0 a_1 a_2).$$

Notice that the coordinates of the point A_2 coincide with the denominator and the numerator of the continued fraction $[a_0; a_1 : a_2]$.

15.2.2 Continued fractions related to broken lines

Let us extend the definition of LLS sequences for sails to the case of broken lines, basing the definition on the expressions of Proposition 15.5. Consider an arbitrary broken line $A_0 \ldots A_n$ satisfying the following condition: any of its edge is not contained in a line passing through the origin O.

Definition 15.7. Define

$$a_{2k} = \det(OA_k, OA_{k+1}), \quad k = 0, \ldots, n;$$
$$a_{2k-1} = \frac{\det(A_k A_{k-1}, A_k A_{k+1})}{a_{2k-2} a_{2k}}, \quad k = 1, \ldots, n.$$

The sequence $(a_0, \ldots, a_{2n})$ is called the *LLS sequence* for the broken line. The number $[a_0; \cdots : a_{2n}]$ is said to be the *continued fraction for the broken line* $A_0 \ldots A_n$.

This definition extends the definition of LSLS sequences to O-broken lines.

Proposition 15.8. *Let a broken line be an O-broken line. Then its LLS sequence coincides with its LSLS sequence.*   □

Since the proof of this proposition is a straightforward calculation, we leave it to the reader.

Finally, we show how the $GL(2, \mathbb{R})$-transformations act on the LLS sequences.

Proposition 15.9. *Consider two broken lines $A_0 \ldots A_n$ and $B_0 \ldots B_n$ whose LLS sequences are $(a_0, \ldots, a_{2n})$ and $(b_0, \ldots, b_{2n})$ respectively. Suppose that there exists a $GL(2, \mathbb{R})$-operator taking the first broken line to the second one. Let the determinant of this operator equal λ. Then we have*

$$\begin{cases} a_{2k} = \lambda b_{2k}, & k = 0, \ldots, n, \\ a_{2k-1} = \frac{1}{\lambda} b_{2k-1}, & k = 1, \ldots, n \end{cases}.$$

Proof. The statement follows directly from formulas of Proposition 15.5, since the area of any parallelogram in the definition is multiplied by λ.   □

15.2.3 Geometry of continued fractions for broken lines

Let us recall the notion of polynomials P_k and Q_k from Chapter 1. The polynomials P_k and Q_k are uniquely defined by the following three conditions: first, they are relatively prime; secondly,

$$\frac{P_k(x_0,\ldots,x_k)}{Q_k(x_0,\ldots,x_k)} = [x_0;x_1 : \cdots : x_k];$$

thirdly,

$$P_k(0,\ldots,0) + Q_k(0,\ldots,0) = 1.$$

Theorem 15.10. *Consider an O-broken line* $A_0\ldots A_n$ *with LLS sequence*

$$(a_0,a_1,\ldots,a_{2n-2}).$$

Let also $A_0 = (1,0)$ *and* $A_1 = (1,a_0)$. *Then*

$$A_n = \big(Q_{2n-1}(a_0,a_1,\ldots,a_{2n-2}),P_{2n-1}(a_0,a_1,\ldots,a_{2n-2})\big).$$

Remark 15.11. Similarly to Theorem 3.1 we have the following continuant expression for Theorem 15.10:

$$A_n = (K_{2n-2}(a_1,\ldots,a_{2n-2}),K_{2n-1}(a_0,a_1\ldots,a_{2n-2})).$$

Proof. We prove this theorem by induction on the number of vertices in the broken line, i.e., in n.

Base of induction. If the broken line has two vertices A_0,A_1 and its LLS sequence is (a_0), then $A_1 = (1,a_0)$.

Step of induction. Assume that the statement holds for all broken lines with k vertices. Let us give a proof for an arbitrary broken line with $k+1$ vertices. Consider a broken line $A_0\ldots A_k$. Let its LLS sequence be $(a_0,\ldots,a_{2k})$.
   Let

$$T = \begin{pmatrix} a_0a_1 + 1 & -a_1 \\ -a_0 & 1 \end{pmatrix}.$$

The transformation T takes A_1 to $(1,0)$ and the line A_2A_1 to the line $x = 1$. Under the transformation T the broken line $A_0\ldots A_k$ is taken to another broken line, which we denote by $B_0B_1\ldots B_k$. Since the determinant of T equals 1, then by Proposition 15.9, the LLS sequence for $B_0B_1\ldots B_k$ coincides with $(a_0,\ldots,a_{2k})$. By the induction assumption, the statement holds for the broken line $B_1\ldots B_k$. Thus we have

$$B_k = \big(Q_{2k-1}(a_2,\ldots,a_{2k}),P_{2k-1}(a_2,\ldots,a_{2k})\big).$$

Letting $B_k = (q,p)$, then

$$A_k = T^{-1}(B_k) = \big(p+a_1q,a_0p+(a_0a_1+1)q\big).$$

Therefore, we have

$$\frac{a_0 p + (a_0 a_1 + 1)q}{p + a_1 q} = a_0 + \cfrac{1}{a_1 + \frac{p}{q}} = \frac{P_{2k+1}(a_0, a_1, \ldots, a_{2k})}{Q_{2k+1}(a_0, a_1, \ldots, a_{2k})}.$$

Now we substitute $p = P_{2k-1}(a_2, \ldots, a_{2k})$ and $q = Q_{2k-1}(a_2, \ldots, a_{2k})$. Let us calculate the value of the sum of the numerator and denominator in the leftmost fraction at the point $O = (0, \ldots, 0)$:

$$((a_0 P_{2k-1} + (a_0 a_1 + 1)Q_{2k-1}) + (P_{2k-1} + a_1 Q_{2k-1}))(O) = (P_{2k-1} + Q_{2k-1})(O) = 1.$$

The first equality holds since we take $a_0 = a_1 = 0$. The second equation holds by definition of P_{2k-1} and Q_{2k-1}. Therefore,

$$P_{2k+1}(a_0, a_1, \ldots, a_{2k}) = a_0 P_{2k-1}(a_2, \ldots, a_{2k}) + (a_0 a_1 + 1)Q_{2k-1}(a_2, \ldots, a_{2k}),$$
$$Q_{2k+1}(a_0, a_1, \ldots, a_{2k}) = P_{2k-1}(a_2, \ldots, a_{2k}) + a_1 Q_{2k-1}(a_2, \ldots, a_{2k}).$$

This concludes the proof for the broken line $A_0 \ldots A_k$. The induction step is complete.   □

Theorem 15.10 extends the geometric interpretation of regular continued fractions to the case of arbitrary continued fractions in the following way.

Corollary 15.12. *Consider a broken line $A_0 \ldots A_n$ with $A_0 = (1, 0)$, $A_1 = (1, a_0)$, and $A_n = (x, y)$. Let $\alpha = [a_0; a_1 : \cdots : a_{2n}]$ be the corresponding continued fraction for this broken line. Then*

$$\frac{y}{x} = \alpha.$$

For the case of an infinite value for the continued fraction for the broken line, $x/y = 0$.   □

Remark 15.13. Supposing that the sequence of numbers $(a_0, a_1, \ldots, a_{2n})$ in Corollary 15.12 contains only positive integers, the broken line $A_0 \ldots A_n$ coincides with the sail of the angle defined by rays $y = 0$ and $y = \alpha x$ in the first quadrant of the plane. In particular, $A_0 \ldots A_n$ is in the boundary of a convex set and its LLS sequence coincides with the LLS sequence of the sail for the angle $\angle A_0 O A_n$.

From Corollary 15.12 we get the following statement.

Corollary 15.14. *Consider two broken lines $A_0 \ldots A_n$ and $B_0 \ldots B_m$ with LLS sequences $(a_0, a_1, \ldots, a_{2n})$ and $(b_0, b_1, \ldots, b_{2m})$ respectively. Let $B_0 = A_0$ and the vector $A_0 A_1$ coincide with the vector $\frac{a_0}{b_0} B_0 B_1$. Suppose that*

$$[a_0; \cdots : a_{2n}] = [b_0; \cdots : b_{2m}].$$

Then the points A_n, B_m, and the origin O are contained in one line.

Proof. Consider a transformation in the group $SL(2, \mathbb{R})$ taking the point A_0 to $(1, 0)$ and A_1 to some point on the line $x = 1$. It takes both broken lines to some broken lines with $B_0 = A_0 = (1, 0)$ and the points A_1 and B_1 to some points on the line $x = 1$. By Proposition 15.9 this operator does not change the continued fraction. Now we

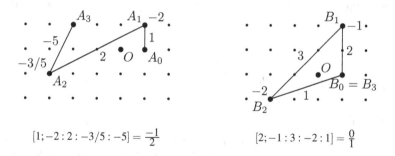

$$[1; -2 : 2 : -3/5 : -5] = \frac{-1}{2}$$

$$[2; -1 : 3 : -2 : 1] = \frac{0}{1}$$

Fig. 15.4 Two examples of broken lines with three edges.

are in a position to apply Corollary 15.10: the images of the points A_n, B_m, and the origin are on a line. Therefore, the points A_n, B_m, and O are on a line. □

Knowing the continued fraction for a broken line, it is possible to say whether the corresponding broken line is closed.

Corollary 15.15. (On necessary and sufficient conditions for broken lines to be closed.) *Consider a broken line $A_0A_1 \ldots A_n$ with the LLS sequence $(a_0, a_1, \ldots, a_{2n})$. Let $A_0 = (1,0)$ and A_0A_1 be on the line $x = 1$. Then the broken line is closed if and only if*

$$Q_{2n+1}(a_0, a_1, \ldots, a_{2n}) = 1 \quad and \quad P_{2n+1}(a_0, a_1, \ldots, a_{2n}) = 0.$$

The condition $P_{2n+1} = 0$ is equivalent to the following one:

$$[a_0; a_1 : \cdots : a_{2n}] = 0.$$

Notice also that for an arbitrary broken line the conditions are almost the same, as seen from Proposition 15.9.

Example 15.16. Let us write the conditions for a broken line consisting of three edges. Suppose that the LLS sequence of this broken line is $(a_0, a_1, a_2, a_3, a_4)$. Then the conditions are introduced by the following system:

$$\begin{cases} a_0a_1a_2a_3a_4 + a_0a_1a_2 + a_0a_1a_4 + a_0a_3a_4 + a_2a_3a_4 + a_0 + a_2 + a_4 = 0, \\ a_1a_2a_3a_4 + a_1a_2 + a_1a_4 + a_3a_4 + 1 = 1. \end{cases}$$

In Fig. 15.4 we show two examples of broken lines with three edges.

We conclude this section with the following open problem.

Problem 8. Consider a broken line L and a proper Euclidean transformation T (the origin may not be preserved by T). Find the relation between the elements of the LLS sequences of the broken lines L and $T(L)$.

Due to Proposition 15.9 it is sufficient to solve this problem only for translations by a vector.

15.2.4 Proof of Theorem 4.16

First let us prove the following statement.

Proposition 15.17. *Consider an integer angle $\angle AOB$ with vertex at the origin O and with edges passing through $A = (1,0)$ and $B = (1,\alpha)$, respectively, where α satisfies $0 < \alpha < 1$. Let the odd regular continued fraction for α be*

$$\alpha = [0; a_1 : \cdots : a_{2n-2}].$$

Then the LLS sequence of the angle $\angle AOB$ is

$$(a_2, \ldots, a_{2n-2}).$$

Proof. The congruence is provided by the multiplication by

$$\begin{pmatrix} 1 & a_1 \\ 0 & 1 \end{pmatrix}.$$

$\square$

We have finally collected all necessary ingredients to to prove Theorem 4.16.

Proof of Theorem 4.16. Recall that we are given by two linearly independent integer vectors

$$OA = (p,q) \quad \text{and} \quad OB = (r,s).$$

The proof for the statement of Theorem 4.16 is given by the study of numerous straightforward cases of various signs for p, q, r, s, and $\det(OA, OB)$. Let us study the case $p, q, r, s > 0, \det(OA, OB) < 0$ (we omit the studies of all the other cases here).

In this case two sequences of integers

$$(a_0, a_1, \ldots, a_{2m}) \quad \text{and} \quad (b_0, a_1, \ldots, b_{2n})$$

are defined as the sequences of elements of odd regular continued fractions of rational numbers

$$\frac{q}{p} = [a_0; a_1 : \ldots : a_{2m}] \quad \text{and} \quad \frac{s}{r} = [b_0; a_1 : \ldots : b_{2n}].$$

Denote by E the coordinate vector $(1,0)$. Consider the broken line L that is a concatenation of the following two sails. First we take the sail of the angle $\angle AOE$ (in the case where the last edge of this sail is not vertical we formally add the infinitesimal edge EE with vertical direction and 0 integer length) and the sail for the

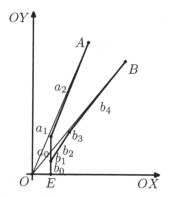

Fig. 15.5 The broken line corresponding to the concatenation of the sails for $\angle AOE$ and $\angle EOB$.

angle $\angle EOB$ (similarly another infinitesimal edge EE is added in the case where the first edge of the sail of the angle is not vertical). See Fig. 15.5.

The broken line L has the following properties:

— its first vertex is in the ray OA and the last vertex is in the ray OB;

— the direction of its first edge is inwards (with respect to the angle).

Then the angle is integer congruent to the angle $\angle EOC$ with $C = (1, \alpha)$ where $|\alpha|$ is defined by the LLS sequence of the above broken line as

$$\alpha = [\varepsilon a_{2m} : \varepsilon a_{2m-1} : \cdots : \varepsilon a_1 : \varepsilon a_0 : 0 : \delta b_0 : \delta b_1 : \cdots : \delta b_{2n}].$$

In case of $p, q, r, s > 0, \det(OA, OB) < 0$, the first part of the broken line L will be the sail (considered as a broken line) of $\angle AOE$ passed clockwise. Thus the elements of its LLS sequence are reversed and negative to the values of the LLS sequence for $\angle AOE$ (i.e., $\varepsilon = -1$). Note that in the case where $q/p < 1$ we end up with an infinitesimal (zero length) vertical vector which additionally brings two elements: $\lfloor p/q \rfloor$ for the angle with the vertical line passing through E and 0, indicating that we stay at E. Then we switch to the second sail. Both sails are starting vertically (or asymptotically vertically in the case where a_1 or b_1 are zeroes), hence the angle between the edges corresponding to a_0 and b_0 is zero. By that reason we should add a zero element to the LLS sequence for L here. Further we go back following the sail of the angle $\angle EOB$ (in this case $\delta = 1$), which is described by the continued fraction

$$[b_0 : b_1 : \cdots : b_{2n}]$$

(here again we formally add $b_0 = 0$ and $b_1 = \lfloor s/r \rfloor$ if $r/s < 1$). Therefore the LLS sequence of the broken line L is

$$(-a_{2m}, -a_{2m-1}, \ldots, -a_1, -a_0, 0, b_0, b_1, \ldots, b_{2n}).$$

Now let us define α as

$$\alpha = [-a_{2m} : -a_{2m-1} : \cdots : -a_1 : -a_0 : 0 : b_0 : b_1 : \cdots : b_{2n}]$$

and let the regular continued fraction for $|\alpha|$ be as follows

$$|\alpha| = [c_0; c_1 : \cdots : c_{2k}].$$

From Proposition 15.17 we have

$$\text{ltan} \angle EOC = \begin{cases} [c_0; c_1 : c_2 : \cdots : c_{2k}], & \text{if } c_0 \neq 0; \\ [c_2; \cdots : c_{2k}], & \text{if } c_0 = 0. \end{cases}$$

Therefore, the LLS sequence of $\angle AOB \cong \angle EOC$ is either $(c_0, c_1, \ldots, c_{2k})$ in the case where $c_0 \neq 0$ or $S = (c_2, \ldots, c_{2k})$ in the case where $c_0 = 0$. This concludes the proof of the case $p, q, r, s > 0, \det(OA, OB) < 0$.

The cases for the other choices of signs for p, q, r, s and $\det(OA, OB)$ are considered similarly, we omit them here.   □

15.3 Areal and angular densities for differentiable curves

Let us go even further and consider a differentiable curve instead of a broken line. In this situation it helps to think of curves as broken lines with infinitesimally small segments. In some sense the LLS sequence "splits" into a pair of functions related to the odd and even elements of the LLS sequence. Further, we call these two functions *areal* and *angular densities*. In this section we discuss some basic properties of areal and angular densities, in particular we shall show that the areal density is the inverse to the velocity of a body that moves according to the second Kepler law.

15.3.1 Notions of real and angular densities

Throughout this section we suppose that all curves are of class C^2 and have the arc-length parameterization.

Definition 15.18. Let γ be an arc-length parameterized C^2-class curve. By definition the *areal density* and the *angular density* at some point t are

$$A(t) = \lim_{\varepsilon \to 0} \frac{\det(O\gamma(t), O\gamma(t+\varepsilon))}{\varepsilon} = \det(O\gamma(t), \dot{\gamma}(t))$$

and

$$B(t) = \lim_{\varepsilon \to 0} \frac{\det(\gamma(t)\gamma(t-\varepsilon), \gamma(t)\gamma(t+\varepsilon))}{\varepsilon \det(O\gamma(t-\varepsilon), O\gamma(t)) \cdot \det(O\gamma(t), O\gamma(t+\varepsilon))}.$$

We continue with the following geometric interpretations for the function A.

Proposition 15.19. (Areal density and the Kepler's second law.) *Suppose that a body moves along the curve* γ *with velocity* $1/A$. *Then the sector area velocity of the body is constant and equals* 1. □

The proposition follows directly from the definition.

Let us show that the value of A^2B at a point t coincides with the signed curvature at t (which we denote by $\kappa(t)$).

Proposition 15.20. *Let the vectors* $O\gamma(t)$ *and* $\dot{\gamma}(t)$ *be noncollinear at some moment* t. *Then the following holds:*

$$A^2(t)B(t) = \kappa(t).$$

Proof. By definition we have

$$A^2(t)B(t) =$$
$$\lim_{\varepsilon \to 0} \left(\left(\frac{\det(O\gamma(t), O\gamma(t+\varepsilon))}{\varepsilon} \right)^2 \frac{\det(\gamma(t)\gamma(t-\varepsilon), \gamma(t)\gamma(t+\varepsilon))}{\varepsilon \det(O\gamma(t-\varepsilon), O\gamma(t)) \cdot \det(O\gamma(t), O\gamma(t+\varepsilon))} \right)$$
$$= \lim_{\varepsilon \to 0} \frac{\det(\gamma(t)\gamma(t-\varepsilon), \gamma(t)\gamma(t+\varepsilon))}{\varepsilon^3}.$$

Let us rewrite the curve γ in coordinates, i.e., $\gamma(t) = (x(t), y(t))$. Since the parameter t is arc-length we have

$$|\gamma'(t)| = \sqrt{(x'(t))^2 + (y'(t))^2} = 1 \quad \text{and} \quad \kappa(t) = x'(t)y''(t) - y'(t)x''(t).$$

Let us consider the Taylor series expansions

$$\gamma(t+\varepsilon) = \gamma(t) + \varepsilon\gamma'(t) + \frac{\varepsilon^2}{2}\gamma''(t) + \text{terms of higher order},$$
$$\gamma(t-\varepsilon) = \gamma(t) - \varepsilon\gamma'(t) + \frac{\varepsilon^2}{2}\gamma''(t) + \text{terms of higher order}.$$

Let us substitute

$$\gamma(t)\gamma(t+\varepsilon) = \left(\varepsilon x' + \frac{\varepsilon^2}{2}x'', \varepsilon y' + \frac{\varepsilon^2}{2}y'' \right) + \text{terms of higher order},$$
$$\gamma(t)\gamma(t-\varepsilon) = \left(-\varepsilon x' + \frac{\varepsilon^2}{2}x'', -\varepsilon y' + \frac{\varepsilon^2}{2}y'' \right) + \text{terms of higher order}.$$

Hence we have

$$\lim_{\varepsilon \to 0} \frac{\det(\gamma(t)\gamma(t-\varepsilon)\gamma(t)\gamma(t+\varepsilon))}{\varepsilon^3} = \lim_{\varepsilon \to 0} \frac{\varepsilon^3(x'y'' - y'x'') + \text{4th order terms}}{\varepsilon^3}$$
$$= \lim_{\varepsilon \to 0} \frac{\varepsilon^3\kappa(t) + \text{4th order terms}}{\varepsilon^3}$$
$$= \kappa(t).$$

Therefore, $A^2(t)B(t) = \kappa(t)$. □

Let us prove the theorem on existence of a curve with a given angular density. In some sense this theorem is a smooth analogue of the algorithm to construct a broken line by the elements of the corresponding continued fraction of Section 15.2.1. It is interesting to observe that to reconstruct a curve it is necessary to know the areal density (which corresponds only to odd elements in the LLS sequences for broken lines).

Theorem 15.21. *Suppose that we know the areal density $A(t)$ smoothly depending on a parameter t in some neighborhood of t_0, the starting position $\gamma(t_0)$, and the origin O.*

— If $|A(t_0)| > |O\gamma(t_0)|$, then there is no finite curve satisfying the above conditions

— If $|O\gamma(t_0)| > |A(t_0)| > 0$, then the curve γ is uniquely defined in some neighborhood of the point $\gamma(t_0)$.

Proof. In polar coordinates (r, φ) with center at O, the curve γ is defined by the following system of differential equations:

$$\begin{cases} r^2 \dot{\varphi} = A, \\ \dot{r}^2 + r^2 \dot{\varphi}^2 = 1. \end{cases}$$

This system is equivalent to the union of the following two systems:

$$\begin{cases} \dot{\varphi} = \frac{A}{r^2}, \\ \dot{r} = \sqrt{1 - \frac{A^2}{r^2}}, \end{cases} \quad \text{and} \quad \begin{cases} \dot{\varphi} = \frac{A}{r^2}, \\ \dot{r} = -\sqrt{1 - \frac{A^2}{r^2}}. \end{cases}$$

By the main theorem of ordinary differential equations (e.g., see [10]), each of these two systems has a finite solution if and only if $|r| > |A| > 0$. $\square$

15.3.2 Curves and broken lines

There are many interesting questions related to the convergence of broken lines to smooth curves in the context of areal and angular densities. Let us first show one particular result in this direction.

Let $\gamma(t)$ be a curve with arc-length parameter $t \in [0, T]$ and densities $A(t)$ and $B(t)$. In addition we suppose that the vectors $O\gamma(t)$ and $\dot{\gamma}(t)$ are linearly independent for all admissible values of t. For a positive integer n we denote by γ_n the broken line $V_{0,n} \ldots V_{n,n}$ with vertices

$$V_{i,n} = \gamma\left(\frac{i}{n}T\right).$$

Suppose that the LLS sequence of γ_n is $(a_{0,n}, \ldots, a_{2n,n})$. Consider the following two functions A_n and B_n:

$$A_n(t) = a_{2\lfloor nt/T \rfloor + 1, n} \quad \text{and} \quad B_n(t) = a_{2\lfloor nt/T \rfloor, n}.$$

These functions satisfy the following convergence theorem.

Theorem 15.22. *Let γ be a C^2-curve with arc-length parameterization. Then the sequence of functions (A_n) converges pointwise to the function A, and the sequence of functions (B_n) converges pointwise to the function B.*

Proof. This follows directly from the definition of density functions and the properties of LLS sequences shown in Proposition 15.5.   □

We continue with two open problems. The first one is in some sense an inverse problem to the statement of Theorem 1.16.

Problem 9. Let a sequence of broken lines L_i converge pointwise to a C^2-curve γ. Study the relations between LLS sequences of the broken lines L_i and their "limits" from one side and the density functions of γ from the other side.

Notice that it is not clear what kind of limits for the elements of LLS sequence one should use here.

Finally, it is interesting to have some criterion for the curve to be closed, as was done for broken lines in Corollary 15.15.

Problem 10. What are the conditions on the functions $A(t)$ and $B(t)$ for the resulting curve γ to be closed?

15.3.3 Some examples

Let us write down the areal and angular densities for straight lines, ellipses, and logarithmic spirals.

Lines. Let O be the origin and γ the line $x = a$. Then we have

$$A(t) = a \quad \text{and} \quad B(t) = 0.$$

Ellipses and their centers. Consider the ellipse $\frac{x^2}{a^2} + \frac{y^2}{b^2} = 1$ with $a \geq b > 0$ and let O be the origin, i.e., at the symmetry center of the ellipse. Then we have the following expressions for the areal and angular densities:

$$A(t) = \frac{ab}{\sqrt{a^2 \sin^2 t + b^2 \cos^2 t}} \quad \text{and} \quad B(t) = \frac{1}{ab\sqrt{a^2 \sin^2 t + b^2 \cos^2 t}}.$$

Notice that the ratio of the functions $A(t)$ and $B(t)$ is constant:

$$\frac{A(t)}{B(t)} = a^2 b^2.$$

Ellipses and their foci. Consider again the ellipse $\frac{x^2}{a^2} + \frac{y^2}{b^2} = 1$ with $a \geq b > 0$. Now let O be the point $(-\sqrt{a^2 - b^2}, 0)$, which is one of the foci of the ellipse. Then the areal and the angular densities are as follows:

$$A(t) = \frac{ab + b\sqrt{a^2 - b^2}\cos t}{\sqrt{a^2 \sin^2 t + b^2 \cos^2 t}}$$

and

$$B(t) = \frac{a}{b\sqrt{a^2 \sin^2 t + b^2 \cos^2 t}\left(a + \cos t\sqrt{a^2 - b^2}\right)^2}.$$

Remark on Keplerian planetary motion. Consider the Sun and some planet that moves around the Sun. Then the planet moves according to Kepler's laws:

I. The orbit of the planet is an ellipse with the Sun at one of the two foci of the ellipse.
II. The motion has constant sector area velocity.
III. The square of the orbital period of the planet is proportional to the cube of the semimajor axis of its orbit.

The trajectory of the planet is an ellipse defined by the equation $\frac{x^2}{a^2} + \frac{y^2}{b^2} = 1$ with $a \geq b > 0$ in some Euclidean coordinates (according to Kepler's first law), in addition, the Sun is one of the foci. Then according to Kepler's second law, the planet moves with velocity $\lambda/A(t)$ for all t. The constant λ is defined from Kepler's third law:

$$\frac{T^2}{a^3} = \frac{T_e^2}{a_e^3},$$

where by T we denote the orbital period of our planet, and by T_e and a_e, respectively, the orbital period and the semimajor axis for Earth. Denote by L the length of the ellipse for the planet, i.e.,

$$L = 4a \int\limits_0^{\pi/2} \sqrt{1 - \left(1 - \frac{b^2}{a^2}\right)\cos^2 t}\, dt.$$

Since $T = |\lambda| \int_0^L |1/A(t)| dt$, we have

$$\lambda = \pm \frac{T_e}{\int_0^L |1/A(t)| dt} \left(\frac{a}{a_e}\right)^{\frac{3}{2}}.$$

Similar situations hold for parabolic and hyperbolic motions.

Logarithmic spirals. Consider the logarithmic spiral

$$\{(ae^{bt}\cos t, ae^{bt}\sin t) \mid t \in \mathbb{R}\}.$$

Then the areal and angular densities are expressed as follows:

$$A(t) = \frac{ae^{bt}}{\sqrt{b^2+1}} \quad \text{and} \quad B(t) = \frac{e^{-3bt}\sqrt{b^2+1}}{a^3}.$$

Observe that

$$A^3(t)B(t) = \frac{1}{b^2+1},$$

meaning that for logarithmic spirals the function A^3B is constant.

Remark 15.23. When we consider lines, ellipses, and spirals we have certain relations between densities A and B: the functions A, A/B, and A^3B respectively are constant. Notice that if A^2B is a constant function, then the curvature is constant, and hence we get circles. It is natural to study what kind of curves are defined by certain relations between the densities. For instance, the following question remains open: *what curves do we get if AB (or simply B) is constant?*

15.4 Exercises

Exercise 15.1. Find a proof of Proposition 15.8.

Exercise 15.2. Draw a broken line on five vertices whose LLS sequence satisfies (a) all elements are positive; (b) all elements are negative; (c) even elements are positive and odd elements are negative; (d) odd elements are positive and even elements are negative.

Exercise 15.3. Calculate the areal and angular densities for (a) straight lines; (b) ellipses; (c) logarithmic spirals.

Chapter 16
Extended Integer Angles and Their Summation

Let us start with the following question. Suppose that we have arbitrary numbers a, b, and c satisfying

$$a + b = c.$$

How do we calculate the continued fraction for c knowing the continued fractions for a and b? It turns out that this question is not a natural question within the theory of continued fractions. The simplest algorithm works as follows:
— first calculate the rational number representations for a and b;
— add them according to basic school arithmetic;
— write the continued fraction for the sum.
One can hardly imagine any law to write the continued fraction for the sum directly. The main obstacle here is that the summation of rational numbers does not have a geometric explanation in terms of the integer lattice. In this chapter we propose to consider a "geometric summation" of continued fractions, which we consider a summation of integer angles.

In Section 16.1 we start with the notion of extended integer angles. These angles are the integer analogues of Euclidean angles of the type $k\pi + \varphi$ for arbitrary integers k. We classify extended angles by writing normal forms representing all of them. Finally, we define the M-sums of extended angles and integer angles. Further, in Section 16.2, we show how the continued fractions of extended angles are expressed in terms of the corresponding normal forms. Finally, in Section 16.3 we give a proof of Theorem 6.9(i) on the sum of integer angles in integer triangles, which is based on the techniques introduced in this chapter.

16.1 Extension of integer angles. Notion of sums of integer angles

We start with a few general definitions. An integer affine transformation of the plane is said to be *proper* if it preserves the orientation of the plane. We say that two

© Springer-Verlag GmbH Germany, part of Springer Nature 2022
O. N. Karpenkov, *Geometry of Continued Fractions*,
Algorithms and Computation in Mathematics 26,
https://doi.org/10.1007/978-3-662-65277-0_16

sets S_1 and S_2 are *proper integer congruent* if there exists a proper integer affine transformation of $\mathbb{R}^2$ taking the set S_1 to the set S_2, which we write as $S_1 \cong S_2$.

16.1.1 Extended integer angles and revolution number

16.1.1.1 Equivalence classes of integer oriented broken lines and the corresponding extended angles

We say that an integer oriented broken line $A_n A_{n-1} A_{n-2} \ldots A_0$ is *inverse* to the integer oriented broken line $A_0 A_1 A_2 \ldots A_n$.

Definition 16.1. Consider an integer point V. Two integer oriented V-broken lines l_1 and l_2 (see Definition 15.1) are said to be *equivalent* if their first and their last vertices coincide respectively and the closed broken line generated by l_1 and the inverse of l_2 is contractible in $\mathbb{R}^2 \setminus \{V\}$.

Definition 16.2. The equivalence class of integer oriented V-broken lines containing a broken line $A_0 A_1 \ldots A_n$ is said to be the *extended integer angle* for the broken line $A_0 A_1 \ldots A_n$ with *vertex* V (or for short, *extended angle*). We denote it by $\angle(V, A_0 A_1 \ldots A_n)$.

The set of extended angles is invariant with respect to the action of (proper) integer affine transformations. Therefore, *(proper) integer congruence* for extended angles is well defined.

16.1.1.2 Revolution numbers for extended angles.

Let v be an arbitrary vector, and V an integer point. Consider the ray

$$r = \{V + \lambda \bar{v} \mid \lambda \geq 0\}.$$

Let AB be an arbitrary (oriented from A to B) segment not contained in r. Suppose also that V is not in AB. We denote by $\#(r, AB)$ the number

$$\#(r, AB) = \begin{cases} 0, & AB \text{ does not intersect } r, \\ \frac{1}{2} \operatorname{sgn}\!\left(A(A+v)B\right), & AB \cap r \in \{A, B\}, \\ \operatorname{sgn}\!\left(A(A+v)B\right), & AB \cap r \in AB \setminus \{A, B\}, \end{cases}$$

and call it the *intersection number* of the ray r and the segment AB.

Definition 16.3. Let $A_0 A_1 \ldots A_n$ be a V-broken line, and let r be a ray $\{V + \lambda v \mid \lambda \geq 0\}$. We call the number

$$\sum_{i=1}^{n} \#(r, A_{i-1} A_i)$$

the *intersection number* of the ray r the V-broken line $A_0A_1 \ldots A_n$ and denote it by $\#(r, A_0A_1 \ldots A_n)$.

Definition 16.4. Consider an arbitrary extended angle $\angle(V, A_0A_1 \ldots A_n)$. Define

$$r_+ = \{V + \lambda VA_0 \mid \lambda \geq 0\} \qquad \text{and} \qquad r_- = \{V - \lambda VA_0 \mid \lambda \geq 0\}.$$

The number

$$\frac{1}{2}\left(\#(r_+, A_0A_1 \ldots A_n) + \#(r_-, A_0A_1 \ldots A_n)\right)$$

is called the *revolution number* for the extended angle $\angle(V, A_0A_1 \ldots A_n)$. We denote it by $\#(\angle(V, A_0A_1 \ldots A_n))$. Additionally, we put by definition $\#(\angle(V, A_0)) = 0$.

Example 16.5. Let $O = (0,0)$, $A = (1,0)$, $B = (0,1)$, and $C = (-1,-1)$. Then

$$\#(\angle(O,A)) = 0, \qquad \#(\angle(O,AB)) = \frac{1}{4},$$
$$\#(\angle(O,ABCA)) = 1, \ \#(\angle(O,ACB)) = -\frac{3}{4}.$$

Let us show that the definition of the revolution number is well defined.

Proposition 16.6. *The revolution number of any extended angle is well defined.*

Proof. Consider an arbitrary V-broken line and the corresponding extended angle $\angle(V, A_0A_1 \ldots A_n)$. Let

$$r_+ = \{V + \lambda VA_0 | \lambda \geq 0\} \quad \text{and} \quad r_- = \{V - \lambda VA_0 | \lambda \geq 0\}.$$

By definition, every segment of the broken line $A_0A_1 \ldots A_n$ is at unit integer distance from V. Hence this broken line does not contain V, and the rays r_+ and r_- do not contain edges of the broken line.

Suppose that

$$\angle(V, A_0A_1 \ldots A_n) = \angle(V', A_0'A_1' \ldots A_m').$$

Let us show that

$$\#(\angle(V, A_0A_1 \ldots A_n)) = \#(\angle(V', A_0'A_1' \ldots A_m')).$$

From the definition we have $V = V'$, $A_0 = A_0'$, $A_n = A_m'$, and the broken line $A_0A_1 \ldots A_nA_{m-1}' \ldots A_1'A_0'$ is contractible in $\mathbb{R}^2 \setminus \{V\}$. This implies that

$$\#(\angle(V, A_0A_1 \ldots A_n)) - \#(\angle(V,' A_0'A_1' \ldots A_m'))$$
$$= \frac{1}{2}\left(\#(r_+, A_0A_1 \ldots A_nA_{m-1}' \ldots A_1'A_0') + \#(r_-, A_0A_1 \ldots A_nA_{m-1}' \ldots A_1'A_0')\right)$$
$$= 0 + 0 = 0$$

(we leave to the reader the proof of the second equality; see Problem 16.2). Hence,

$$\#(\angle(V, A_0A_1 \ldots A_n)) = \#(\angle(V', A_0'A_1 \ldots A_m')).$$

Therefore, the revolution number of any extended angle is well defined.   □

Remark. Notice that the revolution number of a closed V-broken line coincides with the degree of a point V with respect to the V-broken line.

Let us formulate one important property of the revolution number.

Proposition 16.7. *The revolution number of extended angles is invariant under proper integer congruences.*   □

Finally, we formulate a nice expression for the rotation number of a closed integer broken line with integer vertices not passing through the origin introduced in a recent preprint [86] by A. Higashitani and M. Masuda. Recall that the *rotation number* about the integer point V of a closed broken line L not passing through V is the degree of the projection of L to the unit circle along the rays with vertex at V. We denote the rotation number by $\mathrm{Rot}_V(L)$. Notice that a rotation number for V-broken lines coincides with the revolution number of the corresponding extended angles.

Proposition 16.8. *Consider a closed integer V-broken line L and enumerate all of its integer points $A_1,\ldots,A_d$ (not only vertices). In addition, set $A_0 = A_d$ and $A_{d+1} = A_1$. Then we have*

$$\mathrm{Rot}_V(L) = \frac{1}{4}\sum_{i=1}^{d}\varepsilon_i + \frac{1}{12}\sum_{i=1}^{d}a_i,$$

where $\varepsilon_i = \det(VA_i, VA_{i+1})$ and a_i is defined from the equation

$$\varepsilon_{i-1}VA_{i-1} + \varepsilon_i VA_{i+1} + a_i VA_i = 0.$$

                                                                                    □

For the proof of this proposition we refer to [86] and [226].

16.1.1.3 Zero integer angles

In the next theorem we use zero integer angles and their trigonometric functions. Let A, B, and C be three integer points on a straight line. Suppose that B is distinct from A and C and that it is not between A and C. We say that the integer angle $\angle ABC$ with vertex at B is *zero*. Further, we put by definition

$$\mathrm{lsin}(\angle ABC) = 0, \quad \mathrm{lcos}(\angle ABC) = 1, \quad \mathrm{ltan}(\angle ABC) = 0.$$

Denote by larctan(0) the angle $\angle AOA$, where $A = (1,0)$ and O is the origin.

Type I:	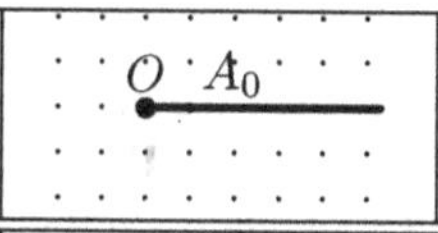	$\angle(O,A_0) = 0\pi + \text{larctan}(0)$, LSLS sequence is ().
Type II$_1$:	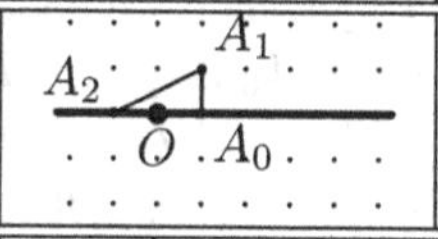	$\angle(O,A_0A_1A_2) = \pi + \text{larctan}(0)$, LSLS sequence is $(1,-2,1)$.
Type III$_2$:	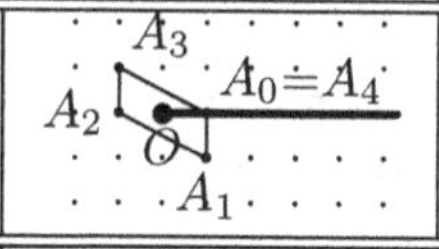	$\angle(O,A_0A_1A_2A_3A_4) = -2\pi + \text{larctan}(0)$, LSLS sequence is $(-1,2,-1,2,-1,2,-1)$.
Type IV$_1$:	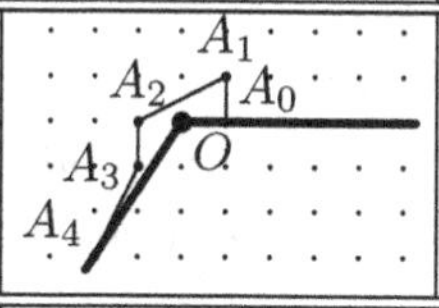	$\angle(O,A_0A_1A_2A_3A_4) = \pi + \text{larctan}\frac{3}{2}$, LSLS sequence is $(1,-2,1,-2,1,1,1)$.
Type V$_1$:	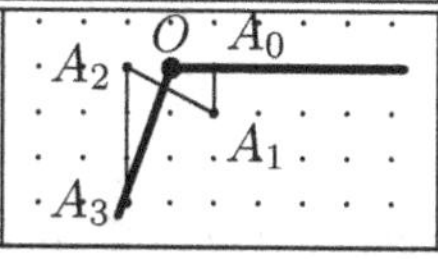	$\angle(O,A_0A_1A_2A_3) = -\pi + \text{larctan}3$, LSLS sequence is $(-1,2,-1,2,3)$.

Fig. 16.1 Examples of normal forms.

16.1.2 On normal forms of extended angles

While working with an abstract definition of extended angles it is useful to keep in mind particular broken lines that characterize the corresponding equivalence classes, which we shall call the *normal forms* of these broken lines.

Denote a sequence

$$\underbrace{(a_0,\ldots,a_n,a_0,\ldots,a_n, \quad \ldots, \quad a_0,\ldots,a_n}_{k \text{ times}},b_0,\ldots,b_m).$$

by

$$\big((a_0,\ldots,a_n)^k,b_0,\ldots,b_m\big).$$

Definition 16.9. (Normal forms of extended angles.) Consider an integer oriented O-broken line $A_0A_1\ldots A_s$, where O is the origin. Let $A_0 = (1,0)$ and (if $s > 0$) let A_1 be on the straight line $x = 1$.
(I) We say that the extended angle $\angle(O,A_0)$ is *of Type* **I** and denote it by $0\pi + \text{larctan}(0)$ (or 0, for short). The empty sequence is said to be *characteristic* for $0\pi + \text{larctan}(0)$.

If the LSLS sequence of the broken line $A_0A_1\ldots A_s$ coincides with one of the following sequences (we call it the *characteristic sequence* for the corresponding angle), then:

(II$_k$) If $\big((1,-2,1,-2)^{k-1},1,-2,1\big)$, where $k \geq 1$, then we denote the angle Φ_0 by $k\pi + \mathrm{larctan}(0)$ (or $k\pi$, for short) and say that Φ_0 is *of Type* **II$_k$**;

(III$_k$) If $\big((-1,2,-1,2)^{k-1},-1,2,-1\big)$, where $k \geq 1$, then we denote the angle Φ_0 by $-k\pi + \mathrm{larctan}(0)$ (or $-k\pi$, for short) and say that Φ_0 is *of Type* **III$_k$**;

(IV$_k$) If $\big((1,-2,1,-2)^{k},a_0,\ldots,a_{2n}\big)$, where $k \geq 0$, $n \geq 0$, $a_i > 0$, for $i = 0,\ldots,2n$, then we denote the angle Φ_0 by $k\pi + \mathrm{larctan}([a_0;a_1 : \cdots : a_{2n}])$ and say that Φ_0 is *of Type* **IV$_k$**;

(V$_k$) If $\big((-1,2,-1,2)^{k},a_0,\ldots,a_{2n}\big)$, where $k > 0$, $n \geq 0$, $a_i > 0$, for $i = 0,\ldots,2n$, then we denote the angle Φ_0 by $-k\pi + \mathrm{larctan}([a_0;a_1 : \cdots : a_{2n}])$ and say that Φ_0 is *of Type* **V$_k$**.

See examples of several normal forms in Fig. 16.1.

Theorem 16.10. *For every extended angle Φ there exists a unique normal angle of Definition 16.9 that is proper integer congruent to Φ. (This angle is called the normal form of Φ.)*

We start the proof with the following lemma.

Lemma 16.11. *Consider integers m, $k \geq 1$, and $a_i > 0$ for $i = 0,\ldots,2n$.*
(i) Suppose that the LSLS sequences for the extended angles Φ_1 and Φ_2 are respectively

$$\big((1,-2,1,-2)^{k-1},1,-2,1,-2,a_0,\ldots,a_{2n}\big) \quad and$$
$$\big((1,-2,1,-2)^{k-1},1,-2,1,m,a_0,\ldots,a_{2n}\big).$$

Then Φ_1 is proper integer congruent to Φ_2.
(ii) Suppose that the LSLS sequences for the extended angles Φ_1 and Φ_2 are respectively

$$\big((-1,2,-1,2)^{k-1},-1,2,-1,m,a_0,\ldots,a_{2n}\big) \quad and$$
$$\big((-1,2,-1,2)^{k-1},-1,2,-1,2,a_0,\ldots,a_{2n}\big).$$

Then Φ_1 is proper integer congruent to Φ_2.

Proof. We start the proof with the first statement of the lemma. Without loss of generality we assume that the vertices of the extended angles Φ_1 and Φ_2 are at the origin, say

$$\Phi_1 = \angle(O,A_0\ldots A_{2k+n+1}),$$
$$\Phi_2 = \angle(O,B_0\ldots B_{2k+n+1}).$$

Additionally we assume that $A_0 = B_0 = (1,0)$ and the points A_1 and B_2 are on the lines $x = 1$ and $(m+1)y = x$ respectively. These conditions together with the LSLS sequences uniquely identify the vertices

$$\begin{cases} A_{2l} &= \big((-1)^l,0\big), & \text{for } l < k-1, \\ A_{2l+1} &= \big((-1)^l,(-1)^l\big), & \text{for } l < k-1, \\ A_{2k} &= \big((-1)^k,0\big), \\ A_{2k+1} &= \big((-1)^k,(-1)^k a_0\big), \end{cases}$$

and

$$\begin{cases} B_{2l} = ((-1)^l, 0), & \text{for } l < k-1, \\ B_{2l+1} = ((-1)^l(-m-1), (-1)^l), & \text{for } l < k-1, \\ B_{2k} = ((-1)^k, 0), \\ B_{2k+1} = ((-1)^k, (-1)^k a_0). \end{cases}$$

Hence $A_{2k} = B_{2k}$ and $A_{2k+1} = B_{2k+1}$. Notice that the remaining parts of both LSLS sequences (i.e., $(a_0, \ldots, a_{2n})$) coincide. Therefore, the point A_l coincides with the point B_l for $l > 2k$.

On the one hand, it is clear that the integer oriented broken line

$$A_0 \ldots A_{2k}(=B_k) B_{k-1} B_{k-2} \ldots B_0$$

is contractible. On the other hand, $A_l = B_l$ for $l > 2k$. Therefore, the angle Φ_1 is proper integer congruent to Φ_2. This concludes the proof of Lemma 16.11(i).

The proof of Lemma 16.11(ii) is similar, and we leave it as an exercise for the reader. $\square$

Proof of Theorem 16.10. Uniqueness of normal forms. Let us first show that the extended angles listed in Definition 16.9 are pairwise proper integer noncongruent.

Consider the revolution numbers of the extended angles listed in Definition 16.9:

Types	I	$\text{II}_k (k \geq 0)$	$\text{III}_k (k \geq 0)$	$\text{IV}_k (k \geq 0)$	$\text{V}_k (k > 0)$
Revolution numbers	0	$1/2(k+1)$	$-1/2(k+1)$	$1/4 + 1/2k$	$1/4 - 1/2k$

Therefore, the revolution numbers of extended angles distinguish the types of the angles.

For Types **I**, **II**$_k$, and **III**$_k$ the proof of uniqueness is complete, since any such type contains exactly one extended angle.

Now we prove that any two normal forms of Type **IV**$_k$ are not proper integer congruent for any integer $k \geq 0$. Consider the extended angle

$$\Phi = k\pi + \text{larctan}([a_0; a_1 : \cdots : a_{2n}])$$

of Definition 16.9. Assume that the O-broken line for the angle Φ is $A_0 A_1 \ldots A_m$, where $m = 2|k| + n + 1$. Notice that the LSLS sequence for this broken line is characteristic for Φ. We consider the two cases of odd and even k separately.

If k is even, then the integer angle $\angle A_0 O A_m$ is proper integer congruent to the integer angle $\text{larctan}([a_0; a_1 : \cdots : a_{2n}])$. The integer arctangent is a proper integer affine invariant for Φ. This invariant distinguishes all the extended angles of Type **IV**$_k$ with even k.

Suppose now that k is odd. Set $B = O + \overline{A_0 O}$. The integer angle $\angle BVA_m$ is proper integer congruent to $\text{larctan}([a_0; a_1 : \cdots : a_{2n}])$. The integer arctangent is a proper integer affine invariant for the extended angle Φ, distinguishing extended angles of Type **IV**$_k$ with odd k.

The proof of uniqueness of normal forms of Type V_k repeats the proof for normal forms of Type IV_k.

Therefore, the extended angles listed in Definition 16.9 are not proper integer congruent.

Existence of normal forms. Now we prove that every extended angle is proper integer congruent to one of the normal forms.

Consider an arbitrary extended angle $\Phi = \angle(V, A_0 A_1 \ldots A_n)$. If $\#(\Phi) = k/2$ for some integer k, then Φ is proper integer congruent to an angle of one of the Types **I—III**. If $\#(\Phi) = 1/4$, then the extended angle Φ is proper integer congruent to the extended angle defined by the sail of the integer angle $\angle A_0 V A_n$ of Type IV_0.

Suppose now that $\#(\Phi) = 1/4 + k/2$ for some positive integer k. Choose the broken line defining Φ whose LSLS sequence is of the following form:

$$((1, -2, 1, -2)^{k-1}, 1, -2, 1, m, a_0, \ldots, a_{2n}), \tag{16.1}$$

where $a_i > 0$, for $i = 0, \ldots, 2n$. Notice that it is always possible to construct such a broken line: first, construct the broken line for

$$((1, -2, 1, -2)^{k-1}, 1, -2, 1).$$

Secondly, take the sail of the corresponding angle. The union of the broken line and the sail is a broken line whose LSLS sequence is exactly as in (16.1). By Lemma 16.11 the extended angle Φ defined by such an LSLS sequence is proper integer congruent to the extended angle of Type IV_k defined by the sequence

$$((1, -2, 1, -2)^{k-1}, 1, -2, 1, -2, a_0, \ldots, a_{2n}),$$

which is in the list of normal forms.

Finally, we consider the case $\#(\Phi) = 1/4 - k/2$ for some positive integer k. As in the previous case there exists a broken line defining Φ whose LSLS sequence is

$$((-1, 2, -1, 2)^{k-1}, -1, 2, -1, m, a_0, \ldots, a_{2n}),$$

where $a_i > 0$, for $i = 0, \ldots, 2n$. By Lemma 16.11 the extended angle defined by this sequence is proper integer congruent to the extended angle of Type V_k defined by the sequence

$$((-1, 2, -1, 2)^{k-1}, -1, 2, -1, 2, a_0, \ldots, a_{2n}),$$

which is again in the list of normal forms.

This completes the proof of Theorem 16.10.

Let us reformulate Theorem 16.10 in the following way.

Corollary 16.12. *Two extended angles are proper integer congruent if and only if they have the same normal form.*

16.1.3 Trigonometry of extended angles. Associated integer angles

Having the list of normal forms, we can extend the trigonometric functions to the case of extended integer angles.

Definition 16.13. Consider an extended angle Φ with the normal form $k\pi+\varphi$ for some integer (possible zero) angle φ and for an integer k.
(*i*) The integer angle φ is said to be *associated* with the extended angle Φ.
(*ii*) The numbers $\operatorname{ltan}(\varphi)$, $\operatorname{lsin}(\varphi)$, and $\operatorname{lcos}(\varphi)$ are called the *integer tangent*, the *integer sine*, and the *integer cosine* of the extended angle Φ.

There exists a canonical embedding of the set of all integer angles into the set of all extended angles, since every sail is an integer broken line at distance one from the vertex of the angle.

Definition 16.14. Consider an arbitrary integer angle φ. The angle

$$0\pi + \operatorname{larctan}(\operatorname{ltan}\varphi)$$

is said to be *corresponding* to the angle φ, which we denote by $\overline{\varphi}$.

From Theorem 16.10 it follows that for every integer angle φ there exists a unique extended angle $\overline{\varphi}$ corresponding to φ. Therefore, two integer angles φ_1 and φ_2 are proper integer congruent if and only if the corresponding extended angles $\overline{\varphi}_1$ and $\overline{\varphi}_2$ are proper integer congruent.

16.1.4 Opposite extended angles

As in Euclidean geometry, in integer geometry every extended angle possesses an opposite angle.

Definition 16.15. Consider an extended angle

$$\Phi = \angle(V, A_0 A_1 \ldots A_n).$$

The angle

$$\angle(V, A_n A_{n-1} \ldots A_0)$$

is said to be *opposite* to Φ, which we denote by $-\Phi$.

Let us write the normal form for the opposite extended angle.

Proposition 16.16. *For every extended angle $\Phi \cong k\pi+\varphi$ we have*

$$-\Phi \cong (-k-1)\pi + (\pi - \varphi).$$

$\square$

16.1.5 Sums of extended angles

In the next definition we introduce sums of integer and extended angles. A sum of two angles is not uniquely defined, in contrast to Euclidean geometry.

Definition 16.17. Consider arbitrary extended angles Φ_i, where $i = 1, \ldots, l$ for $l > 1$. Let the normal forms of the Φ_i have the characteristic sequence

$$(a_{0,i}, a_{1,i}, \ldots, a_{2n_i,i}) \qquad \text{for } i = 1, \ldots, l.$$

Let $M = (m_1, \ldots, m_{l-1})$ be an arbitrary $(l-1)$-tuple of integers. Construct an extended angle Φ with characteristic sequence

$$\left(a_{0,1}, a_{1,1}, \ldots, a_{2n_1,1}, m_1, a_{0,2}, \ldots, a_{2n_2,2}, m_2, \ldots, m_{l-1}, a_{0,l}, \ldots, a_{2n_l,l}\right).$$

The normal form of Φ is called the *M-sum of extended angles* Φ_i $(i = 1, \ldots, l)$ and is denoted by

$$\sum_{M,i=1}^{l} \Phi_i, \quad \text{or equivalently by} \quad \Phi_1 +_{m_1} \Phi_2 +_{m_2} \cdots +_{m_{l-1}} \Phi_l.$$

It is clear that M-sums of extended angles are defined in a unique way up to the set choice of the set M.

Example 16.18. Let $\Phi = 0\pi + \mathrm{larctan}\, 1$. Then

$$\Phi +_{-3} \Phi = \pi + \mathrm{larctan}\, 1,$$
$$\Phi +_{-2} \Phi = \pi + \mathrm{larctan}\, 0,$$
$$\Phi +_{-1} \Phi = 0\pi + \mathrm{larctan}\, 1,$$
$$\Phi +_{0} \Phi = 0\pi + \mathrm{larctan}\, 2,$$
$$\Phi +_{1} \Phi = 0\pi + \mathrm{larctan}\, \tfrac{3}{2}.$$

It is interesting to observe that the M-sum of extended angles is neither associative nor commutative.

Proposition 16.19. *The M-sum of extended angles is nonassociative.*

Proof. For example, let

$$\Phi_1 = 0\pi + \mathrm{larctan}\, 2,$$
$$\Phi_2 = 0\pi + \mathrm{larctan}\, \tfrac{3}{2},$$
$$\Phi_3 = 0\pi + \mathrm{larctan}\, 5.$$

Then

$$\Phi_1 +_{-1} \Phi_2 +_{-1} \Phi_3 = \pi + \mathrm{larctan}\, 4,$$
$$\Phi_1 +_{-1} (\Phi_2 +_{-1} \Phi_3) = 2\pi + \mathrm{larctan}\, 0,$$
$$(\Phi_1 +_{-1} \Phi_2) +_{-1} \Phi_3 = 0\pi + \mathrm{larctan}\, 1.$$

$\square$

Proposition 16.20. *The M-sum of extended angles is noncommutative.*

Proof. For example, let

$$\Phi_1 = 0\pi + \text{larctan } 1 \quad \text{and} \quad \Phi_2 = 0\pi + \text{larctan } \frac{5}{2}.$$

Then

$$\Phi_1 +_1 \Phi_2 = 0\pi + \text{larctan } \frac{12}{7} \neq 0\pi + \text{larctan } \frac{13}{5} = \Phi_2 +_1 \Phi_1.$$

$\square$

Remark 16.21. The definition of *M*-sums of extended angles is canonically lifted to the case of classes of proper integer congruences of extended angles.

16.1.6 Sums of integer angles

We conclude this section with the definition of *M*-sums of integer angles.

Definition 16.22. Consider integer angles α_i, where $i = 1, \ldots, l$ for some $l \geq 2$. Let $\overline{\alpha}_i$ be the corresponding extended angles for α_i, and let $M = (m_1, \ldots, m_{l-1})$ be an arbitrary $(l-1)$-tuple of integers. The integer angle φ associated with the extended angle

$$\Phi = \overline{\alpha}_1 +_{m_1} \overline{\alpha}_2 +_{m_2} \cdots +_{m_{l-1}} \overline{\alpha}_l.$$

is called the *M-sum of integer angles* α_i $(i = 1, \ldots, l)$ and denoted by

$$\sum_{M,i=1}^{l} \alpha_i, \quad \text{or equivalently by} \quad \alpha_1 +_{m_1} \alpha_2 +_{m_2} \cdots +_{m_{l-1}} \alpha_l.$$

Remark 16.23. The definition of *M*-sums of integer angles is also canonically lifted to the case of classes of proper integer congruences of integer angles.

16.2 Relations between extended and integer angles

Recall that $\lfloor r \rfloor$ denotes the maximal integer not greater than r.

Theorem 16.24. *Consider an extended angle $\Phi = \angle(V, A_0A_1 \ldots A_n)$. Suppose that the normal form for Φ is $k\pi + \varphi$ for some pair (k, φ). Let $(a_0, a_1, \ldots, a_{2n-2})$ be the LSLS sequence for the integer oriented broken line $A_0A_1 \ldots A_n$. Suppose that*

$$[a_0; a_1 : \cdots : a_{2n-2}] = \frac{p}{q}.$$

Then the following hold:

$$\varphi \cong \begin{cases} \text{larctan } 1, & \text{if } p/q=\infty, \\ \text{larctan } \frac{p}{q}, & \text{if } p/q\geq 1, \\ \text{larctan } \frac{|p|}{|q|-\lfloor(|q|-1)/|p|\rfloor|p|}, & \text{if } 0<p/q<1, \\ 0, & \text{if } p/q=0, \\ \pi - \text{larctan } \frac{|p|}{|q|-\lfloor(|q|-1)/|p|\rfloor|p|}, & \text{if } -1<p/q<0, \\ \pi - \text{larctan}(-\frac{p}{q}), & \text{if } p/q\leq -1. \end{cases}$$

Proof. Without loss of generality we assume that V is the origin O, $A_0 = (1,0)$, and

$$A_0 + \frac{1}{a_0}\, \text{sgn}(A_0 O' A_1) A_0 A_1 = (1,1).$$

(One can get this after a certain integer affine transformation of the plane.)

By Theorem 3.1 the coordinates of the point A_n are (q,p). This directly implies the statement of the theorem for the cases $p>q>0$, $p/q = 0$, and $p/q = \infty$.

Suppose now that $q>p>0$. Consider the integer angle $\varphi = \angle A_0 P A_n$. Let $B_0 \ldots B_m$ be the sail for φ. Direct calculation shows that the point

$$D = B_0 + \frac{B_0 B_1}{l\ell(B_0 B_1)}$$

coincides with the point $(1+\lfloor(q-1)/p\rfloor, 1)$. Consider the proper integer linear transformation T satisfying the following two conditions: first, T takes the point $A_0 = B_0$ to itself, and second, T takes the point D to $(1,1)'$. These two conditions uniquely identify T as

$$T = \begin{pmatrix} 1 & -\lfloor(q-1)/p\rfloor \\ 0 & 1 \end{pmatrix}.$$

The transformation T takes the point $A_n = B_m$, with coordinates (q,p), to the point with coordinates $(q - \lfloor(p-1)/p\rfloor p, p)'$. Therefore,

$$\varphi = \text{larctan } \frac{p}{q - \lfloor(q-1)/p\rfloor p}.$$

The proof for the case $p > 0$ and $q < 0$ repeats the proof described above after switching to the corresponding adjacent angles.

Finally, the case with $p < 0$ is centrally symmetric with respect to all the cases considered above. This concludes the proof of Theorem 16.24.   □

16.3 Proof of Theorem 6.9(i)

Now we return to the proof of the first statement of the theorem on sums of integer tangents for integer angles in integer triangles:

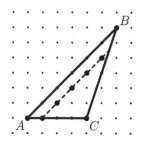

Fig. 16.2 For the given triangle $\triangle ABC$ we have $1\ell_1(AB;C) = 5$.

Theorem 6.9(i). On sums of integer tangents of angles in integer triangles.
Let $(\alpha_1, \alpha_2, \alpha_3)$ be an ordered triple of angles. There exists a triangle with consecutive angles integer congruent to α_1, α_2, and α_3 if and only if there exists $i \in \{1,2,3\}$ such that the angles $\alpha = \alpha_i$, $\beta = \alpha_{i+1(\mathrm{mod}\,3)}$, $\gamma = \alpha_{i+2(\mathrm{mod}\,3)}$ satisfy the following conditions:
(a) for $\xi =]\operatorname{ltan}\alpha, -1, \operatorname{ltan}\beta[$ the following holds $\xi < 0$, or $\xi > \operatorname{ltan}\alpha$, or $\xi = \infty$;
(b) $]\operatorname{ltan}\alpha, -1, \operatorname{ltan}\beta, -1, \operatorname{ltan}\gamma[= 0$.

16.3.1 Two preliminary lemmas

We begin the proof of Theorem 6.9(i) with two preliminary lemmas.

Definition 16.25. Let $\triangle ABC$ be an integer triangle. Consider all integer points in the interior of $\triangle ABC$ and on its sides lying at unit integer distance from the side AB. We denote their number by $1\ell_1(AB;C)$ (see Fig. 16.2).

Recall that all integer points at unit integer distance from AB described in the definition are contained in one straight line parallel to AB. Besides that,

$$0 < 1\ell_1(AB;C) < 1\ell(AB).$$

Now we prove the following lemma.

Lemma 16.26. *For every integer triangle $\triangle ABC$ the following holds:*

$$\overline{\angle CAB} +_{1\ell(AB)-1\ell_1(AB;C)-1} \overline{\angle ABC} +_{1\ell(BC)-1\ell_1(BC;A)-1} \overline{\angle BCA} = \pi$$

(where by $\overline{\varphi}$ we denote the extended angle corresponding to φ).

Proof. Consider an arbitrary integer triangle $\triangle ABC$. Suppose that the pair of vectors (BA, BC) defines the positive orientation of the plane (otherwise, we consider the triangle $\triangle ACB$). Set

$$D = A + CB \quad \text{and} \quad E = A + CA$$

(see Fig. 16.4 on page 210).

Since $CADB$ is a parallelogram, we have

$$\triangle BAD \cong \triangle ABC,$$

and hence

$$\angle BAD \cong \angle ABC \quad \text{and} \quad 1\ell_1(BA;D) = 1\ell_1(AB;C).$$

Since $EABD$ is a parallelogram, we get

$$\triangle AED \cong \triangle BAD \cong \triangle ABC.$$

Therefore,

$$\angle DAE \cong \angle BCA \quad \text{and} \quad 1\ell_1(DA;E) = 1\ell_1(BC;A).$$

Let $A_0 \ldots A_n$ be the sail of the angle $\angle CAB$ with the corresponding LLS sequence

$$(a_0, \ldots, a_{2n-2}).$$

Further, let $B_0 B_1 \ldots B_m$ be the sail of $\angle BAD$ (where $B_0 = A_n$) with the corresponding LLS sequence

$$(b_0, \ldots, b_{2m-2}).$$

Finally, let $C_0 C_1 \ldots C_l$ be the sail of $\angle DAE$ (where $C_0 = B_m$) with the corresponding LLS sequence

$$(c_0, \ldots, c_{2l-2}).$$

Consider now the A-broken line

$$A_0 \ldots A_n B_1 B_2 \ldots B_m C_1 C_2 \ldots C_l.$$

The LSLS sequence for this broken line is

$$(a_0, \ldots, a_{2n-2}, t, b_0, \ldots, b_{2m-2}, u, c_0, \ldots, c_{2l-2}),$$

where t and u are certain integers. By the definition of the M-sum of extended angles this sequence defines the extended angle

$$\overline{\angle CAB} +_t \overline{\angle BAD} +_u \overline{\angle DAE},$$

which is equal to π by construction.

Now let us compute t. Denote by A_n' the integer point in the segment $A_{n-1}A_n$ that is next to A_n. Consider the set of integer points at unit integer distance from AB and lying in the half-plane with AB at the boundary and containing the point D. This set coincides with the following set (see an example in Fig. 16.3):

$$\{A_{n,k} = A_n + A_n'A_n + kAA_n \mid k \in \mathbb{Z}\}.$$

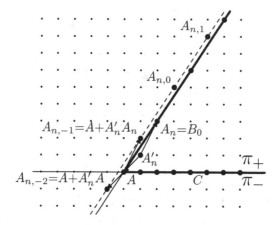

Fig. 16.3 Integer points $A_{n,t}$.

Denote by π_+ the open half-plane bounded by the straight line AC and containing the point B. Let also π_- denote the complement of π_+ in the plane. Since

$$A_{n,-2} = A + A'_n A,$$

the integer points $A_{n,k}$ for k less than or equal to -2 are in the closed half-plane π_-. Since

$$A_{n,-1} = A + A'_n A_n,$$

the integer points $A_{n,k}$ for k greater than or equal to -1 are in the open half-plane π_+.

The intersection of the parallelogram $AEDB$ and the open half-plane π_+ contains exactly $\mathrm{l}\ell(AB)$ points of the described set: these points are the points $A_{n,k}$ with $-1 \leq k \leq \mathrm{l}\ell(AB) - 2$.

Since

$$\triangle BAD \stackrel{\hat{}}{\cong} \triangle ABC,$$

the number of points $A_{n,k}$ in the closed triangle $\triangle BAD$ is $\mathrm{l}\ell_1(AB;C)$: these points are exactly the points $A_{n,k}$ with k satisfying

$$\mathrm{l}\ell(AB) - \mathrm{l}\ell_1(AB;C) - 1 \leq k \leq \mathrm{l}\ell(AB) - 2.$$

Denote by k_0 the integer

$$\mathrm{l}\ell(AB) - \mathrm{l}\ell_1(AB;C) - 1.$$

The point A_{n,k_0} is contained in the segment $B_0 B_1$ of the sail of the integer angle $\angle BAD$ (see Fig. 16.4). Since

$$\angle BAD \stackrel{\hat{}}{\cong} \angle ABC,$$

we have

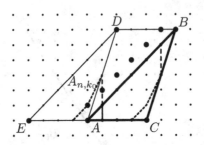

Fig. 16.4 The point A_{n,k_0}.

$$t = \text{sgn}(A_{n-1}AA_n)\,\text{sgn}(A_nAB_1)\,\text{sgn}(A_{n-1}A_nB_1)\,l\sin\angle A_{n-1}A_nB_1$$
$$= 1\cdot 1\cdot \text{sgn}(A_{n-1}A_nA_{n,k_0})\,l\sin\angle A_{n-1}A_nA_{n,k_0}$$
$$= \text{sign}(k_0)|k_0| = k_0 = 1\ell(AB) - 1\ell_1(AB;C) - 1.$$

For the same reason,

$$u = 1\ell(DA) - 1\ell_1(DA;E) - 1 = 1\ell(BC) - 1\ell_1(BC;A) - 1.$$

Therefore,

$$\overline{\angle CAB} +_{1\ell(AB)-1\ell_1(AB;C)-1} \overline{\angle ABC} +_{1\ell(BC)-1\ell_1(BC;A)-1} \overline{\angle BCA} = \pi.$$

The proof is complete.   □

Lemma 16.27. *Let α, β, and γ be nonzero integer angles. Suppose that*

$$\overline{\alpha} +_u \overline{\beta} +_v \overline{\gamma} = \pi.$$

Then there exists a triangle with three consecutive integer angles proper integer congruent to α, β, and γ respectively.

Proof. Let

$$O = (0,0), \qquad A = (1,0), \quad \text{and} \quad D = (-1,0).$$

Choose the integer points

$$B = (q_1,p_1) \quad \text{and} \quad C = (q_2,p_2)$$

with integers p_1, p_2 and positive integers q_1, q_2 such that

$$\angle AOB = \text{larctan}(\text{ltan}\,\alpha) \quad \text{and} \quad \angle AOC = \overline{\angle AOB} +_u \overline{\beta}.$$

From the construction, the pair of vectors (OB, OC) defines the positive orientation, and $\angle BOC \hat{\cong} \beta$. Since

$$\overline{\alpha} +_u \overline{\beta} +_v \overline{\gamma} = \pi \quad \text{and} \quad \overline{\alpha} +_u \overline{\beta} \stackrel{\triangle}{\cong} \angle AOC,$$

we have $\angle COD \stackrel{\triangle}{\cong} \gamma$.

Define

$$B' = (q_1 p_2, p_1 p_2) \quad \text{and} \quad C' = (q_2 p_1, p_1 p_2)$$

and consider the triangle $\triangle B'OC'$. The coincidence of angles

$$\angle B'OC' = \angle BOC$$

implies

$$\angle B'OC' \stackrel{\triangle}{\cong} \beta.$$

Since $\beta \neq 0$, the points B' and C' are distinct and the straight line $B'C'$ does not coincide with the straight line OA. Since the second coordinates of both points B' and C' equal $q_1 q_2$, the straight line $B'C'$ is parallel to the straight line OA. Therefore, by Proposition 5.23 we have

$$\angle C'B'O \cong \angle AOB' = \angle AOB \cong \alpha \quad \text{and} \quad \angle OC'B' \cong \angle C'OD = \angle COD \cong \gamma.$$

So, the consecutive integer angles of the triangle $\triangle B'OC'$ are integer congruent to α, β, and γ.   □

16.3.2 Conclusion of the proof of Theorem 6.9(i)

Let α, β, and γ be nonzero integer angles satisfying the conditions (a) and (b) of Theorem 6.9(i). From the second condition

$$]\text{ltan}(\alpha), -1, \text{ltan}(\beta), -1, \text{ltan}(\gamma)[= 0,$$

it follows that

$$\overline{\alpha} +_{-1} \overline{\beta} +_{-1} \overline{\gamma} = k\pi.$$

Since all three tangents are positive, we have either $k = 1$ or $k = 2$.

By the first condition,

$$]\text{ltan}\,\alpha, -1, \text{ltan}\,\beta[< 0 \quad \text{or} \quad]\text{ltan}\,\alpha, -1, \text{ltan}\,\beta[> \text{ltan}\,\alpha.$$

Hence

$$\overline{\alpha} +_{-1} \overline{\beta} = 0\pi + \varphi,$$

for some integer angle φ, and hence $k \leq 1$. Together with the above, this implies that $k = 1$. Therefore, by Lemma 16.27 there exists a triangle with three consecutive integer angles integer congruent to α, β, and γ, respectively.

Let us prove the converse statement. First we prove condition (b) of Theorem 6.9(i). We do it by reductio ad absurdum. Suppose that there exists a triangle

$\triangle ABC$ with consecutive integer angles

$$\alpha = \angle CAB, \qquad \beta = \angle ABC, \quad \text{and} \quad \gamma = \angle BCA,$$

such that

$$\begin{cases} \,]\operatorname{ltan}\alpha, -1, \operatorname{ltan}\beta, -1, \operatorname{ltan}\gamma[\neq 0, \\ \,]\operatorname{ltan}\beta, -1, \operatorname{ltan}\gamma, -1, \operatorname{ltan}\alpha[\neq 0, \\ \,]\operatorname{ltan}\gamma, -1, \operatorname{ltan}\alpha, -1, \operatorname{ltan}\beta[\neq 0. \end{cases}$$

This system of inequalities and Lemma 16.26 imply that at least two of the integers

$$1\ell(AB) - 1\ell_1(AB;C) - 1, \qquad 1\ell(BC) - 1\ell_1(BC;A) - 1, \quad \text{and} \quad 1\ell(CA) - 1\ell_1(CA;B) - 1$$

are nonnegative.

Without loss of generality we suppose that

$$\begin{cases} 1\ell(AB) - 1\ell_1(AB;C) - 1 \geq 0, \\ 1\ell(BC) - 1\ell_1(BC;A) - 1 \geq 0. \end{cases}$$

On the one hand, since all integers of the continued fraction

$$r =]\operatorname{ltan}(\alpha), 1\ell(AB) - 1\ell_1(AB;C) - 1, \operatorname{ltan}(\beta), 1\ell(BC) - 1\ell_1(BC;A) - 1, \operatorname{ltan}(\gamma)[$$

are nonnegative and the last one is positive, either $r > 0$ or $r = \infty$. On the other hand, by Lemma 16.26 and by Theorem 16.24, we have that $r = 0/-1 = 0$. We arrive at a contradiction. Hence the second condition always holds.

Now we prove that condition (a) holds. Consider a triangle $\triangle ABC$ with consecutive integer angles

$$\alpha = \angle CAB, \qquad \beta = \angle ABC, \quad \text{and} \quad \gamma = \angle BCA.$$

We have proven the second condition, so without loss of generality we assume that

$$]\operatorname{ltan}(\alpha), -1, \operatorname{ltan}(\beta), -1, \operatorname{ltan}(\gamma)[= 0.$$

Since

$$\overline{\alpha} +_{-1} \overline{\beta} +_{-1} \overline{\gamma} = \pi,$$

we have

$$\overline{\alpha} +_{-1} \overline{\beta} = 0\pi + \varphi$$

for some integer angle φ. The last expression for φ implies the first condition of the theorem.

This concludes the proof of Theorem 6.9(i). $\square$

16.4 Exercises

Exercise 16.1. Find the revolution number of the O-broken line with vertices

$$A_0 = (1,0), \quad A_1(-1,2), \quad A_2 = (0,-1), \quad A_3 = (1,1), \quad A_4 = (1,-2).$$

Find the normal form of the extended angle $\angle(O, A_0 A_1 A_2 A_3 A_4)$.

Exercise 16.2. Let $A_0 A_1 \ldots A_n$ be a closed V-broken line. Suppose that it is contractible in $\mathbb{R}^2 \setminus \{V\}$. Prove that for every ray r with vertex at V we have

$$\#(r, A_0 A_1 \ldots A_n) = 0.$$

Exercise 16.3. Find a proof of Lemma 16.11(ii).

Exercise 16.4. (a). Prove that every extended angle possesses a unique opposite extended angle.
(b). Prove that the M-sum is uniquely defined by the ordered sequence of extended angles and the set of integers M.

Exercise 16.5. Calculate the normal forms for the following angles

$$(2\pi + \operatorname{larctan} 1) +_{-1} (3\pi + \operatorname{larctan} \tfrac{5}{2});$$
$$(4\pi + \operatorname{larctan} \tfrac{4}{3}) +_0 (3\pi + \operatorname{larctan} \tfrac{7}{3});$$
$$(-5\pi + \operatorname{larctan} 119) +_2 (3\pi + \operatorname{larctan} \tfrac{5}{3}).$$

Exercise 16.6. Find an example of a triple of integer angles that are not the angles of an integer triangle.

Exercise 16.7. Is it true that for an arbitrary pair of integer angles (α, β) there exists an integer angle γ such that these three angles are angles of some integer triangle?

Chapter 17
Integer Angles of Polygons and Global Relations for Toric Singularities

In Chapter 6 we proved a necessary and sufficient criterion for a triple of integer angles to be the angles of some integer triangle. In this chapter we prove the analogous statement for the integer angles of convex polygons. Further, we discuss an application of these two statements to the theory of complex projective toric surfaces. We refer the reader to the general theory of toric surfaces in the works of V.I. Danilov [49], G. Ewald [61], W. Fulton [66], T. Oda [162], and A. Trevisan [214]. In this book we do not consider the multidimensional case (we refer the interested reader to the paper of H. Tsuchihashi [215]).

In Section 17.1 we formulate and prove a theorem on integer angles of convex polygons. After a brief introduction of the main notions and definitions of complex projective toric surfaces (Section 17.2) we discuss two problems related to singular points of toric varieties using integer geometry techniques.

17.1 Theorem on angles of integer convex polygons

We start with a theorem on necessary and sufficient conditions for integer angles to be the angles of some convex integer polygon.

Theorem 17.1. *Let $(\alpha_1, \ldots, \alpha_n)$ be an arbitrary ordered n-tuple of nonzero integer angles. Then the following two conditions are equivalent:*
(i) there exists a convex n-gon with consecutive integer angles $\alpha_1, \ldots, \alpha_n$;
(ii) there exists a sequence of integers $(m_1, \ldots, m_{n-1})$ such that

$$\overline{\pi - \alpha_i} +_{m_1} \cdots +_{m_{n-1}} \overline{\pi - \alpha_n} = 2\pi.$$

Proof. Condition (i) $\Rightarrow$ Condition (ii). Suppose that there exists an integer convex polygon $A_1 A_2 \ldots A_n$ with prescribed consecutive integer angles $\alpha_1, \ldots, \alpha_n$. We assume that the pair of vectors $(A_2 A_3, A_2 A_1)$ defines the positive orientation of the plane (otherwise, we consider the polygon $A_n A_{n-1} \ldots A_1$). Let

© Springer-Verlag GmbH Germany, part of Springer Nature 2022
O. N. Karpenkov, *Geometry of Continued Fractions*,
Algorithms and Computation in Mathematics 26,
https://doi.org/10.1007/978-3-662-65277-0_17

$$B_1 = O + A_n A_1 \quad \text{and} \quad B_i = O + A_{i-1} A_i \text{ for } i = 2, \ldots, n.$$

Define

$$\beta_i = \begin{cases} \angle B_i O B_{i+1}, & \text{if } i = 1, \ldots, n-1, \\ \angle B_n O B_1, & \text{if } i = n. \end{cases}$$

Consider the broken line that is the union of all the consecutive sails for angles $\beta_1, \ldots, \beta_n$. This broken line defines an extended angle proper integer congruent to $2\pi + 0$. The LSLS sequence for this broken line contains exactly $n-1$ new elements (together with all the elements of the LLS sequences for the sails of the angles $\beta_1, \ldots, \beta_n$). Denoting these numbers by $m_1, \ldots, m_{n-1}$, we get

$$\overline{\beta_1} +_{m_1} \cdots +_{m_{n-1}} \overline{\beta_n} = 2\pi.$$

From the definition of β_i ($i = 1, \ldots, n$) it follows that $\beta_i \cong \pi - \alpha_i$. Hence

$$\overline{\pi - \alpha_i} +_{m_1} \cdots +_{m_{n-1}} \overline{\pi - \alpha_n} = 2\pi.$$

Therefore, the first condition implies the second condition.

Condition (ii) $\Rightarrow$ *Condition* (i). We assume that there exists $M = (m_1, \ldots, m_{n-1})$ such that

$$\overline{\pi - \alpha_i} +_{m_1} \cdots +_{m_{n-1}} \overline{\pi - \alpha_n} = 2\pi.$$

This equation implies the existence of integer points $B_1 = (1, 0)$, $B_i = (x_i, y_i)$, for $i = 2, \ldots, n-1$, and $B_n = (-1, 0)$ such that

$$\angle B_i O B_{i-1} \cong \pi - \alpha_{i-1}, \text{ for } i = 2, \ldots, n, \quad \text{and} \quad \angle B_1 O B_n \cong \pi - \alpha_n.$$

Denote by P the integer point

$$O + \sum_{i=1}^{n} OB_i.$$

Since $\alpha_i \neq 0$, the angle $\pi - \alpha_i$ is not contained in a line ($i = 1, \ldots, n$). Hence, the origin O is an interior point of the convex hull of the points B_i for $i = 1, \ldots, k$. Therefore, there exists an integer s such that the integer triangle $\triangle B_s P B_{s+1}$ contains the origin O and in addition, O is not in the edge $B_s B_{s+1}$ (here $B_{n+1} = B_1$ and $B_0 = B_n$). This implies that

$$O = \lambda_1 OP + \lambda_2 OB_i + \lambda_3 OB_{i+1},$$

where $\lambda_1 > 0$, $\lambda_2 \geq 0$, and $\lambda_3 \geq 0$ are rational numbers. So there exist positive integers c_i ($i = 1, \ldots, n$) such that

$$\sum_{i=1}^{n} (c_i OB_i) = 0. \tag{17.1}$$

Consider $A_0 = O$ and $A_i = A_{i-1} + c_i OB_i$ for $i = 2, \ldots, n$. Since the numbers $c_1, \ldots, c_i$ are integers, the broken line $A_0 A_1 \ldots A_n$ is integer. From equation (17.1) the broken

Fig. 17.1 Two integer noncongruent polygons with proper integer congruent integer angles.

line is closed (i.e., $A_0 = A_n$). By construction, the integer angle at A_i of this broken line is proper integer congruent to α_i $(i = 1, \ldots, n)$. Since $c_i > 0$ for $i = 1, \ldots, n$ and since all the vectors OB_i are in counterclockwise order, the broken line is a convex polygon. Therefore, condition (ii) holds. $\square$

Remark 17.2. On the one hand, Theorem 17.1 is an extension of Theorem 6.9(i). On the other hand, the direct generalization of Theorem 6.9(ii) is false: *the integer angles do not uniquely determine the proper integer affine homothety types of integer convex polygons.* See an example in Fig. 17.1.

A completely satisfactory description of integer congruence classes of lattice convex polygons has not yet been found. It is known only that the number of convex polygons with lattice area bounded from above by n grows exponentially in $n^{1/3}$, as n tends to infinity (see [8] and [19]).

17.2 Toric surfaces and their singularities

17.2.1 Definition of toric surfaces

We start with the definition of complex projective toric surfaces associated to integer convex polygons.

Definition 17.3. Consider an integer convex polygon $P = A_0 A_1 \ldots A_n$. Let the intersection of this (closed) polygon with the integer lattice $\mathbb{Z}^2$ consist of the points $B_i = (x_i, y_i)$ for $i = 0, \ldots, m$. We enumerate the points in such a way that $B_i = A_i$ for $i = 0, \ldots, n$. Denote by Ω the following set in the complex projective space $\mathbb{C}P^m$:

$$\left\{ \left(t_1^{x_0} t_2^{y_0} t_3^{-x_0 - y_0} : t_1^{x_1} t_2^{y_1} t_3^{-x_1 - y_1} : \cdots : t_1^{x_m} t_2^{y_m} t_3^{-x_m - y_m} \right) \big| t_1, t_2, t_3 \in \mathbb{C} \setminus \{0\} \right\}.$$

The closure of the set Ω in the natural topology of $\mathbb{C}P^m$ is called the *complex projective toric surface associated with the polygon P* and denoted by X_P.

Example 17.4. For instance, consider the following two polygons:

$$P = \text{[figure]} \quad \text{and} \quad Q = \text{[figure]}.$$

Then $X_P = \mathbb{C}P^2$, and X_Q is the conic in $\mathbb{C}P^3$ defined by the equation

$$x_1 x_3 = x_0 x_4.$$

Notice that the integer area coincides with the degree of the resulting surface. This is a general fact of toric geometry.

Remark 17.5. One can generalize the described construction to the case of poly-hedra of arbitrary dimensions. Notice that in the multidimensional case this construction may have nonisolated singular points. In this situation there is a standard procedure to resolve them (via certain ς-processes). The resulting variety is the toric variety corresponding to the polyhedron.

Remark 17.6. One of the main properties of toric surfaces (and toric varieties in general) is as follows. Every toric surface admits a huge group of isomorphisms. While here we discuss only toric surfaces, similar statements hold for toric varieties of arbitrary dimension. Consider a toric surface X_P, parameterized as $X_P(t_0, t_1, t_2)$. Consider the multiplicative group of complex numbers $\mathbb{C}^* = \mathbb{C} \setminus \{0\}$. Every toric surface X_P admits a natural action of the group $\mathbb{C}^* \times \mathbb{C}^*$. An arbitrary element (a, b) of the group $\mathbb{C}^* \times \mathbb{C}^*$ acts on the toric surface X_P as follows:

$$f_{(a,b)} : X_P \to X_P, \qquad f_{(a,b)}\big(X_P(t_0, t_1, t_2)\big) = X_P(a t_0, b t_1, t_2).$$

We use this action below to find singular points of toric surfaces.

17.2.2 Singularities of toric surfaces

We begin with the following definition.

Definition 17.7. Algebraic singularities of complex projective toric surfaces are called *toric singularities*.

Let us first collect general properties of complex projective toric surfaces. For $i = 0, \ldots, m$, set

$$\tilde{A}_i = (0 : \cdots : 0 : 1 : 0 : \cdots : 0),$$

where 1 stands in the ith place.

Theorem 17.8. (*i*) *The set X_P is a complex projective complex-two-dimensional surface with isolated algebraic singularities;*
(*ii*) *the complex projective toric surface contains the points $\tilde{A}_i$ for $i = 0, \ldots, n$ (where $n+1$ is the number of vertices of the convex polygon);*
(*iii*) *the points of $X_P \setminus \{\tilde{A}_0, \tilde{A}_1, \ldots, \tilde{A}_n\}$ are nonsingular;*
(*iv*) *the point $\tilde{A}_i$ is singular if and only if $\alpha_i \not\cong$ larctan 1, where α_i is the angle of the polygon P at vertex A_i;*
(*v*) *the algebraic singularity at $\tilde{A}_i$ ($0 \le i \le n$) is uniquely determined by the integer affine type of the nonoriented sail of α_i.*   □

Example 17.9. Consider the polygon

$$P = \text{}$$

and the corresponding toric surface X_P. The surface $X_P \subset \mathbb{C}P^3$ is defined by the equation

$$x_0 x_1 x_2 = x_3^3.$$

Its singularities are the points

$$(1:0:0:0), \quad (0:1:0:0), \quad \text{and} \quad (0:0:1:0).$$

In the appropriate affine charts these three singularities are defined by the equation

$$xy = z^3.$$

For the proof of Theorem 17.8 we refer to classical textbooks on toric geometry ([49], [61], [66], [162], etc.). Nevertheless, we would like to outline some ideas and remarks related to the proof. The first item is a classical statement of algebraic geometry.

Outline of the proof of item (ii). Consider an arbitrary vertex A_i of the polygon P. Let l_i be some strictly supporting line for the polygon at A_i (i.e., $l_i \cap P = \{A_i\}$). Consider a vector v orthogonal to l_i and directed away from the polygon. Assume that $v = (\lambda_1, \lambda_2)$. Consider a curve in $\mathbb{R}P^m \subset \mathbb{C}P^m$ with real parameter t:

$$\left\{ \left(t^{\lambda_1 x_0 + \lambda_2 y_0} : t^{\lambda_1 x_1 + \lambda_2 y_1} : \cdots : t^{\lambda_1 x_m + \lambda_2 y_m} \right) \mid t \in \mathbb{R} \right\}.$$

It is clear that this curve is a subset of X_P. By construction of v, the largest value of the exponent will be at the place i corresponding to A_i. Thus the limit point of this curve is exactly the point $\tilde{A}_i$ as t tends to infinity. Therefore, the point $\tilde{A}_i$ belongs to the toric surface X_P. For a similar reason, the points $\tilde{A}_{n+1}, \ldots, \tilde{A}_m$ are not in X_P.

Outline of the proof of item (iii). To detect all singular points of X_P we use the action of the group $\mathbb{C}^* \times \mathbb{C}^*$ introduced in Remark 17.6. Notice that the group takes singular points to singular points and nonsingular points to nonsingular points. There are two-dimensional, one-dimensional, and zero-dimensional orbits of the action of $\mathbb{C}^* \times \mathbb{C}^*$ on X_P. Since all the singular points of X_P are isolated, they are all contained in the zero-dimensional orbits, i.e., at points where $\mathbb{C}^* \times \mathbb{C}^*$ acts trivially. It can be only the points with a single nonzero coordinate

$$(0 : \cdots : 0 : 1 : 0 : \cdots : 0).$$

The toric surface X_P contains only singular points with ones at places $0, \ldots, n$. That is exactly the points $\tilde{A}_0, \ldots, \tilde{A}_n$.

Remarks on items (iv) and (v). As stated in Theorem 17.8(v), singularities of toric surfaces are in one-to-one correspondence with integer affine types of nonoriented sails of integer angles. Therefore, toric singularities are classified by the LLS sequences of corresponding integer angles up to switching the order of the sequence. We call the corresponding LLS sequence the *continued fractions* of the toric singularity. Therefore, the pair of rational numbers

$$(\operatorname{ltan}\alpha, \operatorname{ltan}\alpha^t)$$

for an arbitrary integer angle α is a complete invariant of the corresponding toric singularities, which we call the *sail pair* of the singularity.

The continued fractions for toric singularities are slightly different from the Hirzebruch—Jung continued fractions for toric singularities (see [93] and [88]). The latter ones have all negative integer elements except for the first one, which is a positive integer. It is interesting to note that Hirzebruch—Jung continued fractions generate a sequence of σ-processes that resolve the corresponding singularity. The relations between regular continued fractions and Hirzebruch—Jung continued fractions are described in [182].

17.3 Relations on toric singularities of surfaces

In this subsection we study global conditions on complex projective toric surface singularities associated to *n*-gons (in algebraic language, *n* is understood as the *Euler characteristic* of the corresponding toric surface; see, for instance, [214]). In Section 17.3.1 we fix an arbitrary *n* and study whether a certain *n*-tuple of toric singularities is realizable as a set of singular points of a toric surface defined by an *n*-gon. Further, in Section 17.3.2, we fix an arbitrary *m*-tuple of toric singularities and construct an *n*-gon (with large enough *n*) whose singularities are exactly the singularities of the *m*-tuple.

17.3.1 Toric singularities of n-gons with fixed parameter n

We begin this subsection with triangles. The corresponding toric surfaces have at most three toric singularities. The relations on these singularities are described by a reformulation of Theorem 6.9(i).

Corollary 17.10. *Consider three arbitrary complex-two-dimensional toric singularities with sail pairs (a_i, b_i), where $i = 1, 2, 3$. Then the following two statements are equivalent:*

(i) *There exists a complex projective toric surface of Euler characteristic equal to 3 whose singular points have sail pairs (a_i, b_i) for $i = 1, 2, 3$.*

(ii) *There exist a permutation* $\sigma \in S_3$ *and rational numbers* $c_i \in \{a_i, b_i\}$ *for* $i = 1, 2, 3$ *such that the following conditions hold:*

— *the continued fraction* $]c_{\sigma(1)}, -1, c_{\sigma(2)}[$ *is either negative, greater than* $c_{\sigma(1)}$, *or equal to* ∞;

— $]c_{\sigma(1)}, -1, c_{\sigma(2)}, -1, c_{\sigma(3)}[= 0$. □

Remark. Let us say one more time that one should use only odd continued fractions for c_1, c_2, and c_3 in the statement of the above proposition.

When $n > 3$ we have the following weaker statement. It follows directly from Theorem 17.1.

Corollary 17.11. *Consider n arbitrary complex-two-dimensional toric singularities with sail pairs (a_i, b_i), where $i = 1, \ldots, n$. Then* **(i)** $\Rightarrow$ **(ii)**, *where*

(i) *There exists a complex projective toric surface of Euler characteristic n whose singular points have sail pairs (a_i, b_i), for $i = 1, \ldots, n$.*

(ii) *There exist rational numbers $c_i \in \{a_i, b_i\}$ for $i = 1, \ldots, n$ and a sequence of integers $(m_1, \ldots, m_{n-1})$ such that the following condition holds:*

$$]c_1, m_1, c_2, m_2, \ldots, m_{n-1}, c_n[= 0.$$

□

The following problem is still open.

Problem 11. Find a necessary and sufficient condition for existence of a complex projective toric surface of Euler characteristic n in terms of sail pairs of its singularities and a sequence of integers M (as in Corollary 17.11).

Notice that all integers of M are greater than -2. In addition, at least one of the elements of M is equal to -1.

17.3.2 Realizability of a prescribed set of toric singularities

Finally, we discuss the realizability of toric surfaces with a prescribed set of toric singularities.

Proposition 17.12. *For every collection (with multiplicities) of complex-two-dimensional toric algebraic singularities there exists a complex projective toric surface with this collection of singularities.*

In the proof of Proposition 17.12 we use the following lemma.

Lemma 17.13. *For every collection of integer angles α_i $(i = 1, \ldots, n)$ there exist an integer $k \geq n-1$ and a k-tuple of integers $M = (m_1, \ldots, m_k)$ such that*

$$\overline{\alpha_1} +_{m_1} \cdots +_{m_{n-1}} \overline{\alpha_n} +_{m_n} \text{larctan } 1 +_{m_{n+1}} \cdots +_{m_k} \text{larctan } 1 = 2\pi.$$

Proof. Consider any collection of integer angles α_i $(i = 1, \ldots, n)$ and set

$$\Phi = \overline{\alpha_1} +_1 \overline{\alpha_2} +_1 \cdots +_1 \overline{\alpha_n}.$$

It is clear that one of the LSLS sequences for Φ is obtained by adding the LLS sequences of the angles α_i and the number 1 taken $n-1$ times. All the elements of this LSLS sequence are positive integers, and hence Φ is of the form $\varphi + 0\pi$ for some integer angle φ.

If $\varphi \cong$ larctan 1, then we have

$$\Phi +_{-2} \text{larctan } 1 +_{-2} \text{larctan } 1 +_{-2} \text{larctan } 1 = 2\pi.$$

Then $k = n+2$, and $M = (1, \ldots, 1, -2, -2, -2)$.

Suppose now that $\varphi \ncong$ larctan 1. Then the following holds:

$$\overline{\varphi} +_{-1} \overline{\pi - \varphi} +_{-2} \text{larctan } 1 +_{-2} \text{larctan } 1 = 2\pi.$$

Consider the sail for the angle $\pi - \varphi$. Suppose that the sequence of all its consecutive integer points (not only vertices) is $(B_0, \ldots, B_s)$, where the order coincides with the order of the sail. Then we have

$$\angle B_i O B_{i+1} \cong \text{larctan } 1 \quad \text{for every } i = 1, \ldots, s.$$

Put $b_i = \text{lsin } \angle B_i O B_{i+1}$ for $i = 1, \ldots, s$. We get

$$\begin{aligned}
\overline{\varphi} &+_{-2} \text{larctan } 1 +_{-2} \text{larctan } 1 +_{-2} \text{larctan } 1 \\
&= \overline{\alpha_1} +_1 \overline{\alpha_2} +_1 \cdots +_1 \overline{\alpha_n} \\
&+_{-1} \text{larctan } 1 +_{b_1} \text{larctan } 1 +_{b_2} \cdots +_{b_s} \text{larctan } 1 \\
&+_{-2} \text{larctan } 1 +_{-2} \text{larctan } 1 +_{-2} \text{larctan } 1 = 2\pi.
\end{aligned}$$

Therefore, $k = n+s+3$, and

$$M = (\underbrace{1, 1, \ldots, 1, 1}_{(n-1) \text{ times}}, -1, b_1, \ldots, b_s, -2, -2, -2).$$

The proof of Lemma 17.13 is complete.   □

Proof of Proposition 17.12. Consider an arbitrary collection of surface toric algebraic singularities. Suppose that they are enumerated by integer angles α_i $(i = 1, \ldots, n)$. By Lemma 17.13 there exist $k \geq n-1$ and $M = (m_1, \ldots, m_k)$ such that

$$\overline{(\pi - \alpha_1)} +_{m_1} \cdots +_{m_{n-1}} \overline{(\pi - \alpha_n)} +_{m_n} \text{larctan } 1 +_{m_{n+1}} \cdots +_{m_k} \text{larctan } 1 = 2\pi.$$

Then by Theorem 17.1 there exists a convex polygon $P = A_0 \ldots A_k$ with integer angles proper integer congruent to the angles α_i $(i = 1, \ldots, n)$, and the last $k-n+1$ angles equal larctan 1.

By Theorem 17.8(c) all singularities of the toric surface X_P are contained in the set $\{\tilde{A}_0,\tilde{A}_1,\ldots,\tilde{A}_k\}$. By Theorem 17.8(d) the points $\tilde{A}_i$ corresponding to larctan 1 are also nonsingular. The collection of toric singularities at the remaining points coincides with the given collection. This concludes the proof of Proposition 17.12.

□

Example 17.14. Let us construct a projective toric surface having a unique toric singularity with the sail pair $(7/5, 7/3)$. Consider $\alpha \mathrel{\hat{=}} \text{larctan}(7/5)$. First of all, we draw the angle $\pi - \alpha$ and its adjacent angle $\pi - (\pi - \alpha)$ (which is α). Further, we subdivide the half-plane in the complement to the union of α and $\pi - \alpha$ in a standard way into two angles β and γ (such that the resulting LSLS sequence ends in $(-2, 1, -2, 1)$).

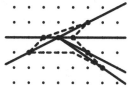

Secondly, we subdivide the angle $\alpha = \pi - (\pi - \alpha)$ into angles integer congruent to larctan 1, shown here:

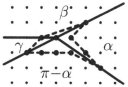

Finally, we construct the hexagon P all of whose angles are adjacent to the angles of the obtained decomposition.

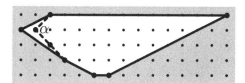

The toric surface X_P has a unique singular point; its sail pair is $(7/5, 7/3)$.

17.4 Exercises

Exercise 17.1. Consider an arbitrary toric surface X_P. Characterize all points of X_P where the action of $\mathbb{C}^* \times \mathbb{C}^*$ on X_P is   (a) constant;   (b) one-dimensional;   (c) two-dimensional.

Exercise 17.2. Write the toric surfaces for the following polygons

$$P = \boxed{\triangle}, \qquad Q = \boxed{\triangle}, \qquad \text{and} \qquad R = \boxed{\diamondsuit}.$$

Enumerate all their singularities and find the sail pairs of the corresponding singularities.

Exercise 17.3. Find a toric surface with a unique toric singularity whose sail pair is $(5/3, 5/2)$.

Exercise 17.4. Prove that the sequence M in Corollary 17.11 has at least one negative integer. Show that all negative numbers are equal to -1.

Exercise 17.5. Find an example of an integer quadrangle whose sequence M has exactly one negative integer.

Part II
Multidimensional Continued Fractions

In this second part we study multidimensional versions of several classical theorems on continued fractions. For this purpose we use multidimensional continued fractions in sense of Klein—Voronoi. The material of this part is relatively recent, so we accompany it with related open problems and conjectures. In addition, we provide links to classical problems and conjectures, such as the Littlewood conjecture, the lonely runner problem, White's problem, etc. In conclusion, we briefly discuss other generalizations of continued fractions.

As in the planar case, we begin with a discussion of basic features of multidimensional integer geometry. In Chapter 18 we introduce the main integer objects and present the corresponding invariants. Further, in Chapter 19 we study a more specific question related to emptiness of integer simplices, pyramids, and tetrahedra. In particular, we formulate and proof White's theorem on three-dimensional empty integer triangles.

In Chapter 20 we introduce multidimensional continued fractions in the sense of Klein, which are special polyhedral surfaces related to simplicial cones in real spaces. We study both finite and infinite Klein's continued fractions. We prove that for a generic cone, the corresponding continued fraction has a good polyhedral structure (using the multidimensional Kronecker's approximation theorem). Finally, we study the structure of two-dimensional compact faces of continued fractions in the sense of Klein.

In the next two chapters, we investigate the periodic structure of algebraic multidimensional continued fractions. The structure of such continued fractions is periodic due to the action of the Dirichlet groups. We study some classical aspects concerning Dirichlet groups in Chapter 21. In particular, we describe several technical details related to the calculation of the bases of those groups, and the questions of lattice reduction (e.g., the simplest LLL-algorithm). Further, in Chapter 22 we discuss periodic multidimensional continued fractions. Finally, we prove a generalized version of Lagrange's theorem and explain the relation of multidimensional continued fractions to the Oppenheim conjecture.

We discuss the multidimensional Gauss—Kuzmin statistics in Chapter 23. Firstly, we extend the Möbius measure of the projective plane to higher dimensions. Secondly, we formulate general results regarding the generalized Gauss—Kuzmin statistics on relative frequencies of polygonal faces in continued fractions in the sense of Klein. Finally, we give examples of explicit calculation for certain integer types of faces. We conclude with several open problems on the distribution of faces in multidimensional continued fractions.

In Chapter 24 we bring together several techniques to construct multidimensional continued fractions. There are two different algorithmic approaches to this issue. The first approach deals with a face-by-face construction of the Klein's polyhedral continued fraction. In the second approach one makes an approximation of a sail and conjectures which faces of the approximation are the faces of the continued fraction. We discuss the corresponding algorithm and study one particular example.

Chapter 25 is dedicated to multidimensional Gauss's reduction theory. We show that the Hessenberg matrices play the role of reduced matrices. Further we discuss a complete geometric invariant of conjugacy classes of integer invertible matrices in the simplest case that all eigenvalues are distinct. If a matrix has a real spectrum, then the corresponding invariant is its Klein's continued fraction. For the remaining cases we use multidimensional continued fractions in the sense of Klein—Voronoi as invariants, which naturally extends Klein's continued fractions to the case of matrices with arbitrary spectra. After that we present the technique of multidimensional Gauss reduction. Our next goal is to show how to solve Diophantine equations related to Markov–Davenport characteristics, the main ingredient here is algebraic periodicity induced by the Dirichlet group. We conclude this chapter with an exhaustive study of the new case of the family of three-dimensional operators having two complex conjugate eigenvalues. This part of material has never been published before.

We investigate questions concerning best approximation of multidimensional simplicial cones in Chapter 26. In particular we discuss the relation to the maximal commutative subgroup approximation. Further we show that simultaneous approximation can be also considered in the context of maximal commutative subgroup approximation.

Finally, in Chapter 27 we expose a collection of different generalizations of multidimensional continued fractions mentioning interesting theorems related to them. We do not pretend that this collection is complete. The aim of this chapter is to show the diversity of methods that are used to study the extensions of classical problems of continued fraction theory.

Chapter 18
Basic Notions and Definitions of Multidimensional Integer Geometry

In this chapter we generalize integer two-dimensional notions and definitions of Chapter 2. As in the planar case, our approach is based on the study of integer invariants. Furthe,r we use them to study the properties of multidimensional continued fractions. First, we introduce integer volumes of polytopes, distances, and angles. Then we express volumes of polytopes, integer distances, and integer angles in terms of integer volumes of simplices. Finally, we show how to compute integer volumes of simplices via certain Plücker coordinates in the Grassmann algebra (this formula is extremely useful for the computation of multidimensional integer invariants of integer objects contained in integer planes). We continue this with a discussion of the Ehrhart polynomials, which one can consider a multidimensional generalization of Pick's formula in the plane. This chapter is concluded with the classification of elementary integer-regular polygons.

18.1 Basic integer invariants in integer geometry

Let us expand integer geometry to the multidimensional case. The main object of integer geometry is the lattice $\mathbb{Z}^n$ in $\mathbb{R}^n$, i.e., the lattice of vectors all of whose coordinates are integers.

18.1.1 Objects and the congruence relation

We say that a vector (point) is *integer* if all its coordinates are integers. A plane in $\mathbb{R}^n$ is called *integer* if its integer vectors form a sublattice whose rank equals the dimension of the plane. We say that a polyhedron is *integer* if all its vertices are integer. A cone in $\mathbb{R}^n$ is said to be *integer* if its vertex is integer and each of its edges contains integer points distinct from the vertex.

© Springer-Verlag GmbH Germany, part of Springer Nature 2022
O. N. Karpenkov, *Geometry of Continued Fractions*,
Algorithms and Computation in Mathematics 26,
https://doi.org/10.1007/978-3-662-65277-0_18

The integer *congruence relation* is defined as in the two-dimensional case by the group of affine transformations preserving the set of integer points, i.e., $\mathrm{Aff}(n,\mathbb{Z})$. This group is a semidirect product of $\mathrm{GL}(n,\mathbb{Z})$ and the group of translations on integer vectors. We use "$\cong$" to indicate that two objects are *integer congruent*.

18.1.2 Integer invariants and indices of sublattices

As in the planar case described in Chapter 2, the integer invariants in the multidimensional case are easily defined in terms of indices of certain sublattices. We first give a geometric description of the index of sublattices in terms of the number of integer points in certain parallelepipeds.

Proposition 18.1. *The index of a sublattice generated by integer vectors $v_1,\dots,v_k$ in an integer k-dimensional plane equals the number of all integer points P satisfying*

$$AP = \sum_{i=1}^{n} \lambda_i v_i \quad \text{with} \quad 0 \le \lambda_i < 1, \quad i \in \{1,\dots,n\},$$

where A is an arbitrary integer point.

Proof. The proof of this statement is similar to the proof of the planar one. Let H be a subgroup of $\mathbb{Z}^2$ generated by $v_1,\dots,v_k$. Define

$$\mathrm{Par}(v_1,\dots,v_k) = \left\{ \sum_{i=1}^{n} \lambda_i v_i \ \middle|\ 0 \le \lambda_i < 1, \quad i = 1,\dots,n \right\}.$$

First, we prove that for every integer vector g there exists an integer point P lying in the set $\mathrm{Par}(v_1,\dots,v_k)$ such that $AP \in gH$. Let

$$g = \sum_{i=1}^{n} \mu_i v_i.$$

Then consider

$$P = A + \sum_{i=1}^{n} (\mu_i - \lfloor \mu_i \rfloor) v_i.$$

Since $0 \le \mu_i - \lfloor \mu_i \rfloor < 1$ for $i = 1,\dots,n$, the point P is inside the parallelogram and $AP \in gH$.

Second, we show the uniqueness of such a point P. Suppose that for two integer points $P_1, P_2 \in \mathrm{Par}(v_1,\dots,v_k)$ we have $AP_1 \in gH$ and $AP_2 \in gH$. Hence the vector P_1P_2 is an element in H. The only element of H of type

$$\alpha v + \beta w \quad \text{with} \quad 0 \le \alpha, \beta < 1$$

is the zero vector. Hence $P_1 = P_2$.

Therefore, the integer points of $\mathrm{Par}(v_1, \ldots, v_k)$ are in one-to-one correspondence with the cosets of H in $\mathbb{Z}^2$.   □

18.1.3 Integer volume of simplices

Let us start with the notion of integer volume for integer simplices. We will define the integer volumes for polyhedra in the next subsection.

Definition 18.2. An *integer volume* of an integer simplex $A_1 \ldots A_n$ is the index of the sublattice generated by the vectors $A_2A_1, \ldots, A_nA_1$ in the integer lattice of the space $\mathrm{Span}(A_2A_1, \ldots, A_nA_1)$. Denote it by $\mathrm{lV}(A_1 \ldots A_n)$.
For the points $A_1 \ldots A_n$ that are contained in some $(n-2)$-dimensional plane we say that $\mathrm{lV}(A_1 \ldots A_n) = 0$.

Example 18.3. Consider a three-dimensional simplex $S \subset \mathbb{R}^4$ with vertices

$$s_1 = (2,3,0,1), \quad s_2 = (1,4,2,4), \quad s_3 = (1,0,0,4), \quad s_4 = (1,0,0,1).$$

The integer volume of S is 6.

We postpone the calculation of the volume of simplices for a while (see Example 18.32 below).

18.1.4 Integer angle between two planes

First, we give the definition of an integer angle for two linear subspaces.

Definition 18.4. Consider two integer linear spaces L_1 and L_2 that are not contained one in another. The *integer angle* between L_1 and L_2 is the index of the sublattice generated by all integer vectors of $L_1 \cup L_2$ in the integer lattice of the space $\mathrm{Span}(L_1, L_2)$. Denote it by $\mathrm{l}\alpha(L_1, L_2)$.
In case one space is a subspace of the other we agree to say that the integer angle between them is zero.

In order to calculate integer angles via integer volumes one can use the following proposition.

Proposition 18.5. *Consider two integer linear spaces L_1 and L_2 that are not contained one in another. Let the sets of independent integer vectors*

$$\{u_1, \ldots, u_k, w_1, \ldots, w_m\}, \quad \{v_1, \ldots, v_l, w_1, \ldots, w_m\}, \quad and \quad \{w_1, \ldots, w_m\}$$

form bases in L_1, L_2, and $L_1 \cap L_2$ respectively. Then we have

$$\mathrm{l}\alpha(L_1, L_2) = \frac{\mathrm{lV}(u_1, \ldots, u_k, v_1, \ldots, v_l, w_1, \ldots, w_m)\,\mathrm{lV}(w_1, \ldots, w_m)}{\mathrm{lV}(u_1, \ldots, u_k, w_1, \ldots, w_m)\,\mathrm{lV}(v_1, \ldots, v_l, w_1, \ldots, w_m)}.$$

Proof. First, let us change the basis (w_i) of $L_1 \cap L_2$ to the basis $(\overline{w}_i)$ that generates the integer sublattice in $L_1 \cap L_2$. The value of the formula stays unchanged, since the numerator and the denominator are both divided by $lV^2(w_1, \ldots, w_m)$.

Second, we extend the basis $(\overline{w}_i)$ with integer vectors $\overline{u}_i$ to the basis of $(\overline{u}_i, \overline{w}_i)$ that generates the integer lattice of L_1. The value of the formula stays unchanged, since the numerator and the denominator are both divided by the same quantity $lV(u_1, \ldots, u_k, \overline{w}_1, \ldots, \overline{w}_m)$.

Third, we extend the basis $(\overline{w}_i)$ with integer vectors $\overline{v}_i$ to the basis of $(\overline{v}_i, \overline{w}_i)$ that generates the integer lattice of L_2. For similar reasons the value of the formula stays unchanged.

From the definition we have

$$
\begin{cases}
lV(\overline{u}_1, \ldots, \overline{u}_k, \overline{v}_1, \ldots, \overline{v}_k, \overline{w}_1, \ldots, \overline{w}_m) = l\alpha(L_1, L_2), \\
lV(\overline{u}_1, \ldots, \overline{u}_k, \overline{w}_1, \ldots, \overline{w}_m) = 1, \\
lV(\overline{v}_1, \ldots, \overline{v}_k, \overline{w}_1, \ldots, \overline{w}_m) = 1, \\
lV(\overline{w}_1, \ldots, \overline{w}_m) = 1.
\end{cases}
$$

For these vectors the statement of the proposition is clearly holds. This concludes the proof.   □

Finally, we give a definition of the integer angle between arbitrary (affine) planes.

Definition 18.6. Consider two integer planes π_1 and π_2. Let L_1 and L_2 be the spaces of vectors corresponding to π_1 and π_2. The *integer angle* between π_1 and π_2 is the index $l\alpha(L_1, L_2)$. We denote it by $l\alpha(\pi_1, \pi_2)$.

18.1.5 Integer distance between two disjoint planes

Definition 18.7. Consider two disjoint integer planes π_1 and π_2. Let L_1 and L_2 be the spaces of vectors corresponding to π_1 and π_2, and let $a = A_1 A_2$ be a vector such that $A_1 \in \pi_1$ and $A_2 \in \pi_2$. The *integer distance* between π_1 and π_2 is the index of the sublattice generated by all integer vectors of $L_1 \cup L_2 \cup \text{Span}(a)$ in the integer lattice of the space $\text{Span}(L_1, L_2, a)$. Denote it by $ld(\pi_1, \pi_2)$.

In the case of an intersecting subspace of another we agree to say that the integer distance between them is zero.

Remark 18.8. One of the simplest useful invariants in three-dimensional integer geometry is the distance between an integer point p and an integer plane π, i.e., the index of the lattice generated by the integer vectors joining p and all integer points of π in the whole integer lattice of $\mathbb{Z}^3$. Integer distance has a nice geometric description. Draw all integer planes parallel to π. One of them contains the point p (let us call it π_p). The integer distance $ld(p, \pi)$ is the number of integer planes in the region bounded by the lines π_p and π plus one. Let us illustrate by the following example:

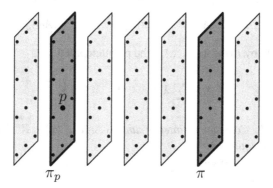

There are three integer planes parallel to π and lying between π_p and π. Hence, $\mathrm{ld}(p,\pi) = 3+1 = 4$.

Let us present a formula relating integer distances and integer volumes.

Proposition 18.9. *Consider two disjoint integer planes π_1 and π_2. Let L_1 and L_2 be the spaces of vectors corresponding to π_1 and π_2, and let a be a vector with one integer endpoint in π_1 and one integer endpoint in π_2. Suppose that the sets of independent integer vectors*

$$\{u_1,\ldots,u_k,w_1,\ldots,w_m\}, \quad \{v_1,\ldots,v_l,w_1,\ldots,w_m\}, \quad and \quad \{w_1,\ldots,w_m\}$$

form bases in L_1, L_2, and $L_1 \cap L_2$ respectively. Then we have

$$\mathrm{ld}(\pi_1,\pi_2) = \frac{\mathrm{IV}(a,u_1,\ldots,u_k,v_1,\ldots,v_l,w_1,\ldots,w_m)\,\mathrm{IV}(w_1,\ldots,w_m)}{\mathrm{IV}(u_1,\ldots,u_k,w_1,\ldots,w_m)\,\mathrm{IV}(v_1,\ldots,v_l,w_1,\ldots,w_m)}.$$

$\square$

The proof of Proposition 18.9 is similar to the proof of Proposition 18.5, so we skip it here.

In this section we have defined several invariants of planes and pairs of planes. In general, many other nice invariants for different arrangements of integer planes (for instance, for triples of integer lines whose span is 5-dimensional) are defined in a similar way.

18.2 Integer and Euclidean volumes of basis simplices

Here we study simplices of maximal dimension whose edges generate an integer lattice. We prove Proposition 18.11, which is analogues to Proposition 2.15.

Definition 18.10. An integer polyhedron is called *empty* if it does not contain any integer points other than its vertices.

Denote by $V(A_0A_1\ldots A_n)$ the Euclidean volume of a simplex $A_0A_1\ldots A_n$. Recall that

$$V(A_0 A_1 \ldots A_n) = \frac{1}{n!} \left| \det(A_0 A_1, A_0 A_2, \ldots, A_0 A_n) \right|.$$

We also denote by $P(A_0; A_1, \ldots, A_n)$ the parallelepiped

$$\left\{ A_0 + \sum_{i=1}^{n} \lambda_i A_i A_0 \ \middle| \ 0 \le \lambda_i \le 1 \right\}.$$

Proposition 18.11. *Consider an integer simplex $A_0 A_1 \ldots A_n$ in $\mathbb{R}^n$. Then the following statements are equivalent:*
  (a) $P(A_0; A_1, \ldots, A_n)$ *is empty;*
  (b) $\mathrm{IV}(A_0 A_1 \ldots A_n) = 1$;
  (c) $V(A_0 A_1 \ldots A_n) = \frac{1}{n!}$.

Remark 18.12. In Section 2 we showed that every empty triangle has integer area 1 (and Euclidean area $1/2$). This is no longer true in the multidimensional case; for instance, see Example 18.33. We describe the situation in the multidimensional case in the next chapter.

Proof. **(a) $\Rightarrow$ (b).** Let a parallelepiped $P(A_0; A_1, \ldots, A_n)$ be empty. Therefore, by Proposition 18.1 there is only one coset for the subgroup generated by vectors $A_1 A_0, \ldots, A_n A_0$. Hence, the vectors $A_1 A_0, \ldots, A_n A_0$ generate the integer lattice, and $\mathrm{IV}(A_0 A_1 \ldots A_n) = 1$.

(b) $\Rightarrow$ (c). Let $\mathrm{IV}(A_0 A_1 \ldots A_n) = 1$. Hence the vectors $A_1 A_0, \ldots, A_n A_0$ generate the integer lattice. Thus every integer point is an integer combination of them, so for the jth coordinate vector e_j we have

$$e_j = \sum_{j=1}^{n} \lambda_{ij} A_i A_0,$$

with integers λ_{ij} for $1 \le i, j \le n$. Denote by M_1 the matrix where at the jth place of the ith column is λ_{ij}. Let also M_2 be the matrix whose ith column is filled with the coordinates of the vector $A_0 A_i$ for $i = 1, \ldots, n$. Then

$$M_1 \cdot M_2 = \mathrm{Id},$$

where Id is the identity matrix. Since these matrices are integer, both their determinants equal either 1 or -1. Hence the Euclidean volume of the simplex $A_0 A_1 \ldots A_n$ coincides with the Euclidean volume of the coordinate simplex and equals $\frac{1}{n!}$.

(c) $\Rightarrow$ (a). Consider an integer simplex $A_0 A_1 \ldots A_n$ of Euclidean volume $\frac{1}{n!}$. Suppose that the corresponding parallelepiped $P(A_0; A_1, \ldots, A_n)$ has an integer point distinct from its vertices. Then there exists an integer simplex such that one of its facets (of codimension 1) coincides with a facet of $A_0 A_1 \ldots A_n$, and the height of this facet is smaller than the corresponding height in $A_0 A_1 \ldots A_n$. Hence there exists an integer simplex whose Euclidean volume is smaller than $\frac{1}{n!}$, which is impossible (since the determinant of n integer vectors is an integer). $\qquad\square$

Proposition 18.13. *For an arbitrary simplex S of full dimension we have the following formula:*

$$\text{IV}(S) = V(P(S)).$$

Proof. Let us prove this statement by induction on the integer volume of simplices.

Base of induction. The statement for simplices of integer volume 1 follows directly from Proposition 18.11.

Step of induction. Let the statement hold for all simplices of integer volume less then N ($N > 1$). Let us prove it for an arbitrary simplex $S = v_0 v_1 \dots v_n$ of lattice volume N.

Consider a polyhedron $P(S)$. By Proposition 18.11, since $\text{IV}(S) > 1$, there is an integer point in $P(S)$ distinct from the vertices. Without loss of generality we suppose that this point is not on any facet parallel to $v_1, \dots, v_n$. Hence

$$d(v_0, v_1 \dots v_n) > 1.$$

Denote this value by d_0.

Consider an integer point $\bar{v}_0$ at unit integer distance to the plane of the facet $v_1 \dots v_n$. By Proposition 18.9 we get

$$\frac{\text{IV}(v_0 v_1 \dots v_n)}{\text{IV}(\bar{v}_0 v_1 \dots v_n)} = d_0.$$

It is clear that for Euclidean volumes we have the same relation:

$$\frac{V(P(v_0 v_1 \dots v_n))}{V(P(\bar{v}_0 v_1 \dots v_n))} = d_0.$$

Since $d_0 > 0$, we have $\text{IV}(\bar{v}_0 v_1 \dots v_n) < N$, and hence we are in a position to apply the induction assumption:

$$\text{IV}(v_0 v_1 \dots v_n) = d_0 \text{IV}(\bar{v}_0 v_1 \dots v_n) = d_0 V(P(\bar{v}_0 v_1 \dots v_n)) = V(P(v_0 v_1 \dots v_n)).$$

This concludes the proof.   □

18.3 Integer volumes of polyhedra

In this subsection we write down an expression for the integer volume of simplices via certain Euclidean volumes. Further, we use the additivity of Euclidean volume to get a natural extension of lattice volume to the case of integer polyhedra. Finally, we discuss how to decompose an arbitrary convex integer polyhedron into integer empty simplices.

18.3.1 Interpretation of integer volumes of simplices via Euclidean volumes

We begin with a definition of determinants of sublattices.

Definition 18.14. Consider a sublattice L of the integer lattice $\mathbb{Z}^n$. The Euclidean volume of the simplex whose edges generate L is called the *determinant* of L and denoted by $\det L$.

Let us say a few words about the determinant being well defined. Recall that L can be mapped by a linear (not necessarily integer) mapping to $\mathbb{Z}^k \subset \mathbb{R}^k$. In Proposition 18.11 we showed that all simplices whose edges generate $\mathbb{Z}^k$ have the same Euclidean volume. Hence all simplices whose edges generate L also have the same Euclidean volume.

In the following theorem we establish a relation between the integer volume of a simplex S and the Euclidean volume of the corresponding parallelepiped $P(S)$.

Theorem 18.15. *Let S be a k-dimensional integer simplex in $\mathbb{R}^n$ and let $L(S)$ denote the integer lattice of the k-dimensional integer plane containing S. Then we have*

$$\mathrm{lV}(S) = \frac{V(P(S))}{\det(L(S))}.$$

Proof. Consider a linear map T sending L to $\mathbb{Z}^k \subset \mathbb{R}^k$. We have

$$\mathrm{lV}(T(S)) = \mathrm{lV}(S), \qquad V(T(P(S))) = \frac{V(P(S))}{\det(L(S))}, \quad \text{and} \quad \det(T(L(S))) = 1.$$

Hence the statement of Theorem 18.15 follows directly from a similar statement for simplices of full dimension:

$$\mathrm{lV}(T(S)) = V(T(P(S))),$$

which appears in Proposition 18.13. $\square$

18.3.2 Integer volume of polyhedra

Using the result of Theorem 18.15 we extend the definition of integer area to general polyhedra.

Definition 18.16. Consider an integer polyhedron P of dimension n. The *integer volume* of P equals

$$n! \frac{V(P)}{\det(L(P))},$$

which we denote by $\mathrm{lV}(P)$.

As in the two-dimensional case, we have the additivity property for lattice area.

Proposition 18.17. *The integer volume of polyhedra is additive, i.e., if an integer polyhedron P is a disjoint union of integer polyhedra $P_1, \ldots, P_k$ then*

$$\mathrm{IV}(P) = \sum_{i=1}^{k} \mathrm{IV}(P_k).$$

Proof. By definition the integer volumes of polyhedra of dimension n are proportional to their Euclidean volumes. Therefore, the additivity of Euclidean volume implies additivity of integer volumes.   □

18.3.3 Decomposition into empty simplices

Let us formulate two theorems on the existence of decompositions into simplices. The first holds in arbitrary dimensions, while the second is specific to the three-dimensional case.

Theorem 18.18. *For every convex integer polyhedron P (not contained in a hyperplane) there exists a decomposition into integer empty simplices.*

Proof. Let us give a sketch of the proof. The proof is based on induction on the number of integer points in $P \in \mathbb{R}^n$.

Base of induction. If P has only $n+1$ integer points inside, then it is an empty simplex.

Step of induction. Suppose now that the statement holds for every $k < m$. Let us prove it for m. Consider an integer polyhedron P. Suppose that it contain points $p_1, \ldots, p_m$. Let $f_1, \ldots, f_s$ be all the $(n-2)$-dimensional faces of P. We fix one integer point of P, say some p_i, and consider all pyramids with vertex at this point and bases in faces $f_1, \ldots, f_s$. These pyramids subdivide P into integer polyhedra. If P contains at least $n+2$ points, then it is possible to choose p_i such that this subdivision contains at least two smaller polyhedra (each of them contains a smaller number of integer points). By induction all the smaller polyhedra are decomposable; hence P is decomposable itself.   □

The decomposition of Theorem 18.18 usually contains triangles of distinct integer types that are not necessarily integer congruent to the coordinate simplex. In the paper [95] of J.-M. Kantor and K.S. Sarkaria, a stronger result for three-dimensional polyhedra is given.

Theorem 18.19. *For every convex integer polyhedron P in $\mathbb{R}^3$ there exists a decomposition of 4P into integer tetrahedra congruent to the basis tetrahedron.*   □

(Recall that nP is the polyhedron whose vertices have all the coordinates n-times greater than the coordinates of P.)

Idea of the proof. By Theorem 18.18 it is enough to find a proof for the list of empty tetrahedra, which is known due to White's theorem (see Corollary 19.3 below). It turns out that for every empty tetrahedron $T \in \mathbb{R}^3$, the tetrahedron $4T$ admits a decomposition into integer tetrahedra congruent to the basis tetrahedron. We skip here the technical details related to explicit constructions of these decompositions.

$\square$

18.4 Lattice Plücker coordinates and calculation of integer volumes of simplices

In order to calculate integer volumes of integer simplices we embed the space of all k-dimensional sublattices into the Grassmann algebra. Further, we express the integer volume of a simplex in terms of certain Plücker coordinates of this embedding.

18.4.1 Grassmann algebra on $\mathbb{R}^n$ and k-forms

The *Grassmann algebra* (or *exterior algebra*) of the vector space $\mathbb{R}^n$ is the algebra over $\mathbb{R}$ with multiplication (usually denoted by $\wedge$), generated by elements 1, $e_1, \ldots, e_n$, where $e_1, \ldots, e_n$ is the basis of $\mathbb{R}^n$. The relations defining the algebra are

$$e_i \wedge e_j = -e_j \wedge e_i;$$
$$e_i \wedge 1 = 1 \wedge e_i = e_i;$$
$$1 \wedge 1 = 1.$$

This algebra is denoted by $\Lambda(\mathbb{R}^n)$, with the operation $\wedge$ called the *wedge* product.

Let r be a positive integer. Denote by $\Lambda^k(\mathbb{R}^n)$ the vector space generated by the elements $e_{i_1} \wedge \cdots \wedge e_{i_k}$, where $i_1, \ldots, i_k \in \{1, \ldots, n\}$. In addition we put $\Lambda^0 \mathbb{R}^n = \mathbb{R}$. The space $\Lambda^k(\mathbb{R}^n)$ is called the *kth exterior power* of $\mathbb{R}$ and its elements are called *k-forms* on $\mathbb{R}$.

Proposition 18.20. *(i) The dimension of the kth exterior power of $\mathbb{R}^n$ is as follows:*

$$\dim \Lambda^k(\mathbb{R}^n) = \begin{cases} \binom{n}{k}, & \text{if } k \leq n, \\ 0, & \text{if } k > n. \end{cases}$$

(ii) Any element of the exterior algebra can be written as a sum of k-forms, so we have

$$\Lambda(\mathbb{R}^n) = \Lambda^0(\mathbb{R}^n) \oplus \Lambda^1(\mathbb{R}^n) \oplus \cdots \oplus \Lambda^n(\mathbb{R}^n).$$

$\square$

Let us define a natural basis of the space $\Lambda(\mathbb{R}^n)$ for an arbitrary $k \leq n$. Let I and J be two strictly increasing sequences of k elements of the set $\{1, \ldots, n\}$. We say that

$I > J$ if there exists $s \leq k$ such that $i_t = j_t$ for $t < s$ and $i_s > j_s$. We call this ordering *lexicographic*.

Definition 18.21. For a strictly increasing sequence I of k elements of the set $\{1,\ldots,n\}$ we define

$$\omega_I = e_{i_1} \wedge \cdots \wedge e_{i_k}.$$

The basis of $\Lambda^k(\mathbb{R}^n)$ consisting of all k-forms of type ω_I whose indices are ordered lexicographically is called the *lexicographic basis* of $\Lambda(\mathbb{R}^n)$.

Example 18.22. The dimension of the space $\Lambda^2(\mathbb{R}^4)$ is 6. The lexicographic basis of $\Lambda^2(\mathbb{R}^4)$ is as follows:

$$\left(e_1 \wedge e_2, \quad e_1 \wedge e_3, \quad e_1 \wedge e_4, \quad e_2 \wedge e_3, \quad e_2 \wedge e_4, \quad e_3 \wedge e_4\right).$$

18.4.2 Plücker coordinates

The set of all k-dimensional subspaces of $\mathbb{R}^n$ is called the *linear Grassmannian*, and denoted by $\mathrm{Gr}(k, \mathbb{R}^n)$. Denote by $P(\Lambda^k(\mathbb{R}^n))$ the projectivization of the rth exterior power $\Lambda^k(\mathbb{R}^n)$, i.e., we consider all forms up to a multiplication by a nonzero constant.

Definition 18.23. The *Plücker embedding* of a Grassmannian into projective space is the map

$$\phi : \mathrm{Gr}(k, \mathbb{R}^n) \to P(\Lambda^k(\mathbb{R}^n)),$$
$$\mathrm{Span}(v_1, \ldots, v_k) \to P(\{v_1 \wedge \ldots \wedge v_k\}).$$

It is clear that the form $v_1 \wedge \cdots \wedge v_k$ up to a nonzero multiplicative factor does not depend on the choice of basis in $\mathrm{Span}(v_1, \ldots, v_k)$.

Let $e_1, \ldots, e_n$ be a basis of $\mathbb{R}^n$ and let $\omega_1, \ldots, \omega_N$ be the corresponding lexicographic basis in $\Lambda^k(\mathbb{R}^n)$.

Definition 18.24. Consider a k-dimensional linear subspace L in $\mathbb{R}^n$ with basis $v_1, \ldots, v_k$. Let

$$v_1 \wedge \cdots \wedge v_k = \sum_{i=1}^{N} \lambda_i \omega_i.$$

The *Plücker coordinates* of L are the projective coordinates $(\lambda_1 : \ldots : \lambda_N)$.

To compute the Plücker coordinates one should calculate the determinants of all $(k \times k)$ minors of the $(k \times n)$ matrix whose elements are coordinates of vectors $v_1, \ldots, v_k$.

Example 18.25. Consider a two-dimensional linear subspace of $\mathbb{R}^3$ generated by vectors $v_1 = (1, 1, 1)$ and $v_2 = (1, 2, 5)$. We have

$$v_1 \wedge v_2 = (e_1 + e_2 + e_3) \wedge (e_1 + 2e_2 + 5e_3)$$
$$= \det \begin{pmatrix} 1 & 1 \\ 1 & 2 \end{pmatrix} e_1 \wedge e_2 + \det \begin{pmatrix} 1 & 1 \\ 1 & 5 \end{pmatrix} e_1 \wedge e_3 + \det \begin{pmatrix} 1 & 1 \\ 2 & 5 \end{pmatrix} e_2 \wedge e_3.$$

Hence the Plücker coordinates of this linear subspace are

$$\left(\det \begin{pmatrix} 1 & 1 \\ 1 & 2 \end{pmatrix} : \det \begin{pmatrix} 1 & 1 \\ 1 & 5 \end{pmatrix} : \det \begin{pmatrix} 1 & 1 \\ 2 & 5 \end{pmatrix} \right) = (1 : 4 : 3).$$

18.4.3 Oriented lattices in $\mathbb{R}^n$ and their lattice Plücker embedding

In this subsection we define a modification of the Plücker embedding that is useful in the study of lattices. We say that a lattice is a k-lattice if it has k generators and they span a k-dimensional subspace in $\mathbb{R}^n$. A k-lattice is said to be *oriented* if we fix an orientation of the span. Denote the space of all oriented k-lattices by $\mathrm{Lat}(k, \mathbb{R}^n)$, which we call the *oriented lattice Grassmannian*.

Definition 18.26. The *lattice Plücker embedding* of an oriented lattice Grassmannian in the space $\Lambda^k(\mathbb{R}^n)$ is the map

$$\phi : \mathrm{Lat}(k, \mathbb{R}^n) \to \Lambda^k(\mathbb{R}^n),$$
$$\langle v_1, \ldots, v_k \rangle \to v_1 \wedge \cdots \wedge v_k,$$

where $v_1, \ldots, v_k$ is a basis of the lattice $\langle v_1, \ldots, v_k \rangle$ defining the positive orientation.

Remark 18.27. It is clear that the lattice Plücker embedding is well defined (since different generators of the lattice define the same k-form up to sign, which is also preserved for generators of positive orientation).

We define lattice Plücker coordinates in a similar way.

Definition 18.28. Consider a k-lattice L in $\mathbb{R}^n$ with generators $v_1, \ldots, v_k$ defining the basis of positive orientation. Let

$$v_1 \wedge \cdots \wedge v_k = \sum_{i=1}^{N} \lambda_i \omega_i.$$

The *lattice Plücker coordinates* of L are the coordinates $(\lambda_1, \ldots, \lambda_N)$.

Remark. Notice that the lattice Plücker coordinates of lattices are affine and the arbitrary Plücker coordinates of subspaces are projective.

Example 18.29. As in Example 18.25, we consider a two-dimensional linear subspace of $\mathbb{R}^3$ generated by vectors $v_1 = (1, 1, 1)$ and $v_2 = (1, 2, 5)$. Its lattice Plücker coordinates are $(1, 4, 3)$.

18.4.4 Lattice Plücker coordinates and integer volumes of simplices

If L is a sublattice of the integer lattice $\mathbb{Z}^n$, then all its lattice Plücker coordinates are integers. Let us now show how to write the integer volume of integer simplices in terms of lattice Plücker coordinates.

Theorem 18.30. *Let $s_1 s_2 \ldots s_{k+1}$ be a simplex in $\mathbb{R}^n$ and let $(p_1, \ldots, p_N)$ be the lattice Plücker coordinates for the lattice generated by vectors $s_i s_{k+1}$ for $i = 1, \ldots, k$. Then*

$$\mathrm{lV}(s_1 s_2 \ldots s_{k+1}) = \gcd(p_1, \ldots, p_N).$$

Denote by v_i the vector $s_i s_{k+1}$ for $i = 1, \ldots, k$. We begin with the following lemma.

Lemma 18.31. *The number $\gcd(p_1, \ldots, p_N)$ is a $\mathrm{GL}(n, \mathbb{Z})$-invariant.*

Proof. Let

$$v_1 \wedge \cdots \wedge v_k = \sum_{i=1}^{N} p_i \omega_i$$

(as usual, $(\omega_1, \ldots, \omega_N)$ is the lexicographic basis associated to a chosen basis $(e_1, \ldots, e_n)$). Consider an arbitrary $A \in \mathrm{GL}(n, \mathbb{Z})$. We have

$$A(v_1 \wedge \cdots \wedge v_k) = A\left(\sum_{i=1}^{N} p_i \omega_i \right) = \sum_{i=1}^{N} p_i A(\omega_i).$$

Since all k-forms $A(\omega_i)$ have integer coefficients and all the numbers p_i are divisible by $\gcd(p_1, \ldots, p_N)$, the greatest common divisor for the lattice Plücker coordinates for the form $A(v_1 \wedge \cdots \wedge v_k)$ (denote them by $(q_1, \ldots, q_N)$) is divisible by $\gcd(p_1, \ldots, p_N)$ as well.

For the same reason, applying the inverse operator A^{-1} to the k-form $A(\omega_i)$, we have that $\gcd(q_1, \ldots, q_N)$ is divisible by $\gcd(p_1, \ldots, p_N)$. Hence

$$\gcd(q_1, \ldots, q_N) = \gcd(p_1, \ldots, p_N).$$

Therefore, $\gcd(p_1, \ldots, p_N)$ is a $\mathrm{GL}(n, \mathbb{Z})$-invariant. $\square$

Proof of Theorem 18.30. The statement holds for all lattices in the plane spanned by $e_1, \ldots, e_k$. In this case the first lattice Plücker coordinate is the Euclidean volume of the parallelogram $P(s_1, s_2, \ldots, s_{k+1})$, which coincides with the integer volume of the simplex $s_1 s_2 \ldots s_{k+1}$ (see Corollary 18.15). All the other lattice Plücker coordinates are zero.

An arbitrary k-dimensional plane can be taken to the coordinate k-plane by some $\mathrm{GL}(n, \mathbb{Z})$ transformation. By Lemma 18.31, the $\gcd(p_1, \ldots, p_N)$ is a $\mathrm{GL}(n, \mathbb{Z})$ invariant. So in this case it is also equal to the integer volume of the simplex. $\square$

Example 18.32. Let us calculate the volume of the three-dimensional tetrahedron $S = s_1 s_2 s_3 s_4$ of Example 18.3. Recall that

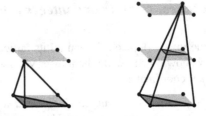

Fig. 18.1 The empty tetrahedra T_1 and T_2.

$$s_1 = (2,3,0,1), \quad s_2 = (1,4,2,4), \quad s_3 = (1,0,0,4), \quad s_4 = (1,0,0,1).$$

This tetrahedron defines a three-dimensional sublattice in $\mathbb{Z}^4$ generated by vectors $v_1 = s_1 s_4$, $v_2 = s_2 s_4$, and $v_3 = s_3 s_4$. So, we get

$$v_1 \wedge v_2 \wedge v_3 = (e_1 + 3e_2) \wedge (4e_2 + 2e_3 + 3e_4) \wedge 3e_4$$
$$= 0 \cdot e_1 \wedge e_2 \wedge e_3 + 12 \cdot e_1 \wedge e_2 \wedge e_4 + 6 \cdot e_1 \wedge e_3 \wedge e_4 + 18 \cdot e_2 \wedge e_3 \wedge e_4.$$

The lattice Plücker coordinates are $(0, 12, 6, 18)$, and their greater common divisor is 6. Hence the integer volume of S equals 6.

18.5 Ehrhart polynomials as generalized Pick's formula

In this section we discuss one of the generalizations of Pick's formula. We start with the following example, showing that there is no straightforward combinatoric generalization of Pick's formula.

Example 18.33. Consider the following two tetrahedra (see in Fig. 18.1):

$$T_1 = \mathrm{conv}((0,0,0),(1,0,0),(0,1,0),(0,0,1));$$
$$T_2 = \mathrm{conv}((0,0,0),(1,0,0),(0,1,0),(1,1,2)).$$

These tetrahedra are both empty, but they have distinct areas:

$$V(T_1) = 1/6 \quad \text{and} \quad V(T_2) = 1/3.$$

The emptiness of the tetrahedra means that they have only four integer points as vertices, and hence they have the same number of integer points in all the corresponding faces.

The following estimate relates the number of integer vertices in the polyhedron and its Euclidean volume.

Theorem 18.34. (S.V. Konyagin, K.A. Sevastyanov [126].) *The number of vertices of an n-dimensional convex integer polyhedron P is bounded from above by* $c(n)V(P)^{\frac{n-1}{n+1}}$, *where* $c(n)$ *is a constant not depending on P.* □

So the generalization should have a different nature. Let us say a few words on Ehrhart polynomials, which to some extent generalize Pick's formula.

Definition 18.35. Let P be a d-dimensional convex integer polytope in $\mathbb{R}^n$, and let t be a real variable. The function

$$L(P,t) = \#(tP \cap \mathbb{Z}^n)$$

is called the *Ehrhart polynomial*.

Actually, the function $L(P,t)$ is a polynomial only to a certain extent.

Theorem 18.36. (E. Ehrhart [60].) *Consider an arbitrary convex polyhedron P in* $\mathbb{R}^n$. *Let us restrict the variable t to integers. Then the Ehrhart polynomial of P is a polynomial of degree d.* □

So for the integer variable t there exist rational numbers $a_0, \ldots, a_d$ such that

$$L(P,t) = a_d t^d + a_{d-1} t^{d-1} + \cdots + a_0.$$

Let us discuss the lattice-geometric meaning of several coefficients.

Proposition 18.37. *Let P be an integer polyhedron. Then*

$$a_0 = 1;$$
$$a_{d-1} = \frac{1}{2 \cdot (d-1)!} \sum_F \mathrm{lV}(F);$$
$$a_d = \frac{\mathrm{lV}(P)}{d!}.$$

The sum in the formula for a_{d-1} *is taken over all faces of P of codimension 1.* □

The other coefficients of the Ehrhart polynomial do not have a simple combinatorial interpretation now. We refer the interested reader to a recent book by M. Beck and S. Robins [22].

Remark 18.38. Let us study the Ehrhart polynomials for integer polygons in the plane. Consider a polygon P, let I be the number of integer points in the interior of P, and let E be the number of integer points in the union of edges of P. Then we have

$$L(P,t) = A(P) \cdot t^2 + E/2 \cdot t + 1.$$

By Definition 18.35 we have

$$L(P,1) = \#(P \cap \mathbb{Z}^2) = I + E.$$

The last two equations (for $t = 1$) imply Pick's formula:

$$A(P) = I + E/2 - 1.$$

Example 18.39. Let us write the Ehrhart polynomials for the tetrahedra T_1 and T_2 in Example 18.33:

$$L(T_1,t) = \tfrac{1}{6}x^3 + x^2 + \tfrac{11}{6}x + 1;$$
$$L(T_2,t) = \tfrac{1}{3}x^3 + x^2 + \tfrac{5}{3}x + 1.$$

18.6 Integer-regular polyhedra

We conclude this chapter with the classification of integer-regular polyhedra in arbitrary dimensions.

18.6.1 Definition of integer-regular polyhedra

In this subsection we collect several general notions related to integer-regular polyhedra. Recall that a *convex n-dimensional polyhedron* in $\mathbb{R}^n$ is the convex hull of a finite set of points that are not contained in a hyperplane of $\mathbb{R}^n$.

Definition 18.40. A point v of a polyhedron P is a *vertex* if for every pair of distinct to v points (v_1, v_2) of the polyhedron P the segment $v_1 v_2$ does not contain v.
A convex k-dimensional polyhedron Q is a *k-dimensional face* of a convex polyhedron P if the following three conditions hold:
— all vertices of Q are vertices of P;
— the affine plane spanned by Q does not have vertices of P other than vertices of Q;
— the polyhedron P is contained in one of the (multidimensional) half-planes with respect to some affine hyperplane containing Q.

Note that in our notation vertices are 0-dimensional faces (and edges are 1-dimensional faces).

We continue with the definition of supporting planes that is coming from theory of convex polyhedra.

Definition 18.41. A hyperplane π is *supporting* for a polyhedron P if P is contained in one of the half-planes with respect to π and the hyperplane π contains at least one point of P.

We continue with the definition of flags in $\mathbb{R}^n$.

Definition 18.42. A *flag* $F = (\pi_0, \pi_1, \ldots, \pi_{n-1})$ in $\mathbb{R}^n$ is a collection of a point π_0, a line π_1, and planes π_k for $k = 2, \ldots, n$ satisfying the condition

$$\pi_0 \subset \pi_1 \subset \cdots \subset \pi_n.$$

Definition 18.43. A flag $F = (\pi_0, \pi_1, \ldots, \pi_{n-1})$ in $\mathbb{R}^n$ is said to be a *flag of a polyhedron P* if there exists a sequence of faces of P:

$$Q_0 \subset Q_1 \subset \cdots \subset Q_n,$$

where the dimension of Q_i is i for $i = 0, \ldots, n$, such that

$$Q_i \subset \pi_i \quad \text{for } i = 0, \ldots, n-1.$$

A polyhedron is *integer* if all its vertices are integer points.

Now we are in position to define integer regular polyhedra.

Definition 18.44. An integer n-dimensional polyhedron in $\mathbb{R}^n$ is said to be *integer-regular* if for any pair of its flags there exists an integer affine transformation of $\mathbb{R}^n$ sending the first flag to the second one.

Finally let us define elementary integer polyhedra.

Definition 18.45. We say that a point $v = (tx_1, \ldots, tx_n)$ is a *t-dilate* of the point $w = (x_1, \ldots, x_n)$, we write $v = tw$.
An integer polyhedron P_1 with vertices $(tv_1, \ldots, tv_k)$ is said to be the *t-dilate* of the polyhedron P_2 with vertices $(v_1, \ldots, v_k)$, denote it by $P_1 = tP_2$.
An integer polyhedron is *elementary* if it is not t-dilate of an integer polyhedron for every integer $t > 1$.

18.6.2 Schläfli symbols

Consider an arbitrary regular n-dimensional Euclidean polyhedron P. Let $F = (\pi_0, \ldots, \pi_{n-1})$ be one of its flags, with corresponding faces $Q_i \subset \pi_i$.

We consider Q_i as a solid homogeneous body and denote by O_i the center of mass of Q_i (for $i = 0, \ldots, n-1$). Let also O_n denote the center of mass for P itself. The n-dimensional tetrahedron $O_0 O_1 \ldots O_{n-1} O_n$ is called the *chamber* of a regular polyhedron P corresponding to the given flag.

Denote by r_i (for $i = 0, \ldots, n-1$) the reflection about the $(n-1)$-dimensional plane that spans the center of masses $O_n, \ldots, O_{i+1}, O_{i-1}, \ldots, O_0$ (except the i-th one), see Fig. 18.2. These reflections are called *basic*.

We have the following classical statement.

Proposition 18.46. *The group of Euclidean symmetries of an Euclidean regular polyhedron P is generated by the reflections $r_0, r_1, \ldots, r_{n-1}$.*
For $i = 1, \ldots, n-1$ the angle between the invariant hyperplanes of the symmetries r_{i-1} and r_i equals π/a_i, for some integer $a_i \geq 3$. □

Definition 18.47. The expression $\{a_1, \ldots, a_{n-1}\}$ is said to be the *Schläfli symbol* for a (flag of) the regular polygon P. For simplicity we write a^s for the substring $a, a, \ldots, a$ of the length s in the Schläfli symbol.

Since all the flags of a regular polyhedron are congruent, the Schläfli symbol for different faces is the same, hence we can treat it as a the Schläfli symbol for the polyhedron itself.

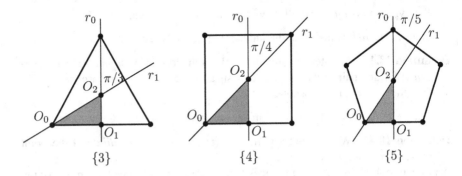

Fig. 18.2 Basic reflections and the Schläfli symbols for regular polygons.

18.6.3 Euclidean regular polyhedra

Let us list all regular convex Euclidean polyhedra (up to Euclidean transformation and homotheties).

List of regular Euclidean polyhedra.

Dimension 1: here any segment is regular, its Schläfli symbol is $\{\}$.

Dimension 2: for every integer $m \geq 3$ we have a regular m-gon, its Schläfli symbol is $\{m\}$.

Dimension 3: Starting from dimension three we have only finitely many regular polyhedra in each dimension. In dimension 3 we have

- $\{3, 3\}$: a regular tetrahedron;
- $\{3, 4\}$: a regular octahedron;
- $\{4, 3\}$: a regular cube;
- $\{3, 5\}$: a regular icosahedron;
- $\{5, 3\}$: a regular dodecahedron.

Dimension 4: In this dimension we have some exotic regular polyhedra. The list here is as follows:

- $\{3, 3, 3\}$: a regular tetrahedron;

- $\{4,3,3\}$: a regular cube;
- $\{3,3,4\}$: a regular generalized octahedron (the other names for this polyhedron is cross polytope, or hyperoctahedron);
- $\{3,4,3\}$: a regular 24-cell (or hyperdiamond, or icositetrachoron);
- $\{3,3,5\}$: a regular 600-cell (or hypericosahedron, or hexacosichoron);
- $\{5,3,3\}$: a regular 120-cell (or hyperdodecahedron, or hecatonicosachoron).

Dimension n (n>4): Surprisingly, in higher dimensions the situation stabilises. There are only three regular Euclidean polyhedra in such cases:

- $\{3^{n-1}\}$: a regular tetrahedron;
- $\{4,3^{n-2}\}$: a regular cube;
- $\{3^{n-2},4\}$ a regular generalized octahedron.

Remark 18.48. In the one- and two-dimensional cases the statement is straightforward. Three-dimensional regular polyhedra were already known in ancient times. The cases of dimensions higher than three were studied by Schläfli in [188].

18.6.4 Preliminary integer notation

In this subsection we generate several lists of integer polyhedra that will be later used in Theorem 18.49.

First of all, denote by e_i the i-th unit coordinate vector for all admissible positive integers i.

The segment. Denote by $\{\}^{\mathbb{Z}}$ the unit integer segment.

Tetrahedra. For any $n > 1$ and any integer p we denote by $\{3^{n-1}\}^{\mathbb{Z}}_p$ the n-dimensional tetrahedron with the vertices:

$$V_0 = O; \qquad V_i = O + e_i, \quad \text{for } i = 1, \ldots, n-1,$$

forming the unit coordinate tetrahedron in the hyperplane $x_n = 0$ and the vertex

$$V_n = (p-1) \sum_{k=1}^{n-1} e_k + p e_n.$$

Cubes. Recall that in general any cube C is generated by one of its vertices P and an n-tuple of linearly independent vectors v_i, namely

$$C = \left\{ P + \sum_{i=1}^{n} \alpha_i v_i \,\middle|\, 0 \leq \alpha_i \leq 1, i = 1, \ldots, n \right\}.$$

We denote by

- $\{4,3^{n-2}\}^{\mathbb{Z}}_1$ for any $n \geq 2$: the integer cube with vertex at the origin and generated by all basis vectors;

- $\{4,3^{n-2}\}^{\mathbb{Z}}_2$ for any $n \geq 2$: the integer cube with vertex at the origin and generated by the first $n-1$ basis vectors and the vector

$$e_1 + e_2 + \ldots + e_{n-1} + 2e_n;$$

- $\{4,3^{n-2}\}^{\mathbb{Z}}_3$ for any $n \geq 3$: the integer cube with a vertex at the origin and generated by the vectors:

$$e_1 \quad \text{and} \quad e_1 + 2e_i \quad \text{for } i = 2, \ldots, n.$$

Generalized octahedra. Denote by

- $\{3^{n-2},4\}^{\mathbb{Z}}_1$ for any $n \geq 2$: the integer generalized octahedron with the vertices $O \pm e_i$ for $i = 1, \ldots, n$. This polyhedron is the convex hull of the unit coordinate tetrahedron together with its centrally symmetric (about the origin) copy;

- $\{3^{n-2},4\}^{\mathbb{Z}}_2$ for any positive n: the integer generalized octahedron whose vertices are the vertices of $\{3^{n-3},4\}^{\mathbb{Z}}_1$:

$$O \pm e_i \quad \text{for } i = 1, \ldots, n-1,$$

and extra two vertices

$$O \pm (e_1 + e_2 + \ldots + e_{n-1} + 2e_n);$$

- $\{3^{n-2},4\}^{\mathbb{Z}}_3$ for any positive n: the integer generalized octahedron with the vertices

$$O, \quad O - e_1, \quad O - e_1 - e_i \quad \text{for } i = 2, \ldots, n, \quad \text{and} \quad e_i \quad \text{for } i = 2, \ldots, n.$$

Two particular 24-cells. Here we have the following two polyhedra:

- $\{3,4,3\}^{\mathbb{Z}}_1$: the 24-cell with 8 vertices of the form

$$O \pm 2(e_2 + e_3 + e_4), \quad O \pm 2(e_1 + e_2 + e_4),$$
$$O \pm 2(e_1 + e_3 + e_4), \quad O \pm 2e_4,$$

and 16 vertices of the form

$$O \pm (e_2 + e_3 + e_4) \pm (e_1 + e_2 + e_4) \pm (e_1 + e_3 + e_4) \pm e_4.$$

- $\{3,4,3\}^{\mathbb{Z}}_2$: the 24-cell with 8 vertices of the form

$$O \pm 2(e_1 + e_2 + e_3 + e_4), \quad O \pm 2(e_1 - e_2 + e_3 + e_4),$$
$$O \pm 2(e_1 + e_2 - e_3 + e_4), \quad O \pm 2(e_1 + e_2 + e_3 - e_4),$$

and 16 vertices of the form

$$O \pm (e_1 + e_2 + e_3 + e_4) \pm (e_1 - e_2 + e_3 + e_4)$$
$$\pm (e_1 + e_2 - e_3 + e_4) \pm (e_1 + e_2 + e_3 - e_4).$$

18.6.5 Integer-regular polyhedra in arbitrary dimensions

The following theorem was proved in [100] (see also [149, 154] for further development).

Theorem 18.49. *Any elementary integer regular polyhedron is integer congruent to exactly one of the following polyhedra:*
Dimension 1: *the unit segment* $\{\}^{\mathbb{Z}}$.

Dimension 2: *As we have seen in Proposition 2.26, we have 6 polygons in dimension 2:*

- *the triangles* $\{3\}_1^{\mathbb{Z}}$ *and* $\{3\}_2^{\mathbb{Z}}$;
- *the squares* $\{4\}_1^{\mathbb{Z}}$ *and* $\{4\}_2^{\mathbb{Z}}$;
- *the octagons* $\{6\}_1^{\mathbb{Z}}$ *and* $\{6\}_2^{\mathbb{Z}}$.

Dimensions 3, 5, 6, 7, 8, . . .: *these dimensions are rather regular. Here we have the following three classes of integer regular polygons:*

- *the tetrahedra* $\{3^{n-1}\}_i^{\mathbb{Z}}$ *where positive integers i are divisors of $n+1$;*
- *the generalized octahedra* $\{3^{n-2}, 4\}_i^{\mathbb{Z}}$, *for* $i = 1, 2, 3$;
- *the cubes* $\{4, 3^{n-2}\}_i^{\mathbb{Z}}$, *for* $i = 1, 2, 3$.

Dimension 4: *Finally, as a rule dimension 4 is exceptional. Here we have four combinatorially different integer-regular types of polyhedra:*

- *the tetrahedra* $\{3, 3, 3\}_1^{\mathbb{Z}}$ *and* $\{3, 3, 3\}_5^{\mathbb{Z}}$;
- *the generalized octahedra* $\{3, 3, 4\}_i^{\mathbb{Z}}$, *for* $i = 1, 2, 3$;
- *the 24-cells* $\{3, 4, 3\}_1^{\mathbb{Z}}$ *and* $\{3, 4, 3\}_2^{\mathbb{Z}}$;
- *the cubes* $\{4, 3, 3\}_i^{\mathbb{Z}}$, *for* $i = 1, 2, 3$.

Fig. 18.3 illustrates the adjacency diagram of the elementary integer-regular polyhedra. The list of integer-regular polyhedra in dimension 3 consists of 9 items; they are shown in Fig. 18.4.

18.7 Exercises

Exercise 18.1. Calculate the integer distance and the integer angle between the following two planes in $\mathbb{R}^4$:

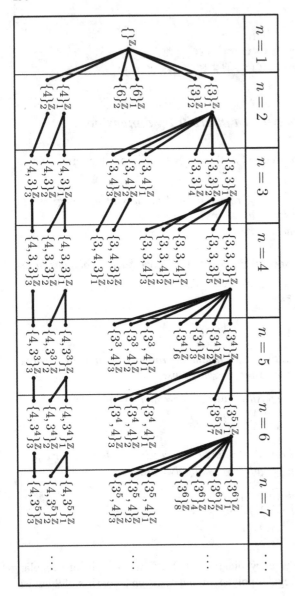

Fig. 18.3 Adjacency of the elementary integer-regular polyhedra.

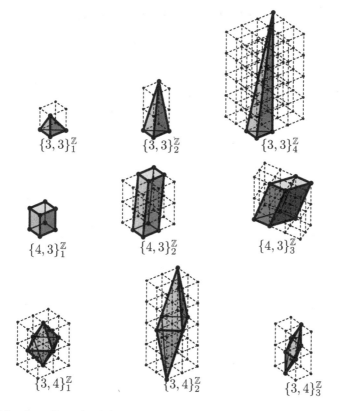

Fig. 18.4 Nine three-dimensional elementary integer-regular polyhedra.

$$\begin{cases} x = 0, \\ y = 0, \end{cases} \quad \text{and} \quad \begin{cases} x+y = 0, \\ z+t = 4. \end{cases}$$

Exercise 18.2. Find some integer invariants of arrangements of three lines in $\mathbb{R}^n$.

Exercise 18.3. Prove the statements of Proposition 18.20.

Exercise 18.4. Prove that any two integer k-dimensional planes in $\mathbb{R}^n$ are integer congruent.

Exercise 18.5. Find the Plücker coordinates of the sublattice generated by the following vectors:

$$v_1 = (1,2,3,0,0), \quad v_2 = (1,1,1,1,1), \quad \text{and} \quad v_3 = (0,0,3,2,1).$$

Exercise 18.6. Find and prove the expression for the leading coefficient of the Ehrhart polynomial.

Exercise 18.7. Calculate the Ehrhart polynomial for the tetrahedron

$$T = \mathrm{conv}((0,0,0),(2,0,0),(0,3,0),(1,2,3)).$$

Chapter 19
On Empty Simplices, Pyramids, Parallelepipeds

Recall that an integer polyhedron is called *empty* if it does not contain integer points other than its vertices. In this chapter we give the classification of empty tetrahedra and the classification of pyramids whose integer points are contained in the base of pyramids in $\mathbb{R}^3$. Later in the book we essentially use the classification of the mentioned pyramids for studying faces of multidimensional continued fractions. In particular, the describing of such pyramids simplifies the deductive algorithm of Chapter 24 in the three-dimensional case. We continue with two open problems related to empty objects in lattices. The first one is a problem of description of empty simplices in dimensions greater than 3. The second is the lonely runner conjecture. We conclude this chapter with a proof of a theorem on the classification of empty tetrahedra.

19.1 Classification of empty integer tetrahedra

Recall that in the two-dimensional case there is a unique class of empty triangles (see Proposition 2.15 and Remark 2.16). In this section we examine the situation in the three-dimensional case. We start with White's theorem on geometric properties of empty tetrahedra.

Let $ABCD$ be a tetrahedron with enumerated vertices. Denote by $P(ABCD)$ the following parallelepiped:

$$\{A + \alpha AB + \beta AC + \gamma AD \mid 0 \le \alpha \le 1, 0 \le \beta \le 1, 0 \le \gamma \le 1\}.$$

Theorem 19.1. (White's theorem [222].) *Let $ADBA'$ be an empty tetrahedron. Then all integer points in $P(ADBA')$ except the vertices are contained in one of the planes passing through centrally symmetric edges distinct from the edges of the tetrahedron.*

© Springer-Verlag GmbH Germany, part of Springer Nature 2022
O. N. Karpenkov, *Geometry of Continued Fractions*,
Algorithms and Computation in Mathematics 26,
https://doi.org/10.1007/978-3-662-65277-0_19

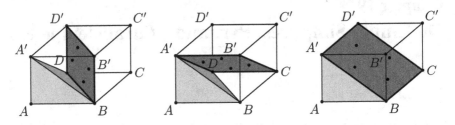

Fig. 19.1 If $ADBA'$ is empty, then all integer points are in one of the dark gray three planes. We give an example with four points inside.

Remark 19.2. For every parallelepiped there exist exactly three planes as in the theorem. See an example with four inner integer points in Fig. 19.1.

Theorem 19.1 leads to a complete classification of empty integer tetrahedra with one marked vertex.

Corollary 19.3. (*i*) **Classification of marked empty tetrahedra.** *An integer empty tetrahedron with a marked vertex is integer congruent to exactly one of the following marked tetrahedra:*

— *a tetrahedron with vertices* $(0,0,0)$, $(1,0,0)$, $(1,0,1)$, *and* $(1,1,0)$;

— $T_{1,r}^{\xi}$ *of the list "T-W," where* $r \geq 2$, $0 < \xi \leq r/2$, *and* $\gcd(\xi, r) = 1$ *(see Fig. 19.2).* *The point* $(0,0,0)$ *is a marked vertex for all the tetrahedra in the list.*

(*ii*) **Classification of empty tetrahedra.** *Any empty tetrahedron is integer congruent to an empty marked tetrahedron. The tetrahedra* T_{1,r_1}^{ξ} *and* T_{1,r_2}^{v} *(without marked vertex) are integer congruent if and only if* $r_1 = r_2$ *and for* $r_1 > 1$, *at least one of the following four equations holds:*

$$(\xi) \equiv (\pm v \mod r_1)^{\pm 1}.$$

Remark 19.4. Classifications of empty tetrahedra and empty marked tetrahedra coincide only for $r = 1, 2, 3, 4, 5, 6, 8, 10, 12, 24$. (This is the case when the square of every element of the multiplicative group $(\mathbb{Z}/m\mathbb{Z})^*$ equals ± 1. Hence all the summands in the multiplicative group are either $\mathbb{Z}/4\mathbb{Z}$ or $\mathbb{Z}/2\mathbb{Z}$. There are 15 such groups. Only nine of them satisfy the condition. In addition we have the case $r = 1$.)

We give all the proofs for the statements of this section in Section 19.4.

19.2 Classification of completely empty lattice pyramids

We say that a pyramid is *marked* if the vertex of the pyramid is specified. (So every tetrahedron corresponds to four marked pyramids, while all the other pyramids are in one-to-one correspondence with marked pyramids.) A pyramid in $\mathbb{R}^3$ is called

The list "T-W"	Parameters	Coordinates of the face	Integer-affine type of the face	Id
$T^{\xi}_{a,r}$	$a\geq 1,\ r\geq 2,$ $0<\xi\leq r/2,$ $\gcd(\xi,r)=1$	$(\xi,r-1,-r),$ $(a+\xi,r-1,-r),$ $(\xi,r,-r)$	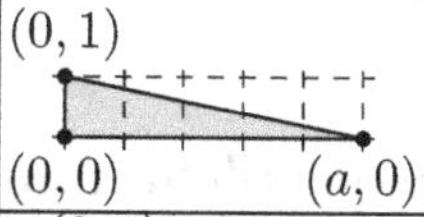 $(0,1)$ $(0,0)\quad (a,0)$	r
U_b	$b\geq 2$	$(1,b-1,2),$ $(2,-1,2),$ $(0,-1,2)$	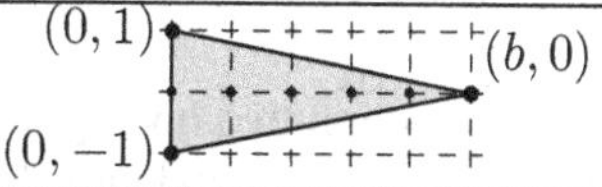 $(0,1)\quad (b,0)$ $(0,-1)$	2
V		$(-2,1,2),$ $(-1,-1,2),$ $(1,2,2)$	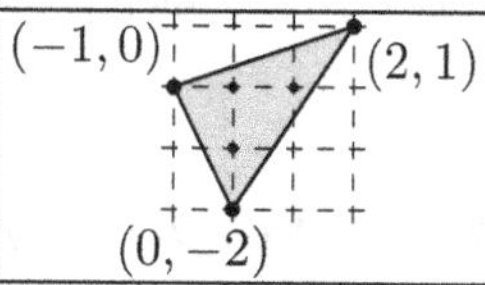 $(-1,0)\quad (2,1)$ $(0,-2)$	2
W		$(0,2,3),$ $(1,1,3),$ $(2,3,3)$	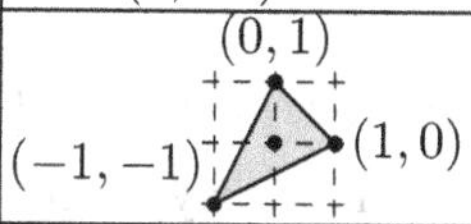 $(0,1)$ $(-1,-1)\quad (1,0)$	3

Fig. 19.2 The T-W list of empty pyramids. The origin $(0,0,0)$ is the marked vertex for all the pyramids.

integer if its vertex is an integer point and its base is an integer polygon. An integer pyramid is called *completely empty* if every integer point inside the pyramid is either contained in the base or coincides with the vertex of the pyramid.

An integer pyramid is called *one-story* (*multistory*) if the integer distance from the vertex to the base is 1 (≥ 1). One-story integer empty pyramids can have an arbitrary polygon as a base, since there is no geometric obstacle to constructing arbitrary pyramids. The multistory case is more interesting, as described in the following theorem.

Theorem 19.5. *([98]) Every multistory completely empty convex three-dimensional marked pyramid is either triangular or quadrangular. Every triangular multistory completely empty pyramid is integer congruent to exactly one of the marked pyramids from the list "T-W" (see Fig. 19.2). Every quadrangular multistory completely empty pyramid is integer congruent to exactly one of the following pyramids:*

	Coords. of the vertex	Coordinates of the base	Integer-affine type of the base	Id
$M_{a,b}$ $b\geq a\geq 1$	$(0,0,0)$	$(-1,0,2),$ $(-a-1,1,2),$ $(-1,2,2),$ $(b-1,1,2)$	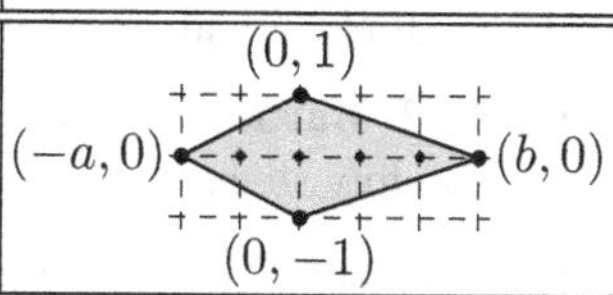 $(0,1)$ $(-a,0)\quad (b,0)$ $(0,-1)$	2

□

We skip the complete technical proof Theorem 19.5 (for a complete proof we refer to [98]).

19.3 Two open problems related to the notion of emptiness

Let us briefly mention two open problems related to arrangements of integer points in the interior of base parallelepipeds of integer sublattices.

19.3.1 Problem on empty simplices

In this subsection we say a few words about the following problem.

Problem 12. Classify empty integer simplices in dimension n.

From Proposition 2.15 it follows that all empty triangles are integer congruent to the coordinate triangle ($n = 2$). In Corollary 19.3 all distinct empty three-dimensional tetrahedra are listed ($n = 3$). For the rest of the cases ($n \geq 4$), the problem is open. In the case $n = 4$, we would like to mention one interesting recent result. To formulate it we introduce a notion of the width of integer polyhedra.

Definition 19.6. The width of an integer polyhedron P in $\mathbb{R}^n$ is the minimal integer distance between two parallel integer hyperplanes containing P in the middle.

Let us illustrate this definition with several examples.

Example 19.7. The widths of the following triangles,

are respectively 1, 2, and 3. These triangles have minimal possible integer area among the triangles of the given width: the areas are 1, 3, and 7 respectively.

Example 19.8. The width of every empty three-dimensional tetrahedron is 1.

The last example means that every empty tetrahedron is contained between two neighboring parallel integer planes (see Fig. 19.3).

Almost all four-dimensional empty integer simplices have the following property.

Theorem 19.9. (M. Barile et al. 2011 [20].) *Up to possibly a finite number of exceptions, every empty simplex in dimension 4 has width 1 or 2.* □

To conclude this subsection we observe once more that the complete classification of empty simplices in $\mathbb{R}^4$ is still not known, while the situation in higher dimensions is entirely open.

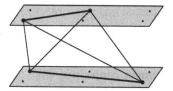

Fig. 19.3 All empty three-dimensional tetrahedra are of width 1.

19.3.2 Lonely runner conjecture

In this subsection we discuss the *lonely runner conjecture*, which is also known as *view-obstruction problem for n-dimensional cubes*. We start with a general formulation of the conjecture.

Conjecture 13. (Lonely runner conjecture.) Suppose k runners having distinct constant speeds start at a common point and run laps on a circular track with circumference 1. Then for any given runner, there is a time at which that runner is at arc distance at least $1/k$ away from every other runner.

This conjecture was introduced by J.M. Wills [223] in 1967. Later it was considered by T.W. Cusick (in [43], [45], and [46]) as the geometric view-obstacle problem. The cases $k = 2, 3$ are easy exercises. The conjecture holds for $k = 4$ (U. Betke and J.M. Wills [23], T.W. Cusick [45]), $k = 5$ (T.W. Cusick and C. Pomerance [48], see also in [24]), $k = 6$ (T. Bohman et al. [25]; see also in [24]), and $k = 7$ (J. Barajas and O. Serra [17] and [18]). For $k \geq 8$ it is not known whether this conjecture is true.

Let us describe the question in geometric terms. Without loss of generality we study the conjecture only for the first runner. We say that the starting point is the zero point of a circle. It is usual to put the speed of the first runner equal to zero. We associate each runner that travels with nonzero speed with a point on a circle. The configuration space of all $(k-1)$-tuples on the circle is the $(k-1)$-dimensional torus T^{k-1} (which is the direct product of $k-1$ copies of circles $S^1 = \mathbb{R}/\mathbb{Z}$). We mark the point $(0, \ldots, 0)$ on this torus as the first runner (who is actually not running but standing). While the time goes from zero to infinity, the configuration of $(k-1)$-runners travels linearly in the torus T^{k-1}. The question is whether this trajectory is always far from the middle point of the torus $(1/2, 1/2, \ldots, 1/2)$.

To simplify the picture it is good to consider the universal covering space of T^{k-1}, which is $\mathbb{R}^{k-1}$. Here the standing runner is associated with an integer lattice $\mathbb{Z}^{k-1}$. Now we reformulate the lonely runner conjecture as a view-obstacle problem.

Lonely runner conjecture as a view-obstacle problem. A hunter is standing in the middle (origin) of an n-dimensional garden. At each vertex of the shifted lattice $\mathbb{Z}^{k-1} + (1/2, 1/2, \ldots, 1/2)$ a tree grows. All the trees are hypercubes of size $1 - 2/k$ centered at the vertex of the shifted lattice and with sides parallel to the coordinate axes. Suppose the hunter is not willing to shoot in the directions having at least one zero coordinate. Then he will shoot into some tree. It is supposed that the bullet has zero size.

Finally, we give a discrete lattice version of the lonely runner conjecture.

Discrete version of lonely runner conjecture. Does there exist a full-rank sublattice of $\mathbb{Z}^{k-1}$ (suppose it is generated by $v_1,\ldots,v_{k-1}$) such that the hypercube

$$\Big\{\sum_{i=1}^{k-1}\lambda_i v_i \ \Big| \ \frac{1}{k}\leq \lambda_i \leq \frac{k-1}{k}, \ i=1,2,\ldots,k-1\Big\}$$

does not contain integer points.

Remark 19.10. Let us say a few words about the inverse question of meeting of all the runners. Suppose k runners having distinct constant speeds start at a common point and run laps on a circular track. Then for every $\varepsilon > 0$ and $T > 0$, there is a time $t > T$ at which all the runners are at arc distance at most ε away from each other. We leave this as an exercise to the reader.

19.4 Proof of White's theorem and the empty tetrahedra classification theorems

We start with a proof of White's theorem. Further, we deduce from it the classification of empty tetrahedra. Before proving White's theorem, we introduce some necessary definitions and prove several preliminary statements.

19.4.1 IDC-system

We begin with some necessary definitions.

Definition 19.11. Let A, B, C, and D be integer points that span the entire space $\mathbb{R}^3$ (the collection of points is ordered). Let us construct a system of coordinates related to these points. Let b, c, and d be the integer distances from the points B, C, and D to the planes of the faces ACD, ABD, and ACD, respectively. Consider the point A as the origin. Set the coordinates of B, C, and D equal to $(b,0,0)$, $(0,c,0)$, and $(0,0,d)$ respectively. The coordinates of the other points in $\mathbb{R}^3$ are defined by linearity. We say that this system of coordinates is an *integer-distance coordinate system with respect to ABCD* or (*IDC-system* for short).

We call the points with integer coordinates in an IDC-system the *IDC-nodes*. We say that an IDC-node is *integer* if it is an integer point of $\mathbb{R}^3$. Notice that every integer point in the old system is an IDC-node. The converse is true if and only if the vectors AB, AC, and AD generate the integer lattice. The coordinates of integer

IDC-nodes coincide with the integer distances to the coordinate planes in the IDC-system.

19.4.2 A lemma on sections of an integer parallelepiped

If a and b are two integer vectors, denote by $L_{a,b}$ the integer sublattice generated by a and b. Consider the set of all cosets in $\mathbb{Z}^3/L_{a,b}$. Every left coset of this quotient group is contained in a plane parallel to the plane spanned by the vectors a and b. In addition, every such plane contains only finitely many left cosets, their number being equal to the index of the sublattice $L_{a,b}$ in the integer sublattice in the subspace $\mathrm{Span}(a,b)$.

Lemma 19.12. *Consider an integer parallelepiped $P = ABCDA'B'C'D'$ and a plane π parallel to ABCD. Let π intersect the parallelepiped (by some parallelogram). Then the following two statements hold.*

(i) The section of the parallelepiped P by the plane π contains representatives of all left cosets of $\mathbb{Z}^3/L_{a,b}$ that are in π.

(ii) Every left coset of $\mathbb{Z}^3/L_{a,b}$ contained in π intersects the parallelepiped P, and the intersection is described by one of the following cases:
   (a) a point inside the parallelogram $\pi \cap P$;
   (b) two points on opposite edges of $\pi \cap P$;
   (c) four points coincide with the vertices of $\pi \cap P$.   □

We leave the proof of this lemma as a simple exercise for the reader.

19.4.3 A corollary on integer distances between the vertices and the opposite faces of a tetrahedron with empty faces

Corollary 19.13. *All integer distances from the vertices of an integer (three-dimensional) tetrahedron with empty faces to the opposite faces are equal.*

Proof. Consider an integer tetrahedron $OABC$ with empty faces. Suppose that

$$\mathrm{ld}(A, OBC) = n.$$

Let us show that $\mathrm{ld}(B, OAC) = \mathrm{ld}(C, OAB) = n$.

Consider the parallelepiped $P(OABC)$. Since the triangles OBC, OAB, and OAC are empty, the corresponding faces of $P(OABC)$ are empty as well. Hence all faces of the parallelepiped are empty. Consider all the integer planes parallel to OBC that divide the parallelepiped into two nonempty parts. Since $\mathrm{ld}(A, OBC) = n$, the number of these planes equals $n - 1$.

By Lemma 19.12 every such integer plane contains exactly one left coset of the lattice $\mathbb{Z}^3/L_{OB,OC}$. Since the faces of the parallelepiped are empty, the intersection of this class with a parallelepiped is one point in the interior. Hence the parallelepiped $P(OABC)$ contains exactly $n-1$ integer points.

Suppose that $\text{ld}(B, OAC) = m$, and $\text{ld}(B, OAC) = k$. Then for similar reasons the parallelepiped contains $m-1$ and $k-1$ interior integer points, respectively. Therefore, we have $k = m = n$.   □

19.4.4 Lemma on one integer node

For the proof of Theorem 19.1 we need the following lemma. Let $\text{ld}(B, ACD) = r$ for some $r > 1$. Consider the IDC-system with respect to $ADBA'$ (with integer-distance coordinates (x, y, z)). Denote by B', C, C', and D' the vertices with coordinates $(0, r, r)$, $(r, r, 0)$, (r, r, r), and $(r, 0, r)$, respectively.

Lemma 19.14. *There is a unique integer node in the interior of the intersection of the plane $x + y + z = r + 1$ and the parallelepiped (here we restrict ourselves to the case $r > 1$).*

Proof. Since $\text{ld}(A, A'B'CD) = r$ and the parallelogram $A'B'CD$ is contained in the plane $x + y + z = r$, for every integer n, the plane $x + y + z = n$ is integer. So the plane $x + y + z = r + 1$ is also integer.

Consider a parallelogram obtained from $A'B'CD$ by the shifting on the vector $(0, 0, 1)$ in the IDC-system. This parallelogram is contained in the plane $x + y + z = r + 1$. The edges of the shifted parallelogram are on the planes of faces of the parallelepiped and they do not contain vertices. Hence there exists a unique point in the interior of the shifted parallelogram, denoted by $K(x_0, y_0, z_0)$. Therefore, we have at most one point satisfying the conditions of the lemma.

Let us show that $K(x_0, y_0, z_0)$ is inside the parallelepiped. The shifted parallelogram is defined by the following four inequalities and one equation:

$$0 \le x \le r, \qquad 0 \le y \le r, \quad \text{and} \quad x + y + z = r + 1.$$

Since the point $(0, 0, r + 1)$ is not integer ($r + 1$ is not divisible by r), we have $z_0 \le r$.

Suppose that $z_0 \le 0$. In this case the vector KC is integer; hence the point $K' = A + KC$ is an integer node (see Fig. 19.4). Notice that the point K' is in the tetrahedron $ADBA'$ and it is distinct from the vertices of the tetrahedron. So $ADBA'$ is not empty, contradicting the conditions of the lemma. Therefore, we obtain $z_0 > 0$.

Hence we get $0 \le x_0 \le r, 0 \le y_0 \le r$, and $0 \le z_0 \le r$. Since the edges of the shifted parallelogram do not pass through the vertices of the parallelepiped, the point K is in the interior of the parallelepiped.   □

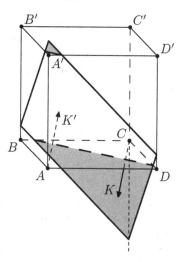

Fig. 19.4 The points K and K' are the integer nodes.

19.4.5 Proof of White's theorem

We prove the statement of the theorem by induction on the number of points inside the parallelepiped (or equivalently on the value of $\mathrm{ld}(A', ABD)$).

Base of induction. Suppose that $\mathrm{ld}(A', ABD) = 2$. Then the unique integer node inside the parallelepiped is the center of symmetry of the parallelepiped. It is in the intersection of all planes passing through opposite edges.

Step of induction. Suppose that the statement of the theorem holds for all empty tetrahedra with integer distance from the vertices to the opposite edges less than r. Let us prove the theorem for an arbitrary empty tetrahedron $A'ABD$ with

$$\mathrm{ld}(A', ABD) = r$$

By Lemma 19.14 there exists a unique integer node in the plane $x + y + z = r + 1$ inside the parallelepiped $P(A'ABD)$, denoted by $A''(x_0, y_0, z_0)$ in the IDC-system with respect to the tetrahedron $ABDA'$. Without loss of generality, we suppose that $z_0 \geq x_0$ and $z_0 \geq y_0$. The tetrahedron $ADBA''$ is an empty integer tetrahedron, since all the nodes inside $ADBA''$ except A'' are inside $ADCA'$ as well. We get $\mathrm{ld}(A'', ADB) < r$, and hence by induction we have that all integer nodes of $P(ADBA'')$ are either in the plane $A''CD$, in the plane $A''BC$, or in the plane $B''BD$ (where $B'' = A'' + AB$). Consider the intersections of these planes with the plane $z = 1$. One of these intersections should contain an integer node inside the parallelepiped $P(ADBA'')$.

Case 1. The integer nodes are in the plane $A''CD$. The plane $A''CD$ is defined by the equation

$$x + \frac{r - x_0}{z_0} z = r.$$

The intersection of the plane $A''CD$ with the plane $z = 1$ is the line $(x_1, t, 1)$, where t is a linear parameter of the line and x_1 is a constant equal to $r - \frac{r - x_0}{z_0}$. This plane contains integer nodes only if x_1 is integer. Let us estimate it (assuming that $y_0 \leq z_0$). On the one hand,

$$x_1 = r - \frac{r - x_0}{z_0} = r - \frac{r - (r + 1 - z_0 - y_0)}{z_0} = r + \frac{1}{z_0} - 1 - \frac{y_0}{z_0}$$
$$\geq r - 2 + \frac{1}{z_0} > r - 2.$$

On the other hand, $x_1 < r$. Since x_1 is an integer and $r - 2 < x_1 < r$, we have $x_1 = r - 1$. Therefore, there exist integer nodes with coordinates $(r - 1, t, 1)$, and hence by Lemma 19.12 one of them (with coordinates $(r - 1, t_0, 1)$, for some integer t_0) is contained in the parallelepiped $P(ADBA')$. So the rest of the integer nodes of the parallelepiped $P(ADBA')$ have the coordinates $(r - k, (k \cdot t_0 \mod r), k)$, and hence they are all in the parallelogram $A'B'CD$.

Case 2. The integer nodes are in the plane $A''BC$. The plane $A''BC$ is defined by the equation

$$y + \frac{r - y_0}{z_0} z = r.$$

The intersection of the plane $A''BC$ with the plane $z = 1$ is the line $(t, y_1, 1)$, where t is a linear parameter of the line and y_1 is a constant equal to $r - \frac{r - y_0}{z_0}$. This plane contains integer nodes only if x_1 is integer. Let us estimate it (assuming that $x_0 \leq z_0$). On the one hand,

$$y_1 = r - \frac{r - y_0}{z_0} = r - \frac{r - (r + 1 - z_0 - x_0)}{z_0} = r + \frac{1}{z_0} - 1 - \frac{x_0}{z_0}$$
$$\geq r - 2 + \frac{1}{z_0} > r - 2.$$

On the other hand, $y_1 < r$. Since y_1 and $r - 2 < y_1 < r$, we have $y_1 = r - 1$. Therefore, there exist integer nodes with coordinates $(t, r - 1, 1)$, and hence by Lemma 19.12 one of them (with coordinates $(t_0, r - 1, 1)$, for some integer t_0) is contained in the parallelepiped $P(ADBA')$. So the rest of the integer nodes of the parallelepiped $P(ADBA')$ have the coordinates $((k \cdot t_0 \mod r), r - k, k)$, and hence they are all in the parallelogram $A'B'CD$.

The integer nodes are not in the plane $B''BD$. Let us show that the case of the plane $B''BD$ is an empty case. This plane is defined by the equation

$$x + y + \frac{r - y_0 - x_0}{z_0} z = r.$$

The intersection of the plane $B''BD$ with the plane $z = 1$ is the line $\left(\frac{a+t}{2}, \frac{a-t}{2}, 1\right)$, where t is a linear parameter for the line and a is a constant equivalent to $r - \frac{r - x_0 - y_0}{z_0}$.

Such a line contains integer nodes only if a is integer. On the one hand.

$$a = r - \frac{r - y_0 - x_0}{z_0} = r - \frac{z_0 - 1}{z_0} > r - 1.$$

On the other hand, we have $a < r$. Therefore, there are no integer nodes in the intersection of the parallelepiped $P(ADBA')$ (and of the parallelepiped $P(ADBA'')$ as well) with the plane $z = 1$, which is impossible.

Therefore, the first two cases enumerate all realizable positions of integer nodes. In both cases the integer nodes are in one of the planes mentioned in the formulation of the theorem. Hence, the statement holds for an arbitrary pyramid $ABCDA'$ with $\mathrm{ld}(A', ABD) = r$. The proof of theorem is completed by induction. □

Remark 19.15. Notice that the third case in the proof is empty, since we assume that the third coordinate of A'' is not less than the first and the second coordinates.

19.4.6 Deduction of Corollary 19.3 from White's theorem

Proof of Corollary 19.3(i). We prove three statements: on completeness of the list, on emptiness of the tetrahedra in the list, and on pairwise noncongruence of the listed tetrahedra.

Completeness of the list. Considering an integer empty tetrahedron $A'ABD$ with a marked vertex A', let $\mathrm{ld}(A', ABD) = r$. If $r = 1$, then we get a tetrahedron integer congruent to $(0,0,0)$, $(1,0,0)$, $(1,0,1)$, and $(1,1,0)$.

Suppose that $r > 1$. Then consider a point K a unit integer distance from the plane ABD lying in the parallelepiped $P(ABDA')$. By White's theorem, K is in one of the three diagonal planes. Without loss of generality, we assume that K is in $A'B'CD$. So K has coordinates $(\xi, r - 1, 1)$ in the IDC-system related to $ABDA'$. Let us rewrite the coordinates of the point in the basis AB, AC, and AK (notice that it is a basis of the integer lattice). We have

$$A = (0,0,0), \quad B = (1,0,0), \quad C = (0,1,0), \quad \text{and} \quad A' = (-\xi, 1 - r, r).$$

Hence $ABDA'$ is integer congruent to $T_{1,r}^{\xi}$.

The parameter ξ is relatively prime to r, since there would otherwise exist integer points on the edges distinct from the vertices. The tetrahedra $T_{1,r}^{\xi}$ and $T_{1,r}^{r-\xi}$ are integer congruent. Therefore, we restrict the consideration of the tetrahedra to the case $0 < \xi \le r/2$.

Emptiness of the tetrahedra. All the tetrahedra are empty, since all integer points in $P(ABDA')$ are contained in one of the diagonal planes that does not intersect the tetrahedron in the interior of the parallelepiped.

Any two marked tetrahedra in the list are not integer congruent. Let us introduce the following integer invariant to distinguish the tetrahedra. Consider an
arbitrary marked tetrahedron $ABDA'$ with vertex A' and the related three-sided angle
with vertex at A' and a section ABD. By Lemma 19.14 this plane contains a unique
integer point (denoted by K) on the plane parallel to ABD at integer distance $r+1$
from A'. Integer distances to the faces of the three-sided angle are 1, ξ, and $r-\xi$ for
some ξ. The unordered collection $[1,\xi,r-\xi]$ is an integer invariant of the marked
tetrahedron $A'ABD$. This invariant (together with $\mathrm{ld}(A',ABD)=r$) distinguishes all
distinct tetrahedra in the list of the corollary. $\square$

Proof of Corollary 19.3(ii). We prove the same three statements as in the proof of
Corollary 19.3(i).

Completeness of the list. Emptiness of the tetrahedra. By Corollary 19.3(i) every
empty tetrahedron is integer congruent to at least one of the marked tetrahedra in
the list of Corollary 19.3(i). We have already shown that all the tetrahedra in this
list are empty.

Integer congruence of tetrahedra in the list. The affine group of symmetries of
tetrahedra S_4 includes the affine group of symmetries of marked tetrahedra S_3. For
each of the four angles of a tetrahedron we construct the integer point uniquely
defined by the conditions of Lemma 19.14. Direct calculation shows that the integer 4-tuples of distances from these points to four planes of the faces of an empty
tetrahedron are as follows

$$(1,1,\xi,r-\xi),\quad(1,1,\xi,r-\xi),\quad(v,r-v,1,1),\quad\text{and}\quad(v,r-v,1,1),$$

where r is the volume of the tetrahedron and the pair (ξ,v) satisfies $\xi \cdot v \equiv (1$
$\mathrm{mod}\ r)$. So this unordered 4-tuple of unordered 4-tuples is an invariant of empty
tetrahedra. The tetrahedra $T_{1,r}^{\xi}$, $T_{1,r}^{v}$, $T_{1,r}^{r-\xi}$, and T_{r}^{r-v} in the list of Corollary 19.3(i)
are integer congruent; all the other pairs of empty tetrahedra in the list have distinct
4-tuples of unordered 4-tuples of distances, and hence they are not congruent. $\square$

19.5 Exercises

Exercise 19.1. Find triangles of width 4 with the smallest possible integer area.

Exercise 19.2. Prove that the width of all empty 3-dimensional tetrahedra is 1.

Exercise 19.3. Find proofs for the lonely runner conjecture for 2 and 3 runners.

Exercise 19.4. Prove the statement of Remark 19.10.

Exercise 19.5. Write the Ehrhart polynomial for all empty tetrahedra.

Exercise 19.6. Prove the equivalence of the lonely runner conjectures in the classical, view-obstacle, and discrete forms.

Chapter 20
Multidimensional Continued Fractions in the Sense of Klein

In the first part of this book we studied continued fractions from the geometric point of view of sails and their integer invariants. In this chapter we present a natural generalization of sails to the multidimensional case by F. Klein.

20.1 Background

In 1839, C. Hermite [84] posed the problem of generalizing ordinary continued fractions to the higher-dimensional case. Since then, there have been many different definitions generalizing different properties of ordinary continued fractions. In this book we focus on the geometric generalization proposed by F. Klein in [119] and [120]. Multidimensional continued fractions in the sense of Klein have many relations with other branches of mathematics. For example, in [215], H. Tsuchihashi described the relationship between periodic multidimensional continued fractions and multidimensional cusp singularities. M.L. Kontsevich and Yu.M. Suhov studied the statistical properties of random multidimensional continued fractions in [124], and later they were studied by the author in [101]. O.N. German [74] and J.-O. Moussafir [155] discussed the connection between the sails of multidimensional continued fractions and Hilbert bases. The relations to approximation theory of maximal commutative subgroups is discussed by A. Vershik and the author in [109]. The combinatorial topological generalization of Lagrange's theorem was obtained by E.I. Korkina in [127] and its algebraic generalization by G. Lachaud [134]. The book [13] of V.I. Arnold is a good survey of geometric problems and theorems associated with one-dimensional and multidimensional continued fractions in the sense of Klein (see also his articles [9], [11], and [12]).

Originally, F. Klein introduced his multidimensional continued fractions in the context of studying decomposable forms with integer coefficients, which we discuss later in more general settings in Chapter 25. Periodicity of multidimensional sails plays an important role in the study of algebraic irrationalities, since the pe-

© Springer-Verlag GmbH Germany, part of Springer Nature 2022
O. N. Karpenkov, *Geometry of Continued Fractions*,
Algorithms and Computation in Mathematics 26,
https://doi.org/10.1007/978-3-662-65277-0_20

riodic structure is defined by the induced action of the group of units in the orders of algebraic fields. Later, we give a multidimensional analogue of Lagrange's theorem, in Chapter 22. Periods of sails are complete invariants of conjugacy classes of $SL(n, \mathbb{Z})$ matrices in multidimensional Gauss's reduction theory (see Chapter 25). In Chapter 23 we present statistical properties of multidimensional sails. Most of the interesting questions related to multidimensional sails appear already in the three-dimensional case, so within this book we pay the most attention to sails in $\mathbb{R}^3$. We observe that some three-dimensional theorems have quite simple generalizations to the case of higher dimensions, while some others become hard open problems even in the four-dimensional case.

We say a few words about the other generalizations of multidimensional continued fractions in Chapter 27.

20.2 Some notation and definitions

20.2.1 A-hulls and their boundaries

We begin with several preliminary definitions.

Consider a set $S \subset \mathbb{R}^n$. We denote by $\overline{S}$ the closure of S. A point $p \in S$ is called *interior* if there exists a ball in $\mathbb{R}^n$ with center at p that is completely contained in S. The union of all interior points of S is called the *interior* of S. The *boundary* of S is the set of points in $\overline{S}$ not belonging to the interior of S. We denote it by ∂S. An element of the boundary of S is called a *boundary point* of S.

A *simplicial cone* in $\mathbb{R}^n$ is the convex hull of k rays with the same vertex and linearly independent directions. (It immediately follows that $k \leq n$.) The *vertex* of a simplicial cone is the common vertex of the rays defining the cone. The rays are called the *edges* of the cone. The convex hull of any subset of rays defining the cone C is called a *face* of C. A simplicial cone is called *integer* if its vertex is integer. A simplicial cone is called *rational* if its vertex is integer and all its edges contain integer points distinct from the vertex.

Definition 20.1. Consider an integer simplicial cone C of dimension d with center at the origin in $\mathbb{R}^n$. An *A-hull* of the cone C is the convex hull of all integer points lying in the closure C except the origin. We denote this set by A-hull(C).

Now we formulate the classical concept of the sail, which is the cornerstone of the theory of multidimensional continued fractions in the sense of Klein.

Definition 20.2. The boundary of the set A-hull(C) is called the *sail* of this cone.

20.2.2 Definition of multidimensional continued fraction in the sense of Klein

Let us give the classical definition of multidimensional continued fractions. Consider a set of $n+1$ hyperplanes of $\mathbb{R}^{n+1}$ passing through the origin in general position (i.e., whose union is homeomorphic of the union of all coordinate hyperplanes). The complement to the union of these hyperplanes consists of $N = 2^{n+1}$ open simplicial cones $C_1, \ldots, C_N$.

Definition 20.3. The set of all sails for all the cones $C_1, \ldots, C_N$ is called the n-*dimensional continued fraction* associated to the given $n+1$ hyperplanes in $\mathbb{R}^{n+1}$ passing through O in general position.

In this chapter we investigate combinatorial and topological properties of sails. First we study the case of finite multidimensional continued fractions. Further we prove a generalized Kronecker's approximation theorem (which is actually interesting by itself) and deduce from it general theorems on the polyhedral structure of sails. Finally, we discuss the situation around two-dimensional faces of sails in more detail.

20.2.3 Face structure of sails

Let us first recall several preliminary general definitions.

Definition 20.4. Let S be a convex set in $\mathbb{R}^n$. A hyperplane π is said to be *supporting* if the following two conditions hold:
(i) the set S is entirely contained in one of the two (multidimensional) half-planes defined by π;
(ii) $\pi \cap \overline{S} \neq \emptyset$.

To define faces of sails we use the following rather abstract definition.

Definition 20.5. Assume that π is a supporting hyperplane for a convex set S and that S is not a subset of π. Then we say that the set $F_\pi = \pi \cap S$ is a *face* of S. The dimension of F_π is the *dimension* of a face. If F_π is a point or a segment (including rays), we say that the face is either a *vertex* or an *edge*.

Remark 20.6. Notice that it is not necessary that the faces of A-hulls coincide with the faces of their closures. One of the illustrations of this is the cone of Example 20.20 below, shown in Fig. 20.4. The shaded area is not a face of A-hull(C), but it is a face of its closure. The face of the A-hull(C) is the dashed ray.

Let us now introduce the notion of faces for sails.

Definition 20.7. Consider a simplicial n-dimensional cone C. A subset F in the sail of C is called a *face of the sail* if F is a face of $\overline{\text{A-hull}(C)}$.

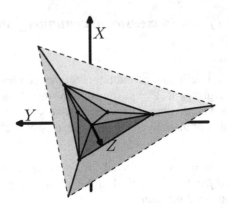

Fig. 20.1 A sail of a finite continued fraction.

In the most interesting cases the set A-hull(C) is closed (see Theorem 20.24). So the following rather tautological proposition is of interest in studying the faces of sails.

Proposition 20.8. *Consider a simplicial n-dimensional cone C in $\mathbb{R}^n$. Every face of the set* A-hull(C) *is a convex hull of a subset of $\mathbb{Z}^n$.*   □

20.3 Finite continued fractions

Consider a rational simplicial cone C in $\mathbb{R}^{n+1}$ with vertex at the origin. The sail of C has finitely many compact faces of all dimensions. In fact, all compact faces are contained in the pyramid $OA_1A_2 \ldots A_{n+1}$, where A_i is the first integer point on the ith edge distinct from O. We call sails of rational cones *finite*. If a multidimensional continued fraction has a finite sail, then all the other sails of this continued fraction are also finite; we call this continued fraction *finite*.

Example 20.9. In Fig. 20.1 we give an example of a sail for a simplicial cone defined by the three vectors $(3,2,3)$, $(1,-4,3)$, and $(-2,1,3)$.

We recall that the classification of two-dimensional cones (i.e., planar angles) and therefore the corresponding geometric one-dimensional sails is provided by LLS sequences for both the finite and infinite cases in Chapter 4. The problem of describing multidimensional (in particular two-dimensional) finite sails is relatively new and currently has no consistent answer. Let us study the two-dimensional sails that have a unique compact face. It is clear that this face is triangular, since the cone has three edges.

Theorem 20.10. *Let the sail for a simplicial cone C in $\mathbb{R}^3$ have a unique compact face. Then this face falls into one of the following two cases.*

(*i*). *If the integer distance from the origin to the triangular compact face is one, then every integer triangle is realizable as this face. Two sails whose unique compact faces are at integer distance one are integer congruent if and only if their triangular faces are integer congruent.*

(*ii*). *Suppose that the face is at an integer distance to the origin greater than* 1. *Then the integer congruence classes of corresponding sails are in one-to-one correspondence with the list "T-W" of completely empty pyramids shown in Fig.* 19.2 *(here each pyramid is associated to a simplicial cone centered at a marked vertex of the pyramid and having edges containing edges of the pyramid). The compact faces of the sails are exactly the bases of such pyramids.*

Proof. If the integer distance from the face to the vertex is one, then the statement is straightforward. If the distance is greater than one, then the statement follows directly from Theorem 19.5 on the classification of integer multistory completely empty marked pyramids.   □

Later we formulate a more general theorem for two-dimensional faces at distance greater than 1 (see Corollary 20.36).

For sails with the number of compact faces greater than 1 almost nothing is known. The following general problem remains open.

Problem 14. Describe all finite two-dimensional sails (and the corresponding continued fractions).

The first tasks here are to investigate sails having two, three, four, etc., compact faces.

20.4 On a generalized Kronecker's approximation theorem

Kronecker's approximation theorem states the following: *if θ is an arbitrary irrational number, then the sequence of numbers $\{k\theta\}$ (for $k > 0$) is dense in the unit interval.* In this subsection we formulate and prove a multidimensional generalization of Kronecker's approximation theorem.

We begin with a preliminary definition of the operation of addition for sets.

20.4.1 Addition of sets in $\mathbb{R}^n$

Let S and T be two subsets of $\mathbb{R}^n$. We define the *sum* of S and T as follows:

$$S \oplus T = \{p + q \mid p \in S, q \in T\}.$$

We show an example of the sum of a triangle and a square in Fig. 20.2.

The sum operation is invariant with respect to affine transformations.

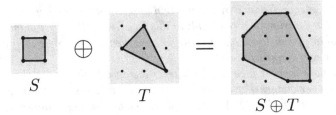

Fig. 20.2 Addition of two convex polygons.

Proposition 20.11. *The sum of two convex sets is convex.*

Proof. The proof is straightforward. Consider two convex sets S and T. Let $u, v \in S \oplus T$. Then $u = s_1 + t_1$ and $v = s_2 + t_2$, where $s_1, s_2 \in S$ and $t_1, t_2 \in T$. Hence for every $\lambda \in [0, 1]$ we have

$$\lambda u + (1 - \lambda)v = \lambda(s_1 + t_1) + (1 - \lambda)(s_2 + t_2) = (\lambda s_1 + (1 - \lambda)s_2) + (\lambda t_1 + (1 - \lambda)t_2).$$

The two summands in the last expression are in S and T respectively. Hence the sum is in $S \oplus T$. Hence for any two points of $S \oplus T$ the segment connecting these points is also in $S \oplus T$. Therefore, the set $S \oplus T$ is convex.   □

20.4.2 Integer approximation spaces and affine irrational vectors

Let us give some preliminary definitions. We call a subspace L *integer* if it contains an integer sublattice of full rank in L.

Definition 20.12. Let $u \in \mathbb{R}^n$. We call the intersection of all integer vector subspaces of $\mathbb{R}^n$ containing u the *integer approximation space of* u. It is denoted by R_u.

Definition 20.13. A vector $u \in \mathbb{R}^n$ is called *affinely irrational* if its endpoint is not contained in any integer affine hyperplane of R_u.

Example 20.14. (*i*) Consider the vector $v = (\sqrt{2}, \sqrt{3})$. The integer approximation space of v is two-dimensional; the vector v is affinely irrational.

(*ii*) The integer approximation space of $(1, \sqrt{2})$ is also two-dimensional (i.e., it is $\mathbb{R}^2$). Nevertheless, this vector is not affinely irrational, since its end is contained in the integer line $x = 1$.

(*iii*) Now let $u = (\sqrt{2}, \sqrt{2})$. Then R_u is the line $y = x$, and u is affinely irrational.

(*iv*) Finally, for the vector $w = (1, 1)$ we have that R_w is the line $y = x$. The integer affine hyperplanes in R_w are just integer points. The vector w is integer itself, so w is not affinely irrational.

Further, we use the following proposition.

Proposition 20.15. *Let $u \in \mathbb{R}^n$. Then for every $\lambda \in \mathbb{R} \setminus S$, where S is a countable set, the vector λu is affinely irrational.*

Proof. The set S is formed by the intersections of the integer affine hyperplanes in R_u with the line $\ell = \{\lambda u \mid u \in \mathbb{R}\}$. No such integer hyperplane π contains the line ℓ, since otherwise, π would contain R_u; hence π intersects ℓ in at most one point. Since there are countably many integer planes, S is countable. $\qquad\square$

20.4.3 Formulation of the theorem

We are interested in the following generalization of the Kronecker's approximation theorem (see [156] and [136]).

Theorem 20.16. (Multidimensional Kronecker's Approximation Theorem.)
Consider an arbitrary vector u in $\mathbb{R}^n$.

(i) Let $L_u = R_u \cap \mathbb{Z}^n$. Then $L_u \oplus \{ku \mid k \in \mathbb{R}\}$ is dense in R_u.

(ii) For every $x \in R_u$ and every $\varepsilon > 0$, the set $B(x, \varepsilon) \oplus \{ku \mid k \in \mathbb{R}\}$ contains infinitely many integer points.

(iii) For every $x \in R_u$ and every $\varepsilon > 0$, the set $B(x, \varepsilon) \oplus \{ku \mid k \in \mathbb{R}_+\}$ contains infinitely many integer points.

(iv) There exists an integer sublattice T of rank $n - \dim R_u$ such that

$$\overline{\mathbb{Z}^n \oplus \{ku \mid k \in \mathbb{Z}\}} = R_u \oplus T.$$

(v) If the vector u is affinely irrational, we have that for every $x \in R_u$ and every $\varepsilon > 0$, the set $B(x, \varepsilon) \oplus \{ku \mid k \in \mathbb{Z}_+\}$ contains infinitely many integer points.

Example 20.17. Consider a vector $u = [\sqrt{2}, \sqrt{3}, \sqrt{2} + \sqrt{3}]$. Then the space R_u is a plane $x_1 + x_2 = x_3$ and the vector u is affinely irrational. Hence all five statements of the theorem hold.

20.4.4 Proof of the Multidimensional Kronecker's approximation theorem

In the proof of Theorem 20.16 we use the following lemma.

Lemma 20.18. *Consider $u \in \mathbb{R}^n$ and let $\dim R_u = d$. Suppose that the integer lattice $L_u = R_u \cap \mathbb{Z}^n$ has an integer basis $(e_1, \ldots, e_d)$.*

(i) For any nonzero λ, we have $R_u = R_{\lambda u}$.

(ii) Consider an affinely irrational vector u. Let

$$w = \lambda_0 u + \sum_{i=1}^{d} \lambda_i e_i,$$

where the numbers $\lambda_0, \lambda_1, \ldots, \lambda_n$ are rational, and additionally $\lambda_0 \neq 0$. Then $R_u = R_w$ and the vector w is affinely irrational.

(iii) Let $u = (a_1, \ldots, a_{d-1}, 1)$, in coordinates related to the basis $e_1, \ldots, e_d$, denote $\tilde{u} = (a_1, \ldots, a_{d-1}, 0)$. Then $R_{\tilde{u}} = \mathrm{Span}(e_1, \ldots, e_{d-1})$ and $\tilde{u}$ is affinely irrational.

Proof. Lemma 20.18(i) holds, since the linear spaces defined by u and λu coincide.

Let us prove Lemma 20.18(ii). Let u be affinely irrational and let w be as in the statement of the lemma. Let us first prove that $R_w = R_u$. A hyperplane of R_u containing w is integer if and only if a hyperplane shifted by $u - w$ containing w is integer. Hence $R_w = R_u$ and w is affinely irrational.

Finally we prove Lemma 20.18(iii). Consider an arbitrary integer subspace R containing $\tilde{u}$. Then the span of R and e_d contains u. Hence R_u is in the span of $R_{\tilde{u}}$ and e_d. Therefore, $R_{\tilde{u}}$ coincides with the span of $e_1, \ldots, e_{d-1}$.

Now we prove that $\tilde{u}$ is affinely irrational by reductio ad absurdum. Suppose that it is contained in some integer hyperplane of $\mathrm{Span}(e_1, \ldots, e_{d-1})$ defined by a linear equation $f = 0$. Then u satisfies two independent linear equations $f = 0$ and $x_d = 0$. Hence u is contained in some integer plane π of codimension 2 in R_u. Hence u is contained in the integer hyperplane $\mathrm{Span}(\pi, O)$ (where O is the origin) of R_u, and hence R_u is not the integer approximation space of u. We arrive at a contradiction. The proof is complete.   □

Proof of Theorem 20.16. Let us prove Theorem 20.16 by induction in dimension of R_u.

Base of induction. The statement for the case $\dim R_u = 1$ follows directly from the classical Kronecker's approximation theorem (recall that here u is an irrational number).

Step of induction. Let all the statements of the theorem hold for all vectors whose integer approximation space is $(d-1)$-dimensional. We first prove statement (v) for an arbitrary affinely integer vector u with $\dim R_u = d$ by reductio ad absurdum. Suppose that for some $x \in R_u$, there exists $\varepsilon > 0$ such that the set

$$B(x, \varepsilon) \oplus \{ku \mid k \in \mathbb{Z}_+\}$$

contains only finitely many integer points. Consider $x_0 = x + Nu$ for a sufficiently large N such that the set

$$B(x_0, \varepsilon) \oplus \{ku \mid k \in \mathbb{Z}_+\}$$

does not contain any integer point. For a pair of positive integers $k_1 > k_2 > 0$, denote by $\Delta(k_1, k_2)$ the vector

$$\{x_0 + k_1 u\} - \{x_0 + k_2 u\},$$

where for a vector $r = (r_1, \ldots, r_d)$, its *quotient part* is denoted by $\{r\}$, i.e.,

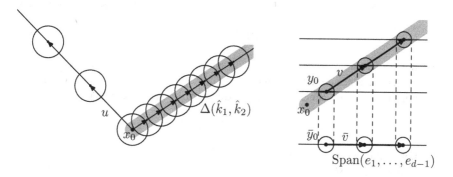

Fig. 20.3 Reduction of the dimension.

$$\{r\} = (r_1 - \lfloor r_1 \rfloor, \ldots, r_d - \lfloor r_d \rfloor).$$

It is clear that $\Delta(k_1, k_2)$ belongs to the unit coordinate cube. Since this cube is compact, there exist integers $\hat{k}_1 > \hat{k}_2 > 0$ such that

$$|\Delta(\hat{k}_1, \hat{k}_2)| < \frac{\sqrt{3}}{2}\varepsilon.$$

Notice that $\Delta(\hat{k}_1, \hat{k}_2) \neq 0$, since $(\hat{k}_1 - \hat{k}_2)u \notin \mathbb{Z}^n$. For every nonnegative m, the ball

$$B(x_0, \varepsilon) \oplus \{m\Delta(\hat{k}_1, \hat{k}_2)\}$$

does not contain integer points, since this ball is obtained from the ball

$$B(x_0, \varepsilon) \oplus \{(\hat{k}_1 - \hat{k}_2)u\}$$

by a shift on the integer vector

$$\lfloor x_0 + \hat{k}_1 u \rfloor - \lfloor x_0 + \hat{k}_2 u \rfloor,$$

where $\lfloor r \rfloor$ is the *vector floor function*, which is equal to $r - \{r\}$. Hence the set

$$B(x_0, \varepsilon) \oplus \{m\Delta(\hat{k}_1, \hat{k}_2) \mid m \in \mathbb{Z}_+\}$$

does not contain integer points. By construction, this set contains an $(\varepsilon/2)$-tubular neighborhood of a ray with vertex at x_0 and direction $\Delta(\hat{k}_1, \hat{k}_2)$, i.e., the set

$$T = B(x_0, \varepsilon/2) \oplus \{t\Delta(\hat{k}_1, \hat{k}_2) \mid t \in \mathbb{R}_+\};$$

see Fig. 20.3 (left). Now we are interested in the points of intersections of the ray $\{x_0 + t\Delta(\hat{k}_1, \hat{k}_2) \mid k \in \mathbb{R}_+\}$ with the integer planes parallel to the span of the vectors $e_1, \ldots, e_{d-1}$. Let the first and second intersections with these planes be y_0 and $y_0 + v$.

Then all the intersections are in the set $\{y_0 + kv \mid k \in \mathbb{Z}_+\}$. By construction, the set

$$B(y_0, \varepsilon/2) \oplus \{kv \mid k \in \mathbb{Z}_+\}$$

does not contain any integer point. Let $\bar{y}_0$ and $\bar{v}$ be the point and the vector whose first $d-1$ coordinates coincide with the coordinates of y_0 and v, respectively, and the last coordinate is zero. Since the last coordinate of y_0 and v are integers, the set

$$B(\bar{y}_0, \varepsilon/2) \oplus \{k\bar{v} \mid k \in \mathbb{Z}_+\}$$

does not contain any integer point.

By Lemma 20.18(ii) it follows that $R_u = R_{\Delta(\hat{k}_1, \hat{k}_2)}$. Further, by Lemma 20.18(i), we get $R_{\Delta(\hat{k}_1, \hat{k}_2)} = R_v$. Since the last coordinate of R_v equals 1, we are in the situation of Lemma 20.18(iii). Hence the vector $\bar{v}$ is affinely irrational and

$$R_{\bar{v}} = \mathrm{Span}(e_1, \ldots, e_{d-1}).$$

Hence y_0 is in $R_{\bar{v}}$. Then by the induction assumption for $(d-1)$-dimensional integer approximation spaces, the set

$$B(\bar{y}_0, \varepsilon/2) \oplus \{k\bar{v} \mid k \in \mathbb{Z}_+\}$$

contains infinitely many integer points. We arrive at a contradiction. Therefore, the statement of Theorem 20.16(v) is true for an arbitrary vector u with $\dim R_u = d$.

Theorem 20.16(iii) is an easy corollary of Theorem 20.16(v): namely, for an arbitrary u take a scalar λ such that λu is affinely irrational (due to Proposition 20.15), and hence Theorem 20.16(iii) for u follows from Theorem 20.16(v) for λu.

Theorem 20.16(i) and (ii) follow directly from Theorem 20.16(iii), and Theorem 20.16(iv) follows from Theorem 20.16(i).

All these prove the correctness of the step of induction and conclude the proof of the multidimensional Kronecker's approximation theorem.   □

20.5 Polyhedral structure of sails

In this section we study topological and combinatorial structures of sails. We first show that all sails are homeomorphic to real spaces. Then we discuss the integer polyhedral structure of sails.

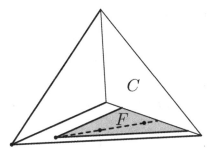

Fig. 20.4 The intersection of a face F with the closure of the A-hull of C.

20.5.1 The intersection of the closures of A-hulls with faces of corresponding cones

We start with the following proposition for arbitrary simplicial cones.

Proposition 20.19. *Consider a simplicial n-dimensional cone C in $\mathbb{R}^n$ with vertex at the origin. Let F be a face of C or else $F = C$. Then we have*

$$\overline{\text{A-hull}(C)} \cap F = \overline{\text{A-hull}(F)} \oplus F.$$

If the dimension of the lattice contained in $\text{Span}\, F$ equals the dimension of F, we have

$$\overline{\text{A-hull}(F)} \oplus F = \overline{\text{A-hull}(F)}.$$

Example 20.20. In Fig. 20.4, we give an example of a three-dimensional simplicial cone C in $\mathbb{R}^3$ with a two-dimensional face F containing a one-dimensional integer sublattice. The integer points of the face F are shown by black dots. The set $\overline{\text{A-hull}(F)}$ is the integer ray containing all the integer points of F. It is shown dashed in the figure. The intersection of the set $\overline{\text{A-hull}(C)}$ with the face F is colored in gray.

Proof. First, we prove that

$$\overline{\text{A-hull}(F)} \oplus F \subset \overline{\text{A-hull}(C)} \cap F.$$

Let $x \in \overline{\text{A-hull}(F)} \oplus F$. This means that $x = p + v$, where p is a point of the closure of the A-hull of F and v is a vector contained in the face F. If $v = 0$, then x coincides with p, which is in $\overline{\text{A-hull}(F)} \subset \overline{\text{A-hull}(C)}$. Suppose now that $v \neq 0$. For every positive integer N we choose an integer point $p_N = (a_1(N), \ldots, a_n(N))$ in $C \setminus F$ such that

$$\left| \frac{v}{|v|} - \frac{p_N - p}{|p_N - p|} \right| < \frac{1}{N}.$$

(It is always possible to choose such point, to be sufficiently far from the origin, which we leave as an exercise for the reader.) Then the segment with endpoints p

and p_N is contained in the A-hull of C. Set

$$\tilde{v}_N = \frac{|v|}{|p_N - p|}(p_N - p).$$

By construction we have

$$p + v_N \in \overline{\text{A-hull}(C)}.$$

Then

$$v = \lim_{N \to \infty} \tilde{v}_N.$$

Hence

$$p + v = p + \lim_{N \to \infty} \tilde{v}_N \in \overline{\text{A-hull}(C)} \cap F.$$

Therefore, we get

$$\overline{\text{A-hull}(F)} \oplus F \subset \overline{\text{A-hull}(C)} \cap F.$$

Second, we prove that

$$\overline{\text{A-hull}(C)} \cap F \subset \overline{\text{A-hull}(F)} \oplus F.$$

Let x be not in $\overline{\text{A-hull}(F)} \oplus F$. Then there exists a supporting plane $P^{d-1} \subset \text{Span}(F)$ of the sum $\overline{\text{A-hull}(F)} \oplus F$ that separates x and $\overline{\text{A-hull}(F)} \oplus F$ (by Proposition 20.11 the sum $\overline{\text{A-hull}(F)} \oplus F$ is convex), where d is the dimension of the plane $\text{Span}(F)$. In addition, this hyperplane can be chosen such that it does not contain any vector of F. Then the intersection $P^{d-1} \cap F$ is a $(d-1)$-dimensional simplex. Denote its vertices by $v_1, \ldots, v_d$. Then, first, the pyramid $Ov_1 \ldots v_d$ intersects $\overline{\text{A-hull}(F)} \oplus F$ only in its base $v_1 \ldots v_d$, and second, the pyramid contains x in the interior. Let us choose the vertices $v_{d+1} \ldots v_n$ on the edges of the cone C that are not adjacent to the face F sufficiently close to the origin such that all the integer points other than the origin (if any) of the pyramid $Ov_1 \ldots v_n$ are contained in the base $v_1 \ldots v_n$. This implies that the set $\overline{\text{A-hull}(C)}$ intersects the pyramid only in its base. Since x is not in the base $v_1 \ldots v_n$, it is not contained in $\overline{\text{A-hull}(C)} \cap F$.

So we have proved the first equality of the proposition.

Suppose now that the dimension of the lattice contained in F equals the dimension of F itself. Let us prove

$$\overline{\text{A-hull}(F)} \oplus F = \overline{\text{A-hull}(F)}.$$

It is sufficient to prove that for an arbitrary point x in the interior of $\overline{\text{A-hull}(F)}$ and vector $v \in F$ the ray with vertex at x and direction v is in $\overline{\text{A-hull}(F)}$. Since the integer lattice is of full rank in F, for any edge e_k of F there exist a sufficiently large constant C_k and a sequence $x_m(k)$ of distinct integer points inside F with distances to this edge not greater than C_k whose norm tends to infinity. Then, starting from some m, the ray with vertex at x and direction v intersects all simplices $(x_m(1), \ldots, x_m(d))$, where $d = \dim F$. Hence this ray is in the convex hull of x and all the elements $x_m(1), \ldots, x_m(d)$ for $m > 0$. Therefore, this ray is a subset of $\overline{\text{A-hull}(F)}$.   □

20.5.2 Homeomorphic types of sails

The homeomorphic types of sails are described by the following theorem.

Theorem 20.21. *Every sail of an n-dimensional simplicial cone in $\mathbb{R}^n$ is homeomorphic to $\mathbb{R}^{n-1}$.*

Let us prove the following lemma.

Lemma 20.22. *Let C be an n-dimensional simplicial cone in $\mathbb{R}^n$ and let x be an interior point of its A-hull. Then the shifted cone $\{x\} \oplus C$ is contained in the interior of A-hull(C).*

Proof. Since x is an interior point of A-hull(C), there exists a ball with center at x contained in A-hull(C). Consider a point $x_1 \neq x$ in this ball such that $x - x_1 \in C$. Then the cone $\{x\} \oplus C$ is in the interior of the cone $\{x_1\} \oplus C$. By Proposition 20.19 in the case of $F = C$ we have

$$\overline{\text{A-hull}(C)} = \overline{\text{A-hull}(C)} \cap C = \overline{\text{A-hull}(C)} \oplus C.$$

Hence,

$$\{x_1\} \oplus C \subset \overline{\text{A-hull}(C)}.$$

Therefore, the cone $\{x\} \oplus C$ is contained in the interior of the A-hull of the cone C.   □

Proof of Theorem 20.21. Let C be a simplicial n-dimensional cone in $\mathbb{R}^n$. Consider an arbitrary interior point x of the A-hull of C and take an $(n-1)$-dimensional sphere S_x with center at x that is completely contained in the A-hull. Denote by C_x the closed cone $C \oplus \{x\}$. Project the sail to the sphere S_x along the radial rays with vertex x. Let us show that the projection is bijective with the set $S_x \setminus C_x$ (see Fig. 20.5).

First, by Lemma 20.22 the cone C_x is in the interior of the A-hull of C, and hence there are no boundary points of A-hull (i.e., of the sail) that project to $S_x \cap C_x$.

Second, we show the surjectivity. Each ray with vertex at x and intersecting $S_x \setminus C_x$ intersects the complement to the cone, which is not in the closure of the A-hull. Hence this ray contains at least one point of the boundary of the A-hull (i.e., of the sail). Therefore, the projection of the sail to the set $S_x \setminus C_x$ is surjective.

Finally, we prove the injectivity. Consider a ray r with center at x and intersecting the set $S_x \setminus C_x$. Let x_r be the last boundary point of the closure of the A-hull, i.e., all the other points of the closure of the A-hull are in between x and x_r (see Fig. 20.5). Denote by P_r the convex hull of the set $\{x_r\} \cup S_x$. The set $\overline{\text{A-hull}(C)}$ is convex and contains the point x_r and the sphere S_x. Therefore,

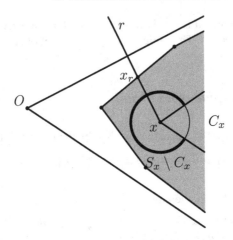

Fig. 20.5 The set $S_x \setminus C_x$ and a point x_r.

$$P_r \subset \overline{\text{A-hull}(C)}.$$

Hence all points of the ray r between x and x_r are interior points of the set A-hull(C), and hence they are not on the sail (which is the boundary of $\overline{\text{A-hull}(C)}$). Therefore,

$$\partial\left(\overline{\text{A-hull}(C)}\right) \cap r = \{x_r\}.$$

Therefore, the projection is injective.

We have shown that there is a one-to-one projection sending the sail to $S_x \setminus C_x$. On the one hand, the projection of the sail is continuous. On the other hand, the inverse to the projection is also continuous (this follows directly from the convexity of the sail; we leave this as an exercise for the reader). The set $S_x \setminus C_x$ is homeomorphic to an open $(d-1)$-dimensional disk (and hence homeomorphic to $\mathbb{R}^{d-1}$), and therefore, the sail is homeomorphic to $\mathbb{R}^{d-1}$. $\qquad\square$

20.5.3 Combinatorial structure of sails for cones in general position

In this book we mostly discuss sails for cones that are either integer (i.e., each edge has an integer point distinct from the origin) or in general position. The combinatoric structure of the sails for integer cones is rather simple and was discussed in Section 20.3 above. Let us examine the situation for sails in general position.

20.5.3.1 Theorem on closeness of sails for cones in general position

First, we give the definition of a cone in general position.

Definition 20.23. A simplicial n-dimensional cone in $\mathbb{R}^n$ is said to be *in general position* if
    — none of its faces contains integer points;
    — each of its edges contains a vector whose integer approximation space coincides with $\mathbb{R}^n$.

It turns out that for cones in general position the set A-hull(C) is closed.

Theorem 20.24. *Let a simplicial n-dimensional cone C be in general position. Then the set* A-hull(C) *is closed.*

We begin the proof of Theorem 20.24 with two lemmas.

Lemma 20.25. *Let a simplicial n-dimensional cone C be in general position. Then*

$$\overline{\text{A-hull}(C)} \cap \partial C = \emptyset.$$

Proof. Consider an arbitrary face F of the cone C. From Proposition 20.19 we have

$$\overline{\text{A-hull}(C)} \cap F = \overline{\text{A-hull}(F)} \oplus F = \emptyset \oplus F = \emptyset.$$

Hence the intersection of $\overline{\text{A-hull}(C)}$ with the union of all faces is empty as well.   □

Lemma 20.26. *Let C be a simplicial n-dimensional cone and let F be a face of the sail of C. Consider an interior point p of C contained in F. Suppose that π is a supporting plane of the set* $\overline{\text{A-hull}(C)}$ *passing through p. Then we have the following:*
(i) If $\pi \cap C$ is not compact, then C is not in general position.
(ii) If $\pi \cap C$ is compact, we have $p \in$ A-hull(F).

Proof. Lemma 20.26(i). We prove the statement by contradiction. Assume that C is in general position. The hyperplane π divides the cone into two connected components, one of which, say C_1, does not contain integer points in the interior (since it is a supporting plane of the of $\overline{\text{A-hull}(C)}$). By construction, the set C_1 is not compact, and hence its intersection with one of the edges of the cone C is a ray. We assume that this edge contains a vector u whose approximation space is $\mathbb{R}^n$. Now consider any interior point x of the set C_1. On the one hand, there exists a positive ε such that the tubular neighborhood of the edge shifted to x:

$$B(x, \varepsilon) \oplus \{ku \mid k \in \mathbb{R}_+\},$$

is completely contained in C_1. Therefore, by construction,

$$B(x, \varepsilon) \oplus \{ku \mid k \in \mathbb{R}_+\} \cap \mathbb{Z}^n = \emptyset.$$

On the other hand, by the multidimensional Kronecker's approximation theorem (Theorem 20.16(iii)), this tubular neighborhood contains infinitely many integer points. We arrive at a contradiction. Hence Lemma 20.26(i) holds.

Lemma 20.26(ii). Let $\pi \cap C$ be compact. Consider a sequence of points $p_i \in$ A-hull(C) that converges to p. By definition, A-hull(C) is the convex hull of a subset of integer points. Hence by Carathéodory's theorem, for every i there exists an integer simplex $\Delta_i \subset$ A-hull(C) containing the point p_i. Since π is a supporting plane, the distance from one of the vertices of Δ_i to the plane π is smaller than $|p - p_i|$. Since $\mathbb{Z}^n$ is discrete and $\pi \cap C$ is compact, there exists N such that for every $i > N$ every simplex Δ_i has an integer vertex in $\pi \cap C$. Since $\pi \cap C$ is compact, there are only finitely many integer points in $\pi \cap C$. Hence we can choose an infinite subsequence $\tilde{p}_i$ in the sequence p_i all of whose simplices $\tilde{\Delta}_i$ have the same integer vertex v_1 in the plane π.

If $v_1 = p$, then we are done. In case $v_1 \neq p$, for every $\tilde{\Delta}_i$ we can choose a vertex $w_i \neq v_1$ such that the distance from w_i to the plane π tends to zero as i tends to infinity. This is due to the compactness of $\pi \cap C$. Therefore, there exists $\tilde{N}$ such that all w_i belong to $\pi \cap C$ for $i > N$. Since the set

$$(\pi \cap C) \cap \mathbb{Z}^n$$

is finite, we can choose an infinite subsequence $\hat{\Delta}_i$ of $\tilde{\Delta}_i$ whose vertices w_j coincide with an integer point of π, say with v_2.

So each simplex $\hat{\Delta}_i$ contains an integer edge $v_1 v_2$ in the plane π. We continue iteratively constructing an infinite subsequence of Δ_i with a common face $v_1 \ldots v_k \subset \pi$. Either at some step k we have $p \in v_1 \ldots v_k$, or we reach $k = n$, and therefore, there exists an infinite subsequence of (Δ_i) all of whose elements contain the simplex $v_1 \ldots v_n$ in π as a subset, and the last vertex is at positive integer distance from π. In the latter case,

$$p \in \overline{v_1 \ldots v_n} = v_1 \ldots v_n.$$

Therefore, the point p is in A-hull(C).

$\square$

Proof of Proposition 20.24. Consider an arbitrary point p of the sail. As we know from Lemma 20.25, p is an interior point of the cone C. Let π be a supporting plane containing p. By Lemma 20.26(i), since C is in general position, the set $\pi \cap C$ is compact. Then by Lemma 20.26(ii) we have $p \in$ A-hull(F). Therefore, the set A-hull(F) contains its boundary. $\square$

20.5.3.2 Structural theorem on cones in general position

Let us give a classical definition of extremal points and extremal rays.

Definition 20.27. A vertex in a convex set is said to be *extremal* if it is not contained in the interior of any segment in the set.

An edge (or a ray) contained in a convex set is said to be *extremal* if every segment of this convex set either does not intersect the ray (edge) or at least one of its endpoints is in the ray.

Remark 20.28. Notice that not every extremal vertex of a convex set S is a vertex of S. For example, consider the union of the half-disk $\{x^2 + y^2 \leq 1 \mid y \geq 0\}$ and the square with vertices $(1,0)$, $(1,-2)$, $(-1,-2)$, and $(-1,0)$. Here the points $(\pm 1, 0)$ are extremal vertices and yet they are not vertices (i.e., 0-dimensional faces).

Now we formulate the main structural theorem on cones in general position (see also [136]).

Theorem 20.29. *Let a simplicial n-dimensional cone C be in general position. Then we have*

(*i*) *all faces of the sail for C are compact integer polyhedra;*

(*ii*) *the set of all vertices of the sail is discrete;*

(*iii*) *the sail does not contain rays;*

(*iv*) *each vertex of the sail is adjacent to only finitely many extremal edges.*

Proof. We begin with Theorem 20.29(*iii*), proving it by contradiction. Let a cone C satisfy the conditions of the theorem and let its sail contain a ray r. Consider any support hyperplane of $\overline{\text{A-hull}(C)}$ containing this ray. This hyperplane divides the cone into two connected components, one of which, say C_1, does not have integer points in the interior (since it is a supporting plane of $\overline{\text{A-hull}(C)}$). By construction, C_1 is not compact, and hence by Lemma 20.26(i) the cone C is not in general position. We have arrived at a contradiction. Hence Theorem 20.29(*iii*) holds.

Let us prove Theorem 20.29(*i*). Suppose, to the contrary a face F is not compact. Hence there exists a sequence of points p_i in F, that increase in norm to infinity. Consider the sequence of points $\overline{p}_i$ where

$$\overline{p}_i = p_1 + \frac{p_1 p_i}{|p_1 p_i|}.$$

All the elements of this sequence are points in the unit sphere centered at p_1. Since the sphere is compact, there exists at least one accumulation point p in the sphere. Hence the ray with vertex at p_1 and direction $p_1 p$ is entirely contained in $\overline{F} = F$. Therefore, by Theorem 20.29(*iii*) the cone C is not in general position. This contradicts the condition of the theorem. Hence the face F is compact.

By Proposition 20.8 the face F is the convex hull of a subset of $\mathbb{Z}^n$. Compactness of F implies the finiteness of $\mathbb{Z}^n \cap F$. Therefore, F is a compact integer polyhedron. This concludes the proof of Theorem 20.29(*i*).

Let us prove Theorem 20.29(*iv*). Suppose the statement is false, and a vertex p is adjacent to an infinite number of extremal edges. Since $\mathbb{Z}^n$ is discrete, the length of such edges is not universally bounded. Hence the closure of the union of all these extremal edges contains a ray. Since all edges are extremal, this ray is in the sail. This contradicts Theorem 20.29(*iii*).

Finally, Theorem 20.29(*ii*) holds, since the integer lattice is discrete.

$\square$

20.5.4 A-hulls and quasipolyhedra

To conclude this section we mention just one result on the quasipolyhedral structure of sails.

Definition 20.30. A convex set in $\mathbb{R}^d$ is called *quasipolyhedral* if its intersection with every polytope is polyhedral.

In some of the literature, quasipolyhedral sets are called *generalized polyhedra*. They were studied for the first time by V.L. Klee in [118] in the framework of separation theory. Let us list several properties describing quasipolyhedral sets.

Theorem 20.31. *A convex set P is quaspolyhedral if and only if the following three conditions are satisfied:*
— *P is closed;*
— *the set of extremal points is locally finite;*
— *for every extremal point p, the set of all edges and extremal rays meeting at p is finite.* $\square$

Let us apply this theorem to sails.

Corollary 20.32. *Consider a simplicial n-dimensional cone C in $\mathbb{R}^n$. The closure of the A-hull of C is a quasipolyhedral set if every only if any face of C containing a nonzero integer point spans an integer affine space (i.e., a space with full-rank integer sublattice in it).* $\square$

For more information we refer to [136].

20.6 Two-dimensional faces of sails

We conclude this chapter with a collection of some results and questions on two-dimensional faces of continued fractions. We distinguish two essentially different situations: questions related to faces that are a unit integer distance from the origin, and questions related to faces that are integer distance from the origin greater than one.

20.6.1 Faces with integer distance to the origin equal one

All two-dimensional faces of the sails of continued fractions are convex integer polygons. If the integer distance to the origin is 1, then the only restriction for a polygon P to be a face of an n-dimensional continued fraction is whether it is possible to inscribe P in some $(n+1)$-gon Q (not necessarily integer!) such that the convex hull of all integer points in Q coincides with P.

Proposition 20.33. *Every n-gon is realizable as a face of an m-dimensional continued fraction if $m \geq n-1$.*

Proof. It is clear that every integer n-gon P can be inscribed in an m-gon Q, for $m \geq n$, such that the convex hull of all integer points in Q coincides with P. We leave the details of the proof to the reader. □

So the main question arises when $m < n-1$.

Problem 15. Which n-gons are not realizable as faces of an m-dimensional continued fraction?

The answer to this question is not understood even in the case $m = 2$.

Let us introduce a family of examples of quadrangles that are not realizable as faces of two-dimensional continued fractions.

Proposition 20.34. *For arbitrary integers $b \geq a \geq 1$, the quadrangle with vertices $(-1,0)$, $(-a-1,1)$, $(-1,2)$, $(b-1,1)$ cannot be a face of a two-dimensional continued fraction.*

Proof. Suppose that the statement is false. Let there exist a two-dimensional continued fraction that has one of the compact faces (say F) integer equivalent to a quadrangle with vertices $(-1,0)$, $(-a-1,1)$, $(-1,2)$, $(b-1,1)$ for some integers $b \geq a \geq 1$. Let π be the plane of the face. Let us introduce integer coordinates on the plane π such that the coordinates of the vertices of the face F are exactly $(a,0)$, $(0,1)$, $(-b,0)$, and $(0,-1)$. Notice that a point in this plane is integer (as a point in $\mathbb{R}^3$) if and only if it is integer in the new coordinates in π.

Every two-dimensional simplicial cone defining a two-dimensional continued fraction has exactly three faces of dimension 2. So the intersection of the edges of the cone with the plane π is a triangle, which we denote by T. Since F is a face of the corresponding continued fraction we have, first, that the face F is contained in the interior of the triangle T and, second, that the set $T \setminus F$ does not contain any integer points. Notice that the points $(1,1)$ and $(1,-1)$ are not in F, and the point $(1,0)$ is in F. In addition, the points $(1,0)$, $(1,1)$, and $(1,-1)$ are contained in a straight line. This implies that the open angle (denoted by φ) with vertex $O(0,0)$ and edges passing through the points $(1,1)$ and $(1,-1)$ respectively, contains at least one vertex of the triangle T (see Fig. 20.6).

For two adjacent angles to φ and for the opposite angle φ, a similar statement holds. Therefore, each of these four nonintersecting angles has a vertex of T, which is impossible, since T is a triangle. We have arrived at a contradiction. □

All the questions discussed in this subsection are related to the two-dimensional case. Similar questions are also interesting for the multidimensional case as well, though it is clear that the complexity rises with the dimension.

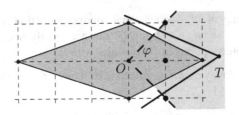

Fig. 20.6 One of the vertices of T is in the shaded (open) angle.

20.6.2 Faces with integer distance to the origin greater than one

We showed in the previous section that every two-dimensional face is realizable as a face of some continued fraction (probably of a large dimension). It turns out that the majority of such faces are a unit integer distance from the origin. There is a complete description of all integer faces that occur at integer distance greater than one.

Theorem 20.35. *Every two-dimensional bounded face an integer distance greater than one from the origin forms a multistory completely empty pyramid.* □

Therefore, the list of such faces is a sublist of all multistory two-dimensional continued fractions (see Theorem 19.5).

Corollary 20.36. (*i*) *The list of faces at integer distance greater than* 1 *for k-dimensional continued fractions for $k > 2$ coincides with the list of bases of multistory marked pyramids.*

(*ii*) *If $k = 2$, the list of all admissible faces coincides with the list of bases of triangular multistory marked pyramids.*

Proof. Existence of all admissible faces follows directly from Theorem 19.5 and Proposition 20.33.

Quadrangular faces are not realizable in the two-dimensional case due to Proposition 20.34. □

We conclude this chapter with the following open question.

Problem 16. Which three-dimensional polytopes are realizable as three-dimensional faces of sails at integer distance greater than 1.

20.7 Exercises

Exercise 20.1. Prove the one-dimensional Kronecker's approximation theorem: if θ is an arbitrary irrational number, then the sequence of numbers $\{k\theta\}$ (for $k > 0$) is dense in the unit interval.

Exercise 20.2. Construct the sail for the cone with vertex at the origin and generated by integer vectors $(1,1,1)$, $(-1,2,4)$, and $(2,1,4)$.

Exercise 20.3. (a) What is the sum of a round disk and a quadrangle?
(b) Suppose the sum of a convex k-gon and a convex l-gon is an m-gon. For which triples (k,l,m) is this possible?
(c) Suppose the sum of a convex k-dimensional polyhedron and an l-dimensional polyhedron is an m-dimensional polyhedron. For which triples (k,l,m) is this possible?

Exercise 20.4. Let C be a simplicial cone with the vertex at the origin and let F be one of its faces. Consider a point p in F and a vector v of F. Prove that for every positive integer N there exists an integer point $p_n = (a_1(N),\ldots,a_n(N))$ in $C \backslash F$ such that

$$\left| \frac{v}{|v|} - \frac{p_N - p}{|p_N - p|} \right| < \frac{1}{N}.$$

Exercise 20.5. Every n-gon is realizable as a face of an m-dimensional continued fraction if $m \geq n-1$.

Exercise 20.6. For an arbitrary n construct an n-gon that is realizable as a face of a two-dimensional continued fraction.

Chapter 21
Dirichlet Groups and Lattice Reduction

Let us study the structure of Dirichlet groups, which is actually the main reason for the periodicity of algebraic multidimensional sails (see Chapter 22). The simplest case of two-dimensional Dirichlet groups was studied in Chapter 12 in the first part of this book. In this chapter we study the general multidimensional case.

We start with Dirichlet's unity theorem on the structure of the group of units in orders in Section 21.1. Further, in Section 21.2 we describe the relation between Dirichlet groups and groups of units. In Sections 21.3 and 21.4 we show how to calculate bases of the Dirichlet groups and positive Dirichlet groups respectively. Finally, in Section 21.5 we briefly discuss the LLL-algorithm on lattice reduction which helps to decrease the computational complexity in many problems related to lattices (including the calculation of Dirichlet group bases).

21.1 Orders, units, and Dirichlet's Unit Theorem

In this subsection we give some preliminary material on the field theory related to the structure of the units in orders.

Definition 21.1. Let K be a field of algebraic numbers and $\mu_1, \ldots, \mu_m$ an arbitrary finite sequence of numbers in K. The set M of all linear combinations $c_1\mu_1 + \cdots + c_m\mu_m$ with integer coefficients $c_1, \ldots, c_m$ is called a *module* of the field K. The numbers $\mu_1, \ldots, \mu_m$ are called the *generators of the module M*.

Every module M in an algebraic field contains a basis, i.e., a sequence $\mu_1, \ldots, \mu_m$ that generates M as linear combinations with integer coefficients such that only the trivial combination represents zero. It is clear that the number of elements in the basis coincides with the number of linearly independent elements (over $\mathbb{Q}$) in M.

Definition 21.2. Let K be a field of algebraic numbers of degree n and let M be a module in K. If M contains n linearly independent (over $\mathbb{Q}$) numbers, then it is said to be *complete*, otherwise it is called *incomplete*.

© Springer-Verlag GmbH Germany, part of Springer Nature 2022
O. N. Karpenkov, *Geometry of Continued Fractions*,
Algorithms and Computation in Mathematics 26,
https://doi.org/10.1007/978-3-662-65277-0_21

Definition 21.3. A complete module in a field of algebraic numbers K is an *order* in K if it is a ring containing 1.

Example 21.4. Let K be an algebraic number field. The set of all numbers in K whose minimal polynomial has integer coefficients is an order in K. This order is often called *maximal*, since every other order of K is contained in this order.

Definition 21.5. Let $D \subset K$ be an arbitrary order. A number ε in D is a *unit* if ε^{-1} is contained in D.

Theorem 21.6. (Dirichlet's unit theorem.) *Let K be a field of algebraic numbers of degree $n = s + 2t$. Consider an arbitrary order D in K. Then D contains units $\varepsilon_1, \dots, \varepsilon_r$ for $r = s + t - 1$ such that every unit ε in D has a unique decomposition of the form*

$$\varepsilon = \xi \varepsilon_1^{a_1} \cdots \varepsilon_r^{a_r},$$

where $a_1, \dots, a_r$ are integers and ξ is a root of 1 contained in D. $\square$

We refer to [26] for a proof of this theorem.

21.2 Dirichlet groups and groups of units in orders

Let us describe the relation between Dirichlet groups and groups of units in orders in algebraic number fields.

21.2.1 Notion of a Dirichlet group

Consider an arbitrary integer matrix A with characteristic polynomial irreducible over $\mathbb{Q}$.

Definition 21.7. Denote by $\Gamma(A)$ the set of all integer matrices commuting with A.

(*i*) The *Dirichlet group* $\Xi(A)$ is the subset of invertible matrices in $\Gamma(A)$.

(*ii*) The *positive Dirichlet group* $\Xi_+(A)$ is the subset of $\Xi(A)$ that consists of all matrices with positive real eigenvalues.

The matrices of $\Gamma(A)$ form a ring with standard matrix addition and multiplication. (As a group, $\Gamma(A)$ is isomorphic to $\mathbb{Z}^{n+1}$.)

21.2.2 On isomorphisms of Dirichlet groups and certain groups of units

Consider an arbitrary matrix A with characteristic polynomial χ_A irreducible over $\mathbb{Q}$. Denote by $\mathbb{Q}[A]$ the set of all matrices that are polynomials in A with rational coefficients. It is clear that $\mathbb{Q}[A]$ is a ring contained in $\mathrm{Span}(\mathrm{Id}, A, \ldots, A^{n-1})$. This ring is naturally isomorphic to quotient ring $\mathbb{Q}[x]/\chi_A(x)$.

Consider an eigenvalue ξ of A. It is clear that the field $\mathbb{Q}[\xi]$ is also isomorphic to $\mathbb{Q}[x]/\chi_A(x)$. Hence there exists a natural isomorphism

$$h_{A,\xi} : \mathbb{Q}[A] \to \mathbb{Q}[\xi],$$

sending $p(A)$ to $p(\xi)$ for every element p of the quotient ring $\mathbb{Q}[x]/\chi_A(x)$. Since $\Xi(A)$ is contained in $\mathbb{Q}[A]$, we have an embedding:

$$h_{A,\xi} : \Xi(A) \to \mathbb{Q}[\xi].$$

In the proofs of Proposition 21.8 and Theorem 21.9 we use the following notation:

$$P_A = \{p \in \mathbb{Q}[x]/\chi_A(x) \mid p(A) \in \mathrm{Mat}(n, \mathbb{Z})\}.$$

From the definitions it is clear that

$$h_{A,\xi}\Gamma(A) = \{p(\xi) \mid p \in P_A\}.$$

Proposition 21.8. *Consider an arbitrary integer matrix A with characteristic polynomial irreducible over $\mathbb{Q}$. Let ξ be one of eigenvalues of A. Then the set $h_{A,\xi}(\Gamma(A))$ is an order in $\mathbb{Q}(\xi)$.*

Proof. First, the set P_A described above is closed under addition and multiplication, since if $p_1(A)$ and $p_2(A)$ are integer matrices, then $(p_1 + p_2)(A)$ and $(p_1 p_2)(A)$ are also integer matrices. Hence $h_{A,\xi}(\Gamma(A))$ is closed under addition and multiplication. Since $\mathbb{Q}[A]$ is a ring (even a field), the addition and multiplication in $h_{A,\xi}(\Gamma(A))$ satisfy all ring axioms. Hence $h_{A,\xi}(\Gamma(A))$ is a ring.

Second, as an additive group, $h_{A,\xi}(\Gamma(A))$ is a subgroup of $\mathrm{Mat}(n, \mathbb{Z})$ isomorphic to the free abelian group $\mathbb{Z}^{n \times n}$, which implies that it is a free abelian group with at most n^2 generators, and hence it is a finite module.

Third, this module contains the identity matrix Id.

And finally, the elements $\mathrm{Id}, \xi, \xi^2, \ldots, \xi^{n-1}$ are linearly independent over $\mathbb{Q}$, and hence the set $h_{A,\xi}(\Gamma(A))$ is a complete module. Therefore, by definition, $h_{A,\xi}(\Gamma(A))$ is an order. $\qquad\square$

Theorem 21.9. *Consider an arbitrary integer matrix A with characteristic polynomial irreducible over $\mathbb{Q}$. Let ξ be one of the eigenvalues of A. Then the Dirichlet group $\Xi(A)$ is isomorphic to the multiplicative group of units in the order $h_{A,\xi}(\Gamma(A))$ of the field $\mathbb{Q}(\xi)$. The isomorphism is realized by the map $h_{A,\xi}$ as above.*

Proof. Recall that $h_{A,\xi}$ is a one-to-one map between $\Xi(A)$ and $h_{A,\xi}(\Xi(A))$.

Letting $p(\xi) \in h_{A,\xi}(\Gamma(A)))$ be a unit, i.e., it is invertible, then there exists an element $q \in P(A)$ such that $p(\xi) \cdot q(\xi) = 1$. Hence $q = p^{-1}$ in the quotient ring $\mathbb{Q}[x]/\chi_A(x)$, and

$$p(A) \cdot q(A) = \mathrm{Id}.$$

Therefore, the matrix $p(A)$ is invertible in $\text{Mat}(n,\mathbb{Z})$, and hence it is in $\Xi(A)$.

Suppose now that $p(A) \in \Xi(A)$ has an inverse in $\Xi(A)$ that is $q(A)$. Let us restrict these matrices to the eigenspace corresponding to ξ. We have directly $p(\xi) \cdot q(\xi) = 1$. Thus, $p(\xi)$ is a unit.

Hence, the Dirichlet group $\Xi(A)$ is isomorphic to the multiplicative group of units in the order $h_{A,\xi}(\Gamma(A))$, and the isomorphism is given by $h_{A,\xi}$.   □

From Dirichlet's unit theorem (Theorem 21.6) and Theorem 21.9 we have the following description of the Dirichlet group $\Xi(A)$.

Corollary 21.10. *Consider an arbitrary integer matrix A with characteristic polynomial irreducible over $\mathbb{Q}$. Suppose it has s real and $2t$ complex conjugate eigenvalues. Then there exists a finite abelian group G such that*

$$\Xi(A) = G \oplus \mathbb{Z}^{s+t-1}.$$

□

21.2.3 Dirichlet groups related to orders that do not have complex roots of unity

A polynomial is called *cyclotomic* if its roots are all distinct primitive nth roots of unity for some integer n.

A cyclotomic polynomial is an irreducible polynomial. It is usually denoted by Φ_n. These polynomials form only a very small subset of all polynomials. For instance, for all $n > 2$, the degrees of Φ_n are even, hence if we study matrices with irreducible characteristic polynomials in odd dimensions, then we do not have any orders containing roots of unity other than ± 1. In fact, the degree equals the *Euler totient* $\phi(n)$ (the number of positive integers less than or equal to n that are coprime to n). The sequence of Euler totients $(\phi(n))$ converges to infinity as n tends to infinity. Hence for each even d there are only finitely many cyclotomic polynomials of any fixed degree k.

If the Dirichlet group $\Xi(A)$ related to orders does not have complex roots of unity, then we directly have

$$\Xi(A) = (\mathbb{Z}/2\mathbb{Z})^k \oplus \mathbb{Z}^{s+t-1}$$

for some integer k. Since the matrix $-\text{Id}$ is in $\Xi(A)$, we have $k \geq 1$.

In these cases, for the positive Dirichlet group, we have $\Xi_+(A) = \mathbb{Z}^{s+t-1}$, since all symmetries about the plane have negative eigenvalues and every subgroup of a free abelian group is free abelian.

In the special case that all the eigenvalues of A are real we have the following statement (we call such a matrix a *real spectrum matrix*).

Proposition 21.11. *Consider an arbitrary integer real spectrum matrix* $A \in \mathrm{SL}(n,\mathbb{Z})$ *with characteristic polynomial irreducible over* $\mathbb{Q}$. *Then* $\Xi_+(A) = \mathbb{Z}^{n-1}$.

Proof. Every generator of $\Xi_+(A)$ is a matrix with positive eigenvalues. Therefore, the operator is not cyclic. Hence $\Xi_+(A)$ is a free abelian group. Since for every $B \in \Xi(A)$ we have $B^2 \in \Xi_+(A)$, the rank of $\Xi_+(A)$ is $n-1$. $\square$

21.3 Calculation of a basis of the additive group $\Gamma(A)$

In this subsection we show how to calculate the basis of an additive group $\Gamma(A)$ for an arbitrary integer matrix A with irreducible characteristic polynomial. Let us identify the space $\mathrm{Mat}(n,\mathbb{R})$ with the space $\mathbb{R}^{(n)^2}$ and consider the standard metrics for this space (the distance between two matrices is the Euclidean distance between the corresponding points in $\mathbb{R}^{n^2}$). Then every integer matrix corresponds to some integer point of $\mathbb{R}^{n^2}$. In Proposition 21.12 below we show that the group $\Gamma(A)$ is isomorphic to $\mathbb{Z}^{n+1}$. Further, we construct a basis of a maximal-rank subgroup in $\Gamma(A)$. Finally, we choose a basis in the parallelepiped generated by this basis. So the algorithm can be performed as follows.

Algorithm to calculate a basis of $\Gamma(A)$.

Input data. In the input we have an arbitrary integer matrix A with irreducible characteristic polynomial.

Goal of the algorithm. To calculate a basis of the group $\Gamma(A)$.

Step 1. Since matrices $\mathrm{Id}, A, A^2, \ldots, A^n$ are linearly independent over $\mathbb{Q}$, they form a basis of some sublattice of $\Gamma(A)$.

Step 2. To reduce all the calculations we apply the LLL-algorithm to the basis of Step 1 to get a reduced basis with much smaller coefficients.

Step 3. Calculate a basis of $\Gamma(A)$ starting with the basis of a sublattice of $\Gamma(A)$ constructed in Step 2.

Output. The basis of $\Gamma(A)$.

21.3.1 Step 1: preliminary statements

Proposition 21.12. *For every integer irreducible real spectrum matrix* A *the set* $\Gamma(A)$ *forms an additive group isomorphic to* $\mathbb{Z}^{n+1}$.

Proof. In the diagonal basis, the group of all matrices commuting with A is isomorphic to $\mathbb{R}^{n+1}$ by addition. Hence the set of all integer matrices forms an integer lattice in this $(n+1)$-dimensional subspace. So the group is isomorphic to $\mathbb{Z}^k$ with $k \leq n+1$.

The matrices $\mathrm{Id}, A, A^2, \ldots, A^n$ are linearly independent over $\mathbb{Q}$ since the characteristic polynomial of A is irreducible over $\mathbb{Q}$. Hence, $k = n+1$. □

Corollary 21.13. *The group* $\Gamma(A)$ *is the intersection of the integer lattice* $\mathbb{Z}^{n \times n} \subset \mathrm{Mat}(n, \mathbb{R})$ *with the space* $\mathrm{Span}(\mathrm{Id}, A, A^2, \ldots, A^n)$. □

21.3.2 Step 2: application of the LLL-algorithm

Applying the LLL-algorithm described below to the sublattice generated by the matrices $\mathrm{Id}, A, A^2, \ldots, A^n$, one constructs a reduced basis. This will decrease the Euclidean lengths of basis vectors generating the sublattice (see Section 21.5 for the full-rank lattice algorithm and Remark 21.18 for the non-full rank lattice algorithm). In general it is not necessary to use the LLL-algorithm here, so that one can proceed directly to Step 3.

21.3.3 Step 3: calculation of an integer basis having a basis of an integer sublattice

In Step 2 we end up with a basis $M_1, \ldots, M_{n+1}$ of a full-rank sublattice in the space $\mathrm{Span}(\mathrm{Id}, A, A^2, \ldots, A^n)$. Let us show how to calculate effectively a basis of the integer lattice in the span.

We do this inductively.

Base of induction. We take $\overline{M} = \frac{1}{\lambda}M_1$ as the first vector of the basis, where λ is the greatest common divisor of all the element in M_1.

Step of induction. Suppose we have calculated the integer basis $\overline{M}_1, \ldots, \overline{M}_{k-1}$ in $\mathrm{Span}(M_1, \ldots, M_{k-1})$. Let us extend it with $\overline{M}_k$ to a basis of an integer lattice in $\mathrm{Span}(M_1, \ldots, M_k)$. As $\overline{M}_k$ we choose the only integer vector of the following form:

$$\overline{M}_k = \frac{1}{d_k}M_k + \sum_{i=1}^{k-1} \frac{\lambda_i}{d_i}\overline{M}_i,$$

where d_i is the integer distance from the ith point of the set $\overline{M}_1, \ldots, \overline{M}_{k-1}, M_k$ to the span of the remaining points, and for integers λ_i we have $0 \leq \lambda_i < d_i$. Since

$$\mathrm{ld}(\overline{M}_k, \mathrm{Span}(\overline{M}_1, \ldots, \overline{M}_{k-1})) = 1,$$

the matrices $\overline{M}_1, \ldots, \overline{M}_k$ form a basis of the lattice in $\mathrm{Span}(M_1, \ldots, M_k)$.

The distances d_i are calculated by the formula of Theorem 18.9, and the integer volumes in coordinates are the greatest common divisors of the corresponding modified Plücker coordinates (see Theorem 18.30). The numbers λ_i are found by exhaustive search.

So in $n+1$ steps we obtain a basis of $\Gamma(A)$ that is the integer lattice in the space $\mathrm{Span}(\mathrm{Id}, A, A^2, \ldots, A^n)$ according to Corollary 21.13.

21.4 Calculation of a basis of the positive Dirichlet group $\Xi_+(A)$

From the algorithmic point of view this step is the most complicated. We describe only the idea for one of the simplest algorithms here and give the corresponding references.

Let $\chi(x)$ be the characteristic polynomial of the matrix A and let ξ be one of the roots of $\chi(x)$. Consider the map

$$h_{A,\xi} : \Xi(A) \to \mathbb{Q}[\xi]$$

as in Section 21.2.2. By Theorem 21.9 this map is an isomorphism between the ring $\Xi(A)$ and image $h_{A,\xi}(\Xi(A))$, which is the multiplicative group of units. In the book [26] the authors show how to construct a basis for the units of an order and a number ρ such that the norms of all its elements are bounded from above by ρ (here the norm is the standard norm on the order considered as $\mathbb{R}^{s+t}$). The method of constructing the constant ρ is standard and quite technical, so we omit it. Now according to [26] we construct a basis by enumeration of all vectors of the set $h(\Xi(A))$ inside the ball $B_\rho(O)$, where $B_\rho(O)$ is a ρ-neighborhood of the origin. The preimage (i.e., h^{-1}) of this basis gives us a basis of the group of invertible elements in the ring $\Xi(A)$, and hence it gives a basis of the subgroup $\Xi_+(A)$.

Remark 21.14. The constant ρ is extremely large (it equals the exponent of some polynomial of the coefficients of the matrix A). An effective algorithm for this step can be found in the book [41] by H. Cohen. Using this algorithm one finds the basis of units in polynomial time (with respect to the coefficients of the matrix A).

21.5 Lattice reduction and the LLL-algorithm

Whenever we are facing computational aspects in terms of the coefficients in some lattice basis, it is useful first to choose the basis in a way that it has vectors that (in some standard norm) as small as possible and then to perform the calculation. In 1982, A.K. Lenstra, H.W. Lenstra, and L. Lovász in [141] proposed a notion of a reduced basis and an algorithm to calculate it. Later the algorithm was named the LLL-algorithm in honor of its inventors. We present the original LLL-algorithm below and refer to [27] and [41] for further information. In this subsection we denote the Euclidean length of a vector v by $|v|$.

21.5.1 Reduced bases

Let $b_1, \ldots, b_n$ be a basis for L. Define inductively

$$b_i^* = b_i - \sum_{j=1}^{i-1} \mu_{ij} b_j^*, \quad \text{where} \quad \mu_{ij} = \frac{\langle b_i, b_j^* \rangle}{\langle b_j^*, b_j^* \rangle}$$

(by $\langle v, w \rangle$ we denote the inner product of two vectors). Actually, the vectors $b_1^*, \ldots, b_n^*$ are the resulting vectors in the Gram—Schmidt orthogonalization process (and $\mu_{i,j}$ are the coefficients that arise in the orthogonalization process).

Definition 21.15. A basis $b_1, b_2, \ldots, b_n$ for a lattice L is said to be *reduced* if the following two conditions hold:
  (*i*) Size reduced conditions: for $1 \leq j \leq i \leq n$, $|\mu_{i,j}| \leq \frac{1}{2}$.
  (*ii*) Lovász conditions: for $1 < i \leq n$, $|b_i^* + \mu_{i,i-1} b_{i-1}^*|^2 \geq \frac{3}{4} b_{k-1}^*$.

Letting L be a lattice generated by $b_1, \ldots, b_n$, denote by $d(L)$ the *determinant* of the lattice:

$$d(L) = |\det(b_1, \ldots, b_n)|.$$

Let us estimate the lengths of the vectors in a reduced basis via the orthogonal basis b_i^* and the determinant of L.

Theorem 21.16. (A.K. Lenstra, H.W. Lenstra, and L. Lovász [141].)
Let $b_1, b_2, \ldots, b_n$ be a reduced basis for a lattice L in $\mathbb{R}^n$. Then we have
(*i*) $|b_j|^2 \leq 2^{i-1} |b_i^*|^2$ *for* $1 \leq j \leq i \leq n$,
(*ii*) $d(L) \leq \prod_{i=1}^{n} |b_i| \leq 2^{n(n-1)/4} d(L)$,
(*iii*) $|b_1| \leq 2^{n(n-1)/4} d(L)^{1/n}$.

Proof. From the size reduced conditions and the Lovász conditions, we have

$$|b_i^*|^2 \geq \left(\frac{3}{4} - \mu_{i,i-1}^2 \right) |b_{i-1}^*| \geq \frac{1}{2} |b_{i-1}^*|^2$$

for all admissible i, and therefore,

$$|b_j^*|^2 \leq 2^{i-j} |b_i^*|^2, \quad \text{for} \quad 1 \leq j \leq i \leq n.$$

From the size reduced conditions, it follows that

$$|b_i|^2 = |b_i^*|^2 + \sum_{j=1}^{i-1} \mu_{i,j}^2 |b_j^*|^2 \leq |b_i^*|^2 + \sum_{j=1}^{i-1} 2^{i-j-2} |b_i^*|^2 = (2^{i-2} + 1/2) \leq 2^{i-1} |b_i^*|^2.$$

This implies the first item of Theorem 21.16:

$$|b_j|^2 \leq 2^{j-1}|b_j^*|^2 \leq 2^{i-1}|b_i^*|^2.$$

By definition of the basis $b_1^*, b_2^*, \ldots, b_n^*$, it follows that

$$d(L) = |\det(b_1^*, b_2^*, \ldots, b_n^*)| = \prod_{i=1}^{n} |b_i^*|,$$

the last equality holding since the vectors b_i^* are mutually orthogonal. Now the inequalities of the second and third items of Theorem 21.16 follow directly from the inequalities of the first item and the fact that $|b_i^*| < |b_i|$. □

21.5.2 The LLL-algorithm

One can think of this algorithm as a type of generalization of Euclid's algorithm.

We use the following notation. Letting r be the nearest integer to $|\mu_{k,l}|$, denote by $[b_k]_l$ the vector $b_k - rb_l$.

LLL-algorithm.

Input data. As input we have a basis of the lattice b_i. The additional orthogonalization data $(b_i^*, \mu_{i,j})$ is defined by b_i as above. The algorithm is iterative. At each iteration we replace the basis b_i by a new one. Whenever we change the basis b_i we should update the orthogonalization data $(b_i^*, \mu_{i,j})$.

Goal of the algorithm. To calculate a reduced basis of the lattice generated by the basis b_i.

Parameters and conditions of the iteration process. An iteration has one parameter $k \in \{1, 2, \ldots, n+1\}$. For the first iteration we put $k = 2$. The algorithm terminates when $k = n + 1$. After each iteration we have the value of k and the following conditions satisfied:

(i_k). For $1 \leq j \leq i < k$, $|\mu_{i,j}| \leq \frac{1}{2}$.
(ii_k). For $1 < i < k$, $|b_i^* + \mu_{i,i-1}b_{i-1}^*|^2 \geq \frac{3}{4}b_{k-1}^*$.

Description of the iteration. We have the basis b_i and the parameter $k \in \{1, 2, \ldots, n\}$ (if $k = n+1$, the algorithm terminates). First, if $k > 1$, we replace b_k by $[b_k]_{k-1}$ and respectively update $(b_i^*, \mu_{i,j})$. Now $|\mu_{k,k-1}| \leq \frac{1}{2}$. Second, we choose one of two cases to continue.

Case 1. We choose this case if

$$|b_i^* + \mu_{i,i-1}b_{i-1}^*|^2 < \frac{3}{4}b_{k-1}^* \qquad \text{and} \qquad k \geq 2.$$

We interchange b_{k-1} and b_k and update $(b_i^*, \mu_{i,j})$. Finally, we reduce k by one ($k :=$ $k-1$) and exit the iteration.

Case 2. We choose this case if

$$|b_i^* + \mu_{i,i-1} b_{i-1}^*|^2 \geq \frac{3}{4} b_{k-1}^* \qquad \text{or} \qquad k = 1.$$

Then we consequently replace b_k by $[b_k]_l$ and respectively update $(b_i^*, \mu_{i,j})$ for $l =$ $k-1, k-2, \ldots, 1$. Finally, we increase k by one ($k := k+1$) and exit the iteration.

Output. The algorithm terminates once $k = n+1$. As output, we have a reduced basis of the lattice.

It is important to notice that the algorithm terminates in finitely many iterations. The running time is bounded as follows.

Theorem 21.17. (A.K. Lenstra, H.W. Lenstra, and L. Lovász [141].) *Let $L \subset$ $\mathbb{Z}^n$ be a lattice with basis $b_1, b_2, \ldots, b_n$, and let $B \in \mathbb{R}, B \geq 2$, be such that $|b_i|^2 \leq$ B for $1 \leq i \leq n$. Then the number of arithmetic operations needed by the basis reduction in the described LLL-algorithm is $O(n^4 \log B)$, and the integers on which these operations are performed each have binary length $O(n \log B)$.* □

We are not going to give a proof here, since it is quite technical. The interested reader is referred to the original manuscript [141].

Remark 21.18. The algorithm works also for the case of lattices of rank $l < n$. In this case it terminates when k reaches $l+1$.

21.6 Exercises

Exercise 21.1. Prove that $\Gamma(A)$, considered as a group with matrix addition, is isomorphic to $\mathbb{Z}^{n+1}$.

Exercise 21.2. Find all isomorphism types of Dirichlet groups for matrices in $SL(n, \mathbb{Z})$ whose characteristic polynomials are irreducible over $\mathbb{Q}$, where
   (a) $n = 2$;   (b) $n = 3$;   (c) $n = 4$.

Exercise 21.3. Prove that cyclotomic polynomials are irreducible over $\mathbb{Q}$ and of even degree (except for $(x \pm 1)$).

Exercise 21.4. Find the generators and relation of the Dirichlet group $\Xi(A)$, where

$$\text{(a) } A = \begin{pmatrix} 0 & 1 & 0 \\ 0 & 0 & 1 \\ 1 & 2 & -4 \end{pmatrix}; \qquad \text{(b) } A = \begin{pmatrix} 0 & 1 & 0 \\ 0 & 0 & 1 \\ 1 & -4 & 0 \end{pmatrix}.$$

Chapter 22
Periodicity of Klein polyhedra. Generalization of Lagrange's Theorem

The sails of algebraic multidimensional continued fractions possess combinatorial periodicity due to the action of the positive Dirichlet group on the sails. In case of one-dimensional geometric continued fractions this periodicity is completely described by the periodicity of the corresponding LLS sequences (see Chapter 11). Around twenty years ago, V.I. Arnold posed a series of questions to study periodicity of multidimensional continued fractions in the sense of Klein. At this moment some of his questions have answers, while others remain open. In this chapter we discuss current progress in this direction. The questions related to periodicity of algebraic sails are important in algebraic number theory, since they are in correspondence with algebraic irrationalities. In particular, periods of algebraic sails characterize the groups of units in the corresponding orders.

First, we associate to any matrix with real distinct eigenvalues a multidimensional continued fraction. Second, we discuss periodicity of associated sails in the algebraic case. Further, we give examples of periods of two-dimensional algebraic continued fractions and formulate several questions arising in that context. Then we state and prove a version of Lagrange's theorem for multidimensional continued fractions. Finally, we say a few words about the relation of the Littlewood and Oppenheim conjectures to periodic sails.

22.1 Continued fractions associated to matrices

We begin with the following constructive definition.

Definition 22.1. Consider an $(n+1) \times (n+1)$ matrix A whose eigenvalues are all distinct and real. Take the n-dimensional hyperplanes passing through the origin that are spanned by n linearly independent eigenvectors of A. There are exactly $n+1$ such hyperplanes. These hyperplanes define the n-dimensional continued fraction called the *multidimensional continued fraction associated to A*.

© Springer-Verlag GmbH Germany, part of Springer Nature 2022
O. N. Karpenkov, *Geometry of Continued Fractions*,
Algorithms and Computation in Mathematics 26,
https://doi.org/10.1007/978-3-662-65277-0_22

Let us formulate an algebraic criterion of integer congruence of associated multidimensional continued fractions.

Proposition 22.2. *Let A and B be matrices of* $GL(n+1, \mathbb{R})$ *with distinct real eigenvalues. The continued fractions associated to A and B are integer congruent if and only if there exists a matrix* $X \in GL(n+1, \mathbb{Z})$ *such that* XAX^{-1} *commutes with B.*

Proof. Let A and B be matrices of $GL(n+1, \mathbb{R})$ with distinct real irrational eigenvalues and suppose that their continued fractions are integer congruent. Since the continued fractions are integer congruent, there exists a linear integer lattice-preserving transformation of the space that maps the continued fraction of A to the continued fraction of B. Under such a transformation the matrix A is conjugated by some integer matrix X with unit determinant. All eigenvalues of the matrix XAX^{-1} are distinct and real (since conjugations preserve the characteristic polynomial of the matrix). Since the invariant cones of the first continued fraction map to the invariant cones of the second one, the sets of eigendirections for the matrices B and XAX^{-1} coincide. Hence these matrices are simultaneously diagonalizable in some basis, and therefore, they commute.

Let us prove the converse. Suppose that there exists $X \in GL(n+1, \mathbb{Z})$ such that

$$\tilde{A} = XAX^{-1}$$

commutes with B. Note that the eigenvalues of the matrices A and $\tilde{A}$ coincide. Therefore, all eigenvalues of the matrix $\tilde{A}$ (just as for the matrix B) are real and distinct. Let us consider a basis in which the matrix $\tilde{A}$ is diagonal. Simple verification shows that the matrix B is also diagonal in this basis. Hence the matrices $\tilde{A}$ and B define the same orthant decomposition of $\mathbb{R}^{n+1}$, and thus the same continued fraction as well. $\square$

22.2 Algebraic periodic multidimensional continued fractions

Now we formulate the notion of periodicity of continued fractions associated to algebraic irrationalities. Recall the following general definition.

Definition 22.3. An integer matrix A is called *irreducible* if its characteristic polynomial is irreducible over $\mathbb{Q}$.

We write *RS-matrix* for real spectrum matrices.

It is clear that every irreducible RS-matrix A has distinct eigenvalues. Hence there exists an associated multidimensional continued fraction to A. We work now only with integer matrices of the group $SL(n+1, \mathbb{Z})$.

Definition 22.4. An *n*-dimensional continued fraction associated to an irreducible RS-matrix $A \in SL(n+1, \mathbb{Z})$ is called *an n-dimensional continued fraction of an* $(n+1)$-*algebraic irrationality*. The case $n = 1(2)$ corresponds to *one-* (*two-*) *dimensional continued fractions of quadratic* (*cubic*) *irrationalities*.

Consider an irreducible **RS**-matrix $A \in \text{SL}(n+1, \mathbb{Z})$. From Proposition 21.11 we know that the positive Dirichlet group $\Xi_+(A)$ is isomorphic to $\mathbb{Z}^n$. Each matrix of $\Xi_+(A)$ preserves the integer lattice and the union of all $n+1$ hyperplanes, and hence it preserves the n-dimensional continued fraction. Since all eigenvalues are positive, the sails are mapped to themselves as well. The group $\Xi_+(A)$ acts freely on any sail. The factor of a sail under such a group action is isomorphic to the n-dimensional torus.

Definition 22.5. By *a fundamental domain* of the sail we mean the union of some faces that contains exactly one face from each equivalence class.

In Fig. 22.1 we give an example of the periodic continued fraction associated to the matrix of the two-dimensional golden ratio (later we study this example in more detail; see Example 22.9 below):

$$\begin{pmatrix} 1 & 1 & 1 \\ 1 & 2 & 2 \\ 1 & 2 & 3 \end{pmatrix}.$$

In the top picture we have a fragment of the sail. In the middle there are two triangles of two different orbits (white ones and dark ones). Finally at the bottom we show one of the possible fundamental domains with respect to the action of the positive Dirichlet group.

22.3 Torus decompositions of periodic sails in $\mathbb{R}^3$

As we have already seen in the first part of the book (Chapter 11), the LLS sequence is a complete invariant of one-dimensional sails. The situation in the multidimensional case is much more complicated, since the combinatorial structure of sails is no longer as simple as in the one-dimensional case. A complete invariant is not known even for two-dimensional sails in $\mathbb{R}^3$. Further, in this section we discuss only periodic two-dimensional continued fractions in $\mathbb{R}^3$, recalling that such sails are homeomorphic to $\mathbb{R}^2$ by Theorem 20.21.

Definition 22.6. A class of integer congruences of certain objects in $\mathbb{R}^n$ is called an *integer affine type* of these objects.

Let us discuss the case of $\mathbb{R}^3$. It is not very hard to get an invariant that distinguishes all sails. For instance, *if we know*

— *integer affine types of all faces in the sail;*
— *integer affine types of the pyramids at all vertices of the sail;*
— *additional data about one vertex v_0: integer affine type of the union of the pyramid and the origin,*

then we can uniquely reconstruct the sail, if it exists.

Let us briefly show how to do this.

Algorithm of sail reconstruction.

Input data. As mentioned above.

Goal of the algorithm. To construct the sail (actually some of its finite part, say a certain fundamental domain for periodic sails).

Step 1. From the third condition we construct the pyramid at v_0 of the sail.

Step 2. From the first condition we find all polygonal faces adjacent to v_0.

Inductive step. In previous steps we have found several faces of the sail. By the first condition identify all integer affine types of pyramids that are adjacent to the constructed set of faces (at least in two edges each). By the second condition construct all the faces in new pyramids.

Output. In a finite or countable number of steps we build the whole sail (up to integer congruence).

Remark 22.7. In the case of algebraic irrationalities all the integer affine types of the faces and pyramids would repeat periodically. Hence we need to store only a finite amount of data regarding a single period.

This data set is far from forming a complete invariant of sails. It has many monodromy conditions, so the majority of cases are nonrealizable. So the natural question here is as follows: *how to find whether a certain data set is realizable?*

Recently, V.I. Arnold proposed a weaker invariant to distinguish the periodic sails.

Definition 22.8. A *torus decomposition* corresponding to a periodic two-dimensional sail is the factor-torus face decomposition of the sail equipped with integer affine types of faces and integer distances to them.

V.I. Arnold conjectured that torus decompositions of integer noncongruent sails are distinct. For all checked integer noncongruent sails this conjecture is true. (See Conjecture 18.) Here we should mention that for torus decompositions, the realizability problem is not yet studied.

In the multidimensional case one can define similar torus decompositions. The important question is then to find which decompositions are realizable and which are not realizable.

22.4 Three single examples of torus decompositions in $\mathbb{R}^3$

In this and the next section we study several examples. We construct one of the sails for each example.

Example 22.9. We start with the simplest example that generalizes the regular continued fraction corresponding to the golden ratio $\frac{1+\sqrt{5}}{2}$. Recall that the geometric continued fraction corresponding to the golden ratio is associated to the matrix

$$\begin{pmatrix} 1 & 1 \\ 1 & 2 \end{pmatrix}.$$

All four one-dimensional sails of the one-dimensional continued fraction corresponding to the golden ratio are integer congruent. The associated continued fraction has sails whose LLS sequences are $(\ldots, 1, 1, 1, \ldots)$. This sequence has period (1), which is the simplest possible for one-dimensional sails. The corresponding circle decompositions consist of one vertex and one edge.

The generalization of this one-dimensional continued fraction to the multidimensional case was given by E.I. Korkina in the work [129]. The multidimensional continued fraction associated to the matrix

$$\begin{pmatrix} 1 & 1 & 1 & \cdots & 1 \\ 1 & 2 & 2 & \cdots & 2 \\ 1 & 2 & 3 & \cdots & 3 \\ \vdots & \vdots & \vdots & \ddots & \vdots \\ 1 & 2 & 3 & \cdots & k \end{pmatrix}$$

is called *the generalized golden ratio*.

Remark 22.10. Unfortunately, the generalized golden ratios do not always give the simplest periodic continued fractions in all dimensions, since the characteristic polynomials are not always irreducible. For instance, there are nonconstant factors if the dimension is $k = 4, 7, 10, 12, 13, 16, 17, 19$ for $k \le 20$. Further, we study the case $k = 3$, which corresponds to an irreducible matrix. The case of the simplest matrices for $k = 4$ was studied in [99].

Consider the continued fraction associated to the generalized golden ratio matrix

$$M = \begin{pmatrix} 1 & 1 & 1 \\ 1 & 2 & 2 \\ 1 & 2 & 3 \end{pmatrix}.$$

The torus decomposition corresponding to this matrix is homeomorphic to the following one:

Here the segment AB is identified with the segment DC and the segment AD with the segment BC. In the picture we show only homeomorphic types of the faces (without lattice structure). The integer affine types of the corresponding faces are given in the next picture; both triangles have integer affine types of the simplest triangle:

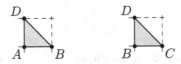

The positive Dirichlet group $\Xi_+(M)$ is generated by the two matrices

$$\begin{pmatrix} 1 & 1 & 1 \\ 1 & 2 & 2 \\ 1 & 2 & 3 \end{pmatrix} \quad \text{and} \quad \begin{pmatrix} 1 & 0 & 1 \\ 0 & 2 & 1 \\ 1 & 1 & 2 \end{pmatrix}.$$

The integer distance from the triangle ABD to the origin equals 2, and from the triangle BCD it equals 1.

This torus decomposition was found by E.I. Korkina [129], G. Lachaud [136], and A.D. Bryuno and V.I. Parusnikov [31] approximately at the same time.

Example 22.11. The second example was studied by A.D. Bryuno and V.I. Parusnikov [31]. They constructed the continued fraction that is associated to the following matrix:

$$M = \begin{pmatrix} 1 & 1 & 1 \\ 1 & -1 & 0 \\ 1 & 0 & 0 \end{pmatrix}.$$

The positive Dirichlet group $\Xi_+(M)$ is generated by the following two matrices:

$$X = M^2, \qquad Y = 2\mathrm{Id} - M^2.$$

The torus decomposition corresponding to this matrix is also homeomorphic to the following one:

Here the segment AB is identified with the segment DC and the segment AD with the segment BC. The triangle ABD has the integer affine type of the simplest triangle. The triangle BCD has the integer affine type of the triangle with vertices $(-1,-1)$, $(0,1)$, and $(1,0)$.

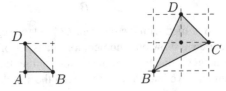

The integer distance from the triangle ABD to the origin equals 2, and from the triangle BCD it equals 1.

Example 22.12. The third example was given by V.I. Parusnikov [172]. This continued fraction is associated to the following matrix:

$$M = \begin{pmatrix} 0 & 1 & 0 \\ 0 & 0 & 1 \\ 1 & 1 & -3 \end{pmatrix}.$$

The positive Dirichlet group $\Xi_+(M)$ is generated by the following two matrices:

$$X = M^2, \qquad Y = 3\text{Id} - 2M^{-1}.$$

The torus decomposition corresponding to this matrix is also homeomorphic to the following one:

Here the segment AB is identified with the segment DC, and the polygonal line AGD with the polygonal line BEC (the point G is identified with the point E). All triangles have the integer affine type of the simplest triangle. The pentagon $BEFDG$ has the integer affine type of the pentagon with vertices $(-1,0)$, $(-1,1)$, $(0,1)$ $(1,0)$, and $(1,-1)$:

The integer distances from the triangle ABC and the pentagon $BEFDG$ to the origin equal 1, from the triangle CDF to the origin equals 2, and from the triangle CFE equals 3.

The continued fractions constructed in Examples 22.9, 22.11, and 22.12 are also known as the continued fractions corresponding to the first, the second, and the third Davenport forms. Several other single torus decompositions for two-dimensional continued fractions were investigated in the works [175], [173], and [174] by V.I. Parusnikov.

For a matrix A we denote by $||A||$ the sum of absolute values of all the coefficients for the matrix. Then we have the following theorem (for more information see [97]).

Theorem 22.13. *Let $A \in SL(3, \mathbb{Z})$ be an irreducible RS-matrix.*
(i) $||A|| \geq 4$.
(ii) If $||A|| = 5$ *(there are* 48 *possible matrices), then the corresponding continued fraction is integer congruent to the generalization of the regular fraction for the golden ratio. This is shown in Example 22.9.*
(iii) For the case $||A|| = 6$ *there are* 912 *different possible matrices. Consider the associated two-dimensional continued fractions:*
    — 480 *continued fractions are integer congruent to the one in Example 22.9;*
    — 240 *of them are integer congruent to the continued fraction of Example 22.12;*
    — 192 *of them are integer congruent to the continued fraction of Example 22.11.*
                                                      □

The classification of two-dimensional continued fractions with norm greater than or equal to seven is unknown.

22.5 Examples of infinite series of torus decomposition

We continue with several infinite series of torus decompositions. The first two infinite series were calculated by E.I. Korkina in [129]. The first of them is shown below.

Example 22.14. The continued fractions of this series are associated to the following matrices for $a \geq 0$:

$$M_a = \begin{pmatrix} 0 & 0 & 1 \\ 1 & 0 & -a-5 \\ 0 & 1 & a+6 \end{pmatrix}.$$

The positive Dirichlet group $\Xi_+(M_a)$ is generated by the following two matrices:

$$X_a = M_a, \qquad Y_a = (M_a - \mathrm{Id})^2.$$

The torus decomposition corresponding to M_a is homeomorphic to the following one:

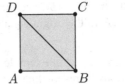

Here the segment AB is identified with the segment DC and the segment AD with the segment BC. The integer distance from the triangle ABD to the origin equals $a + 2$, and from the triangle BCD it equals 1.

Both triangles have integer affine types of the simplest triangle:

Several other examples of infinite series of continued fractions were studied in [96], some of which we now present. The following series generalizes the one from the previous example.

Example 22.15. This series depends on two integer parameters $a, b \geq 0$. Consider the continued fractions associated to the following matrices:

$$M_{a,b} = \begin{pmatrix} 0 & 1 & 0 \\ 0 & 0 & 1 \\ 1 & 1+a-b & -(a+2)(b+1) \end{pmatrix}.$$

The positive Dirichlet group $\Xi_+(M_{a,b})$ is generated by the following two matrices:

$$X_{a,b} = M_{a,b}^{-2}, \qquad Y_{a,b} = M_{a,b}^{-1}(M_{a,b}^{-1} - (b+1)\mathrm{Id}).$$

The torus decomposition corresponding to $M_{a,b}$ is as follows:

Here the segment AB is identified with the segment DC and the segment AD with the segment BC. Both triangles have the same integer affine type of the triangle with vertices $(0,0)$, $(0,1)$, and $(b+1,0)$ ($b = 5$ in the picture):

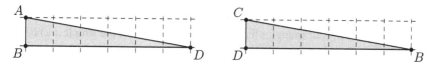

The integer distance from the triangle ABD to the origin equals $a+2$, and from the triangle BCD it equals 1.

Remark. If we substitute $b = 0$ in Example 22.15, then we have a series of matrices whose associated continued fractions are integer congruent to the continued fractions associated to the series of matrices of Example 22.14. We formulate this more precisely in the next proposition.

Proposition 22.16. *The continued fractions associated to the following two matrices are integer congruent (for integers $a \geq 0$):*

$$M_{a,0} = \begin{pmatrix} 0 & 1 & 0 \\ 0 & 0 & 1 \\ 1 & 1+a & -a-2 \end{pmatrix}, \quad M_a' = \begin{pmatrix} 0 & 0 & 1 \\ 1 & 0 & -a-5 \\ 0 & 1 & a+6 \end{pmatrix}.$$

Proof. The matrices $(\mathrm{Id} - M_{a,0})^{-1}$ and M_a' are conjugate by the matrix X in the group $SL(3,\mathbb{Z})$:

$$X = \begin{pmatrix} -1 & -1 & -2 \\ 0 & 0 & -1 \\ 1 & 0 & -1 \end{pmatrix}$$

(i.e., $M_a' = X^{-1}(\mathrm{Id} - M_{a,0})^{-1}X$). Therefore, the corresponding continued fractions are integer congruent by Proposition 22.2.   □

Example 22.17. The series of this example depends on an integer parameter $a \geq 1$. The continued fractions of the series are associated to the following matrices:

$$M_{a,b} = \begin{pmatrix} 0 & 1 & 0 \\ 0 & 0 & 1 \\ 1 & a & -2a-3 \end{pmatrix}.$$

The positive Dirichlet group $\Xi_+(M_a)$ is generated by the following two matrices:

$$X_a = M_a^{-2}, \quad Y_a = \left(2\mathrm{Id} - M_a^{-2}\right)^{-1}.$$

The torus decomposition corresponding to M_a is as follows:

Here the segment AB is identified with the segment DC and the segment AD with the segment BC. All four triangles have integer affine types of the simplest triangle:

The integer distance from the triangles ABD, BDE, and BCE to the origin are equal to $a+2$, $a+1$, and 1, respectively.

Example 22.18. The series of this example depend on two integer parameters $a > 0$ and $b \geq 0$. The continued fractions of the series are associated to matrices

$$M_{a,b} = \begin{pmatrix} 0 & 1 & 0 \\ 0 & 0 & 1 \\ 1 & (a+2)(b+2)-3 & 3-(a+2)(b+3) \end{pmatrix}.$$

The positive Dirichlet group $\Xi_+(M_{a,b})$ is generated by

$$X_{a,b} = \big((b+3)\mathrm{Id} - (b+2)M_{a,b}^{-1}\big)M_{a,b}^{-2} \quad \text{and} \quad Y_{a,b} = M_{a,b}^{-2}.$$

The torus decomposition corresponding to this matrix is homeomorphic to the following one:

Here the segment AB is identified with the segment DC and the polygonal line AFD with the polygonal line BEC (the point F is identified with the point E). The integer affine types of the faces ABF, CGE, CDG, BDF, and $DBEC$ are as follows:

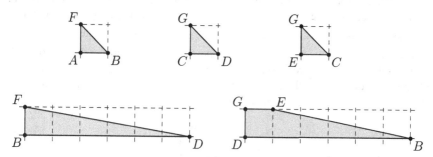

The integer distance from the triangles ABF, BFD, CDG, CEG to the origin equals 1, 1, $2+2a+2b+ab$, and $3+2a+2b+ab$ respectively. The quadrangle $DBEC$ is at unit distance from the origin.

For further details related to such series we refer to [96].

22.6 Two-dimensional continued fractions associated to transpose Frobenius normal forms

Let us consider the family of *transpose Frobenius matrices* (or *transpose companion matrices*) '

$$A_{m,n} := \begin{pmatrix} 0 & 1 & 0 \\ 0 & 0 & 1 \\ 1 & -m & -n \end{pmatrix},$$

where m and n are integers; see Fig 22.2.

Remark 22.19. Suppose that the characteristic polynomial $\chi_{A_{m,n}}(x)$ is irreducible over the field of rational numbers $\mathbb{Q}$. Then the operator of multiplication by the element x in the field $\mathbb{Q}[x]/(\chi_{A_{m,n}}(x))$ in the natural basis $\{1, x, x^2\}$ has the matrix $A_{m,n}$.

Let us observe some basic properties. Denote by Ω the subset of irreducible RS-matrices of this family.

Proposition 22.20. *The set Ω is defined by the inequality*

$$n^2 m^2 - 4m^3 + 4n^3 - 18mn - 27 \leq 0.$$

In addition it is necessary to subtract all the matrices of types: $A_{a,-a}$ and $A_{a,a+2}$, $a \in \mathbb{Z}$ (these are the matrices with reducible characteristic polynomials). □

Notice that the set Ω has the following symmetry.

Proposition 22.21. *The two-dimensional continued fractions for the cubic irrationalities constructed by the matrices $A_{m,n}$ and $A_{-n,-m}$ are integer congruent.* □

Remark 22.22. There are some other integer congruences of continued fractions in Ω. For instance, the matrices $A_{0,-a}$ and $A_{-2a,-a^2}$ have integer congruent continued fractions (since the matrices $A_{0,-a}^2$ and $A_{-2a,-a^2}$ are integer conjugate).

Observe that continued fractions of transpose Frobenius matrices do not cover all possible integer congruence classes of continued fractions for arbitrary RS-matrices in $SL(3, \mathbb{Z})$. For instance, the continued fraction of the following matrix is not integer congruent to that of a Frobenius matrix:

$$A = \begin{pmatrix} 1 & 2 & 0 \\ 0 & 1 & 2 \\ -7 & 0 & 29 \end{pmatrix}$$

(we discuss similar questions later, in Chapter 25). So the following question is of interest.

Problem 17. How often are continued fractions integer congruent to those of the set Ω?

Finally, in Fig. 22.2 we give a table (introduced in [97]) whose squares are filled with torus decompositions of the sails for continued fractions associated to the matrices of Ω. Integer affine types of all the faces of the torus decomposition for one of the sails for $A_{m,n}$ is shown in the square at the intersection of the row with number n and the column with number m. If one of the roots of the characteristic polynomial for the matrix equals 1 or -1, then the corresponding square is marked with the sign $*$ or $\#$ respectively. The squares that correspond to the matrices whose characteristic polynomial has two complex conjugate roots are shaded in light gray.

Remark. Notice that some of the series constructed above are clearly seen in Fig. 22.2.

22.7 Some problems and conjectures on periodic geometry of algebraic sails

We begin with a problem on a complete invariant of sails. In the one-dimensional case the answer is known, since the LLS sequence is a complete invariant of sails. In dimensions two and higher the problem is open (in both the periodic and the general cases). For the periodic case we have the following conjecture and open problem. If the conjecture is true and the problem is solved, then we have a complete invariant for two-dimensional periodic sails.

Conjecture 18. (V.I. Arnold.) Torus decompositions of integer noncongruent sails are distinct.

Problem 19. (V.I. Arnold.) Describe all torus decompositions that are realizable by periodic two-dimensional continued fractions.

Only a few examples are known in this direction. Let us give several trivial examples of torus decompositions that do not correspond to sails.

Example 22.23. The following torus decomposition is not realizable by sails of periodic continued fractions. There is one integer point in the interval AD, which we denote by E:

Here the polygonal line AEB maps to the polygonal line DFC under one of the matrices of the group $\mathrm{SL}(3,\mathbb{Z})$ that preserves the sail. Since AEB is a segment, DFC is also a segment. Therefore, the points B, C, F, and D lie in the same plane. And hence BF is not an edge of some face.

Now we describe the first nontrivial example of torus decomposition, which was given by E.I. Korkina in the work [129].

Example 22.24. Consider the simplest torus triangulation. It consists of two triangles with the simplest integer affine type of the triangle $(0,0)$, $(0,1)$, and $(1,0)$.

The integer distances to both faces are 1. This decomposition is not realizable for periodic sails.

This example (together with many other ones) allows us to formulate the following conjecture.

Conjecture 20. For every torus decomposition for the continued fraction of a cubic irrationality there exists some face with integer distance to the origin greater than one.

On the other hand, every known (periodic) sail has a face with the integer distance to the origin equalling one.

Conjecture 21. For every torus decomposition for the continued fraction of a cubic irrationality there exists some face whose integer distance to the origin equals one.

The following example is a torus decomposition that consists of a single face.

Example 22.25. Consider the torus decomposition consisting of one face with integer affine type of the simplest parallelogram with the vertices $(0,0)$, $(0,1)$, $(1,1)$, and $(1,0)$.

The integer distances to this face can be chosen arbitrarily. This decomposition is not realizable for periodic sails of cubic irrationalities.

It seems that a torus decomposition with a single rectangular face is not realizable in general. Here we conjecture a stronger statement.

Conjecture 22. The torus decomposition for every sail of every cubic irrationality contains a triangular face.

The next question is related to continued fractions of the same cubic extension.

Problem 23. (V.I. Arnold.) Classify continued fractions that corresponds to the same cubic extensions of the field of rational numbers.

Almost nothing is known here (even in the one-dimensional case). For example, even the finiteness of the number of possible continued fractions associated to the same extension is unknown. (For properties of cubic extensions of rational numbers see in the work of B.N. Delone and D.K. Faddeev [58].) Here is a nice example showing that a characteristic polynomial does not distinguish two-dimensional continued fractions.

Example 22.26. The following two matrices having the same characteristic polynomial $x^3 + 11x^2 - 4x - 1$ (and hence the same cubic extension of $\mathbb{Q}$) define integer noncongruent continued fractions:

$$\begin{pmatrix} 0 & 1 & 0 \\ 0 & 0 & 1 \\ 1 & 1 & -2 \end{pmatrix}^3, \quad \begin{pmatrix} 0 & 1 & 0 \\ 0 & 0 & 1 \\ 1 & 4 & -11 \end{pmatrix}.$$

22.8 Generalized Lagrange's Theorem

The combinatorial topological generalization of Lagrange's theorem was announced by E.I. Korkina in [127]; a complete proof of it was given in 2008 by O.N. German and E.L. Lakshtanov in [77]. A slightly weaker algebraic generalization was proposed in 1993 by G. Lachaud; see [134], [135], and [136]. In this chapter we present a version of the combinatorial topological generalization of Lagrange's theorem.

Definition 22.27. Let C be a cone and v a vertex of the sail for C. An *edge star* of v is the union of all edges of the sail adjacent to v; we denote it by St_v.

We say that a star is *regular* if v is in the interior of the cone C.

Consider a star St_v and a point p (not necessary integer) in the complement to St_v. We say that St_v is a *regular star with respect to p* if
— v is in the interior of $conv(St_v, p)$.
— the set $conv(St_v, p) \setminus (conv(St_v) \cup \{p\})$ does not contain integer points.
Define

$$\Gamma(St_v) = \{p \mid St_v \text{ is regular star with respect to } p\}.$$

Theorem 22.28. *Let St_v be a regular star. Then the set $\Gamma(St_v)$ is bounded.*

Proof. We prove the statement by induction on the dimension of the star.

Base of induction. The statement clearly holds for every star (i.e., for one point) in $\mathbb{R}^1$.

Step of induction. Let the statement hold for every star in $\mathbb{R}^{d-1}$; we prove the statement in $\mathbb{R}^d$.

Consider an arbitrary regular star St_v in $\mathbb{R}^d$. Suppose $\Gamma(St_v)$ is not closed. Then there exists a (unit) direction ξ that is an accumulation point of unbounded directions, i.e., for every neighborhood $U \subset S^{d-1}$ of ξ and every $N > 0$ there exists a point $p \in \Gamma(St_v)$ with direction in U and $|vp| > N$. Hence the half-prism

$$conv(St_v) \oplus \{\lambda \xi \mid \lambda \geq 0\}$$

does not contain integer points in the interior except for the points of $conv(St_v)$. There are two different cases to consider here.

The first case is v being inside the half-prism. Then either ξ is rational and the cone contains integer points, or ξ is irrational and then the half-prism contains integer points by the multidimensional Kronecker's approximation theorem (Theorem 20.16). We have arrived at a contradiction.

In the second case we have v on the boundary of the half-prism. This mean that ξ is in the plane of some face St_v^{d-1} of St_v. Let L be the integer hyperplane containing St_v^{d-1}. The direction ξ is also an accumulation points of unbounded directions in $\Gamma(St_v^{d-1})$, since the intersections of pyramids with direction close to ξ with L are also unbounded. This contradicts with induction assumption.

Hence the set $\Gamma(St_v)$ is bounded. $\square$

Corollary 22.29. *The set $\Gamma(\mathrm{St}_v)$ contains finitely many integer points.*   □

Definition 22.30. Let C be a cone in $\mathbb{R}^n$ and let S_C be its sail. A two-sided infinite sequence (v_i) of vertices in the sail S_C in $\mathbb{R}^n$ is called a *chain of S_C* if for every integer k we have

— v_k and v_{k+1} are connected by an edge of the sail;

— the points $v_k, \ldots, v_{k+n-1}$ are affinely independent (i.e., they span the $(n-1)$-dimensional affine plane).

Definition 22.31. Let C be a cone in $\mathbb{R}^n$, S_C its sail, and (v_i) a chain in S_C. Consider an arbitrary subset U of $\mathrm{Aff}(n, \mathbb{Z})$. We say that the chain (v_i) is *U-periodic* if there exists a period t such that for every i there exists $A_i \in U$ satisfying

$$A_i(\mathrm{St}_{v_j}) = (\mathrm{St}_{v_{j+t}}) \quad \text{for } j = i, i+1, \ldots, i+n.$$

Denote by H_n the set of affine transformations $Ax + b$, where A is an integer irreducible RS-matrix in $\mathrm{GL}(n, \mathbb{Z})$ (and hence it has distinct eigenvalues). Let H_n^0 denote the subset of H_n with $b = 0$. Since we have a fixed basis in $\mathbb{R}^n$, we identify linear transformations with matrices in the fixed basis.

Theorem 22.32. **(Geometric generalization of Lagrange theorem.)** *Consider an irrational cone $C \subset \mathbb{R}^n$ and let S_C be its sail. The following two statements are equivalent:*

— *there exists a $\mathrm{GL}(n, \mathbb{Z})$-transformation $A \in H_n^0$ such that $A(C) = C$.*

— *there is an H_n-periodic chain of vertices of the sail S_C.*

Remark 22.33. A stronger similar statement holds for the class of all affine transformations whose corresponding linear transformations satisfy:

— all eigenvalues are distinct from 1;

— all pairs of complex conjugate eigenvalues are distinct and distinct from the absolute values of real eigenvalues.

To avoid technical details we restrict ourselves to the class of affine transformations with irreducible RS-matrix in $\mathrm{GL}(3, \mathbb{Z})$, i.e., to H_n. For more information on the general case we refer to [77].

Lemma 22.34. *Let (v_i) be an H_n-periodic chain of vertices for some sail S in $\mathbb{R}^n$. Then there exists a transformation A in H_n^0 establishing a nontrivial shift of stars, i.e., there exists a positive integer T such that for every k we have*

$$A^T(\mathrm{St}_{v_k}) = \mathrm{St}_{v_{k+T}}.$$

Proof. First, we prove that there exists an affine transformation in H_n establishing a nontrivial shift of the union of all the stars along itself.

Set $B_i = A_{i+1}^{-1} A_i$. Notice that for $k = 1, \ldots, n$ we have $B_i(\mathrm{St}_{v_k}) = \mathrm{St}_{v_k}$, in particular $B_i(v_k) = v_k$. Though it is not clear whether B_i acts on the stars as the identity map or as a symmetry, nevertheless we know that it preserves the set of all elementary vectors corresponding to the edges of the stars. Denoting by r_1 the sum of all primitive vectors corresponding to the edges of the star (St_{v_1}), let $s_1 = v_1 + r_1$. Notice that, first, $B_i(s_1) = s_1$, second, s_1 is not contained in F.

Consider now the $(n-1)$-dimensional plane spanning the face F and the point v, denoted by π. The complement to the plane π contains at least one edge of the star St_{v_1}. Denote by r_2 the sum of all elementary vectors for edges of the star St_{v_1} in the half-space containing this edge, which we denote by π_+. Since B_i is a proper affine transformation preserving the plane π, it preserves the half-space π_+. Thus it preserves the set of all elementary vectors of the star St_{v_1} that are in this half-space, and in particular, the vector r_2. Let $s_2 = v_1 + r_2$. We have $B_i(s_2) = s_2$.

From all of the above, the transformation B_i preserves the $n+1$ points $v_1,\ldots,v_{n-1}$, r_1, and r_2. By the condition on v_i and construction of r_1 and r_2, these points span $\mathbb{R}^n$. Therefore, B_i is the identity transformation and $A_i = A_{i+1}$. So the transformation A_1 (in H_n) acts on the stars as a shift.

Let us find now a linear transformation acting on the stars as a shift. For an arbitrary integer k, we consider a transformation A_1^{-k}. This transformation sends St_{v_k} to St_{v_1} and the set of all admissible vertices $\Gamma(\text{St}_{v_k})$ to $\Gamma(\text{St}_{v_1})$. By Theorem 22.28, $\Gamma(\text{St}_{v_1})$ is bounded, and since it contains only integer points, $\Gamma(\text{St}_{v_1})$ is finite. Let this set have N points. Consider the set

$$\{A_1^k(O) \mid k = 0,\ldots,N\}.$$

All these points are in $\Gamma(\text{St}_{v_N})$, since $A_1^k(\text{St}_{v_N-k}) = \text{St}_k$. Since the stars St_{v_1} and St_{v_N} are integer congruent, the set $\Gamma(\text{St}_{v_N})$ contains exactly N points. Therefore, there exist distinct i and j such that $A_1^i(O) = A_1^j(O)$. Hence $A^{i-j}(O) = O$, meaning that A^{i-j} is a linear transformation in H_n and thus in H_n^0.

The transformation $A^{i-j} \in H_n^0$ establishes a nontrivial shift of the union of all the stars (St_{v_i}) along itself. $\qquad\square$

Proof of Theorem 22.32. Suppose that there exists an $SL(n,\mathbb{Z})$-transformation $A \in H_n^0$ such that $A(C) = C$. Then for an arbitrary vertex of the sail v the chain $(A^n(v))$ with integer parameter n is H_n-periodic.

The converse statement is more complicated. Suppose that there exists an unbounded, in both directions (as an ordered set in $\mathbb{R}^n$), H_n-periodic chain of vertices (v_i) of the sail S_C. Then by Lemma 22.34 there is a transformation A in H_n^0 establishing a nontrivial shift in the sequence of the stars St_{v_i}. Let us prove that $A(C) = C$.

Let us consider an $(n-1)$-dimensional tetrahedron $T = v_1 \ldots v_n$. Notice that T is in the sail, and hence the hyperspace containing it does not contain the origin O. Since $A^m(T) = v_{m+1} \ldots v_{m+n}$ is in C, the tetrahedron T is in the cone $A^{-m}(C)$. Let (e_i) be an eigenbasis of A such that the corresponding eigenvalues (λ_i) form a decreasing sequence. Denote also by (w_i) a basis corresponding to the cone C.

Let $v = \alpha_k e_k + \cdots + \alpha_l e_l$, where $\alpha_k \neq 0$. Then asymptotically, as $m \to +\infty$ we have

$$A^m(v) = \alpha_k \lambda_k^m e_k + \text{smaller terms};$$
$$A^{-m}(v) = \alpha_l \left(\tfrac{1}{\lambda_l}\right)^m e_l + \text{smaller terms}.$$

Therefore, the sequences of rays defined by the directions of $A^m(v)$ and A^{-m} as $m \to +\infty$ converge to the rays with direction e_k and e_l respectively. (Here to define the convergence we consider the topology of the unit sphere. A ray is represented by a point of intersection of this ray with the unit sphere.)

By the above, the set $(A^m(w_i))$ should contain the tetrahedron T for every integer m. Therefore, the limiting cone at $+\infty$ also contains T. Since the tetrahedron T is $(n-1)$-dimensional and the hyperspace of T does not contain the origin, all the rays in the limit has linearly independent directions. Therefore, the rays are distinct and (possibly after some permutations of vectors e_i) the matrix of vectors w_i in the basis (e_i) is lower triangular:

$$\begin{pmatrix} \alpha_{1,1} & 0 & \cdots & 0 & 0 \\ \alpha_{2,1} & \alpha_{1,2} & \cdots & 0 & 0 \\ \vdots & \vdots & \ddots & \vdots & \vdots \\ \alpha_{n-1,1} & \alpha_{n-1,2} & \cdots & \alpha_{n-1,n-1} & 0 \\ \alpha_{n,1} & \alpha_{n,2} & \cdots & \alpha_{n,n-1} & \alpha_{n,n} \end{pmatrix}$$

For the same reasons, all the rays of the limiting cone at $-\infty$ are independent. Hence for every i there exists a vector w_j that is contained in the span of $e_1, \ldots, e_i$. This implies that all the coefficients under the diagonal are zeros. Thus, the matrix is diagonal, and all vectors w_i are eigenvectors of A. Therefore, $A(C) = C$ (recall that by Lemma 22.34 the transformation A is in H_n^0).   □

22.9 Littlewood and Oppenheim conjectures in the framework of multidimensional continued fractions

In this section we briefly show a link between the Littlewood and Oppenheim conjectures and a certain conjecture in the theory of multidimensional continued fractions. Letting v be an arbitrary real number, denote by $\| v \|$ the distance from this number to the set of integers, i.e.,

$$\| v \| = \min_{k \in \mathbb{Z}} |v - k|.$$

Let us formulate the original conjectures.

Conjecture 24. (Littlewood conjecture.) For an arbitrary pair of real numbers (α, β) the following holds:

$$\inf_{m \in \mathbb{Z}_+} m \| m\alpha \| \| m\beta \| = 0.$$

Conjecture 25. (Oppenheim conjecture.) Let $n \geq 3$. Consider n linearly independent linear forms $L_1, \ldots, L_n$ in $\mathbb{R}^n$. Suppose that

$$\inf_{p \in \mathbb{Z}^n \setminus \{0\}} |L_1(p) \cdot L_2(p) \cdots L_n(p)| > 0.$$

Then there exists an irreducible RS-matrix $A \in SL(n, \mathbb{Z})$ such that the planes $L_i(p) = 0$ for $i = 1, \ldots, n$ are the invariant planes of A.

Remark. It is known that for $n = 3$ the Oppenheim conjecture implies the Littlewood conjecture (see, for instance, [39]).

Now we reformulate the Oppenheim conjecture in terms of the geometry of sails. We start with the following definition.

Definition 22.35. Let F be an arbitrary $(n-1)$-dimensional face of a sail in $\mathbb{R}^n$ and let $v_1, \ldots, v_m$ be the vertices of F. Then the *determinant* of F is

$$\det F = \sum_{1 \le i_1 < \cdots < i_n \le m} |\det(v_{i_1}, \ldots, v_{i_n})|.$$

We also say that the *determinant* of an edge star St_v is the Euclidean volume of the convex hull $conv(St_v)$.

Theorem 22.36. (O. German [77].) *Let $n \ge 3$. Consider n linearly independent linear forms $L_1, \ldots, L_n$ in $\mathbb{R}^n$. Denote by C the cone*

$$C = \{p \in \mathbb{R}^n \mid L_i(p) = 0, \ i = 1, \ldots, n\}.$$

Then the following two statements are equivalent:
(i) $\inf_{p \in \mathbb{Z}^n \setminus \{0\}} |L_1(p) \cdot L_2(p) \cdots L_n(p)| > 0$;
(ii) the faces of the sail S_C have uniformly bounded determinants. □

This theorem reduces the Oppenheim conjecture to the following conjecture.

Conjecture 26. Let the faces and edge stars of a cone C have uniformly bounded determinants. Then the sail S_C is algebraically periodic (i.e., it is a sail of some irreducible RS-matrix of $SL(n, \mathbb{Z})$).

Theorem 22.32 implies that to check periodicity it is enough to find an H_n-periodic chain of vertices of the sail.

Finally, we would like to mention here one unsuccessful attempt (which is nevertheless worthy mentioning) to solve these conjectures in [206] and [207]. For further details we refer the interested reader to [75], [76], and [77].

22.10 Exercises

Exercise 22.1. Suppose that a polynomial p with integer coefficients is irreducible over $\mathbb{Q}$. Prove that all the roots of p are distinct.

Exercise 22.2. Check that the Dirichlet groups for Examples 22.9, 22.11, and 22.12 are correctly calculated.

Exercise 22.3. Let $A \in \mathrm{SL}(3, \mathbb{Z})$ be an irreducible RS-matrix. Prove that $\|A\| \geq 5$.

Exercise 22.4. Prove that the torus decomposition consisting of a single square of integer area 2 is not realizable as a sail of a periodic two-dimensional continued fraction.

Exercise 22.5. Consider arbitrary integers m and n. Suppose that the matrix $A_{m,n}$ is an irreducible RS-matrix. Prove that the matrix $A_{-n,-m}$ is also an irreducible RS-matrix and that the continued fraction associated to $A_{-n,-m}$ is integer congruent to the continued fraction associated to $A_{m,n}$.

Exercise 22.6. Consider a cone C and its sail S_C in $\mathbb{R}^n$. Let (v_i) be some chain of the sail S_C. Prove that each $n-1$ consecutive edges of the chain are edges of a face in S_C.

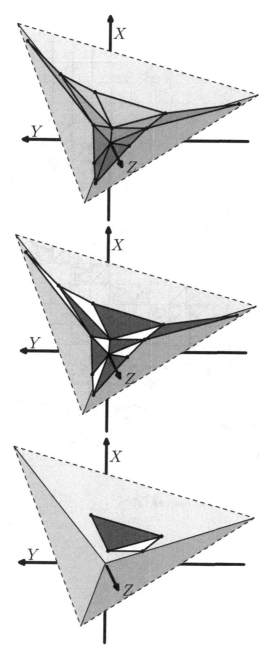

Fig. 22.1 The sail of an algebraic continued fraction (top); its periodic structure (middle); a fundamental domain (bottom).

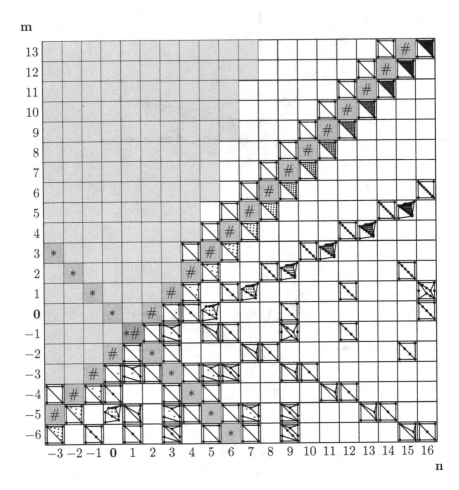

Fig. 22.2 Torus decompositions for matrices $A_{m,n}$.

Chapter 23
Multidimensional Gauss—Kuzmin Statistics

In this chapter we study the distribution of faces in multidimensional continued fractions. The one-dimensional Gauss—Kuzmin distribution was described in Chapter 13, where we discussed the classical approach via ergodic theory and a new geometric approach. Currently, an ergodic approach to the distribution of faces in continued fractions has not been developed. In fact, it is a hard open problem to find an appropriate generalization of the Gauss map suitable to the study of ergodic properties of faces of multidimensional sails. This problem can be avoided, and in fact, the information on the distribution of faces is found via the generalization of the geometric approach via Möbius measures described in the second part of Chapter 13 for the one-dimensional case. We discuss the multidimensional analogue of the geometric approach in this chapter.

23.1 Möbius measure on the manifold of continued fractions

We begin with a description of the structure of a smooth manifold on the set of multidimensional continued fractions. Further, we extend the notions of Möbius forms and measures to the multidimensional case.

23.1.1 Smooth manifold of n-dimensional continued fractions

Denote the set of all continued fractions of dimension n by CF_n. An arbitrary continued fraction is defined by an unordered collection of hyperplanes $(\pi_1, \ldots, \pi_{n+1})$. Denote by l_i, for $i = 1, \ldots, n+1$, the intersection of all the above hyperplanes except for the hyperplane π_i. Obviously, $l_1, \ldots, l_{n+1}$ are *independent* straight lines (i.e., they are not contained in a hyperplane) passing through the origin. These straight lines form an unordered collection of independent straight lines. Conversely, every

© Springer-Verlag GmbH Germany, part of Springer Nature 2022
O. N. Karpenkov, *Geometry of Continued Fractions*,
Algorithms and Computation in Mathematics 26,
https://doi.org/10.1007/978-3-662-65277-0_23

unordered collection of $n+1$ independent straight lines uniquely determines some continued fraction.

Denote the sets of all ordered collections of $n+1$ independent and dependent straight lines by FCF_n and Δ_n, respectively. We say that FCF_n is the space of n-dimensional *framed continued fractions*. Also denote by S_{n+1} the permutation group acting on ordered collections of $n+1$ straight lines. In this notation we have

$$FCF_n = \underbrace{\left(\mathbb{R}P^n \times \mathbb{R}P^n \times \cdots \times \mathbb{R}P^n\right)}_{n+1 \text{ times}} \setminus \Delta_n \quad \text{and} \quad CF_n = FCF_n / S_{n+1}.$$

This implies that the sets FCF_n and CF_n admit natural structures of smooth manifolds induced by the Cartesian product of $n+1$ copies of $\mathbb{R}P^n$. Note also that FCF_n is an $((n+1)!)$-fold covering of CF_n. We call the map $p : FCF_n \to CF_n$ of "forgetting" the order in the ordered collections the *natural projection* of the manifold FCF_n to the manifold CF_n.

23.1.2 Möbius measure on the manifolds of continued fractions

The group $\mathrm{PGL}(n+1, \mathbb{R})$ takes the set of all straight lines passing through the origin of $(n+1)$-dimensional space into itself (for the definition of $\mathrm{PGL}(n+1, \mathbb{R})$ and cross-ratios we refer to Chapter 13). Hence, $\mathrm{PGL}(n+1, \mathbb{R})$ naturally acts on the manifolds CF_n and FCF_n. Furthermore, the action of $\mathrm{PGL}(n+1, \mathbb{R})$ is transitive, i. e., it takes any (framed) continued fraction to any other. Note that for any n-dimensional (framed) continued fraction, the subgroup of $\mathrm{PGL}(n+1, \mathbb{R})$ taking this continued fraction to itself is of dimension n.

Definition 23.1. A form of the manifold CF_n (respectively FCF_n) is said to be a *Möbius form* if it is invariant under the action of $\mathrm{PGL}(n+1, \mathbb{R})$.

Transitivity of the action of $\mathrm{PGL}(n+1, \mathbb{R})$ implies that every two n-dimensional Möbius forms of the manifolds CF_n and FCF_n are proportional.

Definition 23.2. Let M be a smooth enough arcwise connected manifold and let ω be a volume form on it. Denote by μ_ω the measure on M defined as follows: for every open measurable set $S \subset M$,

$$\mu_\omega(S) = \left| \int_S \omega \right|.$$

Definition 23.3. A measure μ of the manifold CF_n (FCF_n) is said to be a *Möbius measure* if there exists a Möbius form ω of CF_n (FCF_n) such that $\mu = \mu_\omega$.

Every two Möbius measures of CF_n (FCF_n) are proportional. The projection p projects the Möbius measures of the manifold FCF_n to the Möbius measures of the manifold CF_n.

23.2 Explicit formulae for the Möbius form

Let us write down explicitly the Möbius forms for the manifold of framed n-dimensional continued fractions FCF_n for arbitrary n.

Consider $\mathbb{R}^{n+1}$ with standard metrics on it. Let π be an arbitrary hyperplane of the space $\mathbb{R}^{n+1}$ with chosen Euclidean coordinates $OX_1 \ldots X_n$, and assume that π does not pass through the origin. We call the set of all collections of $n+1$ ordered straight lines intersecting the plane π the *chart $FCF_{n,\pi}$* of the manifold FCF_n. Let the intersection of π with the ith plane be a point with coordinates $(x_{1,i}, \ldots, x_{n,i})$ on π. For an arbitrary tetrahedron $A_1 \ldots A_{n+1}$ in the plane π, we denote by $V_\pi(A_1, \ldots, A_{n+1})$ its oriented Euclidean volume (with respect to the orientation induced by the coordinates $OX_{1,1} \ldots X_{n,1} X_{1,2} \ldots X_{n,n+1}$ of the chart $FCF_{n,\pi}$).

Remark 23.4. The chart $FCF_{n,\pi}$ is everywhere dense in $(\mathbb{R}^n)^{n+1}$.

Consider the following form in the chart $FCF_{n,\pi}$:

$$\omega_\pi(x_{1,1}, \ldots, x_{n,n+1}) = \frac{\bigwedge\limits_{i=1}^{n+1}\left(\bigwedge\limits_{j=1}^{n} dx_{j,i}\right)}{V_\pi(A_1, \ldots, A_{n+1})^{n+1}}.$$

Proposition 23.5. *The measure μ_{ω_π} is a restriction of a Möbius measure to $FCF_{n,\pi}$.*

Proof. Every transformation of the group $PGL(n+1, \mathbb{R})$ is in one-to-one correspondence with the set of all projective transformations of the plane π. Let us show that the form ω_π is invariant under the action of the transformations (of the everywhere dense set) of the chart $FCF_{n,\pi}$ that are induced by projective transformations of the hyperplane π.

Let us denote by $|\bar{v}|_\pi$ the Euclidean length of a vector $\bar{v}$ in the coordinates $OX_{1,1} \ldots X_{n,n+1}$ of the chart $FCF_{n,\pi}$. Consider an arbitrary point $(x_{1,1}, \ldots, x_{n,n+1})$ of the chart $FCF_{n,\pi}$. Denote by $A_i = A_i(x_{1,i}, \ldots, x_{n,i})$ the point depending on the coordinates of the plane π with coordinates $(x_{1,i}, \ldots, x_{n,i})$, $i = 1, \ldots, n+1$. So for the chosen point, we have a naturally defined tetrahedron $A_1 \ldots A_{n+1}$ in the hyperplane π. Set

$$\overline{f}_{ij} = \frac{\overline{A_j A_i}}{|A_j A_i|_\pi}, \qquad i, j = 1, \ldots, n+1; \quad i \neq j.$$

It is clear that the vectors $\overline{f}_{ij}$ and $\overline{f}_{ji}$ are parallel to the vector $\overline{A_j A_i}$. The collection $(\overline{f}_{ij})$ is a basis in the tangent space to $FCF_{n,\pi}$, which depends continuously on the point in the chart. By dv_{ij} we denote the 1-form corresponding to the coordinate along the vector $\overline{f}_{ij}$ of $FCF_{n,\pi}$.

Let us rewrite the form ω_π in new coordinates:

$$\omega_\pi(x_{1,1},\ldots,x_{n,n+1})$$

$$= \prod_{i=1}^{n+1} \left(\frac{V_\pi(A_i,A_1,\ldots,A_{i-1},A_{i+1},\ldots A_{n+1})}{\prod_{k=1,k\neq i}^{n+1} |\overline{A_k A_i}|_\pi} \right) \cdot \frac{dv_{21} \wedge dv_{31} \wedge \cdots \wedge dv_{n,n+1}}{V_\pi(A_1,\ldots,A_{n+1})^{n+1}}$$

$$= (-1)^{\lfloor \frac{n+3}{4} \rfloor} \cdot \frac{dv_{21} \wedge dv_{12}}{|A_1 A_2|_\pi^2} \wedge \frac{dv_{32} \wedge dv_{23}}{|A_2 A_3|_\pi^2} \wedge \cdots \wedge \frac{dv_{n+1,n} \wedge dv_{n,n+1}}{|A_{n+1} A_n|_\pi^2}.$$

As in the one-dimensional case, the expression

$$\frac{\Delta v_{ij} \Delta v_{ji}}{|A_i A_j|^2}$$

for the infinitesimally small Δv_{ij} and Δv_{ji} is the infinitesimal cross-ratio of the four points A_i, A_j, $A_i+\Delta v_{ji}\overline{f}_{ji}$, and $A_j+\Delta v_{ij}\overline{f}_{ij}$ on the straight line $A_i A_j$. Therefore, the form ω_π is invariant under the action of the transformations (of the everywhere dense set) of the chart $FCF_{n,\pi}$ that are induced by the projective transformations of the hyperplane π. Hence the measure μ_{ω_π} coincides with the restriction of some Möbius measure to $FCF_{n,\pi}$.   □

Corollary 23.6. *A restriction of an arbitrary Möbius measure to the chart $FCF_{n,\pi}$ is proportional to μ_{ω_π}.*   □

Proof. The statement follows from the proportionality of any two Möbius measures.
□

Let us fix an origin O_{ij} for the straight line $A_i A_j$. The integral of the form dv_{ij} (respectively dv_{ji}) for the segment $O_{ij}P$ defines the *coordinate* v_{ij} (v_{ji}) of the point P contained in the straight line $A_i A_j$. As in the one-dimensional case, consider a projectivization of the straight line $A_i A_j$. Denote the angular coordinates by φ_{ij} and φ_{ji}, respectively. In these coordinates,

$$\frac{dv_{ij} \wedge dv_{ji}}{|A_i A_j|_\pi^2} = \frac{1}{4} \cot^2 \left(\frac{\varphi_{ji} - \varphi_{ij}}{2} \right) d\varphi_{ij} \wedge d\varphi_{ji}.$$

Then the following is true.

Corollary 23.7. *The form ω_π is extendible to some form ω_n of FCF_n. In coordinates v_{ij}, the form ω_n is as follows:*

$$\omega_n = \frac{(-1)^{\lfloor \frac{n+3}{4} \rfloor}}{2^{n(n+1)}} \left(\prod_{i=1}^{n+1} \prod_{j=i+1}^{n+1} \cot^2 \left(\frac{\varphi_{ij} - \varphi_{ji}}{2} \right) \right) \cdot \left(\bigwedge_{i=1}^{n+1} \left(\bigwedge_{j=i+1}^{n+1} d\varphi_{ij} \wedge d\varphi_{ji} \right) \right).$$

23.3 Relative frequencies of faces of multidimensional continued fractions

Without loss of generality we choose precisely the form ω_n of Corollary 23.7. Denote by μ_n the projection of the measure μ_{ω_n} to the manifold of multidimensional continued fractions CF_n.

Consider an arbitrary polytope F with vertices at integer points. Denote by $CF_n(F)$ the set of n-dimensional continued fractions that contain the polytope F as a face.

Definition 23.8. The value $\mu(CF_n(F))$ is called the *relative frequency* of a face F.

Let us specify the integer-linear congruence of faces.

Definition 23.9. We say that two faces of a sail are *integer-linear congruent* if there exists an $SL(n, \mathbb{Z})$-transformation sending one face to the other.

Remark 23.10. It some sense the structure of a face includes the origin and the polyhedron of this face. Of course, if two faces are integer-linear congruent, the corresponding polyhedra are integer congruent. The converse is not true. For instance, two empty triangles (that are integer congruent) can be integer distance one, and integer distance two to the origin. Such faces are not integer-linear congruent.

Proposition 23.11. *The frequencies of integer-linear congruent faces coincide.* ☐

The following problem is open for $n \geq 2$.

Problem 27. Find faces of n-dimensional continued fractions with the highest relative frequencies. Is it true that for every constant C there exist only finitely many pairwise integer noncongruent faces whose relative frequencies do not exceed C? Find the corresponding asymptotics (with respect to C tending to infinity).

One of the first sets of problems on statistics of faces of multidimensional continued fractions was proposed by V.I. Arnold. Let us consider the set of three-dimensional integer matrices with rational eigenvalues, denoted by A_3. The continued fraction associated to a matrix of A_3 consists of finitely many faces. Denote by $A_3(m)$ the set of all the matrices of A_3 for which the sums of absolute values of all their elements are not greater than m. The number of such matrices is finite. Let us calculate the number of triangles, quadrangles, etc., among the continued fractions, constructed for the matrices of $A_3(m)$. As m tends to infinity we have a general distribution of the frequencies for triangles, quadrangles, etc. We call the resulting frequencies of integer faces the *frequencies of faces in the sense of Arnold*. The problems of V.I. Arnold include the study of certain properties of such distributions (for instance, *what is more frequent, triangles or quadrangles, what is the average number of integer points inside the faces*, etc.). The majority of these problems are open. For further information we refer to [11] and [12].

Remark 23.12. Let us say a few words about the following particular question of V.I. Arnold: *which faces in two-dimensional sails are more frequent, triangular or quadrangular?* For instance, in the case of $\mathbb{R}^3$, there are two natural models of random polyhedra. The first one is to choose k random points in space and to construct their convex hull. For the second method, one should consider the intersection of k random half-spaces. In the first method, a random convex hull has only triangular faces, while in the second the average number of vertices of faces is four (Schläfli—Saharov theorem). So it is interesting to understand whether two-dimensional sails are "closer" to random polyhedra in the first sense (in this case they have more triangles) or in the second sense (then they should have more quadrangles).

The following conjecture connects the notions of relative frequencies and frequencies in the sense of Arnold.

Conjecture 28. The relative frequencies of faces are proportional to the frequencies of faces in the sense of Arnold.

This conjecture is open in the n-dimensional cases for $n \geq 2$.

23.4 Some calculations of frequencies for faces in the two-dimensional case

23.4.1 Some hints for computation of approximate values of relative frequencies

Consider the space $\mathbb{R}^3$ with the standard metric on it. Let π be an arbitrary plane in $\mathbb{R}^3$ not passing through the origin and with fixed system of Euclidean coordinates $O_\pi X_\pi Y_\pi$. Let $FCF_{2,\pi}$ be the corresponding chart of the manifold FCF_2. For an arbitrary triangle ABC of the plane π, we denote by $S_\pi(ABC)$ its oriented Euclidean area in the coordinates $O_\pi X_1 Y_1 X_2 Y_2 X_3 Y_3$ of the chart $FCF_{2,\pi}$. Denote by $|\bar{v}|_\pi$ the Euclidean length of a vector $\bar{v}$ in the coordinates $O_\pi X_1 Y_1 X_2 Y_2 X_3 Y_3$ of the chart $FCF_{2,\pi}$. Consider the following form in the chart $FCF_{2,\pi}$:

$$\omega_\pi(x_1,y_1,x_2,y_2,x_3,y_3) = \frac{dx_1 \wedge dy_1 \wedge dx_2 \wedge dy_2 \wedge dx_3 \wedge dy_3}{S_\pi((x_1,y_1)(x_2,y_2)(x_3,y_3))^3}.$$

Note that the oriented area S_π of the triangle $(x_1,y_1)(x_2,y_2)(x_3,y_3)$ can be expressed in the coordinates x_i, y_i as follows:

$$S_\pi((x_1,y_1)(x_2,y_2)(x_3,y_3)) = \frac{1}{2}\left(x_3 y_2 - x_2 y_3 + x_1 y_3 - x_3 y_1 + x_2 y_1 - x_1 y_2\right).$$

For the approximate computations of relative frequencies of faces it is useful to rewrite the form ω_π in the dual coordinates (see Remark 23.14 below). Define a triangle ABC in the plane π by three straight lines l_1, l_2, and l_3, where l_1 passes through B and C, l_2 passes through A and C, and l_3 passes through A, and B. Define the straight line l_i ($i = 1,2,3$) in π by the equation (after first making a translation of π in such a way that the origin is taken to some inner point of the triangle)

$$a_i x + b_i y = 1$$

in the variables x and y. Then if we know the 6-tuple of numbers

$$(a_1, b_1, a_2, b_2, a_3, b_3),$$

we can restore the triangle in a unique way.

Proposition 23.13. *In coordinates* $a_1, b_1, a_2, b_2, a_3, b_3$ *the form* ω_π *can be written as follows:*

$$-\frac{8 da_1 \wedge db_1 \wedge da_2 \wedge db_2 \wedge da_3 \wedge db_3}{(a_3 b_2 - a_2 b_3 + a_1 b_3 - a_3 b_1 + a_2 b_1 - a_1 b_2)^3}.$$

$\square$

So, we reduce the computation of relative frequency for the face F (i. e., the value of $\mu_2(CF_2(F))$) to the computation of the measure $\mu_{\omega_2}(p^{-1}(CF_2(F)))$. Consider some plane π in $\mathbb{R}^3$ not passing through the origin. By Corollary 23.7 we have

$$\mu_{\omega_2}(p^{-1}(CF_2(F))) = \mu_{\omega_\pi}(p^{-1}(CF_2(F))) \cap (FCF_{2,\pi}).$$

Finally, the computation should be made for the set $\mu_{\omega_\pi}(p^{-1}(CF_2(F))) \cap (FCF_{2,\pi})$ in dual coordinates a_i, b_i (see Proposition 23.13).

Remark 23.14. In coordinates a_i, b_i the computation of the value of the relative frequency often reduces to the estimation of the integral on the disjoint union of a finite number of six-dimensional Cartesian products of three triangles in coordinates a_i, b_i (see Proposition 23.13). The integration over such a simple domain significantly increases the speed of approximate computations. In particular, the integration can be reduced to the integration over some 4-dimensional domain.

23.4.2 Numeric calculations of relative frequencies

23.4.2.1 Remark on complexity of numeric calculations

Explicit calculation of relative frequencies for the faces seems not to be realizable. Nevertheless, it is possible to make approximations of the corresponding integrals. Usually, the greater the area of the polygon, the smaller its relative frequency. The most complicated approximation calculations correspond to the simplest faces, such as an empty triangle.

In Fig. 23.1 we examine examples of the following faces:
— triangular faces $(0,0,1)$, $(0,1,1)$, $(1,0,1)$, and $(0,0,1)$;
— triangular face $(0,2,1)$, $(2,0,1)$;
— quadrangular face $(0,0,1)$, $(0,1,1)$, $(1,1,1)$, $(1,0,1)$.
For each face we show the plane containing the face.

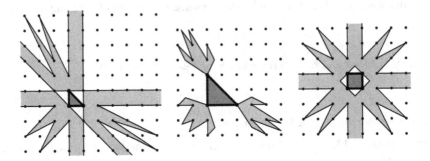

Fig. 23.1 The points painted in light gray correspond to the points at which the rays defining the two-dimensional continued fraction can intersect the plane of the chosen face.

A point is shaded gray if there exists a continued fraction such that one of the rays defining it intersects this plane. The light gray points correspond to the points at which the rays defining the two-dimensional continued fraction intersect the plane of the chosen face.

23.4.2.2 Some numeric results

In Table 23.1 we show the numerical approximations of relative frequencies for 12 faces. In the column "N°" we write a special sign for integer-linear congruence classes of a face. The index denotes the integer distance from the corresponding face to the origin. In the column "face" we draw a picture of the integer-linear congruence class of the face. Further, in the columns "lS" and "ld" we write down the integer areas of faces and integer distances from the planes of faces to the origin respectively. Finally, in the column "μ_2" we show approximate relative frequencies for the corresponding face.

Notice that in the given examples the integer congruence classes of polygons and integer distances to the origin determine the integer-linear congruence type of the face. This is not always the case where we consider greater distances to the origin.

Remark 23.15. The majority of faces of two-dimensional continued fractions are at unit integer distance from the origin. Only three infinite series and three partial examples of faces lie at integer distances greater than one from the origin, as seen in Chapter 19. If the distance to the face is increasing, then the frequency of faces is decreasing on average. The average rate of decrease of the frequency is unknown to the author.

N°	face	IS	Id	μ_2	N°	face	IS	Id	μ_2
I_1		3	1	$1.3990 \cdot 10^{-2}$	VI_1		7	1	$3.1558 \cdot 10^{-4}$
I_3		3	3	$1.0923 \cdot 10^{-3}$	VI_2		7	2	$3.1558 \cdot 10^{-4}$
II_1		5	1	$1.5001 \cdot 10^{-3}$	VII_1		11	1	$3.4440 \cdot 10^{-5}$
III_1		7	1	$3.0782 \cdot 10^{-4}$	VIII_1		7	1	$5.6828 \cdot 10^{-4}$
IV_1		9	1	$9.4173 \cdot 10^{-5}$	IX_1		7	1	$1.1865 \cdot 10^{-3}$
V_1		11	1	$3.6391 \cdot 10^{-5}$	X_1		6	1	$9.9275 \cdot 10^{-4}$

Table 23.1 Some numeric results of calculations of relative frequencies.

23.4.3 Two particular results on relative frequencies

In conclusion, we give two simple statements on relative frequencies of faces.

Proposition 23.16. *Integer congruent polygons P_1 and P_2 at integer distances 1 and 2 to the origin have the same relative frequencies (see, for example, VI_1 and VI_2 of Table 23.1).* □

Denote by A_n the triangle with vertices $(0,0,1)$, $(n,0,1)$, and $(0,n,1)$. Denote by B_n the square with vertices $(0,0,1)$, $(n,0,1)$, $(n,n,1)$, and $(0,n,1)$.

Proposition 23.17. *The following holds:*

$$\lim_{n \to \infty} \frac{\mu(CF_2(A_n))}{\mu(CF_2(B_n))} = 8.$$

□

For further information related to multidimensional Gauss—Kuzmin statistics we refer to the work [124] by M.L. Kontsevich and Yu.M. Suhov and [101] by the author.

23.5 Exercises

Exercise 23.1. Show that relative frequencies of integer-linear congruent faces are equivalent.

Exercise 23.2. Prove Propositions 23.16 and 23.17.

Exercise 23.3. For every face F of Table 23.1 draw the intersection set of the plane spanned by F with all the rays that define two-dimensional continued fractions having F as a face (see examples in Fig. 23.1).

Exercise 23.4. Calculate up to the second digit the relative frequencies for the faces of types VII_1 and X_1 in Table 23.1.

Chapter 24
On the Construction of Multidimensional Continued Fractions

In the first part of this book we saw that the LLS sequences completely determine all possible sails of integer angles in the one-dimensional case. The situation in the multidimensional case is much more complicated. Of course, the convex hull algorithms can compute all the vertices and faces of sails for finite continued fractions, but it is not clear *how to construct (or to describe) vertices of infinite sails of multidimensional continued fractions in general. What integer-combinatorial structures could the infinite sails have?* There is no single example in the case of aperiodic infinite continued fractions of dimension greater than one. The situation is better with periodic algebraic sails, where each sail is characterized by its fundamental domain and the group of period shifts (i.e., the positive Dirichlet group).

In this chapter we show the main algorithms that are used to construct examples of multidimensional continued fractions (finite, periodic, or finite parts of arbitrary sails). We begin with some definitions and background. Further, we discuss one inductive and two deductive algorithms to construct continued fractions. Finally, we demonstrate one of the deductive algorithms on a particular example.

24.1 Inductive algorithm

24.1.1 Some background

A multidimensional periodic algebraic continued fraction is a set of infinite polyhedral surfaces (i.e., sails) that contain an infinite number of faces. As we have already mentioned, the quotient of every sail under the positive Dirichlet group action is isomorphic to the n-dimensional torus. The algebraic periodicity of the polyhedron allows us to reconstruct the whole continued fraction knowing only the fundamental domain (i.e., the union of faces that contains exactly one face from each equivalence class with respect to the action of the corresponding group $\varXi_+$). Moreover, every fundamental domain contains only a finite number of faces of the whole alge-

© Springer-Verlag GmbH Germany, part of Springer Nature 2022
O. N. Karpenkov, *Geometry of Continued Fractions*,
Algorithms and Computation in Mathematics 26,
https://doi.org/10.1007/978-3-662-65277-0_24

braic periodic continued fraction. Hence we are faced with the problem of finding a good algorithm that enumerates all the faces for this domain.

There was no algorithm for constructing multidimensional continued fractions until T. Shintani's work [200] in 1976. The algorithm was further developed by R. Okazaki [166] in 1993. The authors showed the so-called *inductive method* for constructing fundamental domains of multidimensional continued fractions. The algorithm produces the fundamental domain face by face, verifying that each new face does not lie in the same orbit with some face constructed previously. Applying the algorithm, one finds the fundamental domain in finitely many steps. In some sense this is similar to the classical construction of one-dimensional continued fractions.

Using inductive algorithms, E. Thomas and A.T. Vasques obtained several fundamental domains for the two-dimensional case in [211]. Later, E. Korkina in her papers [128], [129], [130] and G. Lachaud in [134] produced an infinite number of fundamental domains for periodic algebraic two-dimensional continued fractions.

24.1.2 Description of the algorithm

For a polytope F and its hyperface E we say that the collection (E,F) is a *hyperflag*. Every $(n-1)$-dimensional face of a sail for an $(n-1)$-dimensional continued fraction is called a *hyperface* of the sail.

Inductive algorithm for constructing the fundamental domains for sails of multidimensional continued fractions.

Input data. We have a matrix A in $SL(n,\mathbb{Z})$ and one of its invariant cones. We start with an empty list of hyperflags L and empty fundamental domain D.

Goal of the algorithm. To construct one of the fundamental domains of the continued fraction for A in the given invariant cone.

Step 1a. Construct a hyperface F_0 of the sail of the cone.

Step 1b. Calculate the face-structure of this face. Add all the hyperflags containing F_0 to the list L as unmarked.

Step 2. Take an unmarked hyperflag (F,E) of the list L and
  — construct the hyperface F' adjacent to the hyperface F along E';
  — check whether F' is not an image of another face F of the list D under the action of some element of the positive Dirichlet group $\Xi_+(A)$. If F' is not an image, then add F' to the list D;
  — mark (F,E) as marked; add (F',E) as marked.

Step 3. Until the list L contains at least one unmarked hyperflag, go to Step 2.

Step 4. Find the face decomposition of all hyperfaces in the list D. It is the closure of one of the fundamental domains.

Output. The face decomposition of the hyperfaces in the list D is the closure of one of the fundamental domains of an $(n-1)$-dimensional continued fraction.

Remark 24.1. The algorithm is relatively slow in Step 2, but the advantage is that one does not need to construct the basis of the positive Dirichlet group. If we work with nonalgebraic sails, then this algorithm constructs a finite domain of a sail (here in Step 2, we only construct an adjacent face).

24.1.3 Step 1a: construction of the first hyperface

24.1.3.1 How to calculate one vertex of the sail

Consider an arbitrary invariant cone C. Let us find some integer point in C. Shift the basis unit parallelepiped completely into the cone C. There exists an integer point lying inside the shifted parallelepiped. The coordinates of this point coincide with integer parts of coordinates for one of the 2^n vertices of this parallelepiped.

Having found some integer point P of the invariant cone C, let us construct some vertex of the sail of C. Consider some integer plane π passing through the origin such that

$$\pi \cap C = O,$$

where O is the origin. Suppose that the integer distance from the point P to this plane is equal to d. Now we look through all the simplices obtained as intersections of the cone C with planes parallel to π at integer distances to the origin equal to $1, \ldots, d$. Suppose that the first simplex containing integer points lies in the plane at integer distance $d' \le d$. The convex hull of all points of this simplex coincides with some faces of the sail. All vertices of this face are vertices of the sail. Choose one of them.

24.1.3.2 Inductive construction of a hyperface

Suppose that we have already constructed a face F_k of the sail of dimension $k-1$ (where $k < n$). Let us construct a face of the sail of dimension k. Let π_1 be an integer plane of dimension $n-2$ that contains F_k, intersects the cone C in a compact set, and does not pass through the origin O. Consider a two-dimensional plane π_2 orthogonal to π_1. Let us project the cone and all its inner integer points along π_1 to π_2, denoting this projection by f. The image of the cone $f(C)$ is a two-dimensional cone. Letting S be the set of all integer points inside the cone, then $f(S)$ is contained in the cone $f(C)$. In addition, $f(S)$ is contained in $f(\mathbb{Z}^n)$, which is a two-dimensional discrete lattice (since π_1 is integer).

Consider a line l in π_2 passing through $f(F_k)$ and assume that this line cuts from the cone $f(C)$ a triangle that contains points of $f(S)$. Take the convex hull of these points in the triangle. One of the two neighboring points of $f(S)$ to $f(F_k)$ in the boundary of this convex hull is in the boundary of the convex hull for $f(S)$ (we call this point P). Let $s \in S$ and $f(s) = P$. Then the span of s and the face F_k is in some face of the sail of dimension k. The face F_{k+1} consists of all integer points in the tetrahedral intersection of the plane spanned by F_k and s with the cone C.

So we get the face F_{k+1}. Proceed further until we get a hyperface F_n.

24.1.4 Step 1b, 4: how decompose the polytope into its faces

Let π be an oriented plane of dimension r, and let $p_0, p_1, \ldots, p_r$ be points in this plane. Define

$$\mathrm{sign}_\pi(p_0, p_1, \ldots, p_r) = \begin{cases} 1 & \text{if } p_0 p_1, \ldots, p_0 p_r \text{ defines a positive orientation,} \\ 0 & \text{if } p_0 p_1, \ldots, p_0 p_r \text{ are dependent,} \\ -1 & \text{if } p_0 p_1, \ldots, p_0 p_r \text{ defines a negative orientation.} \end{cases}$$

Proposition 24.2. *Consider an r-dimensional convex polytope $p_1 \ldots p_s$, and suppose it spans the plane π of dimension r. Consider a subset of indices $i_1 < \ldots < i_r \le s$. Let the tetrahedron $T = p_{i_1} \cdots p_{i_r}$ have nonzero Euclidean (or integer) $(r-1)$-dimensional volume. Then we have:*

(i) The tetrahedron T is in a hyperface if and only if the function

$$\mathrm{sign}_\pi(p_j, p_{i_1}, \ldots, p_{i_r})$$

is either nonnegative or nonpositive simultaneously for all $j \in \{1, \ldots, s\}$.

(ii) If T is in some hyperface, then this hyperface consists of points p_j satisfying the equation

$$\mathrm{sign}_\pi(p_j, p_{i_1}, \ldots, p_{i_r}) = 0.$$

Proof. The condition of the first item means exactly that all points are in one half-plane with respect to the hyperplane containing the tetrahedron $p_{i_1} \cdots p_{i_r}$.

The condition of the second item enumerates the points contained in the plane of T.   □

Remark 24.3. Using Proposition 24.2 one constructs all hyperfaces of a polyhedron P. Further decomposition into faces is done iteratively, applying the hyperface search to already constructed faces.

24.1.5 Step 2: construction of the adjacent hyperface

Suppose that we know a hyperface F and one of its hyperfaces E. Let us construct the hyperface F_E adjacent to F via E.

Let π_1 be an integer plane of dimension $n-2$ containing E and a two-dimensional plane π_2 orthogonal to π_1. Let us project the cone and all its inner integer points along π_1 to π_2, again denoting this projection by f. The image of the cone $f(C)$ is a two-dimensional cone. Letting S be the set of all integer points inside the cone, then $f(S)$ is contained in the cone $f(C)$, and in addition, $f(S)$ is contained in $f(\mathbb{Z}^n)$, which is a two-dimensional discrete lattice (since π_1 is integer). The image of F is the segment AB, and the image of the vertex O of the cone is the point denoted by O'. Consider a point in $f(S) \setminus O'AB$, denoting it by P.

Without loss of generality, we suppose that the points A and P are in one half-plane with respect to the line $O'B$. The line AP cuts from the cone $f(P)$ a triangle containing points of $f(S)$. Take the convex hull of these points in the triangle. One of the two neighboring points of $f(S)$ to B in the boundary of this convex hull is in the boundary of the convex hull for $f(S)$ (we call this point Q). Let $s \in S$ and $f(s) = Q$. Then the span of s and the face E is in a hyperface of the sail adjacent to E. The hyperface F_E consists of all integer points in the tetrahedral intersection of the plane spanned by E and s with the cone C.

24.1.6 Step 2: test of the equivalence class for the hyperface F' to have representatives in the set of hyperfaces D

Consider two ordered bases (e_i^1) and (e_i^2) with matrices M_1 and M_2 in the standard coordinate basis. It is clear that (e_i^1) is integer congruent to (e_i^2) if and only if

$$M_2(M_1)^{-1} \in \mathrm{SL}(n, \mathbb{Z}).$$

Here one needs to check that the elements of $M_2(M_1)^{-1}$ are all integer and that the determinant equals 1.

If in addition we are interested whether these two bases are in the same class with respect to some Dirichlet group $\Xi(A)$, then we should check that the matrix $M_2(M_1)^{-1}$ commutes with the matrix A.

Suppose that we are willing to test whether the equivalence class of a hyperface F' has representatives in the set of faces D. Take an arbitrary n-tuple of independent integer points in F' and the basis (e_i) of $\mathbb{R}^n$ corresponding to it. The test is positive if there exist a face in D and an n-tuple of independent integer points in this face such that the corresponding basis (g_i) is integer congruent to (e_i) and the transformation matrix commutes with the matrix A defining the cone.

24.2 Deductive algorithms to construct sails

24.2.1 General idea of deductive algorithms

In inductive algorithms one constructs faces of the sail one by one inductively. There is another approach to constructing faces of the sail for the cone C. First, one should construct an approximation of the sail itself, and, second, choose the faces of the approximation that are the faces of the sail. This method is especially useful in the algebraic periodic case. Suppose that we are given an integer irreducible real spectrum matrix $A \in \mathrm{SL}(n+1, \mathbb{Z})$. To compute some fundamental domain of a sail of the continued fraction associated to A it is sufficient to do the following:

1. Compute a convex hull approximation of the sail. Namely, take a large enough convenient set of integer points and find its convex hull.

2. Make a conjecture on some fundamental domain. Here we need to guess a set of faces that might form a fundamental domain. We do this by finding a repeatable pattern in the geometry of the faces.

3. Prove the conjecture if possible.

4. If you cannot prove the conjecture, start with **1**, but with a larger convenient set of points.

In this situation the following two questions are relevant:
How can one find a convenient set of integer points for the approximation of the sail?
How can one test whether the conjecture of a fundamental domain of the sail is true?
Let us give answers to these questions.

24.2.2 The first deductive algorithm

In [156] J.O. Moussafir developed an essentially different approach to constructing continued fractions. It works for an arbitrary (not necessarily periodic) continued fraction and computes any bounded part of an infinite polyhedron. The approach is based on deduction. One produces a conjecture on the face structure for a large part of the continued fraction. Then it remains to prove that every conjectured face is indeed a face of the part. This method can be also applied to the case of periodic continued fractions.

The scheme of the first deductive algorithm.

Input data. A cone C.

Goal of the algorithm. Construct a large enough set of faces in C.

Step 1. Find an approximation of the cone and make a face decomposition (better if inside the approximated cone).

Step 2. Determine which faces are the real faces of the sail. Go to Step 1 until there are enough faces.

Step 3. In the algebraic case, find the fundamental domain of the sail.

Output. A large enough set of faces in C.

In the first step we approximate the cone in the following way. The cone is defined by its edges, so we take a good approximation of all the edges, and get a cone approximating our cone.

In the second step we should check that certain polyhedra are faces of the sail for the cone. So we are solving the following problem: *starting with a cone C with vertex at the origin and an integer polyhedron F, find out whether F is a face of the sail for C*. Denote by π_F an integer hyperplane containing F.

(i). Check that the intersection of the cone C with π_F is a simplex, find all the integer points inside it, and check that their convex hull coincides with F.

(ii). Find the integer distance d from the origin to F. Check that all integer planes parallel to π_F and having the origin and F in distinct half-planes (there are exactly $d-1$ such planes) do not have integer points in the intersection with the cone C.

In the third step we find out which faces of the constructed part are shifted by matrices of the group $\Xi_+(A)$ to each other. This was discussed in Section 24.1.5.

As an example, we discuss one particular case of two-dimensional sails later, in Section 24.2.4.

24.2.3 The second deductive algorithm

In this section we describe another deductive construction adapted especially to fundamental domains of periodic continued fractions. The construction involves a method for conjecturing the structure of the fundamental domain and an algorithm testing whether the conjectured domain is indeed fundamental. Usually the number of "false" vertices of this approximation is much smaller than the number of "false" vertices of the approximation produced by the first deductive algorithm.

Note that this algorithm substantially uses the periodicity of a multidimensional continued fraction, and hence it is impossible to apply it to aperiodic continued fractions.

Almost all known simple examples and series of examples of fundamental domains were constructed using this algorithm (see [96]). In particular, the complete list of all two-dimensional periodic continued fractions constructed by matrices of small norm ($|*| \le 6$) was found in [97] (see also Theorem 22.13 above).

Let us briefly itemize the main steps of the algorithm.

The second deductive algorithm to construct fundamental domains.

Input data. A matrix $A \in \mathrm{SL}(n, \mathbb{Z})$ and one of its invariant cones.

Goal of the algorithm. Construct a fundamental domain for the sail of the given matrix A for the given invariant cone.

Step 1. Calculate the basis of the additive group of the ring $\Gamma(A)$.

Step 2. Calculate the basis of the group $\Xi_+(A)$ (using the result of Step 1).

Step 3. Find some vertex of the sail (see Section 24.1.3).

Step 4. Make a conjecture on a fundamental domain of the sail (using the results of Step 2 and Step 3).

Step 5. Test the resulting conjecture from Step 4.

Output. A fundamental domain of an operator A in the given invariant cone.

Remark 24.4. It is assumed that the fundamental domain conjectured in Step 4 and the basis $A_1, \dots, A_n$ of the group $\Xi_+(A)$ satisfy the following conditions:
(i) the closure of the fundamental domain is homeomorphic to the disk;
(ii) the operators with matrices $A_1, \dots, A_n$ define the gluing of this disk to the n-dimensional torus.

We show how to test conjectures in the case of two-dimensional continued fractions in the description of Step 5. The result is partially based on Corollary 20.36 on the classification of two-dimensional faces at the integer distances to the origin greater than one (see also in [98]). For the case of n-dimensional continued fractions for $n \geq 3$, the last step is quite complicated, since the classification of three-dimensional faces at integer distances to the origin greater than one is unknown.

Remark 24.5. Note that all deductive algorithms are not algorithms in the strict sense. One needs to choose some basis of $\Xi_+(A)$ in the right way, produce a good conjecture, and then test it. Even the algorithmic recognition of the period for the given picture of a sail approximation is thought to be a hard problem. That is why this "algorithm" cannot be completed with certainty by some computer program. But on the other hand, this deductive algorithm is effective in practice. All of the examples listed in the article [96] were produced using this algorithm. The examples of this paper generalize and expand almost all known periods of the sails calculated before.

Remark 24.6. Deductive methods can be naturally generalized to the calculations of periods of certain other periodic multidimensional continued fractions related to cones (for more information we refer to the works by H. Minkowski [151] and G.F. Voronoi [220].)

General questions concerning the lattice bases (Steps 1 and 2) were discussed in Chapter 21 for the calculation of a vertex of the sail (Step 3); see Section 24.1.3. Step 5 is general for both deductive algorithms. It will be discussed in the next subsection. So it remains to describe Step 4.

24.2.3.1 Step 4. How to produce a conjecture on the fundamental domain of a sail

Suppose that we know some vertex V of the sail in the invariant cone (from Step 3) and a basis $A_1, \ldots, A_n$ for the group $\Xi_+(A)$ (from Step 2). Let us briefly discuss how to produce a conjecture on a fundamental domain of the sail. First, we compute the set of integer points that contains all vertices of some fundamental domain of the sail. Second, we show how to choose an infinite sequence of *special polyhedron approximations* for the sail. Finally, using a picture of these approximations, we formulate a conjecture on a fundamental domain of the sail.

Proposition 24.7. *Let V be a vertex of the sail of the n-dimensional continued fraction of an $(n+1)$-algebraic irrationality. Then there exists a fundamental domain of the sail such that all vertices of this domain are contained in the convex hull H of the origin and of the 2^n distinct points of the form*

$$V_{\varepsilon_1, \ldots, \varepsilon_n} = \left(\prod_{i=1}^{n} A_i^{\varepsilon_i} \right)(V),$$

where $\varepsilon_i \in \{0,1\}$ for $1 \leq i \leq n$.

Proof. Consider the polyhedral cone C with vertex at the origin and base at the convex polyhedron with vertices $V_{\varepsilon_1, \ldots, \varepsilon_n}$. We take the union of all images of this polyhedral cone under the actions of the operators with matrices

$$A_{m_1, \ldots, m_n} = \prod_{i=1}^{n} A_i^{m_i},$$

for $1 \leq i \leq n$, where $m_i \in \mathbb{Z}$. Obviously this union is equivalent to the union of the whole open invariant cone and the origin. Therefore, every vertex of the sail is obtained from a vertex contained in the cone C by applying an operator with matrix $A_{m_1, \ldots, m_n}$ for some integers m_i. Moreover, the convex hull of all integer points of the given invariant cone contains the convex hull of the vertices of the form $A_{m_1, \ldots, m_n}(V)$. Hence the sail (i.e., the boundary of the convex hull of integer points) is contained in the closure of the complement in the invariant cone to the convex hull of all integer points of the form $A_{m_1, \ldots, m_n}(V)$. This complement is a subset of the union of polyhedra obtained from H by an action of some operator with matrix $A_{m_1, \ldots, m_n}$ (for some integers m_i, $1 \leq i \leq n$). $\qquad\square$

We skip the classical description of the computation of the convex hull for the integer points contained in the polyhedron H. Denote the vertices of this convex hull by V_r for $0 < r \leq N$. Here N is the total number of such points.

Definition 24.8. The convex hull of the finite set of points

$$\{A_{m_1, \ldots, m_n}(V_r) \mid 1 \leq m_i \leq m, \forall i : 0 \leq i \leq n\}$$

is called the *nth special polyhedron approximation* for the sail.

The defined set contains approximately, but fewer than, $m^n N$ points. (Since we calculated some image points for the boundary of H several times, we do not know the exact number of points.) The number N is fixed for the given generators $A_1, \ldots, A_n$ while m varies. We should try to make a conjecture with the least possible m.

Remark 24.9. For all the examples listed in the paper [96] (and for the example of the last section) it was sufficient to take $m = 2$ to produce the corresponding conjectures.

Remark 24.10. The "quality" of the approximation strongly depends on the choice of the basis of $\Xi_+(A)$.

Remark 24.11. Note that though the algorithm works for $n+1$ arbitrary linearly independent matrices of $\Xi_+(A)$, it is not necessary to find generators of $\Xi_+(A)$. Suppose that we know matrices $A_1, \ldots, A_n$ that generate only some full-rank subgroup of the group $\Xi_+(A)$. Let the index of this subgroup be equal to k. Then we are faced with the following two problems. Firstly, the number N will be approximately k times greater than in the previous case. Secondly, one should also find a conjecture on generators of the group $\Xi_+(A)$.

24.2.4 Test of the conjectures produced in the two-dimensional case

In this section we explain how to test conjectures for the case of two-dimensional periodic continued fractions (for more information we refer to [103]). The test consists of seven stages. We prove here that these seven stages are sufficient for the verification of whether the conjecture produced is true. The complexity of these stages is polynomial in the number of all faces.

24.2.4.1 Brief description of the test stages and formulation of the main results

Suppose that we have a conjecture on some fundamental domain D for some sail of a two-dimensional periodic continued fraction associated to some integer irreducible real spectrum matrix A. Let B_1, B_2 be the basis of the group $\Xi_+(A)$. Let p_k (for $k = 0, 1, 2$) be the number of k-dimensional faces of the fundamental domain D. Denote by F_i ($i = 1, \ldots, p_2$) the two-dimensional faces, i.e., polygons. All vertices and edges adjacent to each face are known. It is also conjectured that the fundamental domain D and the basis B_1, B_2 satisfy the following conditions:

(i) the closure of the fundamental domain is homeomorphic to the two-dimensional disk;

(ii) B_1 and B_2 define the gluing of this disk to the n-dimensional torus T^2 (the fundamental domain D is in one-to-one correspondence with this torus).

Test of the conjecture. Our test consists of the following seven stages:
1. test of condition (i);
2. test of condition (ii);
3. calculation of all integer distances from the origin to the two-dimensional planes containing faces F_i and verification of their positivity;
4. test the nonexistence of integer points inside the pyramids with vertices at the origin and bases at F_i (here the integer points of faces F_i are permitted);
5. test the convexity of dihedral angles (for all edges of the fundamental domain);
6. verification that all stars of the vertices are regular;
7. test whether all vertices of D are in the same invariant cone.

Let us give several technical definitions related to conditions 5 and 6.

Definition 24.12. (*i*) A dihedral angle is called *well-placed* if the origin is contained in the corresponding opposite dihedral angle.

(*ii*) Consider an arbitrary polyhedral surface P and its vertex v. The *n-star* of the vertex v in P is the union of all faces of P of dimension not greater than n containing v.

Let us now define the regular stars at vertex v of a fundamental domain. First, we construct a 2-star R_v of the conjectured sail using the algorithm of sail reconstruction (see page 300). Second, let $p : W \to T^2$ be the universal covering of the torus obtained by gluing the fundamental domain via the shift operators. Take any vertex $w \in W$ corresponding to the vertex v. Denote the 2-star at $w \in W$ by R_w. Denote the canonical projection of R_w to R_v by ξ.

If $v \neq (a,0,0)$ for some positive a, then we set $\bar{v}_m = (1/m,0,0)$ for $m \in \mathbb{Z}_+$. (If $v = (a,0,0)$, then we set $\bar{v}_m = (0,1/m,0)$.) A 2-star at v is called *regular* if for the sequence of rays l_m with vertex at the origin and passing through the point $v + \bar{v}_m$ there exists a positive k such that for every $m \geq k$ the following holds: the set

$$\xi^{-1}(l_m \cap R_u)$$

consists of exactly one point.

Let us first formulate a theorem on complexity of the conjecture test.

Theorem 24.13. *The described conjecture test for the fundamental domain D requires every than*

$$C(p_0 + p_1 + p_2)^4$$

additions, multiplications, and comparisons of two integers, where C is a universal constant that does not depend on the p_i. □

We skip the proof of this technical theorem here and refer the interested reader to [103].

Remark 24.14. Note that here we do not take into account the complexity of additions, multiplications, and comparison of two large integers. We think of any such operation as a single operation (as a unit of time). There are some known bounds for the number of digits of the integers that are linear with respect to the coefficients of the matrix A. So the complexity should be multiplied by some polynomial in the coefficients of A.

Theorem 24.15. *Let the set of faces D satisfy the following conditions:*
(1) condition (i);
(2) condition (ii);
(3) positivity of all integer distances from the origin to the two-dimensional planes containing faces F_i;
(4) there are no integer points inside the pyramids with vertices at the origin and bases at F_i (here integer points on the faces F_i are permitted);
(5) all dihedral angles are convex;
(6) all stars of the vertices are regular;
(7) all vertices of D are contained in the same invariant cone.
Then D is a fundamental domain of some sail of the continued fraction associated to the matrix A.

Let us prove that these seven stages are sufficient for the test.

24.2.4.2 Lemma on the injectivity of the face projection

We prove Theorem 24.15 in four lemmas.

First let us give the necessary notation. Let the matrices B_1 and B_2 generate $\Xi_+(A)$. For any integers n and m, we write $B_{n,m}$ for the matrix $B_1^n B_2^m$. We suppose that our domain D satisfies Conditions 1-7 of Theorem 24.15. Let

$$U = \bigcup_{n,m \in \mathbb{Z}} B_{n,m}(D).$$

Consider the two-dimensional unit sphere S^2 centered at the origin O. We denote by π the following map:

$$\pi : \mathbb{R}^3 \setminus O \to S,$$

where every point $x \in \mathbb{R}^3 \setminus O$ maps to the point at the intersection of S^2 and the ray with vertex at the origin and containing x.

Lemma 24.16. *For any face of the polygonal surface U, the map π is welldefined and injective on it.*

Proof. Consider any two-dimensional face F of the surface U. By condition 3, the distance from the origin to the plane containing F is greater than zero. Hence this plane does not contain the origin. Then π is welldefined and injective on F.

Let now E be some edge of U. By conditions 1 and 2, this edge is adjacent to some two-dimensional face and therefore is contained in some plane that does not pass through the origin. Thus the line containing E does not pass through the origin. Hence π is well defined and injective on E.

The injectivity for the vertices is obvious. $\square$

24.2.4.3 Lemma on the finite covering of the fundamental domain

Let $x \in \mathbb{R}^3 \setminus O$. Denote by N_x the tetrahedral angle with vertex at the origin and base with vertices x, $B_1(x)$, $B_1 B_2(x)$, and $B_2(x)$. Notice that

$$\left(\bigcup_{n,m \in \mathbb{Z}} B_{n,m}(N_x) \right) \setminus O$$

is one of eight invariant cones of the continued fraction associated to A that contains x. Note that from conditions 1, 2, and 6 it follows that all points of D are contained in one open invariant cone, which we denote by C.

Lemma 24.17. *Let x be some point of the open invariant cone C. Then the union of all faces of D is contained in a finite union of solid angles of the type $B_{n,m}(N_x)$.*

Proof. By Dirichlet's unit theorem it follows that for every interior point a of the open invariant cone C there exists an open neighborhood satisfying the following condition. The neighborhood can be covered by four solid angles of the type $B_{n,m}(N_x)$ when a belongs to an edge of some $B_{k,l}(N_x)$, by two solid angles when a belongs to the face of some $B_{k,l}(N_x)$, and by one solid angle in the remaining cases. In every case, the neighborhood can be covered by some finite union of solid angles of type $B_{n,m}(N_x)$.

Consider a covering of D by such neighborhoods that correspond to each point of the closure of D. Since the closure of D is closed and bounded in $\mathbb{R}^3$, it is compact. Hence this covering contains some finite subcovering. Therefore, the union of all faces of D is contained in the finite union of solid angles of type $B_{n,m}(N_x)$. $\square$

Corollary 24.18. *Let x be contained in the open invariant cone C. Then the solid angle N_x contains only points from a finite number of fundamental domains of the type $B_{n,m}(D)$.*

Proof. From the last lemma it follows that D is contained in the finite union $\bigcup_{k=1}^{l} B_{n_k,m_k}(N_x)$ (for some positive l). Then the solid angle N_x can contain only points of the fundamental domains $B_{-n_k,-m_k}(D)$ for $1 \le k \le l$. $\square$

24.2.4.4 Lemma on the bijectivity of the projection

Lemma 24.19. *The map π bijectively takes the polygonal surface U to the set $S^2 \cap C$.*

Proof. As was shown above, the surface U is contained in C and is taken to $S^2 \cap C$ under the map π.

Let us introduce the following notation. By condition 2, the action of the operators with matrices B_1 and B_2 determines a gluing of the fundamental domain. After gluing we obtain a torus, denoted by T^2. Let W be the universal covering of T^2. The face decomposition on T^2 lifts to a face decomposition on W. There is a natural two-parameter family (with two integer parameters) of projections $p_{n,m} : W \to U$ that map faces to faces (since the group of shifts $B_{k,l}$ acts on U). Let us choose one of these projections and denote it by p ($p : W \to U$).

Consider the map $\pi \circ p : W \to S^2$. This map does not have branch points at the images of open faces of W, since every face of W bijectively maps to some face of U, and the corresponding face of U injectively maps to $S^2 \cap C$ by Lemma 24.16.

Two faces with a common edge of the universal covering W map to some two faces with common edge of the surface U. Such faces of U generate a well-placed dihedral angle, and hence also injectively map to $S^2 \cap C$. So the map $\pi \circ p$ does not have branch points at the images of open edges.

We now consider some vertex v of W. The edges and faces of W with common vertex v, by condition 6, form a regular 2-star. These edges also map to some edges of U with common vertex. Thus there exists a sequence of points that tends to $\pi \circ p(v)$ (contained in S^2) such that the preimage of every point of the sequence has exactly one preimage in the 2-star of v. Hence $\pi \circ p$ does not have branch points in the branch containing the star at $\pi \circ p(v)$. Therefore, $\pi \circ p$ does not have any branch points at the vertices. So the map $\pi \circ p : W \to S^2 \cap C$ does not have branch points.

Consider an arbitrary point $x \in S^2 \cap C$ and the solid angle N_x corresponding to it. Let x_1 and x_2 be two points of $S^2 \cap N_x$. We will now show that the preimages $(\pi \circ p)^{-1}(x_1)$ and $(\pi \circ p)^{-1}(x_2)$ contain the same number of points. Let us join the points x_1 and x_2 by some curve inside $S^2 \cap N_x$. By Corollary 24.18, we know that the preimage of this curve is contained in a finite number of faces of W. Since there are no branch points in any face (and their number is finite) and there are no boundary faces of W, the number of preimages for $\pi \circ p$ is some (finite) discrete and continuous function on this curve. Therefore, the number of preimages for $\pi \circ p$ of any two points of $S^2 \cap N_x$ is the same. Hence the number of preimages for $\pi \circ p$ of any two points of $S^2 \cap C$ is the same.

From this we conclude that the projection $\pi \circ p$ of the universal covering W (homeomorphic to an open disk) to $S^2 \cap C$ (i.e., homeomorphic to an open disk) is a nonramified covering with finitely many branches. Since the covering (of an open disk by an open disk) is piecewise connected, the number of branches equals one. Since by definition, $p : W \to U$ is surjective and by all of the above it is injective, the maps $p : W \to U$ and $\pi : U \to S^2 \cap C$ are bijective. $\square$

24.2.4.5 Lemma on convexity

Since every ball centered at the origin contains only a finite number of vertices of U (and they do not form a sequence tending to the boundary of the invariant cone C), the polyhedral surface U divides the space $\mathbb{R}^3$ into two connected components. Denote by H the connected component of the complement to U that does not contain the origin.

Lemma 24.20. *The set H is convex.*

Proof. Suppose that some plane passing through the origin intersects the polygonal surface U and does not contain any vertex of U. By Lemma 24.19 such a plane intersects U at some piecewise-connected broken line with an infinite number of edges. The complement of the plane of this broken line consists of two connected components. By assumption all vertices of this broken line are contained in open edges of U. By condition 5 all dihedral angles of U are well-placed. Thus the angle at every vertex of intersection of H with our plane is less than a straight angle. Hence by the previous lemma the intersection is convex.

Consider the set of all planes that pass through the origin, intersect U, and do not contain vertices of U. This set is dense in the set of all planes passing through the origin and intersecting U. Therefore, by continuity it follows that the intersection of H with any plane passing through the origin (and intersecting U) is convex.

Now we prove that the set H is convex. Let x_1 and x_2 be some points of H. Consider the plane that spans x_1, x_2, and the origin. This plane intersects U, since x_1 is in H and the origin is not in H. By the above, the intersection of H with this plane is convex. Hence the segment with endpoints x_1 and x_2 is contained in H. Thus H is convex (by the definition of convexity). □

24.2.4.6 Conclusion of the proof of Theorem 24.15: the main part

So the constructed polygonal surface U has the following properties:
— by Lemma 24.20, U bounds the convex set H;
— by construction, all vertices of U are integer points;
— by Condition 4, the set $C \setminus H$ does not contain integer points.

Therefore, the polygonal surface U is the boundary of the convex hull of all integer points inside C. Thus by definition, U is one of the sails of the continued fraction associated to the matrix A. □

Let us formulate one important conjecture here.

Conjecture 29. Conditions 1–6 imply condition 7.

24.2.5 On the verification of a conjecture for the multidimensional case

Here we briefly outline an idea of how to test a conjecture on a fundamental domain of a multidimensional continued fraction.

Conjecture for the multidimensional case. Suppose that we have a conjecture on some fundamental domain D, and also some basis $B_1, \ldots, B_n$ of the group $\Xi_+(A)$. Also let the fundamental domain and the basis have the following properties:
(i) the closure of the fundamental domain is homeomorphic to the disk;

(ii) the operators with matrices $B_1, \ldots, B_n$ define the gluing of this disk to the n-dimensional torus.

How does one test the conjecture for fundamental domains of multidimensional continued fractions? The verification of conditions (i) and (ii) is straightforward and is omitted. If these conditions hold, we check whether all the n-dimensional faces of the fundamental domain are faces of the sail. This can be done in the following way.

Suppose that the integer distances from the origin to the planes of faces F_i are equal to d_i ($i = 1, \ldots, p$, where p is the number of all n-dimensional faces). Our conjecture is true if and only if for all $i = 1, \ldots, p$ the following conditions hold:
(a) For every integer $d < d_i$ consider the plane parallel to the face F_i with integer distances to the origin equal to d. The intersection of our invariant cone with this plane does not contain any integer point.
(b) For $d = d_i$ the convex hull of all integer points in the intersection coincides with face F_i.
The verification of conditions (a) and (b) is quite complicated from the algorithmic point of view.

We conclude this section with the important inverse question of constructing periodic continued fractions.

Problem 30. (V.I. Arnold.) Does there exist an algorithm to decide whether a given type of fundamental domain is realizable by a periodic continued fraction?

The answer to this question is unknown even for the two-dimensional periodic continued fractions.

24.3 An example of the calculation of a fundamental domain

Let us construct fundamental domains for two-dimensional continued fractions of Example 22.15. Recall that

$$A_{m,n} = \begin{pmatrix} 0 & 1 & 0 \\ 0 & 0 & 1 \\ 1 & -m & -n \end{pmatrix}.$$

Theorem 24.21. *Let $m = b - a - 1$, $n = (a+2)(b+1)$ (where $a, b \geq 0$). Consider the sail for the matrix $A_{m,n}$ containing the point $(0,0,1)$. Let*

$$\begin{array}{ll} A = (1,0,a+2), & B = (0,0,1), \\ C = (b-a-1,1,0), & D = ((b+1)^2, b+1, 1). \end{array}$$

Then the following set of faces forms one of the fundamental domains:
(1) the vertex A;

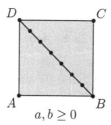

Fig. 24.1 The closure of the fundamental domain of a sail of a fraction associated to the matrix $A_{b-a-1,(a+2)(b+1)}$ (here $b = 6$, and a is arbitrary).

(2) the edges AB, AD, and BD;
(3) the triangular faces ABD and BDC.

   The closure of the fundamental domain is homeomorphic to the square shown in Fig. 24.1 (for the case of an arbitrary a, and $b = 6$).

Proof. Steps 1 and 2. We omit the first and the second steps (these steps are classical; see [41]) and here write down the result. The following two matrices generate the group $\Xi_+(A)$:

$$X_{a,b} = A_{m,n}^{-2}, \quad Y_{a,b} = A_{m,n}^{-1}\left(A_{m,n}^{-1} - (b+1)\mathrm{Id}\right),$$

where Id is the identity element in the group $SL(3,\mathbb{Z})$.

Step 3. We prove that $(0,0,1)$ is a vertex of the sail. Consider the plane passing through A, B, and D:

$$(-1-a)x + (ab+a+b+1)y + z = 1.$$

When the equations (in variables x, y, and z)

$$(-1-a)x + (ab+a+b+1)y + z = \alpha$$

do not have any integer solution for $0 < \alpha < 1$, the integer distance from ABD to the origin is equal to one. There are exactly three integer points (A, B, and D) in the intersection of the plane and the invariant cone. We leave the proof of this as an exercise for the reader.

Step 4. The conjecture of the fundamental domain was produced in the statement of this theorem.

Step 5. It remains to test the conjectured fundamental domain. For the test we need some extra points:

$$E = X_{a,b}^{-1}(B)$$
$$= (1, -ab - a - 2b - 2, a^2b^2 + 2a^2b + 4ab^2 + a^2 + 8ab + 4b^2 + 5a + 7b + 5);$$
$$F = Y_{a,b}(B) = (-a - 2, 1, 0);$$
$$H = X_{a,b}^{-1}(F) = (0, -b - 1, ab^2 + 2ab + 2b^2 + a + 4b + 3).$$

1. (Test of condition (i)). It can be shown in the usual way that the faces have common edge BD, and the edges intersect only at vertices. This implies that all adjacencies are correct, and that only one or two faces are adjacent to each edge. The closure of the boundary is a closed broken line $ABCDA$, homeomorphic to the circle.

2. (Test of condition (ii)). Direct calculations show that the operator with matrix $X_{a,b}$ takes the segment AB to DC (the point A goes to the point D and B to C). The operator with matrix $Y_{a,b}$ takes AD to BC (A goes to B and D to C). Obviously no other points glue together. The Euler characteristic of the obtained surface equals $2 - 3 + 1$, i.e., zero, and the surface is orientable.

3. (Calculation of all integer distances from the origin to the two-dimensional planes containing faces.) The integer distance to the plane of ABD equals

$$\frac{1}{b+1} \cdot \left| \left(\begin{matrix} 1 & 0 & b^2 + 2b + 1 \\ 0 & 0 & b + 1 \\ a + 1 & 1 & 1 \end{matrix} \right) \right| = \frac{b+1}{b+1} = 1.$$

The integer distance from the origin to the plane of BDC equals

$$\frac{1}{b+1} \cdot \left| \left(\begin{matrix} 0 & b^2 + 2b + 1 & b - a - 1 \\ 0 & b + 1 & 1 \\ 1 & 1 & 0 \end{matrix} \right) \right| = \frac{ab + 2b + a + 2}{b+1} = a + 2.$$

4. (Test on nonexistence of integer points inside the pyramids with vertices at the origin and bases at the faces.) Since the integer distance from the origin to the plane containing ABD equals one, the pyramid corresponding to ABD does not contain integer points different from O and the points of the face ABD.

The face BDC is integer-linear congruent to the face with vertices $(1, a + 1, -a - 2)$, $(b + 2, a + 1, -a - 2)$, $(1, a + 2, -a - 2)$ of the the list "T-W" (see Fig. 19.2). The corresponding transformation taking BCD to the face of the list "T-W" is as follows:

$$\left(\begin{matrix} b + 1 & b - a - 1 & b - a \\ 1 & 1 & 1 \\ 0 & -1 & -1 \end{matrix} \right).$$

By Theorem 19.5 the pyramid corresponding to BDC does not contain integer points different from O and the points of the face BDC.

5. (Test on convexity of dihedral angles.) Let us first consider the edge BD. This edge is adjacent to the faces ABD and BDC. The face ABD is contained in the plane

$f_{ABD}(x,y,z) = 0$, and the face BDC is contained in the plane $f_{BDC}(x,y,z) = 0$, where

$$f_{ABD}(x,y,z) = (-1-a)x + (ab+a+b+1)y + z - 1;$$
$$f_{BDC}(x,y,z) = x + (b+1)y - (a+2)z + (a+2).$$

To test that the dihedral angle corresponding to the edge BD is well-placed it is sufficient to verify the following: the point C and the origin O lie in different half-spaces with respect to the plane spanned by the points A, B, and D and the points A and O lie in different half-spaces with respect to the plane spanned by the points C, B, and D. So we need to solve the following system:

$$\begin{cases} f_{ABD}(C) \cdot f_{ABD}(O) < 0, \\ f_{BDC}(A) \cdot f_{BDC}(O) < 0. \end{cases}$$

This system is equivalent to

$$\begin{cases} (a^2 + 3a + 2) \cdot (-1) < 0, \\ (-a^2 - 3a - 1) \cdot (a+2) < 0. \end{cases}$$

Since $a \geq 0$, the inequalities hold. Thus the dihedral angle associated with the edge BD is well-placed. Since the cases of dihedral angles associated to the edges AB (and the faces ADB and AEB) and BC (and the faces BDC and CBF) can be verified in the same way, we omit their descriptions. This concludes the test of Condition 5.

6. (*Verification that all 2-stars of the vertices are regular.*) There is only one vertex in the torus decomposition. Any lift of this point to the universal covering W is adjacent to six edges and six faces. Consider a vertex of the universal covering that maps to the point B. The corresponding 2-star maps to six edges BC, BD, BA, BE, BH, and BF and to six faces BCD, BDA, BAE, BEH, BHF, and BFC, where

$$H = X_{a,b}^{-1}(F) = (0, -b-1, ab^2 + 2ab + 2b^2 + a + 4b + 3).$$

We will check that for every sufficiently small positive ε, a ray l_ε with vertex at the origin and passing through the point $P_\varepsilon = (\varepsilon, 0, 1)$ intersects exactly one of the faces of 2-stars.

First we will check that for every sufficiently small positive ε, the ray l_ε intersects the triangle BCF, or equivalently, that the ray l_ε is contained in the trihedral angle with vertex at the origin O and base in the triangle BCF. The two-dimensional face of the trihedral angle containing B, C, and O can be defined by $f_{ABO} = 0$, and the two-dimensional face of the trihedral angle containing B, F, and O can be defined by $f_{BFO} = 0$; the two-dimensional face of the trihedral angle containing C, F, and O can be defined by $f_{CFO} = 0$, where

$$f_{BCO}(x,y,z) = x + (a+1-b)y;$$
$$f_{BFO}(x,y,z) = x + (a+2)y;$$
$$f_{CFO}(x,y,z) = z.$$

For every sufficiently small positive ε, the ray l_ε is contained in the dihedral angle defined above if the following conditions hold: the points P_ε and F are in the same closed half-space with respect to the plane $f_{BCO} = 0$, the points P_ε and C are in the same closed half-space with respect to the plane $f_{BFO} = 0$, and the points P_ε and B are in the same closed half-space with respect to the plane $f_{CFO} = 0$. Since the points P_ε and B are close to each other for sufficiently small ε, they are in the same closed half-space with respect to the plane $f_{CFO} = 0$. We now check the remaining two conditions:

$$\begin{cases} f_{BCO}(P_\varepsilon) \cdot f_{BCO}(F) \geq 0, \\ f_{BFO}(P_\varepsilon) \cdot f_{BFO}(C) \geq 0, \end{cases} \quad \Leftrightarrow \quad \begin{cases} (-b-1)\varepsilon \geq 0, \\ (b+1)\varepsilon \geq 0. \end{cases}$$

Since $b, \varepsilon \geq 0$, the first inequality does not hold. Thus for every sufficiently small positive ε, the ray l_ε does not intersect the triangle BCF.

The cases of the triangles BCD, BDA, BAE, BEH, and BHF are similar to those described above and are omitted here.

The ray l_ε (for any sufficiently small positive ε) intersects the bijective image of the 2-star of the vertex at exactly one point contained in the edge AB. Therefore, all 2-stars associated to the vertices are regular.

7. (*Test that all the vertices of D are in the same invariant cone.*) The test of the seventh stage for this theorem is trivial, since D contains exactly one vertex.   □

24.4 Exercises

Exercise 24.1. Construct fundamental domains of the matrices in Examples 22.9—22.12 and prove that the corresponding triangulations are correct.

Chapter 25
Gauss Reduction in Higher Dimensions

In this chapter we continue to study integer conjugacy classes of integer matrices in general dimension. Recall that matrices A and B in $SL(n, \mathbb{Z})$ are *integer conjugate* if there exists a matrix C in $GL(n, \mathbb{Z})$ such that

$$B = CAC^{-1}.$$

The natural problem here is as follows: *describe the set of integer conjugacy classes in* $SL(n, \mathbb{Z})$. In this section we give an answer to this question for matrices whose characteristic polynomials are irreducible over the field of rational numbers.

Gauss's reduction theory gives a complete geometric description of conjugacy classes for the case $n = 2$, as we have already discussed in Chapter 11. In the multi-dimensional case the situation is more complicated. It is relatively simple to check whether two given matrices are integer conjugate (see, for instance, [6] and [80]), but to distinguish conjugacy classes is a much harder task. In this chapter we describe a generalization of Gauss's reduction theory to the multidimensional case (see also [107]). We study questions related to the three-dimensional case in more detail.

There exists an alternative algebraic approach to the study of $SL(n, \mathbb{Z})$ conjugacy classes, which we do not touch in this book. There one starts with $GL(n, \mathbb{Q})$ conjugacy classes and splits them into $GL(n, \mathbb{Z})$ conjugacy classes. This reduces the problem to certain problems related to orders of algebraic fields that are defined by the roots of the characteristic polynomial of the corresponding matrices.

25.1 Organization of this chapter

In this chapter we investigate the following five aspects.

© Springer-Verlag GmbH Germany, part of Springer Nature 2022
O. N. Karpenkov, *Geometry of Continued Fractions*,
Algorithms and Computation in Mathematics 26,
https://doi.org/10.1007/978-3-662-65277-0_25

I. Generalized reduced matrices. In the multidimensional Gauss's reduction theory, Hessenberg matrices play the role of reduced matrices. A *Hessenberg matrix* is a matrix that vanishes below the superdiagonal (see [208]); they were introduced by K. Hessenberg in [85]. Hessenberg matrices were essentially used in the QR-algorithm for the eigenvalue problem (see [59], [213], and [167]). Further, in [107] they were considered in the framework of the multidimensional Gauss reduction theory as a multidimensional analogue of reduced matrices.

In Section 25.2 we introduce a natural notion of Hessenberg complexity for Hessenberg matrices. Further we show that *each integer conjugacy class of* $SL(n, \mathbb{Z})$ *has only finitely many distinct Hessenberg matrices with minimal complexity* (Theorem 25.8).

II. Complete invariant of integer conjugacy classes. In Section 25.3 we introduce a complete geometric invariant of a Dirichlet group. It is a periodic multidimensional continued fractions in the sense of Klein—Voronoi (Theorem 25.23). Klein—Voronoi continued fractions naturally generalize Klein continued fractions. Further, we deduce the complete invariants of integer conjugacy classes of $GL(n, \mathbb{Z})$ matrices. We show that the conjugacy classes are represented by periodic shifts of Klein—Voronoi periodic continued fractions (Theorem 25.24).

III. Calculation of reduced matrices. In Section 25.4 we introduce a technique to construct reduced matrices in any integer conjugacy class that has minimal Hessenberg complexity in this class.

IV. Solution of Diophantine equations related to certain decomposable forms. In Section 25.5 we study the integer points at the level sets of the decomposable forms (we have already discussed this question in Chapter 11 in connection with the Markov spectrum). As a technical detail we introduce a notion of multidimensional w-continued fractions similar to its one-dimensional version of Chapter 14.

V. Integer conjugacy classes of $SL(3, \mathbb{Z})$**.** We examine the structure of integer conjugacy classes of $SL(3, \mathbb{Z})$ via Klein—Voronoi continued fractions. The main results concern only the operators with a pair of complex conjugate eigenvectors. It turns out that in this case, Hessenberg matrices distinguish corresponding conjugacy classes asymptotically (Theorem 25.48). The analogous statement does not hold for the case of operators with three real eigenvalues. In general it is much less known for the case of three real eigenvalues.

25.2 Hessenberg matrices and conjugacy classes

In Section 25.2.1 we begin with necessary definitions and notation related to Hessenberg matrices. Further, in Section 25.2.2 we discuss algorithmic aspects of construction of perfect Hessenberg matrices integer congruent to a given one. In Section 25.2.3 we show that every integer conjugacy class with irreducible characteristic polynomial has a finite nonzero number of ς-reduced matrices. Further in Section 25.2.4 we study the set of perfect Hessenberg matrices. We think of this set as a "book" whose "pages" are enumerated by Hessenberg types. Since the matrices from the same page are distinguished by their polynomials, any two matrices on one page are not integer congruent. We conclude this section with a brief discussion of ς-reduced 2-dimensional matrices of the same Hessenberg type.

25.2.1 Notions and definitions

In this subsection we briefly introduce Hessenberg matrices, which we consider as generalized reduced matrices in the multidimensional case.

25.2.1.1 Perfect Hessenberg matrices

A matrix M of the form

$$\begin{pmatrix} a_{1,1} & a_{1,2} & \cdots & a_{1,n-2} & a_{1,n-1} & a_{1,n} \\ a_{2,1} & a_{2,2} & \cdots & a_{2,n-2} & a_{2,n-1} & a_{2,n} \\ 0 & a_{3,2} & \cdots & a_{3,n-2} & a_{3,n-1} & a_{3,n} \\ \vdots & \vdots & \ddots & \vdots & \vdots & \vdots \\ 0 & 0 & \cdots & a_{n-1,n-2} & a_{n-1,n-1} & a_{n-1,n} \\ 0 & 0 & \cdots & 0 & a_{n,n-1} & a_{n,n} \end{pmatrix}$$

is called an (upper) Hessenberg matrix. We say that the matrix M is of Hessenberg type

$$\langle a_{1,1}, a_{2,1} | a_{1,2}, a_{2,2}, a_{3,2} | \cdots | a_{1,n-1}, a_{2,n-1}, \ldots, a_{n,n-1} \rangle.$$

Definition 25.1. A Hessenberg matrix in $\mathrm{SL}(n, \mathbb{Z})$ is said to be *perfect* if for every pair of integers (i, j) satisfying $1 \le i < j+1 \le n$ the following inequalities hold: $0 \le a_{i,j} < a_{j+1,j}$.

In other words, all elements in the first $n-1$ columns of a perfect Hessenberg matrix are nonnegative integers, and in addition, the element $a_{j+1,j}$ is maximal in each of these columns.

25.2.1.2 Reduced Hessenberg matrices

We study only the simplest general case of $\mathrm{SL}(n, \mathbb{Z})$ matrices whose characteristic polynomials are irreducible over $\mathbb{Q}$. Every such matrix is integer conjugate to a perfect Hessenberg matrix, and each conjugacy class of $\mathrm{SL}(n, \mathbb{Z})$ contains infinitely many perfect Hessenberg matrices. To reduce the number of such matrices, we introduce a natural notion of complexity.

Definition 25.2. Consider a Hessenberg matrix $M = (a_{i,j})$. The integer

$$\prod_{j=1}^{n-1} |a_{j+1,j}|^{n-j}$$

is called the *Hessenberg complexity* of M. Denote it by $\varsigma(M)$.

The Hessenberg complexity is in fact equivalent to the volume of the parallelepiped spanned by $v = (1,0,\ldots,0)$, $M(v)$, $M^2(v),\ldots,M^{n-1}(v)$, as we will discuss in Section 25.4.1.

An integer Hessenberg matrix has unit Hessenberg complexity if and only if

$$a_{2,1} = \cdots = a_{n,n-1} = 1.$$

In the literature such matrices are known as *Frobenius matrices* (or companion matrices). The elements of the last column of any Frobenius matrix are exactly the coefficients of the characteristic polynomial multiplied alternately by ± 1.

Example 25.3. Consider the matrix

$$\begin{pmatrix} 1 & 2 & 3 \\ 2 & 3 & 6 \\ 0 & 5 & -1 \end{pmatrix}.$$

This matrix is a perfect Hessenberg matrix of type $\langle 1,2|2,3,5\rangle$. Its Hessenberg complexity is $2^2 \cdot 5 = 20$.

Definition 25.4. A perfect Hessenberg matrix M is said to be ς-*nonreduced* if there exists an integer matrix M' of smaller Hessenberg complexity integer congruent to M. Otherwise, we say that M is ς-*reduced*.

Remark. Let us say a few words about the relation between reduced matrices introduced in Chapter 11 and ς-reduced matrices in $SL(2,\mathbb{Z})$. On the one hand, every reduced hyperbolic (i.e., real spectrum) matrix is a perfect Hessenberg matrix. On the other hand, every ς-reduced matrix is also reduced, although there are certain reduced matrices that are not ς-reduced.

Later, in Theorem 25.8, we show that the number of ς-reduced matrices is finite for every integer conjugacy class. Some classes contain more than one ς-reduced perfect Hessenberg matrix.

Example 25.5. The ς-reduced Hessenberg matrices (with Hessenberg complexity equivalent to 3)

$$M_1 = \begin{pmatrix} 0 & 1 & 2 \\ 1 & 0 & 0 \\ 0 & 3 & 5 \end{pmatrix} \quad \text{and} \quad M_2 = \begin{pmatrix} 0 & 2 & 3 \\ 1 & 1 & 1 \\ 0 & 3 & 4 \end{pmatrix}$$

are integer conjugate.

25.2.2 Construction of perfect Hessenberg matrices conjugate to a given one

In this subsection we show how to construct perfect Hessenberg matrices that are integer congruent to a given one. The main ingredients of the algorithm are in the proof of the following proposition.

Proposition 25.6. *Consider an* $\mathrm{SL}(n,\mathbb{Z})$ *matrix* M *whose characteristic polynomial is irreducible over* $\mathbb{Q}$. *Then for every vector* v *of integer length* 1 *there exists a unique matrix* C *satisfying the following conditions:*

— $C(1,0,\ldots,0) = v$;
— *the matrix* CMC^{-1} *is perfect Hessenberg (we denote this matrix by* $(M|v)$).

In other words, for a given integer operator A, every integer vector of unit integer length is uniquely extended to the basis of the integer lattice in which the operator A has a perfect Hessenberg matrix.

Proof. **Existence.** Consider an $\mathrm{SL}(n,\mathbb{Z})$ matrix M with irreducible characteristic polynomial. Let A denotes the linear operator with matrix M. Consider an integer vector v of unit integer length. For $i = 1,\ldots,n-1$ we define

$$V_i = \mathrm{Span}\left(v, A(v), A^2(v), \ldots, A^{i-1}(v)\right).$$

Since the characteristic polynomial of M is irreducible, the set of all spaces V_i forms a complete flag in $\mathbb{R}^n$. All the vectors $A^j(v)$ are integer vectors; hence every space V_i contains a full-rank integer sublattice.

Let us inductively construct an integer basis $\{\tilde{e}_i\}$ of a vector space $\mathbb{R}^n$ such that:
— for every i the vectors $\tilde{e}_1,\ldots,\tilde{e}_i$ generate the sublattice $\mathbb{Z}^n \cap V_i$;
— the matrix of A is perfect Hessenberg in the basis $\{\tilde{e}_i\}$.

Base of induction. Let $\tilde{e}_1 = v$. Since $1\ell(\tilde{e}_1) = 1$, the vector $\tilde{e}_1$ generates $\mathbb{Z}^n \cap V_1$.

Step of induction. Suppose that we have already constructed $\tilde{e}_1,\ldots,\tilde{e}_k$ forming a basis of $\mathbb{Z}^n \cap V_k$. Let us now calculate $\tilde{e}_{k+1}$. Choose an integer vector g_{k+1} of the space V_{k+1} a unit integer distance to the space V_k. Since $\tilde{e}_1,\ldots,\tilde{e}_k$ generate $\mathbb{Z}^n \cap V_k$, the vectors $\tilde{e}_1,\ldots,\tilde{e}_k, g_{k+1}$ generate the sublattice $\mathbb{Z}^n \cap V_{k+1}$. Since $A(\tilde{e}_k) \in V_{k+1}$, we have

$$A(\tilde{e}_k) = \sum_{i=1}^{k} q_{i,k}\tilde{e}_i + a_{k+1,k}g_{k+1}.$$

For $i = 1,\ldots,k$, we define $b_{i,k}$ and $a_{i,k}$ as integer quotients and remainders of the following equations:

$$q_{i,k} = b_{i,k} \cdot |a_{k+1,k}| + a_{i,k}, \quad \text{where} \quad 0 \le a_{i,k} < a_{k+1,k}.$$

Then we have

$$A(\tilde{e}_k) = |a_{k+1,k}| \left(\text{sign}(a_{k+1,k}) g_{k+1} + \sum_{i=1}^{k} b_{i,k} \tilde{e}_i \right) + \sum_{i=1}^{k} a_{i,k} \tilde{e}_i$$

(where sign is the sing function over the real numbers). Finally, we get

$$\tilde{e}_{k+1} = \text{sign}(a_{k+1,k}) g_{k+1} + \sum_{i=1}^{k} b_{i,k} \tilde{e}_i.$$

The characteristic polynomial of A is irreducible over $\mathbb{Q}$, and hence the integer spaces V_i are not invariant subspaces of A. Therefore, the set $\{\tilde{e}_1, \ldots, \tilde{e}_{k+1}\}$ is a basis of $\mathbb{Z}^n \cap V_{k+1}$. The induction step is complete.

Denote by $\hat{M}$ the matrix of A in the basis $\{\tilde{e}_i\}$. By construction, the matrix $\hat{M}$ is a perfect Hessenberg matrix, and its Hessenberg type is

$$\left\langle a_{1,1}, |a_{2,1}| \Big| a_{1,2}, a_{2,2}, |a_{3,2}| \Big| \cdots \Big| a_{1,n-1}, \ldots, a_{n-1,n-1}, |a_{n,n-1}| \right\rangle.$$

Both M and $\hat{M}$ are matrices of A in two different integer bases, which means they are integer conjugate. The integers $a_{i+1,i}$ are nonzero $(i = 1, \ldots, n-1)$, since the characteristic polynomial of A is irreducible. Therefore, $\varsigma(\hat{M}) > 0$. Denote by C the transition matrix to the basis $\{\tilde{e}_i\}$. Then $C(\tilde{e}_1) = v$ and $\hat{M} = CMC^{-1}$ (where $\hat{M}$ is a perfect Hessenberg matrix).

Uniqueness. Let C_1 and C_2 satisfy the conditions of the theorem. Suppose that C_1 and C_2 are the transition matrices to the bases $\{\tilde{e}_i\}$ and $\{\hat{e}_i\}$ respectively. Then

$$\tilde{e}_1 = C_1(e_1) = v = C_2(e_1) = \hat{e}_1.$$

For $i > 1$ the equality $\tilde{e}_i = \hat{e}_i$ follows from the uniqueness of the choice of the coefficients in the proof of existence. Hence the bases $\{\tilde{e}_i\}$ and $\{\hat{e}_i\}$ coincide. Therefore, $C_1 = C_2$.   □

Now let us briefly summarize the algorithm.

Algorithm to construct perfect Hessenberg matrices.

Input Data: (M, v). Here M is a matrix of a lattice-preserving operator A whose characteristic polynomial is irreducible over q, and v is an integer vector of unit integer length.

Step 1. Put $\tilde{e}_1 = v$.

Inductive Step k. We have already constructed $\tilde{e}_1, \ldots, \tilde{e}_k$. Choose $g_{k+1} \in V_{k+1}$ satisfying $\text{ld}(g_{k+1}, V_k) = 1$ and find integers $q_{i,k}$ for $i = 1, \ldots, k$ and $a_{k+1,k}$ from the decomposition

$$A(\tilde{e}_k) = \sum_{i=1}^{k} q_{i,k} \tilde{e}_i + a_{k+1,k} g_{k+1}.$$

Find $b_{i,k}$ and $a_{i,k}$ $(i = 1, \ldots, k)$ as integer quotients and reminders of the equations

$$q_{i,k} = b_{i,k} \cdot |a_{k+1,k}| + a_{i,k}.$$

Put

$$\tilde{e}_{k+1} = \text{sign}(a_{k+1,k})g_{k+1} + \sum_{i=1}^{k} b_{i,k}\tilde{e}_i.$$

Output Data: the perfect Hessenberg matrix CMC^{-1}, where C is the transition matrix to $\{\tilde{e}_k\}$.

Note that the basis $\{\tilde{e}_i\}$ is constructed in n steps.

Later, in the proof of Theorem 25.31, we use the following corollary.

Corollary 25.7. *Consider an* $\text{SL}(n,\mathbb{Z})$ *operator A with matrix M and let $B \in \Xi(A)$. Then for an arbitrary v we have* $(M|v) = (M|B(v))$.

Proof. Each step of the algorithm produces the same data for v and $B(v)$, due to the fact that A and B commute. Therefore, $(M|v) = (M|B(v))$. ☐

25.2.3 Existence and finiteness of ς-reduced Hessenberg matrices

It turns out that ς-reduced Hessenberg matrices are good for the classification of conjugacy classes in $\text{SL}(n,\mathbb{Z})$. Although they do not form complete invariants of conjugacy classes, they are present in each class and their number is finite. We give an explicit construction of all ς-reduced Hessenberg matrices conjugate to a given one in Section 25.4.3 via Klein—Voronoi continued fractions.

Theorem 25.8. *Let M be an* $\text{SL}(n,\mathbb{Z})$ *matrix. Then the number of ς-reduced Hessenberg matrices integer conjugate to M is finite and greater than zero.*

The proof of this theorem is based on the following proposition.

Proposition 25.9. *Every Hessenberg matrix with positive Hessenberg complexity is identified by its Hessenberg type and characteristic polynomial.* ☐

Proof. Let $M = (a_{i,j})$ be a Hessenberg matrix with positive Hessenberg complexity. The first $n-1$ columns of M are entirely defined by the Hessenberg type of M. The last column is uniquely defined from the characteristic polynomial of M,

$$x^n + c_{n-1}x^{n-1} + \cdots + c_1x + c_0,$$

in the following way. For every k the coefficient c_k is a polynomial in variables $a_{i,j}$ that does not depend on $a_{1,n}, \ldots, a_{k,n}$. This polynomial has a unique monomial containing $a_{k+1,n}$, namely

$$\left(\prod_{j=k+1}^{n-1} a_{j+1,j} \right) a_{k+1,n}.$$

Since $\varsigma(M) \neq 0$, the product in the parentheses is nonzero. Hence $a_{k+1,n}$ is uniquely defined by c_k and the elements $a_{i,j}$ of the first $n-1$ columns. Therefore, all elements of M are defined by the Hessenberg type of M and the characteristic polynomial of M.
$\square$

Proof of Theorem 25.8. **Existence.** By Proposition 25.6 there exist perfect Hessenberg matrices integer conjugate to M. The set of values of Hessenberg complexity is discrete and bounded from below; hence there exists a perfect Hessenberg matrix $\tilde{M}$ integer conjugate to M and with minimal possible Hessenberg complexity. By definition the matrix $\tilde{M}$ is a ς-reduced.

Finiteness. Let c be the Hessenberg complexity of $\tilde{M}$. By definition the Hessenberg complexity of all the other ς-reduced Hessenberg matrices integer conjugate to M equals c. The number of Hessenberg types whose Hessenberg complexity equals c is finite. Notice that the integer conjugate matrices have the same characteristic polynomials. From Proposition 25.9 there exists at most one integer Hessenberg matrix with a given Hessenberg type and a given polynomial. Therefore, the number of ς-reduced Hessenberg matrices integer conjugate to M is finite.
$\square$

25.2.4 Families of Hessenberg matrices with given Hessenberg type

In this subsection we discuss the structure of Hessenberg matrices with given Hessenberg type. We begin with a particular example.

Example 25.10. Let us examine the Hessenberg type $\langle 0,1|1,0,2\rangle$. The set of all matrices of this Hessenberg type is a two-parameter family with parameters m and n:

$$H(\langle 0,1|1,0,2\rangle) = \left\{ \begin{pmatrix} 0 & 1 & 1 \\ 1 & 0 & 0 \\ 0 & 2 & 1 \end{pmatrix} + m \begin{pmatrix} 0 & 0 & 0 \\ 0 & 0 & 1 \\ 0 & 0 & 0 \end{pmatrix} + n \begin{pmatrix} 0 & 0 & 1 \\ 0 & 0 & 0 \\ 0 & 0 & 2 \end{pmatrix} \middle| m, n \in \mathbb{Z} \right\}.$$

Define

$$H_{\langle 0,1|1,0,2\rangle}^{(1,0,1)}(m,n) = \begin{pmatrix} 0 & 1 & n+1 \\ 1 & 0 & m \\ 0 & 2 & 2n+1 \end{pmatrix}.$$

The discriminant of $H_{\langle 0,1|1,0,2\rangle}^{(1,0,1)}(m,n)$ equals

$$-44 - 44n^2 - 56mn - 32n^3 + 32m^3 + 16m^2n^2 + 16mn^2 + 16m^2n - 56n - 8m + 52m^2.$$

The set of matrices with negative discriminant for the given family coincides with the union of the sets of integer solutions of the following quadratic inequalities:

$$2m \leq -n^2 - n - 2 \quad \text{and} \quad 2n \leq m^2 + m.$$

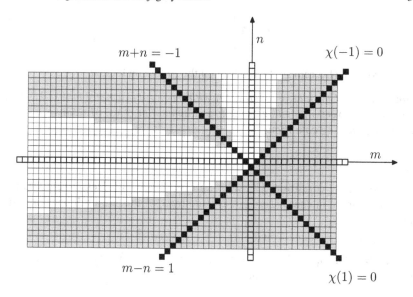

Fig. 25.1 Matrices of Hessenberg type $\langle 0,1|1,0,2\rangle$.

In Fig. 25.1 we represent a matrix $H^{(1,0,1)}_{\langle 0,1|1,0,2\rangle}(m,n)$ by the square in the intersection of the mth column and the nth row. Matrices with reducible characteristic polynomials correspond to black squares. Light gray squares represent matrices with three real eigenvalues. Matrices shown as white squares have a pair of complex conjugate eigenvalues.

It turns out that the general case is similar. Consider a Hessenberg type

$$\Omega = \langle a_{1,1}, a_{2,1} | a_{1,2}, a_{2,2}, a_{3,2} | \cdots | a_{1,n-1}, \ldots, a_{n-1,n-1}, a_{n,n-1} \rangle.$$

Denote by $H(\Omega)$ the set of all Hessenberg matrices in $\mathrm{SL}(n,\mathbb{Z})$ of Hessenberg type Ω.

For $k = 1, \ldots, n-1$ we put

$$v_k(\Omega) = (a_{k,1}, \ldots, a_{k,k+1}, 0, \ldots, 0)$$

and denote by $M_k(\Omega)$ the matrix whose first $n-1$ columns are equal to zero and the last one equals to $v_k(\Omega)$. Denote also by $\sigma(\Omega)$ the $(n-1)$-dimensional simplex with vertices

$$O \text{ (the origin)}, \quad v_1, \quad \ldots, \quad v_{n-1}.$$

Theorem 25.11. (i) The set $H(\Omega)$ is not empty if and only if $\mathrm{lV}(\sigma(\Omega)) = 1$.

(ii) Let $M_0 \in H(\Omega)$. Then $H(\Omega)$ is an integer affine $(n-1)$-dimensional sublattice in the lattice of all integer $(n \times n)$ matrices, i.e.,

$$H(\Omega) = \left\{ M_0 + \sum_{i=1}^{n-1} c_i M_i(\Omega) \,\Big|\, c_1, \ldots, c_{n-1} \in \mathbb{Z} \right\}.$$

The proof of Theorem 25.11 is based on the following lemma.

Lemma 25.12. *Consider an operator A with integer Hessenberg matrix M of type Ω. Let v be the vector standing in the last column of M. Then $M \in \mathrm{GL}(n, \mathbb{Z})$ if and only if the following conditions hold:*
— $\mathrm{lV}(\sigma(\Omega)) = 1$;
— $\mathrm{ld}(v, \mathrm{Span}(\sigma(\Omega))) = 1$.

Proof. **Necessary condition.** Consider an operator A with Hessenberg $\mathrm{GL}(n, \mathbb{Z})$ matrix M in some integer basis $\{g_i\}$. Denote by S_g^{n-1} the $(n-1)$-dimensional simplex with vertices

$$O, \quad g_1, \quad \ldots, \quad g_{n-1}.$$

Since $M \in \mathrm{GL}(n, \mathbb{Z})$, the operator A preserves integer volumes and integer distances. Since $\mathrm{lV}(S_g^{n-1}) = 1$, we have the first condition:

$$\mathrm{lV}(\sigma(\Omega)) = \mathrm{lV}(A(S_g^{n-1})) = \mathrm{lV}(S_g^{n-1}) = 1.$$

Notice that

$$A(\mathrm{Span}(S_g^{n-1})) = \mathrm{Span}(\sigma(\Omega)) \qquad \text{and} \qquad A(g_n) = v.$$

Therefore, we get the second condition:

$$\mathrm{ld}(v, \mathrm{Span}(\sigma(\Omega))) = \mathrm{ld}(g_n, A(\mathrm{Span}(S_g^{n-1}))) = \mathrm{ld}(g_n, \mathrm{Span}(g_1, \ldots, g_{n-1})) = 1.$$

Sufficient condition. Suppose that both conditions of the lemma hold. Then the operator A preserves the integer lattice (generated by $g_1, \ldots, g_n$). Therefore, $M \in \mathrm{SL}(n, \mathbb{Z})$. $\qquad\square$

Proof of Theorem 25.11. (i) Suppose that $\mathrm{lV}(\sigma(\Omega)) = 1$. Choose an integer vector v at unit integer distance to the plane $\mathrm{Span}(\sigma(\Omega))$. By Lemma 25.12 both matrices defined by Ω and vectors $\pm v$ are in $\mathrm{GL}(n, \mathbb{Z})$. One of them is in $\mathrm{SL}(n, \mathbb{Z})$. Conversely, if $H(\Omega)$ contains an $\mathrm{SL}(n, \mathbb{Z})$ matrix, then by Lemma 25.12 the integer volume of $\sigma(\Omega)$ equals 1.

Statement (ii) is straightforward, since the determinant of a matrix is an additive function with respect to the operation of vector addition in the last column. $\qquad\square$

25.2.5 ς-reduced matrices in the 2-dimensional case

Let us say a few words about ς-reduced 2-dimensional matrices in families $H(\Omega)$. We start with the following example.

Example 25.13. Consider the family of matrices $H(\langle 3,4 \rangle)$:

$$\begin{pmatrix} 3 & 3m+2 \\ 4 & 4m+3 \end{pmatrix}.$$

Their Hessenberg complexity equals 4. Almost all matrices of this family have irreducible, over $\mathbb{Q}$, characteristic polynomials with two real roots (except for $m = -1, -2$). For $m = 0, 1, 2, 3, \ldots$ the LLS-periods of the matrices are

$$(2,2), \quad (1,2,1,1), \quad (1,2,1,2), \quad (1,2,1,3), \quad \ldots, \quad (1,2,1,m), \quad \ldots \; .$$

The matrices are ς-reduced for $m \geq 2$. For $m = -3, -4, -5, \ldots$ the corresponding LLS-periods are

$$(4,1), \quad (4,2), \quad (4,3), \quad (4,4), \quad \ldots, \quad (4,-2-m), \quad \ldots \; .$$

Starting from $m \leq -6$ the matrices are ς-reduced.

The example of Hessenberg type $\langle 3,4 \rangle$ shows that almost all matrices of this Hessenberg type are ς-reduced. That is actually the case for all Hessenberg types in $SL(2,\mathbb{Z})$.

Theorem 25.14. *Almost all matrices of a given Hessenberg type in* $SL(2,\mathbb{Z})$ *are* ς-reduced. □

This follows from Theorem 25.31 below and direct calculation of complexities for all vertices of the period; we skip the proof here.

In Theorem 25.48 we prove similar behavior for $SL(3,\mathbb{Z})$ matrices having two complex conjugate eigenvalues (see also Conjecture 31).

25.3 Complete geometric invariant of conjugacy classes

In this section we introduce a geometric complete invariant of integer conjugacy classes: multidimensional continued fractions in the sense of Klein—Voronoi. We start with necessary definitions in Section 25.3.1. In particular, we give a general definition of Klein—Voronoi continued fractions in all dimensions in Section 25.3.1.1 and discuss their algebraic periodicity. Further, in Section 25.3.2 we show that the Klein—Voronoi continued fraction for an operator A is a geometric complete invariant of the corresponding Dirichlet group $\varXi(A)$. Finally, in Section 25.3.3 we discuss that periodic shifts distinguish integer conjugacy classes of $SL(n,\mathbb{Z})$ matrices.

25.3.1 Continued fractions in the sense of Klein—Voronoi

Approximately at the time of the work by F. Klein on continued fractions related to real matrices, G. Voronoi in his dissertation [220] introduced a geometric algorithmic definition that covers cases of matrices having complex conjugate eigenvalues. In [32] and [33] J.A. Buchmann generalized Voronoi's algorithm, making it more convenient for computation of fundamental units in orders. We use ideas of J.A. Buchmann to define the multidimensional continued fraction in the sense of Klein—Voronoi for all cases. Note that if all eigenvalues of an operator are real then the Klein—Voronoi multidimensional continued fraction is a continued fraction in the sense of Klein.

25.3.1.1 General definitions

Consider an operator A in $GL(n, \mathbb{R})$ with distinct eigenvalues. Suppose that it has k real eigenvalues $r_1, \ldots, r_k$ and $2l$ complex conjugate eigenvalues $c_1, \bar{c}_1, \ldots, c_l, \bar{c}_l$, where $k + 2l = n$. Denote by $L_{\mathbb{R}}(A)$ the space spanned by the real eigenvectors.

Denote by $T^l(A)$ the set of all real operators commuting with A such that their real eigenvalues are all units and the absolute values for all complex eigenvalues equal one. In other words:

$$T_A = \{B \in GL(n, \mathbb{R}) \mid AB = BA, \text{spec}(B) \subset S^1, B|_{L_{\mathbb{R}}(A)} = \text{Id}|_{L_{\mathbb{R}}(A)}\},$$

where $\text{spec}(B)$ is the spectrum of B and S^1 is the complex unit circle. In fact, $T^l(A)$ is an abelian group with the operation of matrix multiplication.

For a vector v in $\mathbb{R}^n$, we denote by $T_A(v)$ the orbit of v with respect to the action of the group of operators $T^l(A)$. If v is in general position with respect to the operator A (i.e., it does not lie in invariant planes of A), then $T_A(v)$ is homeomorphic to the l-dimensional torus. For a vector of an invariant plane of A the orbit $T_A(v)$ is also homeomorphic to a torus of positive dimension not greater than l, or to a point.

Example 25.15. Suppose that A is a real spectrum matrix, i.e., $l = 0$. Since all its eigenvectors are real, $T^0(A)$ consists only of the unit operator and $T_A(v) = \{v\}$.

Example 25.16. Now consider the case of a pair complex conjugate eigenvalues, i.e., $l = 1$. The group $T^1(A)$ corresponds to elliptic rotations in the invariant plane of A corresponding to complex eigenvalues. Such rotations are parameterized by the angle of rotation. A general orbit of $T_A(v)$ is an ellipse around the $(n-2)$-dimensional invariant subspace corresponding to real eigenvalues. Every orbit in the invariant subspace spanned by real eigenvalues consists of one point.

Let g_i be a real eigenvector with eigenvalue r_i for $i = 1, \ldots, k$, and let g_{k+2j-1} and g_{k+2j} be vectors corresponding to the real and imaginary parts of some complex eigenvector with eigenvalue c_j for $j = 1, \ldots, l$. Consider the coordinate system corresponding to the basis $\{g_i\}$:

$$OX_1X_2 \ldots X_kY_1Z_1Y_2Z_2 \ldots Y_lZ_l.$$

Denote by π the $(k+l)$-dimensional plane $OX_1X_2 \ldots X_kY_1Y_2 \ldots Y_l$. Let π_+ be the cone in the plane π defined by the equations $y_i \geq 0$ for $i = 1, \ldots, l$. For every v the orbit $T_A(v)$ intersects the cone π_+ in a unique point.

Definition 25.17. A point p in the cone π_+ is said to be π-*integer* if the orbit $T_A(p)$ contains at least one integer point.

Consider all (real) hyperplanes invariant under the action of the operator A. There are exactly k such hyperplanes. In the above coordinates, the ith of them is defined by the equation $x_i = 0$. The complement to the union of all invariant hyperplanes in the cone π_+ consists of 2^k arcwise connected components $C_1(A), \ldots, C_{2^k}(A)$.

Definition 25.18. The convex hull of all π-integer points except the origin contained in an arcwise connected component $C_i(A)$ is called a *factor-sail* of $C_i(A)$. The set of all factor-sails is said to be the *factor-continued fraction* for the operator A. We denote it by $\hat{S}_i(A)$.

Definition 25.19. The union of all orbits $T_A(*)$ in $\mathbb{R}^n$ represented by the points in the factor-sail of $C_i(A)$ is called the *sail* of $T_A(C_i)$, which we denote by $S_i(A)$. The union of all sails is said to be the *continued fraction* for the operator A in the sense of Klein—Voronoi (see Fig. 25.2 below). We denote it by $KVCF(A)$.

In algebraic language we write the factor sail as

$$\hat{S}_i(A) = \partial\left(\operatorname{conv}\left(\{q \in \pi_+ \mid T_A(q) \cap C_i(A) \cap \mathbb{Z}^n \neq \emptyset\} \setminus \{O\}\right)\right),$$

and hence

$$S_i(A) = \bigcup_{p \in \hat{S}_i(A)} T_A(p) \quad \text{and} \quad KVCF(A) = \bigcup_{i=1}^{2^k} S_i(A).$$

Definition 25.20. The intersection of a factor-sail with a hyperplane in π is said to be an m-*dimensional face* of the factor-sail if it is homeomorphic to the m-dimensional disk.

The union of all orbits in $\mathbb{R}^n$ represented by points in some face of the factor-sail is called the *orbit-face* of the operator A.

Integer points of the sail are said to be *vertices* of this sail.

In the simplest possible cases of $k+l = 1$, every factor-sail of A is a point. If $k+l > 1$, then every factor-sail of A is an infinite polyhedral surface homeomorphic to $\mathbb{R}^{k+l-1}$.

25.3.1.2 Algebraic continued fractions

Consider now an operator A in the group $GL(n, \mathbb{Z})$ whose characteristic polynomial is irreducible over $\mathbb{Q}$. Suppose that it has k real roots and $2l$ complex conjugate roots, where $k + 2l = n$.

Further, we consider the Dirichlet group $\Xi(A)$ of all $GL(n, \mathbb{Z})$ operators commuting with A and the corresponding positive Dirichlet group $\Xi_+(A)$ (for more details see Chapter 21). The Dirichlet group $\Xi(A)$ takes a Klein—Voronoi continued fraction to itself and permutes the sails. The positive Dirichlet group $\Xi_+(A)$ consists exactly of operators preserving every sail. By the Dirichlet unit theorem, the group $\Xi(A)$ is homomorphic to $\mathbb{Z}^{k+l-1} \oplus G$, where G is a finite abelian group. The group $\Xi_+(A)$ is homeomorphic to $\mathbb{Z}^{k+l-1}$ and its action on any sail is free. The quotient of any sail by the action of $\Xi_+(A)$ is homeomorphic to the $(n-1)$-dimensional torus.

Definition 25.21. A *fundamental domain* of $KVCF(A)$ is a collection of orbit-faces of $KVCF(A)$ representing exactly one of the different classes of $KVCF(A)/\Xi(A)$. A *fundamental domain* of a sail $S_i(A)$ is a collection of orbit-faces of $S_i(A)$ representing exactly one of the different classes of $S_i(A)/\Xi_+(A)$.

Example 25.22. Let us study an operator A with a Frobenius matrix

$$\begin{pmatrix} 0 & 0 & 1 \\ 1 & 0 & 1 \\ 0 & 1 & 3 \end{pmatrix}.$$

This operator has one real and two complex conjugate eigenvalues. Therefore, the cone π_+ for A is a two-dimensional half-plane. In Fig. 25.2a the half-plane π_+ is shaded in light gray and the invariant plane corresponding to the pair of complex eigenvectors is in dark gray. The vector shown in Fig. 25.2a with endpoint at the origin is an eigenvector of A.

In Fig. 25.2b we show the cone π_+. The invariant plane separates π_+ into two parts. The dots on π_+ are the π-integer points. The boundaries of the convex hulls in each part of π_+ are two factor-sails. One factor-sail is taken to another by the induced action of $-\text{Id}$.

Finally, in Fig. 25.2c we show one of the sails. Three orbit-vertices shown in the figure correspond to the vectors $(1,0,0)$, $(0,1,0)$, and $(0,0,1)$: the large dark points $(0,1,0)$ and $(0,0,1)$ are visible at the corresponding orbit-vertices.

The positive Dirichlet group $\Xi_+(A)$ in our example is homeomorphic to $\mathbb{Z}$, and is generated by A. The group $\Xi(A)$ is homeomorphic to $\mathbb{Z} \oplus \mathbb{Z}/2\mathbb{Z}$ with generators A and $-\text{Id}$. The operator A takes the point $(1,0,0)$ and its orbit-vertex to the point $(0,1,0)$ and the corresponding orbit-vertex. Therefore, every fundamental domain of the continued fraction for the operator A contains one orbit-vertex and one vertex edge. For instance, we can choose the orbit-vertex corresponding to the point $(1,0,0)$ and the orbit-edge corresponding to the "tube" connecting orbit-vectors for the points $(1,0,0)$ and $(0,1,0)$.

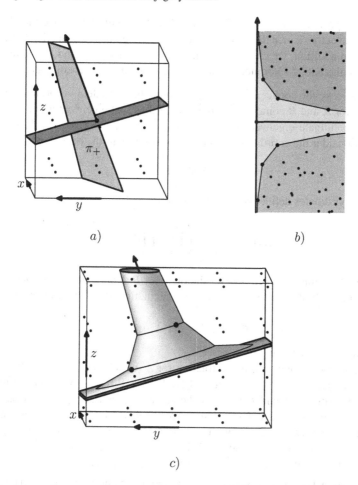

Fig. 25.2 A three-dimensional example (Example 25.22): (a) the cone π_+ and the eigenplane; (b) the continued factor-fraction; (c) a sail of the continued fraction.

25.3.2 Geometric complete invariants of Dirichlet groups

One of the main properties of Klein—Voronoi continued fractions is that they classify all Dirichlet groups.

Theorem 25.23. *Let $A, B \in \mathrm{GL}(n, \mathbb{Z})$ have characteristic polynomials irreducible over $\mathbb{Q}$. Then $\Xi(A) = \Xi(B)$ if and only if $KVCF(A) = KVCF(B)$.*

Remark. If the characteristic polynomial of a matrix is irreducible over $\mathbb{Q}$, then all its eigenvectors are distinct, so all matrices of Proposition 25.23 possess Klein—Voronoi continued fractions.

Proof. Supposing that $\Xi(A) = \Xi(B)$ then A and B commute. Hence they have the same eigenvectors (since they do not have multiple eigenvalues) and the same orbits: $T_A = T_B$. In addition, we choose the same cone π_+ for both A and B. Therefore, by definition, the Klein—Voronoi continued fractions for A and B coincide.

Let us prove the converse statement. Assume that the Klein—Voronoi continued fractions for A and B coincide. Suppose that A has real eigenvectors $g_1,\ldots,g_k$ and complex conjugate eigenvectors $g_{k+2j-1} \pm \sqrt{-1}g_{k+2j}$ for $j = 1,\ldots,l$, where $k + 2l = n$. Consider the coordinate system corresponding to the basis $\{g_i\}$:

$$OX_1X_2\ldots X_kY_1Z_1Y_2Z_2\ldots Y_lZ_l.$$

In these coordinates define the form Φ_A as follows:

$$\Phi_A(x_1,\ldots,x_n) = \left(\prod_{i=1}^{k}x_i\prod_{j=1}^{l}(y_j^2+z_j^2)\right).$$

Similarly we define the form Φ_B for the operator B. Note that A preserves the form Φ_A up to a multiplicative scalar: a simple calculation shows that for every $v \in \mathbb{R}^n$ we have

$$\Phi_A(A(v)) = \det(A)\Phi_A(v).$$

The same is true for B and Φ_B.

The Klein—Voronoi continued fraction for A (which is also $KVCF(B)$) asymptotically coincides with the set $\Phi_A = 0$ (and $\Phi_B = 0$ respectively) at infinity, and hence the matrices A and B have all the same invariant subspaces. In particular, their one-dimensional real eigenspaces corresponding to real eigenvectors and their two-dimensional eigenspaces (we denote them by $\pi_1, \ldots, \pi_l$) defined by pairs of complex conjugate roots coincide. This implies that A and B commute if and only if they commute for the vectors of the invariant planes $\pi_1, \ldots, \pi_l$.

Let us show that operators A and B restricted to a plane π_i ($i = 1,\ldots,l$) commute. For a vertex v of $KVCF(A)$ we set

$$T_v = KVCF(A)\cap\left(v + \langle\pi_1,\ldots,\pi_l\rangle\right).$$

From construction, T_v coincides with the orbit $T_A(v)$. Since the space spanned by all planes $\pi_1,\ldots,\pi_l$ is invariant for both A and B, the set T_v coincides with the orbit $T_B(v)$, and hence $T_A(v) = T_B(v)$.

Consider an arbitrary matrix C of the set T_A. Since all eigenvalues of C have unit modulus and C is diagonalizable in the eigenbasis of A, C preserves Φ_A up to a scalar $\det(C) = \pm 1$. Therefore, $|\Phi_A|$ is constant on $T_A(v)$. For the same reason, $|\Phi_B|$ is constant on $T_A(v) = T_B(v)$. Therefore, by linearity, $\Phi_A = c \cdot \Phi_B$ for some constant $c \neq 0$. Hence, A preserves Φ_B up to a multiplicative scalar.

Consider now the plane π_j for some $1 \leq j \leq l$ and take coordinates (y_j,z_j). It is clear that if an operator preserves $\Phi_B(v)$ up to a multiplicative scalar, then its restriction to the plane π_j preserves the form

$$y_j^2 + z_j^2$$

up to a multiplicative factor. There are two types of matrices that preserve the set of level sets of this form:

$$\begin{pmatrix} a & b \\ -b & a \end{pmatrix} \quad \text{and} \quad \begin{pmatrix} -a & b \\ b & a \end{pmatrix}$$

with arbitrary real parameters a and b. The matrices of the second family have two real eigenvalues in π_j, which is by the above not the case for B. Therefore, both A and B are from the first family. All matrices of the first family commute. Hence A commutes with B in planes π_j for $1 \le j \le l$.

Therefore, A and B are both diagonalizable in a certain complex basis. Hence A and B commute. Therefore, $\Xi(A) = \Xi(B)$. $\qquad\square$

25.3.3 Geometric invariants of conjugacy classes

In this subsection we fix some integer basis in $\mathbb{R}^n$ and identify operators with matrices in this basis. On the one hand, from Theorem 25.23 it follows that the Klein—Voronoi continued fraction for A identifies the Dirichlet group $\Xi(A)$ in a unique way. On the other hand, the operator A defines a shift of $KVCF(A)$ along itself, which we denote by P_A. It is clear that distinct operators of $\Xi(A)$ define inequivalent shifts. So every $A \in GL(n,\mathbb{Z})$ is uniquely identified with a pair $(KVCF(A), P_A)$. The group $GL(n,\mathbb{Z})$ naturally acts on pairs $(KVCF(A), P_A)$ by left multiplication on the first factor and by conjugation on the second factor. We have the following important tautological theorem.

Theorem 25.24. (On geometric complete invariant.) *Two matrices A_1, A_2 of the group $GL(n,\mathbb{Z})$ whose characteristic polynomial is irreducible over $\mathbb{Q}$ are integer conjugate if and only if the pairs $(KVCF(A_1), P_{A_1})$ and $(KVCF(A_2), P_{A_2})$ are in the same $GL(n,\mathbb{Z})$-orbit.* $\qquad\square$

Remark 25.25. The important consequence of this theorem is that all geometric $GL(n,\mathbb{Z})$-invariants of the Klein—Voronoi continued fraction for A (such as integer distances and integer volumes of certain integer point configurations of $KVCF(A)$) are invariants of the conjugacy class of A. Fundamental domains of Klein continued fractions and their invariants in the real spectrum three-dimensional case $(n = k = 3, l = 0)$ were studied in works [130], [129], [134], [136], [96], [97], etc. Currently there is not much known about the construction of Klein—Voronoi continued fractions in the nonreal spectrum case.

25.4 Algorithmic aspects of reduction to ς-reduced matrices

In Section 25.2 above we showed the existence and finiteness of ς-reduced matrices in each integer conjugacy class of $SL(n,\mathbb{Z})$ matrices. The aim of this section is to introduce techniques to construct ς-reduced matrices integer conjugate to a given one. In Section 25.4.1 we give a geometric interpretation of Hessenberg complexity as a volume of a certain simplex, which is called the MD-characteristic. Further, in Section 25.4.2 we use the MD-characteristics to prove that the algorithm of Section 25.2.2 constructs all ς-reduced matrices starting from integer vertices of a Klein—Voronoi continued fraction. The corresponding techniques are outlined in Section 25.4.3.

25.4.1 Markov—Davenport characteristics

In this subsection we characterize the Hessenberg complexity in terms of the Markov—Davenport characteristic.

25.4.1.1 Definition of the MD-characteristic and its invariance under the action of the Dirichlet group

The study of Markov—Davenport characteristics is closely related to the theory of minima of absolute values of homogeneous forms with integer coefficients in n-variables of degree n. One of the first works in this area was written by A. Markov [146] for decomposable quadratic forms in two variables. Further, H. Davenport, in a series of works [52], [53], [54], [55], and [56], made the first steps for the case of decomposable forms for $n = 3$.

Consider $A \in SL(n,\mathbb{Z})$. Denote by $P(A,v)$ the parallelepiped spanned by vectors $v, A(v), \ldots, A^{n-1}(v)$, i.e.,

$$P(A,v) = \left\{ O + \sum_{i=0}^{n-1} \lambda_i A^i(v) \ \middle|\ 0 \le \lambda_i \le 1, i = 0,\ldots,n-1 \right\},$$

where O is the origin.

Definition 25.26. The *Markov—Davenport characteristic* (or the *MD-characteristic*, for short) of an $SL(n,\mathbb{Z})$ operator A is the functional

$$\Delta_A : \mathbb{R}^n \to \mathbb{R} \qquad \text{defined by} \qquad \Delta_A(v) = V(P(A,v)),$$

where $V(P(A,v))$ is the nonoriented volume of $P(A,v)$.

Proposition 25.27. *Consider $A \in SL(n,\mathbb{Z})$ and let $B \in \Xi(A)$. Then for an arbitrary point v we have*

$$\Delta_A(v) = \Delta_A(B(v)).$$

Remark. Proposition 25.27 implies that the MD-characteristic naturally defines a function over the set of all orbits of the Dirichlet group.

Proof. Since $B \in \Xi(A)$, we have the equality $A^n B(v) = BA^n(v)$. Hence the parallelepiped $P(A, B(v))$ coincides with $B(P(A, v))$. Since $B \in SL(n, \mathbb{Z})$, the volume of the parallelepiped is preserved. Therefore,

$$\Delta_A(v) = \Delta_A(B(v)).$$

$\square$

25.4.1.2 Homogeneous forms associated to $SL(n, \mathbb{Z})$ operators

Consider any $SL(n, \mathbb{Z})$ operator A. Suppose that it has k real eigenvectors $g_1, \ldots, g_k$ with eigenvalues $r_1, \ldots, r_k$ and $2l$ complex eigenvectors $g_{k+2j-1} \pm \sqrt{-1} g_{k+2j}$ with complex conjugate eigenvalues $c_1, \bar{c}_1, \ldots, c_l, \bar{c}_l$, where $k + 2l = n$. Consider the system of coordinates

$$OX_1 X_2 \ldots X_k Y_1 Z_1 Y_2 Z_2 \ldots Y_l Z_l$$

corresponding to the basis $\{g_i\}$. A form

$$\alpha \left(\prod_{i=1}^{k} x_i \prod_{j=1}^{l} (y_j^2 + z_j^2) \right)$$

with nonzero α is said to be *associated* to the operator A.

Proposition 25.28. *Let A be an $SL(n, \mathbb{Z})$ operator whose characteristic polynomial has distinct roots. Then the MD-characteristic of A coincides with the absolute value of a form associated to A for a certain nonzero α.*

Proof. Let us consider the formulas of the MD-characteristic of A in the eigenbasis. We assume that the coordinates in this eigenbasis are $(t_1, \ldots, t_n)$. Then for any vector $v = (t_1, \ldots, t_n)$ we have

$$A^j(x) = (r_1^j t_1, \ldots, r_k^j t_k, c_1^j t_{k+1}, \bar{c}_1^j t_{k+2}, \ldots, c_l^j t_{k+2l-1}, \bar{c}_l^j t_{k+2l}).$$

Therefore,

$$\Delta_A(t_1, \ldots, t_n) = \alpha \left| \prod_{i=1}^{k} t_i \prod_{j=1}^{l} (t_{k+2j-1} t_{k+2j}) \right| = \frac{\alpha}{4^l} \left| \prod_{i=1}^{k} x_i \prod_{j=1}^{l} (y_j^2 + z_j^2) \right|.$$

Simple calculations show that $\alpha \neq 0$.

$\square$

25.4.1.3 Hessenberg complexity in terms of the MD-characteristic

Proposition 25.29. *Consider an operator A with Hessenberg matrix $M = (a_{i,j})$ in some integer basis $\{\tilde{e}_i\}$. The Hessenberg complexity $\varsigma(M)$ equals the value of the MD-characteristic $\Delta_A(\tilde{e}_1)$.*

Proof. Denote by V_k the plane spanned by vectors $v, A(v), A^2(v), \dots, A^{k-1}(v)$.

Let us inductively show that

$$A^k(\tilde{e}_1) = \left(\prod_{i=1}^{k} a_{i,i+1}\right) \tilde{e}_{k+1} + v_k, \quad \text{where } v_k \in V_k.$$

Base of induction. We have $A(\tilde{e}_1) = a_{1,2}\tilde{e}_2 + a_{1,1}\tilde{e}_1$.

Step of induction. Suppose that the statement holds for $k = m$, i.e.,

$$A^m(\tilde{e}_1) = \left(\prod_{i=1}^{m} a_{i,i+1}\right) \tilde{e}_{m+1} + v_m, \quad \text{and } v_m \in V_m.$$

Let us prove the statement for $m+1$. Since M is Hessenberg, $A(v_m)$ is in V_{m+1}. Therefore, we have

$$A^{m+1}(\tilde{e}_1) = A\left(\left(\prod_{i=1}^{m} a_{i,i+1}\right) \tilde{e}_{m+1}\right) + A(v_m)$$

$$= \left(\prod_{i=1}^{m+1} a_{i,i+1}\right) \tilde{e}_{m+1} + \left(A(v_m) + \left(\prod_{i=1}^{m} a_{i,i+1}\right)\left(A(\tilde{e}_{m+1}) - a_{m+1,m+2}\tilde{e}_{m+2}\right)\right).$$

The second summand in the last expression is in V_{m+1}. We have completed the induction step.

Therefore,

$$\Delta_A(\tilde{e}_1) = \prod_{i=1}^{n-1} |a_{i+1,i}|^{n-i} = \varsigma(M).$$

□

In the proof of the next theorem we use the following corollary.

Corollary 25.30. *Consider an operator A with matrix M. Let v be any primitive integer vector. Consider the matrix $(M|v)$ constructed by the algorithm of Section 25.2.2. Then we have*

$$\varsigma(M|v) = \Delta_A(v).$$

□

25.4.2 Klein—Voronoi continued fractions and minima of MD-characteristics

In the following theorem we use Klein—Voronoi continued fractions to find minima of MD-characteristics.

Theorem 25.31. *Let A be an* $SL(n, \mathbb{Z})$ *operator with matrix M in some integer basis. Suppose that the characteristic polynomial of A is irreducible over* $\mathbb{Q}$. *Let U be a fundamental domain of the Klein—Voronoi continued fractions for A (see Definition 25.21). Then we have:*

(i) For every ς-*reduced matrix* $\hat{M}$ *integer conjugate to M, there exists* $v \in U$ *such that* $\hat{M} = (M|v)$.

(ii) Let $v \in U$. *The matrix* $(M|v)$ *is* ς-*reduced if and only if the MD-characteristic* Δ_A *attains its minimal value on the set* $\mathbb{Z}^n \setminus \{O\}$ *at the point* v.

Proof. Theorem 25.31(ii) follows directly from Corollary 25.30.

Let us prove Theorem 25.31(i). By Proposition 25.28 there exists a nonzero constant α such that in the system of coordinates $OX_1X_2\ldots X_kY_1Z_1Y_2Z_2\ldots Y_lZ_l$ the MD-characteristic Φ_A is expressed as follows:

$$\alpha \left| \prod_{i=1}^{k} x_i \prod_{i=1}^{l} (y_i^2 + z_i^2) \right|.$$

Suppose that the minimal absolute value of Φ_A on the set of integer points except the origin equals m_0.

Consider a cone π_+ in the plane $z_1 = \cdots = z_l = 0$ and choose the coordinates $(x_1,\ldots,x_k,y_1,\ldots,y_l)$ in it. Consider a projection of $\mathbb{R}^n$ to the cone along the orbits $T_A(*)$. Since the MD-characteristic is constant on $T_A(u)$ for every u, the projection of the MD-characteristic is well defined. Denote it by $\check{\Phi}_A$. In the chosen coordinates of π_+, the function $\check{\Phi}_A$ is written as follows:

$$\alpha \left| \prod_{j=1}^{k} x_j \prod_{j=1}^{l} y_j^2 \right|.$$

This function is convex in every orthant of the cone π_+. Since every factor-sail is the boundary of a certain convex hull in each orthant, all minima of the convex function $\check{\Phi}_A$ restricted to the convex hulls are attained at the boundary, i.e., at π-integer points of factor-sails. Therefore, all integer minima of Φ_A on $\mathbb{Z}^n \setminus \{O\}$ are at vertices of the Klein—Voronoi continued fraction.

By Corollary 25.30, the Hessenberg complexity $\varsigma(M|v)$ coincides with the MD-characteristic $\Delta_A(v)$. Since every matrix integer conjugate to M has a presentation in the form $(M|v)$ and all integer minima of the MD-characteristic are attained at vertices of the Klein—Voronoi continued fraction, every ς-reduced operator $\tilde{M}$ is represented as $(M|v_0)$ for some vertex $v_0 \in KVCF(A)$. By Corollary 25.7, for every

$B \in \Xi(A)$ we have

$$(M|B(v_0)) = (M|v_0).$$

Hence a vector v_0 can be chosen in the fundamental domain U. This concludes the proof.   □

Remark 25.32. In Example 25.5 the ς-reduced Hessenberg matrices

$$M_1 = \begin{pmatrix} 0 & 1 & 2 \\ 1 & 0 & 0 \\ 0 & 3 & 5 \end{pmatrix} \quad \text{and} \quad M_2 = \begin{pmatrix} 0 & 2 & 3 \\ 1 & 1 & 1 \\ 0 & 3 & 4 \end{pmatrix}$$

are integer conjugate but do not coincide. The reason for this is as follows. Consider the Klein—Voronoi continued fraction of A with matrix M_1. It contains integer vertices $p_1 = (1,0,0)$ and $p_2 = (0,1,-1)$. It turns out that p_1 and p_2 are not in the same orbit of the Dirichlet group but have the same MD-characteristic, namely 3. Hence we get two distinct integer conjugate ς-reduced Hessenberg matrices: $M_1 = (M_1|(1,0,0))$ and $M_2 = (M_1|(0,1,-1))$.

25.4.3 Construction of ς-reduced matrices by Klein—Voronoi continued fractions

As we have already proved, the ς-reduced Hessenberg matrix for the operator A is constructed starting from some vertex in a fundamental domain of the Klein—Voronoi multidimensional continued fraction. We use this property to find all ς-reduced Hessenberg matrices in a given integer conjugacy class of matrices.

Algorithm (techniques) to find ς-reduced matrices in integer conjugacy classes.

Input data. An integer matrix M for A.

Goal of the algorithm. To construct all ς-reduced matrices integer conjugate to M.

Step 1. Find a fundamental domain of the Klein—Voronoi continued fraction for the operator A (see Remark 25.33 below).

Step 2. Take all vertices of the fundamental domain constructed in Step 1 and find among them all vertices with minimal value of the MD-characteristic (say $v_1, \ldots, v_k$).

Step 3. By Theorem 25.31(i) and (ii) all ς-reduced matrices integer conjugate to M are $(M|v_1), \ldots, (M|v_k)$. They are all constructed by the algorithm described in Section 25.2.2.

Output. A list of all ς-reduced matrices integer conjugate to M.

Remark 25.33. Currently, Step 1 is the most complicated. The real spectrum case of matrices with all eigenvalues being real has been well studied. For the algorithms of constructing multidimensional continued fractions in this case, see Section 24 and the papers by R. Okazaki [166], T. Shintani [200], J.-O. Moussafir [156], and the author [103]. E. Korkina in [130] and [129], G. Lachaud in [134], [136], A.D. Bruno and V.I. Parusnikov in [31], [174], [175], and [176], and the author in [96] and [97] have produced a large number of fundamental domains for periodic algebraic two-dimensional continued fractions (see also the site [29] by K. Briggs). Some fundamental domains in the three-dimensional case are found, for instance, in [99]. The case with complex conjugate eigenvalues which we study in the next section, is new.

Example 25.34. Let us consider the example of the operator A defined by the matrix

$$\begin{pmatrix} -2 & -4 & -3 \\ 1 & 2 & 2 \\ -1 & -1 & 3 \end{pmatrix}.$$

The characteristic polynomial of this operator has three distinct real roots. Therefore, the Klein—Voronoi continued fraction consists of eight sails. The compositions of operators $-\mathrm{Id}$, A, and $2\mathrm{Id} + A^{-1}$ define integer congruences for all these sails, and hence all ç-reduced operators are written from vertices of one sail. Consider a sail containing the point $(1, 0, 0)$. There are exactly three distinct orbits of the Dirichlet group containing the vertices in this sail. They are defined by the following points:

$$(0,0,1), \quad (1,0,0), \quad \text{and} \quad (3,-1,1).$$

(We skip all calculations of convex hulls corresponding to the sail.) The MD-characteristic of these vectors are respectively 1, 2, and 4. So the minimum of the MD-characteristic (which is 1 in this case) is attained on the vertices of the orbit of the Dirichlet group containing $(0,0,1)$. Therefore, there exists a unique ç-reduced Hessenberg matrix, which is

$$\begin{pmatrix} 0 & 0 & 1 \\ 1 & 0 & 1 \\ 0 & 1 & -3 \end{pmatrix}.$$

The perfect Hessenberg matrices for the vertices $(1,0,0)$ and $(3,-1,1)$ are respectively

$$\begin{pmatrix} 0 & 1 & -1 \\ 1 & 0 & 0 \\ 0 & 2 & -3 \end{pmatrix} \quad \text{and} \quad \begin{pmatrix} 1 & 0 & -1 \\ 2 & 0 & 3 \\ 0 & 1 & -4 \end{pmatrix},$$

and their ç-complexities are 2 and 4.

25.5 Diophantine equations related to the Markov—Davenport characteristic

Let A be an arbitrary $SL(n,\mathbb{Z})$ operator whose characteristic polynomial is irreducible over $\mathbb{Q}$, and let N be an arbitrary integer. In this section we discuss how to solve the Diophantine equation

$$\Delta_A(v) = N. \tag{25.1}$$

We begin with a general definition of w-continued fractions.

25.5.1 Multidimensional w-sails and w-continued fractions

We have already introduced the w-sails of continued fractions in the one-dimensional case in Section 14.2.5. Let us generalize this notion to the multidimensional case.

25.5.1.1 Definition

Consider an operator A in $SL(n,\mathbb{Z})$ whose characteristic polynomial is irreducible over $\mathbb{Q}$. As above, we suppose that A has k real eigenvalues and l pairs of complex conjugate eigenvalues, where $k + 2l = n$. Recall that the complement to the union of all invariant hyperplanes in the cone π_+ consists of 2^k arcwise connected components denoted by $C_1(A), \ldots, C_{2^k}(A)$.

Definition 25.35. We define the n-sails for an arbitrary simplicial cone C with vertex at the origin inductively.

— let the 1-sail be the sail of C.
— suppose that all w-sails for $w < w_0$ are defined. The convex hull of all π-integer points except the origin and except for the π-integer points for all w-factor-sails with $w < w_0$ contained in an arcwise connected component $C_i(A)$ is called the w_0-factor-sail of $C_i(A)$. We denote it by $\hat{S}_{w_0,i}(A)$.

Let us write the w_0-factor-sail of C_i in one formula:

$$\hat{S}_{w_0,i}(A) = \partial\left(\operatorname{conv}\left((\{q \in \pi_+ \mid T_A(q) \cap C_i(A) \cap \mathbb{Z}^n \neq \emptyset\} \setminus \{O\}) \setminus \bigcup_{w=1}^{w_0-1} \hat{S}_{w,i}(A)\right)\right).$$

The set of all w-factor-sails for a given operator is called the w-factor-continued fraction for the operator A.

Definition 25.36. The union of all orbits $T_A(*)$ in $\mathbb{R}^n$ represented by the points in the w-factor-sail of $C_i(A)$ is called the w-sail of $T_A(C_i)$; we denote it by $S_{w,i}(A)$.

The union of all sails is said to be the *w-continued fraction* for the operator A in the sense of Klein—Voronoi. We denote it by $KVCF_w(A)$.

In other words, we have

$$S_{w,i} = \bigcup_{p \in \hat{S}_{w,i}(A)} T_A(p) \quad \text{and} \quad KVCF_w(A) = \bigcup_{i=1}^{2^k} S_{w,i}(A).$$

25.5.1.2 Minima of Markov—Davenport characteristics on *w*-sails

Denote by α_w the minimal absolute value of the Markov—Davenport characteristic $|\Delta_A|$ on $KVCF_w(A)$.

Proposition 25.37. *The sequence* $\alpha_1, \alpha_2, \alpha_3, \ldots$ *is strictly increasing.*

Remark 25.38. The *w*-sails for arbitrary $w > 0$ are not necessarily homothetic to the 1-sail, as it was in the one-dimensional case of Section 14.2.5.

Proof of Proposition 25.37. Consider the coordinate system

$$OX_1 X_2 \ldots X_k Y_1 Z_1 Y_2 Z_2 \ldots Y_l Z_l$$

introduced in Section 25.3.1.1 above. In these coordinates the form Δ_A is written as follows:

$$\Delta_A(x_1, \ldots, x_n) = c \left(\prod_{i=1}^{k} x_i \prod_{j=1}^{l} (y_j^2 + z_j^2) \right),$$

for some nonzero constant c. Consider an arbitrary ray r with vertex at the origin and not contained in any invariant subspace of A. Assume that its direction is $(\lambda_1, \ldots, \lambda_n)$. Then at points of this ray the function Δ_A is as follows:

$$\Delta_A(\lambda_1 t, \ldots, \lambda_n t) = c \left(\prod_{i=1}^{k} \lambda_i \prod_{j=1}^{l} (\lambda_{k+2j-1}^2 + \lambda_{k+2j}^2) \right) t^n. \qquad (25.2)$$

It is clear that the ray r intersects the *w*-sail first and only thereafter intersects the $(w+1)$-sail. Notice that the expression on the right side of Equation 25.2 is proportional to t^n (with respect to t), which increases as t increases. Therefore, the value of Δ_A at the point of intersection of the ray r with the $(w+1)$-sail is strictly greater than the value at the point of intersection with the corresponding *w*-sail.

Notice that every possible limit value of Δ_A on the $(w+1)$-sail is attained at some point of its fundamental domain (since the closure of a fundamental domain is compact). In particular, the infimum of $|\Delta_A|$ at the $(w+1)$-sail is attained as a minimum at some point. Therefore, $\alpha_{w+1} > \alpha_w$. □

Corollary 25.39. *For an arbitrary integer w,* $\alpha_w \geq w$.

Proof. The convexity of a Markov—Davenport form on the intersection of cones C_i with the plane π_+ implies that the minimum α_w is attained at an integer vertex of the $KVCF_w(A)$. Therefore, α_w is a positive integer for every $w \geq 0$. Now the statement of the corollary follows directly from Proposition 25.37.   □

25.5.2 Solution of Equation 25.1

The following general techniques helps to write all the solutions of equation (25.1).

Techniques to solve equation (25.1).

Input Data: Given a pair (A, N), where A is a $GL(n, \mathbb{Z})$ matrix whose characteristic polynomial is irreducible over $\mathbb{Q}$, and N is an integer.

Goal of the algorithm. Find all integer solutions of $\Delta_A(v) = N$.

Step 1. Write the MD-characteristic Δ_A according to Definition 25.26.

Step 2. Find the fundamental domains of all w-sails $C_{w,i}$ for the operator A (for all admissible i) and $w \leq N$; denote their union by U. Notice that $U \cap \mathbb{Z}^n$ is finite.

Step 3. Check all integer points of U whether they satisfy the equation

$$\Delta_A(v) = N.$$

Denote the set of all integer solutions in U by Z_U.

Output Data: the solution of $\Delta(v) = N$ is the set

$$\bigcup_{B \in \Xi(A)} B(Z_U).$$

Remark 25.40. By Corollary 25.39 it is enough to check only the vertices of the w-sails for $w \leq N$.

25.6 On reduced matrices in $SL(3, \mathbb{Z})$ with two complex conjugate eigenvalues

In this section we study the case of Hessenberg $SL(3, \mathbb{Z})$ matrices with two complex conjugate and one real eigenvalue. It turns out that the majority of such matrices of the same hessenberg type are ς-reduced. We start in Section 25.6.1 with some notation and definitions. In Section 25.6.2 we formulate a supplementary theorem on the parabolic structure of the set of nonreal spectrum matrices, whose proof we give in Section 25.6.5. Further, in Section 25.6.3 we formulate the main result on asymptotic uniqueness of ς-reduced matrices, whose proof is in Section 25.6.6. In

Section 25.6.4 we examine some particular examples of families of matrices with fixed Hessenberg type. All the results of this section are new.

25.6.1 Perfect Hessenberg matrices of a given Hessenberg type

For simplicity, we study only $SL(3, \mathbb{Z})$ matrices whose characteristic polynomials are irreducible over $\mathbb{Q}$. There are two main geometrically essentially different cases of such matrices: the *real spectrum matrices* (or *RS-matrices* for short), whose characteristic polynomials have only real eigenvalues, and the *nonreal spectrum matrices* (or *NRS-matrices*) whose characteristic polynomials have a pair of complex conjugate and one real eigenvalue.

For a Hessenberg type Ω we denote the subset of all NRS-matrices in $H(\Omega)$ by $NRS(\Omega)$.

Definition 25.41. Let $\Omega = \langle a_{11}, a_{21} | a_{12}, a_{22}, a_{32} \rangle$. Consider $v = (a_{13}, a_{23}, a_{33})$ such that the determinant of the matrix (a_{ij}) equals 1. Define

$$H_\Omega^v(m, n) = \begin{pmatrix} a_{11} & a_{12} & a_{11}m + a_{12}n + a_{13} \\ a_{21} & a_{22} & a_{21}m + a_{22}n + a_{23} \\ 0 & a_{32} & a_{32}n + a_{33} \end{pmatrix}.$$

It is clear that

$$H(\Omega) = \{ H_\Omega^v(m, n) \mid m \in \mathbb{Z}, n \in \mathbb{Z} \}.$$

In this context, to choose a vector v means to choose the origin O in the plane $H(\Omega)$. So the set $H(\Omega)$ has the structure of a two-dimensional plane (see also Section 25.2.4 above). We denote by OMN the coordinate system corresponding to the parameters (m, n). Notice that the Hessenberg complexity of this family is $a_{12}^2 a_{23}$.

Let $\operatorname{Discr}_\Omega^v(m, n)$ denote the discriminant of the characteristic polynomial of $H_\Omega^v(m, n)$. Then the set $NRS(\Omega)$ is defined by the following inequality in variables n and m:

$$\operatorname{Discr}_\Omega^v(m, n) < 0.$$

Example 25.42. In Fig. 25.3 we show the subset $NRS(\langle 0, 1 | 0, 0, 1 \rangle)$. For this example we choose $v = (0, 0, 1)$.

25.6.2 Parabolic structure of the set of NRS-matrices

The set $NRS(\langle 0, 1 | 0, 0, 1 \rangle)$ in Fig. 25.3 "looks like" the intersection of $\mathbb{Z}^2$ with the union of the convex hulls of two parabolas. Let us formalize this in a general statement.

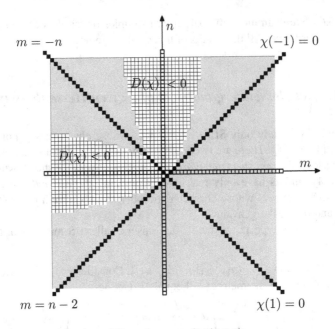

Fig. 25.3 The family of matrices of Hessenberg type $\langle 0,1 | 0,0,1 \rangle$.

Consider the matrix $H_\Omega^v(0,0) = (a_{ij})$ and define b_1, b_2, and b_3 such that the characteristic polynomial of this matrix in the variable t is

$$-t^3 + b_1 t^2 - b_2 t + b_3.$$

Even though in the case of $SL(3,\mathbb{Z})$ we have $b_3 = 1$, nevertheless we write b_3 for generality reasons. For the family $H_\Omega^v(m,n)$ we define the following two quadratic functions:

$$p_{1,\Omega}(m,n) = m - \alpha_1 n^2 - \beta_1 n - \gamma_1;$$
$$p_{2,\Omega}(m,n) = \frac{n}{a_{21}} - \alpha_2 \left(\frac{a_{21}m - a_{11}n}{a_{21}}\right)^2 - \beta_2 \left(\frac{a_{21}m - a_{11}n}{a_{21}}\right) - \gamma_2,$$

where

$$
\begin{cases}
\alpha_1 = -\dfrac{a_{32}}{4a_{21}}, \\[2mm]
\beta_1 = \dfrac{a_{11} - a_{22} - a_{33}}{2a_{21}}, \\[2mm]
\gamma_1 = \dfrac{4b_2 - b_1^2}{4a_{21}a_{32}},
\end{cases}
\qquad
\begin{cases}
\alpha_2 = \dfrac{a_{32}a_{21}}{4b_3}, \\[2mm]
\beta_2 = -\dfrac{b_2}{2b_3}, \\[2mm]
\gamma_2 = \dfrac{b_2^2 - 4b_1 b_3}{4a_{21}a_{32}b_3}.
\end{cases}
$$

In the real plane OMN of the family $H_\Omega^v(m,n)$ we consider the interior of the circle of radius R with center at the origin $(0,0)$ and denote it by $B_R(O)$. For a real number t we set

$$\Lambda_t = \{(m,n) \mid (p_{1,\Omega}(m,n) - t)(p_{2,\Omega}(m,n) - t) < 0\}.$$

Theorem 25.43. *For every positive ε there exists $R > 0$ such that the following inclusions hold:*

$$\Lambda_\varepsilon \backslash B_R(O) \subset NRS(\Omega) \backslash B_R(O) \subset \Lambda_{-\varepsilon} \backslash B_R(O).$$

We give the proof of this theorem in Section 25.6.5 below.

25.6.3 Theorem on asymptotic uniqueness of ς-reduced NRS-matrices

Recall that a ray is said to be *integer* if its vertex is integer and it contains integer points distinct from the vertex.

Definition 25.44. An integer ray in $H(\Omega)$ is said to be an *NRS-ray* if all its integer points correspond to NRS-matrices. A direction is said to be *asymptotic* for the set $NRS(\Omega)$ if there exists an NRS-ray with this direction.

As stated in Theorem 25.43, for every Hessenberg type Ω the set $NRS(\Omega)$ almost coincides with the union of the convex hulls of two parabolas. This implies the following statement.

Proposition 25.45. *Let $\Omega = \langle a_{11}, a_{21} | a_{12}, a_{22}, a_{32} \rangle$. There are exactly two asymptotic directions for the set $NRS(\Omega)$, defined by the vectors $(-1,0)$ and (a_{11}, a_{21}).*

$\square$

Consider a family of Hessenberg matrices H_Ω^v for an appropriate integer vector v.

Definition 25.46. Define

$$R_{1,\Omega,v}^{m,n} = \{H_\Omega^v(m-t,n) \mid t \in \mathbb{Z}_{\geq 0}\};$$
$$R_{2,\Omega,v}^{m,n} = \{H_\Omega^v(m+a_{11}t, n+a_{21}t) \mid t \in \mathbb{Z}_{\geq 0}\}.$$

By $R_{1,\Omega,v}^{m,n}(t)$, or respectively by $R_{2,\Omega,v}^{m,n}(t)$, we denote the tth element in the corresponding family.

Remark 25.47. The families $R_{1,\Omega,v}^{m,n}$ and $R_{2,\Omega,v}^{m,n}$ coincide with the sets of all integer points of certain rays with directions $(-1,0)$ and (a_{11}, a_{21}) respectively. Conversely, from Proposition 25.45 it follows that the set of integer points of every NRS-ray coincides either with $R_{1,\Omega,v}^{m,n}$ or with $R_{2,\Omega,v}^{m,n}$ for some integers m and n.

In Fig. 25.4 we show in dark gray two NRS-rays:

— $R_{1,\langle 1,2|1,1,3\rangle,(0,0,-1)}^{-9,5}$ from the left;

— $R_{2,\langle 1,2|1,1,3\rangle,(0,0,-1)}^{-2,-1}$ from the right.

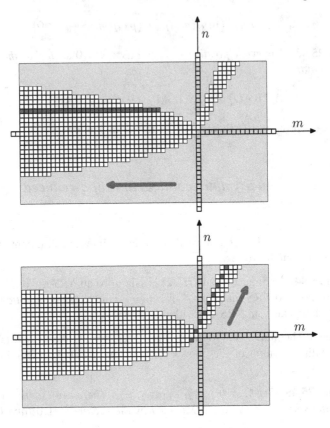

Fig. 25.4 Every NRS-ray contains finitely many ς-nonreduced matrices.

Now we are ready to formulate the main result on asymptotic behavior of NRS-matrices, which we prove later in Section 25.6.6.

Theorem 25.48. (On asymptotic ς-reducibility and uniqueness.) (*i*) *Every NRS-ray* (*as in Fig. 25.4*) *contains only finitely many ς-nonreduced matrices.*
(*ii*) *Every NRS-ray contains only finitely many matrices that are integer conjugate to some other ς-reduced matrix.*

Example 25.49. Every NRS-ray for the Hessenberg type $\langle 0,1|0,0,1\rangle$ contains only ς-reduced perfect matrices. Experiments show that every NRS-ray for $\langle 0,1|1,0,2\rangle$ contains at most one ς-nonreduced matrix (see Fig. 25.5).

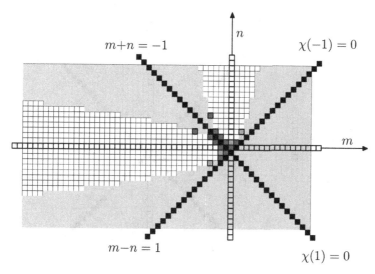

Fig. 25.5 The family of Hessenberg matrices $H^{(0,0,-1)}_{\langle 0,1|1,0,2\rangle}(m,n)$.

25.6.4 Examples of NRS-matrices for a given Hessenberg type

In this subsection we study several examples of families $NRS(\Omega)$ for the Hessenberg types:

$$\langle 0,1|0,0,1\rangle, \quad \langle 0,1|1,0,2\rangle, \quad \langle 0,1|1,1,2\rangle, \quad \text{and} \quad \langle 1,2|1,1,3\rangle.$$

In Figs. 25.5,. 25.6, and 25.7 the dark gray squares correspond to ς-nonreduced matrices. We also fill with gray the squares corresponding to ς-reduced Hessenberg matrices that are the nth powers ($n \geq 2$) of some integer matrices.

Hessenberg perfect NRS-matrices $H^{(1,0,0)}_{\langle 0,1|0,0,1\rangle}(m,n)$. The Hessenberg complexity of all these matrices is 1, and therefore, they are all ς-reduced (see the family in Fig. 25.3 on page 376).

Hessenberg Perfect NRS-Matrices $H^{(0,0,-1)}_{\langle 0,1|1,0,2\rangle}(m,n)$. The Hessenberg complexity of these matrices equals 2. Experiments show that 12 such matrices are ς-nonreduced (see the family in Fig. 25.5). It is conjectured that all other Hessenberg matrices of $NRS(\langle 0,1|1,0,2\rangle)$ are ς-reduced.

Hessenberg Perfect NRS-Matrices $H^{(1,0,1)}_{\langle 0,1|1,1,2\rangle}(m,n)$. The Hessenberg complexity of these matrices equals 2. We have found 12 ς-nonreduced matrices in the family. It is conjectured that all other Hessenberg matrices of $NRS(\langle 0,1|1,1,2\rangle)$ are ς-reduced. See Fig. 25.6.

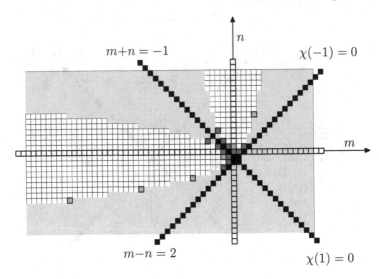

Fig. 25.6 The family of Hessenberg matrices $H_{\langle 0,1|1,1,2\rangle}^{(1,0,1)}(m,n)$.

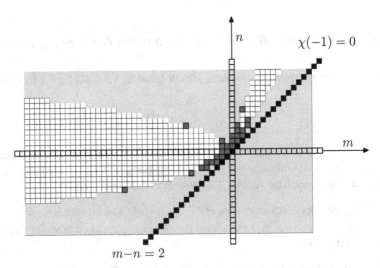

Fig. 25.7 The family of Hessenberg matrices $H_{\langle 1,2|1,1,3\rangle}^{(0,0,-1)}$.

Hessenberg Perfect NRS-Matrices $H_{\langle 1,2|1,1,3\rangle}^{(0,0,-1)}(m,n)$**.** This is a more complicated example of a family of Hessenberg perfect NRS-matrices, with their complexity equaling 12. We have found 27 ς-nonreduced matrices in the family. It is conjectured that all other Hessenberg matrices of $NRS(\langle 1,2|1,1,3\rangle)$ are ς-reduced. See Fig. 25.7.

25.6.5 Proof of Theorem 25.43

We begin the proof with several lemmas, but first a small remark.

Remark. The set $NRS(\Omega)$ is defined by the inequality

$$\mathrm{Discr}_\Omega^\nu(m,n) < 0.$$

In the left part of the inequality there is a polynomial of degree 4 in variables m and n. Notice that the product

$$16a_{21}^2 a_{32}^2 b_3 \left(p_{1,\Omega}(m,n) p_{2,\Omega}(m,n)\right)$$

is a good approximation to $\mathrm{Discr}_\Omega^\nu(m,n)$ at infinity; i.e., the polynomial

$$\mathrm{Discr}_\Omega^\nu(m,n) - 16a_{21}^2 a_{32}^2 b_3 \left(p_{1,\Omega}(m,n) p_{2,\Omega}(m,n)\right)$$

is a polynomial of degree only 2 in the variables m and n.

Lemma 25.50. *The planar curve* $\mathrm{Discr}_{\langle 0,1|0,0,1\rangle}^{(1,0,0)}(m,n) = 0$ *is contained in the domain defined by the inequalities*

$$\begin{cases} (m^2 - 4n + 3)(n^2 + 4m + 3) \geq 0, \\ (m^2 - 4n - 3)(n^2 + 4m - 3) - 72 \leq 0. \end{cases}$$

Remark. Lemma 25.50 implies that the curve $\mathrm{Discr}_{\langle 0,1|0,0,1\rangle}^{(1,0,0)}(m,n) = 0$ is contained in some tubular neighborhood of the curve

$$(m^2 - 4n)(n^2 + 4m) = 0.$$

Proof. Note that

$$\mathrm{Discr}_{\langle 0,1|0,0,1\rangle}^{(1,0,0)}(m,n) = (m^2 - 4n)(n^2 + 4m) - 2mn - 27.$$

Thus, we have

$$\mathrm{Discr}_{\langle 0,1|0,0,1\rangle}^{(1,0,0)}(m,n) - (m^2 - 4n + 3)(n^2 + 4m + 3)$$
$$= -2(n-3)^2 - 2(m+3)^2 - (n+m)^2 \leq 0,$$

and

$$\mathrm{Discr}_{\langle 0,1|0,0,1\rangle}^{(1,0,0)}(m,n) - (m^2 - 4n - 3)(n^2 + 4m - 3) + 72$$
$$= 2(n-3)^2 + 2(m+3)^2 + (n-m)^2 \geq 0.$$

Therefore, the curve $\mathrm{Discr}_{\langle 0,1|0,0,1\rangle}^{(1,0,0)}(m,n) = 0$ is contained in the domain defined in the lemma. $\qquad \square$

Lemma 25.51. *For every* $\Omega = \langle a_{11}, a_{21} | a_{12}, a_{22}, a_{32}\rangle$ *there exists an affine (not necessarily integer) transformation of the plane OMN taking the curve* $\mathrm{Discr}^v_\Omega(m,n) = 0$ *to the curve* $\mathrm{Discr}^{(1,0,0)}_{(0,1|0,0,1)}(m,n) = 0$.

Proof. Let $H^v_\Omega(0,0) = (a_{i,j})$. Note that every matrix $H^v_\Omega(m,n)$ is rational conjugate to the matrix

$$H^{(1,0,0)}_{(0,1|0,0,1)}(m',n'),$$

where

$$\begin{cases} m' = a_{23}a_{32} - a_{11}a_{33} + a_{12}a_{21} - a_{22}a_{33} - a_{11}a_{22} + a_{21}a_{32}m - a_{11}a_{32}n, \\ n' = a_{11} + a_{22} + a_{33} + a_{32}n, \end{cases}$$

by the matrix

$$X^v_\Omega = \begin{pmatrix} 1 & a_{11} & a_{11}^2 + a_{12}a_{21} \\ 0 & a_{21} & a_{11}a_{21} + a_{21}a_{22} \\ 0 & 0 & a_{21}a_{32} \end{pmatrix}.$$

Therefore, the curve $\mathrm{Discr}^v_\Omega(m,n) = 0$ is mapped to the curve $\mathrm{Discr}^{(1,0,0)}_{(0,1|0,0,1)}(m,n) = 0$ bijectively. In *OMN* coordinates this map corresponds to the following affine transformation:

$$\begin{pmatrix} m \\ n \end{pmatrix} \mapsto \begin{pmatrix} a_{21}a_{32}m - a_{11}a_{32}n \\ a_{32}n \end{pmatrix} + \begin{pmatrix} a_{23}a_{32} - a_{11}a_{33} + a_{12}a_{21} - a_{22}a_{33} - a_{11}a_{22} \\ a_{11} + a_{22} + a_{33} \end{pmatrix}.$$

This completes the proof of the lemma.   □

Proof of Theorem 25.43. Consider a family of matrices $H^v_\Omega(-p_{1,\Omega}(0,t) + \varepsilon, t)$ with real parameter t. Direct calculations show that for $\varepsilon \neq 0$ the discriminant of the matrices for this family is a polynomial of the fourth degree in the variable t, and

$$\mathrm{Discr}^v_\Omega(-p_{1,\Omega}(0,t) + \varepsilon, t) = \frac{1}{4}a_{21}a_{32}^5\varepsilon t^4 + O(t^3).$$

Therefore, there exists a neighborhood of infinity with respect to the variable t such that the function $\mathrm{Discr}^v_\Omega(-p_{1,\Omega}(0,t) + \varepsilon, t)$ is positive for positive ε in this neighborhood, and negative for negative ε.

Hence for a given ε there exists a sufficiently large $N_1 = N_1(\varepsilon)$ such that for every $t > N_1$, there exists a solution of the equation $\mathrm{Discr}^v_\Omega(m,n) = 0$ at the segment with endpoints

$$(-p_{1,\Omega}(0,t) + \varepsilon, t) \quad \text{and} \quad (-p_{1,\Omega}(0,t) - \varepsilon, t)$$

of the plane *OMN*.

Now we examine the family in the variable t for the second parabola:

$$H^v_\Omega\left(t - a_{11}p_{2,\Omega}(t,0) - \frac{a_{11}}{\sqrt{a_{11}^2 + a_{21}^2}}\varepsilon, -a_{21}p_{2,\Omega}(t,0) - \frac{a_{21}}{\sqrt{a_{11}^2 + a_{21}^2}}\varepsilon\right).$$

For the same reasons, for a given ε there exists a sufficiently large $N_2 = N_2(\varepsilon)$ such that for every $t > N_2$ there exists a solution of the equation $\mathrm{Discr}_\Omega^\nu(m,n) = 0$ on the segment with endpoints

$$\left(t - a_{11}p_{2,\Omega}(t,0) - \frac{a_{11}}{\sqrt{a_{11}^2 + a_{21}^2}}\varepsilon, -a_{21}p_{2,\Omega}(t,0) - \frac{a_{21}}{\sqrt{a_{11}^2 + a_{21}^2}}\varepsilon\right) \quad \text{and}$$

$$\left(t - a_{11}p_{2,\Omega}(t,0) + \frac{a_{11}}{\sqrt{a_{11}^2 + a_{21}^2}}\varepsilon, -a_{21}p_{2,\Omega}(t,0) + \frac{a_{21}}{\sqrt{a_{11}^2 + a_{21}^2}}\varepsilon\right)$$

of the plane OMN.

The set

$$p_{1,\Omega}(m,n)p_{2,\Omega}(m,n) = 0$$

is a union of two parabolas and therefore has four parabolic branches "approaching infinity." We have shown that for each of these four parabolic branches there exists a branch of the curve $\mathrm{Discr}_\Omega^\nu(m,n) = 0$ contained in the ε-tube of the chosen parabolic branch if we are far enough from the origin.

From Lemma 25.50 we know that $\mathrm{Discr}_{\langle 0,1|0,0,1\rangle}^{\langle 1,0,0\rangle}(m,n) = 0$ is contained in some tubular neighborhood of

$$p_{1,\langle 0,1|0,0,1\rangle}(m,n)p_{2,\langle 0,1|0,0,1\rangle}(m,n) = 0.$$

Then by Lemma 25.51, the curve $\mathrm{Discr}_\Omega^\nu(m,n) = 0$ is contained in some tubular neighborhood of the curve

$$p_{1,\Omega}(m,n)p_{2,\Omega}(m,n) = 0$$

outside some ball centered at the origin. Finally, by Viète's Theorem, the intersection of the curve $\mathrm{Discr}_\Omega^\nu(m,n) = 0$ with each of the parallel lines

$$\ell_t: \quad \frac{a_{11} + a_{21}}{a_{21}}n - m = t$$

contains at most four points. Therefore, there exists a sufficiently large T such that for any $t \geq T$, the intersection of the curve $\mathrm{Discr}_\Omega^\nu(m,n) = 0$ and ℓ_t contains exactly 4 points corresponding to the branches of the parabolas $p_{1,\Omega}(m,n) = 0$ and $p_{2,\Omega}(m,n) = 0$ lying in $\Lambda_{-\varepsilon} \setminus \Lambda_\varepsilon$.

From Lemma 25.50 and Lemma 25.51 it follows that the set

$$\left\{\frac{a_{11} + a_{21}}{a_{21}}n - m < T\right\} \cap \{\mathrm{Discr}_\Omega^\nu(m,n) = 0\}$$

is compact.

Therefore, there exists $R = R(\varepsilon, N_1, N_2, T)$ such that in the complement to the ball $B_R(O)$ we have

$$\Lambda_\varepsilon \subset NRS(\Omega) \subset \Lambda_{-\varepsilon}.$$

The proof of Theorem 25.43 is complete. □

25.6.6 Proof of Theorem 25.48

First we prove that a fundamental domain of a general Klein—Voronoi continued fraction can be chosen from a certain bounded subset of $\mathbb{R}^3$. Then we describe asymptotically the geometric structure of Klein—Voronoi continued fractions. Finally, we prove Theorem 25.48.

25.6.6.1 One general fact on fundamental domains of Klein—Voronoi continued fractions for NRS-matrices of $SL(3,\mathbb{Z})$

Consider an NRS-operator A and any integer point p distinct from the origin O. Denote by $\Gamma_A^0(p)$ the convex hull of the union of two orbits corresponding to the points p and $A(p)$. For every integer k we denote by $\Gamma_A^k(p)$ the set $A^k(\Gamma_A^0(p))$.

Proposition 25.52. *Let p be a nonzero integer point. Then there exists a fundamental domain of the Klein—Voronoi continued fraction for A with all (integer) orbit-vertices contained in the set $\Gamma_A^0(p)$.*

The proof is based on the following lemma. Let

$$\Gamma_A(p) = \bigcup_{k\in\mathbb{Z}} \Gamma_A^k(p).$$

Lemma 25.53. *Let p be a nonzero integer point. Then one of the Klein—Voronoi sails for A is contained in the set $\Gamma_A(p)$.*

Proof. Notice that the set $\Gamma_A(p)$ is a union of orbits. Let us project $\Gamma_A(p)$ to the half-plane π_+. The set $\Gamma_A(p)$ projects to the closure of the complement of the convex hull for the points $\Gamma_A(A^k(p)) \cap \pi_+$ for all integer k in the angle defined by the eigenspaces. Since all points $A^k(p)$ are integer, their convex hull is contained in the convex hull of all points corresponding to integer orbits in the angle. Hence the set $\Gamma_A(p) \cap \pi_+$ contains the projection of the sail. Therefore, the set $\Gamma_A(p)$ contains one of the sails. $\square$

Proof of Proposition 25.52. Since $-\mathrm{Id}$ exchanges the sails, one can choose a fundamental domain entirely contained in one sail. Let this sail contain a point p. By Lemma 25.53 the set $\Gamma_A(p)$ contains this sail. Therefore, all orbit-vertices of a fundamental domain for the Klein—Voronoi continued fraction can be chosen in $\Gamma_A^0(p)$. $\square$

25.6.6.2 Geometry of Klein—Voronoi continued fractions for matrices of $R_{1,\Omega,v}^{m,n}$

Let us prove the following statement.

Proposition 25.54. *Consider an NRS-ray $R^{m,n}_{1,\Omega,v}$. Then there exists a constant $C > 0$ such that for every $t > C$, all orbit-vertices of one of the fundamental domains for the Klein—Voronoi continued fraction of the matrix $R^{m,n}_{1,\Omega,v}(t)$ are contained in the set of all integer orbits corresponding to the integer points in the convex hull of $(1,0,0)$, $(a_{11}, a_{21}, 0)$, and $(-a_{11}, -a_{21}, 0)$.*

We begin with the case of matrices of Hessenberg type $\Omega_0 = \langle 0,1|0,0,1\rangle$. Such matrices form a family $H(\Omega_0)$ with real parameters m and n:

$$H^{(1,0,0)}_{\langle 0,1|0,0,1\rangle}(m,n) = \begin{pmatrix} 0 & 0 & 1 \\ 1 & 0 & m \\ 0 & 1 & n \end{pmatrix}.$$

Put $v_0 = (1,0,0)$.

Lemma 25.55. *Let $R^{m,n}_{1,\Omega_0,v_0}$ be an NRS-ray. Then for every $\varepsilon > 0$ there exists $C > 0$ such that for every $t > C$ the convex hull of the union of two orbit-vertices*

$$T_{R^{m,n}_{1,\Omega_0,v_0}(t)}(1,0,0) \quad and \quad T_{R^{m,n}_{1,\Omega_0,v_0}(t)}(0,1,0)$$

is contained in the ε-tubular neighborhood of the convex hull of three points $(1,0,0)$, $(0,1,0)$, $(0,-1,0)$.

Remark. Lemma 25.55 means that the corresponding domain tends to be flat as t tends to infinity. Recall that $R^{m,n}_{1,\Omega_0,v_0}(t)(1,0,0) = (0,1,0)$ for every t.

Proof. Let us find the asymptotics of eigenvectors and eigenplanes for operators $R^{m,n}_{1,\Omega_0,v_0}(t)$ as t tends to $+\infty$. Denote a real eigenvector of $R^{m,n}_{1,\Omega_0,v_0}(t)$ by $e(t)$. Direct calculations show that there exists $\mu \neq 0$ such that

$$e(t) = \mu\big((1,0,0) + O(t^{-1})\big).$$

Consider the invariant real plane of the operator $R^{m,n}_{1,\Omega_0,v_0}(t)$ (it corresponds to a pair of complex conjugate eigenvalues). Note that this plane is a union of a certain family of closed orbits of $R^{m,n}_{1,\Omega_0,v_0}(t)$. Direct calculations show that any orbit in this family is an ellipse with semiaxes $\lambda g_{max}(t)$ and $\lambda g_{min}(t)$ for some positive real number λ, where

$$g_{max}(t) = (0,t,0) + O(1),$$
$$g_{min}(t) = (0,0,t^{1/2}) + O(t^{-1/2}).$$

The vectors $g_{max}(t) \pm \sqrt{-1} g_{min}(t)$ are two complex eigenvectors of the operator $R^{m,n}_{1,\Omega_0,v_0}(t)$. For the ratio of the lengths of maximal and minimal semiaxes of any orbit we have the following asymptotic estimate:

$$\frac{\lambda|g_{max}(t)|}{\lambda|g_{min}(t)|} = |t|^{1/2} + O(|t|^{-1/2}).$$

First, since

$$(1,0,0) - \frac{1}{\mu}e(t) = O(|t|^{-1}),$$

the minimal semiaxis of the orbit-vertex $T_{R^{m,n}_{1,\Omega_0,v_0}(t)}(1,0,0)$ is asymptotically not greater than $O(t^{-1})$. Therefore, the length of the maximal semiaxis is asymptotically not greater than some function of type $O(|t|^{-1/2})$. Hence, the orbit of the point $(1,0,0)$ is contained in the $(C_1|t|^{-1/2})$-ball of the point $(1,0,0)$, where C_1 is a constant that does not depend on t.

Second, we have

$$(0,1,0) - \frac{1}{t}g_{\max}(t) = O(|t|^{-1}).$$

Thus, the length of the maximal semiaxis of the orbit-vertex $T_{R^{m,n}_{1,\Omega_0,v_0}(t)}(1,0,0)$ is asymptotically not greater than some function $1 + O(t^{-1/2})$. Hence, the length of the minimal semiaxis is asymptotically not greater than some function $O(|t|^{-1/2})$. This implies that the orbit of the point $(0,1,0)$ is contained in the $(C_2|t|^{-1/2})$-tubular neighborhood of the segment with vertices $(0,1,0)$ and $(0,-1,0)$, where C_2 is a constant that does not depend on t.

Therefore, the convex hull of the union of two orbit-vertices

$$T_{R^{m,n}_{1,\Omega_0,v_0}(t)}(1,0,0) \quad \text{and} \quad T_{R^{m,n}_{1,\Omega_0,v_0}(t)}(0,1,0)$$

is contained in the C-tubular neighborhood of the triangle with vertices $(1,0,0)$, $(0,1,0)$, $(0,-1,0)$, where $C = |t|^{-1/2}\max(C_1,C_2)$. This concludes the proof of the lemma. $\square$

Let us now extend the statement of Proposition 25.52 to the general case of Hessenberg matrices.

Corollary 25.56. *Let $\Omega = \langle a_{11}, a_{21} | a_{12}, a_{22}, a_{32} \rangle$ and let $R^{m,n}_{1,\Omega,v}$ be an NRS-ray. Then for every $\varepsilon > 0$ there exists $C > 0$ such that for every $t > C$ the convex hull of the union of two orbit-vertices*

$$T_{R^{m,n}_{1,\Omega,v}(t)}(1,0,0) \quad \text{and} \quad T_{R^{m,n}_{1,\Omega,v}(t)}(a_{11},a_{21},0)$$

is contained in the ε-tubular neighborhood of the convex hull of three points $(1,0,0)$, $(a_{11},a_{21},0)$, $(-a_{11},-a_{21},0)$.

Remark. Recall that $R^{m,n}_{1,\Omega,v}(t)(1,0,0) = (a_{11},a_{21},0)$ for every t.

Proof. Define

$$X = \begin{pmatrix} a_{21}a_{32} & -a_{32}a_{11} & a_{11}a_{22} - a_{21}a_{12} \\ 0 & a_{32} & -a_{11} - a_{22} \\ 0 & 0 & 1 \end{pmatrix}.$$

The operator X defines two linear functions l_1 and l_2 on two variables m and n with coefficients depending only on a_{11}, a_{21}, a_{12}, a_{22}, and a_{32} satisfying

$$H_{\langle 0,1|0,0,1\rangle}^{(1,0,0)}\left(l_1(m,n)-\frac{t}{a_{21}a_{32}}, l_2(m,n)\right)=XH_\Omega(m-t,n)X^{-1}.$$

Therefore, the ray $R_{1,\Omega,v}^{m,n}$ after a change of coordinates and a rescaling is taken to the ray $R_{1,\Omega_0,v_0}^{\tilde{m},\tilde{n}}$ of matrices with Hessenberg type $\Omega_0=\langle 0,1|0,0,1\rangle$ for certain $\tilde{m}$ and $\tilde{n}$.

Lemma 25.55 implies the following. For every $\varepsilon>0$ there exists a positive constant such that for every t greater than this constant the convex hull of the union of two orbit-vertices

$$T_{R_{1,\Omega_0,v_0}^{\tilde{m},\tilde{n}}(t)}(1,0,0)\quad\text{and}\quad T_{R_{1,\Omega_0,v_0}^{\tilde{m},\tilde{n}}(t)}(0,1,0)$$

is contained in the ε-tubular neighborhood of the triangle with vertices $(1,0,0)$, $(0,1,0)$, $(0,-1,0)$.

If we reformulate the last statement for the family of operators in the old coordinates, then we get the statement of the corollary.   □

Proof of Proposition 25.54. Notice that the operator $R_{1,\Omega,v}^{m,n}(t)$ takes the point $(1,0,0)$ to the point $(a_{11},a_{21},0)$. Therefore, the convex hull of the union of two orbit-vertices

$$T_{R_{1,\Omega,v}^{m,n}(t)}(1,0,0)\quad\text{and}\quad T_{R_{1,\Omega,v}^{m,n}(t)}(a_{11},a_{21},0)$$

(we denote it by $W(t)$) coincides with the set $\Gamma_{R_{1,\Omega,v}^{m,n}(t)}^0(1,0,0)$.

From Proposition 25.52 it follows that there exists a fundamental domain for a Klein—Voronoi continued fraction with all its orbit-vertices contained in $W(t)$. Choose a sufficiently small ε_0 such that the ε_0-tubular neighborhood of the triangle with vertices

$$(1,0,0),\qquad(a_{1,1},a_{2,1},0),\quad\text{and}\quad(-a_{11},-a_{21},0)$$

does not contain integer points distinct from the points of the triangle. From Corollary 25.56 it follows that for a sufficiently large t the set $W(t)$ is contained in the ε_0-tubular neighborhood of the triangle. This implies the statement of Proposition 25.54.   □

25.6.6.3 Geometry of Klein—Voronoi continued fractions for matrices of $R_{2,\Omega,v}^{m,n}$

Now let us study the remaining case of the rays of matrices with asymptotic direction (a_{11},a_{21}). We recall that $\Omega_0=\langle 0,1|0,0,1\rangle$.

Proposition 25.57. *Consider an NRS-ray $R_{2,\Omega,v}^{m,n}$. Then there exists a constant $C>0$ such that for every $t>C$ all orbit-vertices of one of the fundamental domains for the Klein—Voronoi continued fraction of the matrix $R_{2,\Omega,v}^{m,n}(t)$ are contained in the set of* *all integer orbits corresponding to the integer points in the convex hull of* $(1,0,0)$, $(-1,0,0)$, *and* $(a_{11},a_{21},0)$. ☐

The proof of this proposition is based on a corollary of the following lemma.

Lemma 25.58. *Let* $R_{2,\Omega_0,v_0}^{m,n}$ *be an NRS-ray. Then for every* $\varepsilon > 0$ *there exists* $C > 0$ *such that for every* $t > C$ *the convex hull of the union of two orbit-vertices*

$$T_{R_{2,\Omega_0,v_0}^{m,n}}(t)(1,0,0) \quad and \quad T_{R_{2,\Omega_0,v_0}^{m,n}}(t)(0,1,0)$$

is contained in the ε*-tubular neighborhood of the convex hull of three points* $(1,0,0)$, $(-1,0,0)$, $(0,1,0)$.

Proof. First, notice that the Klein—Voronoi continued fractions for the operators A and A^{-1} coincide.

Second

$$H_{\langle 0,1|0,0,1 \rangle}^{(1,0,0)}(m,n+t) = XH_{\langle 0,1|0,0,1 \rangle}^{(1,0,0)}(-n-t,-m)X^{-1},$$

where

$$X = \begin{pmatrix} 0 & -1 & -n-t \\ -1 & 0 & -m \\ 0 & 0 & -1 \end{pmatrix}.$$

Thus, in the new coordinates we obtain the equivalent statement for the ray $R_{1,\Omega_0,v_0}^{-n,-m}(t)$. Now Lemma 25.58 follows directly from Lemma 25.55. ☐

Corollary 25.59. *Let* $\Omega = \langle a_{11}, a_{21} | a_{12}, a_{22}, a_{32} \rangle$ *and let* $R_{2,\Omega,v}^{m,n}$ *be an NRS-ray. Then for every* $\varepsilon > 0$ *there exists* $C > 0$ *such that for every* $t > C$ *the convex hull of the union of two orbit-vertices*

$$T_{H_\Omega^v(m+a_{11}t,n+a_{21}t)}(1,0,0) \quad and \quad T_{H_\Omega^v(m+a_{11}t,n+a_{21}t)}(a_{11},a_{21},0)$$

is contained in the ε*-tubular neighborhood of the triangle with vertices* $(1,0,0)$, $(-1,0,0)$, *and* $(a_{11},a_{21},0)$. ☐

Remark. We omit the proofs of Corollary 25.59 and Proposition 25.57, since they repeat the proofs of Corollary 25.56 and Proposition 25.54.

25.6.6.4 Proof of Theorem 25.48

Step 1. Let A be an operator with Hessenberg matrix M in $SL(3,\mathbb{Z})$. By Proposition 25.29 the Hessenberg complexity of the Hessenberg matrix M coincides with the MD-characteristic $\Delta_A(1,0,0)$. Therefore, the Hessenberg matrix M is ς-reduced if and only if the MD-characteristic of A attains the minimal possible absolute value on the integer lattice, except at the origin, exactly at the point $(1,0,0)$.

Step 2. By Theorem 25.31(ii) we know that all minima of the set of absolute values for the MD-characteristic of A are attained at integer points of the Klein—Voronoi sails for A.

Step 3. By Theorem 25.31(i) we can restrict the search of the minimal absolute value of the MD-characteristic to an arbitrary fundamental domains of the Klein—Voronoi continued fraction.

Step 4.1. The case of NRS-rays with asymptotic direction $(-1,0)$. By Proposition 25.54 there exists $T > 0$ such that for every integer $t > T$ all integer points of one of the fundamental domains for $R_{1,\Omega,v}^{m,n}(t)$ are contained in the convex hull of three points $(1,0,0)$, $(a_{11},a_{21},0)$, and $(-a_{11},-a_{21},0)$.

This triangle contains only finitely many integer points, all of which have the last coordinate equal to zero. The value of the MD-characteristic for the points of type $(x,y,0)$ equals

$$(a_{21}x - a_{11}y)a_{32}^2 y^2 t + C,$$

where the constant C does not depend on t, but only on x, y, and the elements $a_{i,j}$ of the matrix $H_\Omega^v(m,n)$.

So, for any point $(x,y,0)$, the MD-characteristics linearly increase with respect to the parameter t. The only exceptions are the points of type $\lambda(1,0,0)$ and $\mu(a_{11},a_{21},0)$ (for integers λ and μ). The values of the MD-characteristics are constant in these points with respect to t.

Since there are finitely many integer points in the triangle $(1,0,0)$, $(a_{11},a_{21},0)$, and $(-a_{11},-a_{21},0)$, for sufficiently large t the MD-characteristic at points of the triangle attains its minimum at $(1,0,0)$ or at $(a_{11},a_{21},0)$. Since $R_{1,\Omega,v}^{m,n}(t)$ takes the point $(1,0,0)$ to the point $(a_{11},a_{21},0)$, the values of the MD-characteristics at $(1,0,0)$ and at $(a_{11},a_{21},0)$ coincide.

Therefore, for sufficiently large t the matrix

$$H_{\langle a_{11},a_{21}|a_{12},a_{22},a_{32}\rangle}^{(1,0,0)}(m-t,n)$$

is always ς-reduced and there are no other ς-reduced matrices integer conjugate to the given one. This implies both statements of Theorem 25.48 for every NRS-ray with asymptotic direction $(-1,0)$.

Step 4.2. The case of NRS-rays with asymptotic direction (a_{11},a_{21}). This case is similar to the case of NRS-rays with asymptotic direction $(-1,0)$, so we omit the proof here.

Proof of Theorem 25.48 is complete. $\square$

25.7 Open problems

In this section we formulate open questions on the structure of the sets of NRS-matrices and briefly describe the situation for RS-matrices.

NRS-matrices. As we have shown in Theorem 25.48, the number of ς-nonreduced matrices in any NRS-ray is finite. Here we conjecture a stronger statement.

Conjecture 31. Let Ω be an arbitrary Hessenberg type. All but a finite number of NRS-matrices of type Ω are ς-reduced.

If the answer to this conjecture is positive, we immediately have the following general question.

Problem 32. Study the asymptotics of the number of ς-nonreduced NRS-matrices with respect to the growth of Hessenberg complexity.

Denote the conjectured number of ς-nonreduced NRS-matrices of Hessenberg type Ω by $\#(\Omega)$. Numerous calculations give rise to the following table for all types with Hessenberg complexity less than 5.

Ω	$\langle 0,1\|0,0,1\rangle$	$\langle 0,1\|1,0,2\rangle$	$\langle 0,1\|1,1,2\rangle$	$\langle 0,1\|1,0,3\rangle$	$\langle 0,1\|1,1,3\rangle$	$\langle 0,1\|1,2,3\rangle$
$\varsigma(\Omega)$	1	2	2	3	3	3
$\#(\Omega)$	0	12	12	6	10	10

Ω	$\langle 0,1\|2,0,3\rangle$	$\langle 0,1\|2,1,3\rangle$	$\langle 0,1\|2,2,3\rangle$	$\langle 1,2\|0,0,1\rangle$	$\langle 0,1\|1,0,4\rangle$	$\langle 0,1\|1,1,4\rangle$
$\varsigma(\Omega)$	3	3	3	4	4	4
$\#(\Omega)$	14	10	10	94	6	8

Ω	$\langle 0,1\|1,2,4\rangle$	$\langle 0,1\|1,3,4\rangle$	$\langle 0,1\|3,0,4\rangle$	$\langle 0,1\|3,1,4\rangle$	$\langle 0,1\|3,2,4\rangle$	$\langle 0,1\|3,3,4\rangle$
$\varsigma(\Omega)$	4	4	4	4	4	4
$\#(\Omega)$	10	8	10	12	8	8

RS-matrices. Now we say a few words about real spectrum matrices (i.e., about $SL(3,\mathbb{Z})$-matrices with three distinct real roots). Mostly we consider the family $H(\langle 0,1\|1,0,2\rangle)$; the situation with the other Hessenberg types is similar.

Recall that

$$H^{(1,0,1)}_{\langle 0,1\|1,0,2\rangle}(m,n) = \begin{pmatrix} 0 & 1 & n+1 \\ 1 & 0 & m \\ 0 & 2 & 2n+1 \end{pmatrix}.$$

This matrix is of Hessenberg type $\langle 0,1\|1,0,2\rangle$, and its Hessenberg complexity equals 2. Hence $H^{(1,0,1)}_{\langle 0,1\|1,0,2\rangle}(m,n)$ is ς-reduced if and only if it is not integer conjugate to some matrix of unit Hessenberg complexity. All such matrices are of Hessenberg type $\langle 0,1\|0,0,1\rangle$.

In Fig. 25.8 we show all matrices $H^{(1,0,1)}_{\langle 0,1\|1,0,2\rangle}(m,n)$ within the square

$$-20 \leq m,n \leq 20.$$

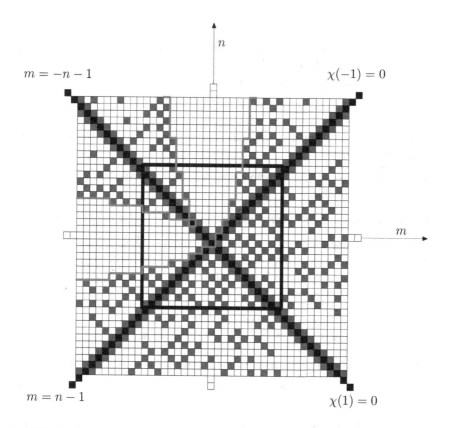

Fig. 25.8 The family of matrices of Hessenberg type $\langle 0,1|1,0,2\rangle$.

The square in the intersection of the mth column with the nth row corresponds to the matrix $H^{(1,0,1)}_{\langle 0,1|1,0,2\rangle}(m,n)$. It is colored in black if the characteristic polynomial has rational roots. The square is colored in gray if the characteristic polynomial is irreducible and there exists an integer vector (x,y,z) with the coordinates satisfying

$$-1000 \le x,y,z \le 1000,$$

such that the MD-characteristic of $H_{\langle 0,1|1,0,2\rangle}(1,0,1)(m,n)$ equals 1 at (x,y,z). All the rest of the squares are white.

If a square is gray, then the corresponding matrix is ς-nonreduced (see Remark 25.29). If a square is white, then we cannot conclude whether the matrix is ς-reduced (since the integer vector with unit MD-characteristics may have coordinates with absolute values greater than 1000).

It is most probable that white squares in Fig. 25.8 represent ς-reduced matrices. We have checked explicitly all white squares for

$$-10 \leq m, n \leq 10,$$

and found that all of them are ς-reduced (such squares are contained inside the big black square shown on the figure).

We show a boundary broken line between the NRS- and RS-squares in gray.

Remark. In Fig. 25.8 the *NRS-domain* is easily visualized. It almost completely consists of white squares. On the other hand, the *RS-domain* contains a relatively large number of black squares. This indicates a significant difference between the RS and NRS cases.

Direct calculations of the corresponding MD-characteristics show that the following proposition holds.

Proposition 25.60. *If an integer $m+n$ is odd, then $H_{\langle 0,1|1,0,2 \rangle}^{(1,0,1)}(m,n)$ is ς-reduced.* $\square$

Remark. On the one hand, Proposition 25.60 implies the existence of rays entirely consisting of ς-reduced matrices. On the other hand, in contrast to the NRS-matrices, there are some rays consisting entirely of ς-nonreduced RS-matrices. For instance, all matrices corresponding to the integer points of the lines

$$(1) \; m = n; \quad (2) \; m = n+2; \quad (3) \; m = -n;$$
$$(4) \; m = -n-2; \quad (5) \; n = 3m-4; \quad (6) \; m = 3n+6$$

are not ς-reduced (we do note state that the list of such lines is complete).

So Theorem 25.48 does not have a direct generalization to the RS-case and we end up with the following problem.

Problem 33. What is the percentage of ς-reduced matrices among matrices of a given Hessenberg type Ω?

It is likely that almost all Hessenberg matrices are ς-reduced (except for some measure-zero subset).

25.8 Exercises

Exercise 25.1. Let π be an integer k-dimensional plane in $\mathbb{R}^n$. Describe the locus of all integer points a unit integer distance to π.

Exercise 25.2. Prove that the Hessenberg matrices

$$\begin{pmatrix} 0 & 1 & 3 \\ 1 & 0 & 0 \\ 0 & 3 & 8 \end{pmatrix} \quad \text{and} \quad \begin{pmatrix} 0 & 2 & 5 \\ 1 & 1 & 2 \\ 0 & 3 & 7 \end{pmatrix}$$

are not integer conjugate but have the same Hessenberg complexity and the same characteristic polynomial.

Exercise 25.3. Let $A \in GL(n, \mathbb{R})$ preserve the level sets of the form $x^2 + y^2$. Prove that there exists α such that either

$$A = \begin{pmatrix} \cos\alpha & \sin\alpha \\ -\sin\alpha & \cos\alpha \end{pmatrix} \quad \text{or} \quad A = \begin{pmatrix} -\cos\alpha & \sin\alpha \\ \sin\alpha & \cos\alpha \end{pmatrix}.$$

Exercise 25.4. Consider $A \in GL(n, \mathbb{Z})$ whose characteristic polynomial is irreducible over $\mathbb{Q}$. Find the generators of the abelian group $T^I(A)$ in the eigenbasis of A.

Exercise 25.5. Consider $A, B \in GL(n, \mathbb{Z})$ whose characteristic polynomial is irreducible over $\mathbb{Q}$. Then A and B commute if and only if the Dirichlet groups of A and B coincide.

Exercise 25.6. If an integer $m+n$ is odd, then $H^{(1,0,1)}_{\langle 0,1|1,0,2\rangle}(m,n)$ is ς-reduced.

Exercise 25.7. Prove that all matrices $H^{(1,0,1)}_{\langle 0,1|1,0,2\rangle}(m,n)$ corresponding to integer points of the lines

$$(1)\ m = n; \quad (2)\ m = n+2; \quad (3)\ m = -n;$$
$$(4)\ m = -n-2; \quad (5)\ n = 3m-4; \quad (6)\ m = 3n+6$$

are ς-reduced. Are there some other similar lines in the plane (m,n)?

Exercise 25.8. How many different nonempty Hessenberg types of complexity n exist for $n = 1, 2, \ldots, 10$?

Exercise 25.9. Prove that the minimum of the Markov—Davenport characteristic on the corresponding w-sail is attained at some vertex of this w-sail.

Exercise 25.10. Find an example of an invariant cone for an irreducible operator whose w-sails are not homothetic to each other.

Chapter 26
Approximation of Maximal Commutative Subgroups

We have already discussed some geometric approximation aspects in the plane in Chapter 14: we have studied approximations, first, of an arbitrary ray with vertex at the origin, and second, of an arrangement of two lines passing through the origin. In this chapter we briefly discuss an approximation problem of maximal commutative subgroups of $GL(n, \mathbb{R})$ by rational subgroups (introduced recently in [109]). In general this problem touches the theory of simultaneous approximation and both subjects of Chapter 14. The problem of approximation of real spectrum maximal commutative subgroups has much in common with the problem of approximation of nondegenerate simplicial cones. So it is clear that multidimensional continued fractions should be a useful tool here. Also we would like to mention that the approximation problem is linked to the so-called limit shape problems (see for instance [217]).

In Section 26.1 we give general definitions and formulate the problem of best approximations of maximal commutative subgroups. Further, in Section 26.2 we discuss the connection of three-dimensional maximal commutative subgroup approximation to the classical case of simultaneous approximation of vectors in $\mathbb{R}^3$.

In this chapter we denote $\sqrt{-1}$ by I.

26.1 Rational approximations of MCRS-groups

In this section we give general definitions and formulate basic concepts of approximation of maximal commutative subgroups. We begin in Section 26.1.1 with the definition of MCRS-groups and say a few words about how they relate to simplicial cones. Further, in Section 26.1.2 we extend the definition of two-dimensional Markov—Davenport forms to the multidimensional case. We define rational subgroups and their "size" in Section 26.1.3. The distance function (discrepancy) between two subgroups is introduced in Section 26.1.4. Then in Section 26.1.5 we

© Springer-Verlag GmbH Germany, part of Springer Nature 2022
O. N. Karpenkov, *Geometry of Continued Fractions*,
Algorithms and Computation in Mathematics 26,
https://doi.org/10.1007/978-3-662-65277-0_26

give the definition of best approximations. Finally, we conclude this section with a link to the classical problem of approximating real numbers by rational numbers.

26.1.1 Maximal commutative subgroups and corresponding simplicial cones

26.1.1.1 Maximal commutative subgroups

For an arbitrary matrix A of $GL(n, \mathbb{R})$, we denote by $C_{GL(n,\mathbb{R})}(A)$ its *centralizer*, i.e., the set of all the elements of $GL(n, \mathbb{R})$ commuting with A.

We say that a matrix is *regular* if all its eigenvalues are distinct (but not necessarily real). In this chapter we discuss only the most studied case of centralizers defined by regular matrices. Notice that if a matrix A is regular, then the centralizer of $C_{GL(n,\mathbb{R})}(A)$ is commutative.

Definition 26.1. Let A be a regular matrix in $GL(n, \mathbb{R})$. We call the centralizer $C_{GL(n,\mathbb{R})}(A)$ the *maximal commutative subgroup* of $GL(n, \mathbb{R})$ (or *MCRS-group* for short).

We say that a subspace is *invariant* for a centralizer $C_{GL(n,\mathbb{R})}(A)$ if it is invariant for every element of $C_{GL(n,\mathbb{R})}(A)$. Let a regular operator A have k real and $2l$ complex eigenvalues ($k + 2l = n$). Then the MCRS-group $C_{GL(n,\mathbb{R})}(A)$ has k one-dimensional and l two-dimensional minimal invariant subspaces. In case $l = 0$, we say that the MCRS-group is *real spectrum*.

Remark 26.2. Notice that if A is a regular matrix in $GL(n, \mathbb{Z})$, then the Dirichlet group $\Xi(A)$ is a subgroup of the centralizer $C_{GL(n,\mathbb{R})}(A)$. We would like just to mention that Dirichlet groups usually play an important role in the study of best approximations of the corresponding MCRS-groups.

26.1.1.2 The space of simplicial cones

An *n-sided simplicial cone* in $\mathbb{R}^n$ is the convex hull of the union of n unordered rays with vertex at the origin. We consider only the cones that are not contained in a hyperplane. Denote by $Simpl_n$ the space of all n-sided simplicial cones.

Let us describe a relation between real spectrum maximal commutative subgroups from one side and n-sided simplicial cones from the other. There exists a natural 2^n-fold covering of the space of all real spectrum MCRS-groups by the space $Simpl_n$: the cones map to MCRS-groups whose eigendirections are the extremal rays of the cones. Therefore, approximation problems within the space $Simpl_n$ can be studied in terms of the MCRS-groups.

Remark 26.3. The space $Simpl_n$ is not compact for $n > 1$. It admits a transitive right action of $GL(n, \mathbb{R})$ and possesses an invariant measure μ_n. This measure is usually called the *Möbius measure*, and is similar to the Möbius measure for the space of all multidimensional continued fractions discussed in Chapter 23.

26.1.2 Regular subgroups and Markov—Davenport forms

Consider an arbitrary MCRS-group $\mathfrak{A}$ in $\mathrm{GL}(n,\mathbb{R})$ and denote its eigenlines (both real and conjugate) by $l_1,\ldots,l_n$. For $i = 1,\ldots,n$, denote by L_i a nonzero linear form over the space $\mathbb{C}^n$ that attains zero values at all vectors of the complex lines l_j for $j \neq i$. Let $\det(L_1,\ldots,L_n)$ be the determinant of the matrix having in the kth column the coefficients of the form L_k for $k = 1,\ldots,n$ in the dual basis.

Definition 26.4. We say that the form

$$\frac{\prod_{k=1}^{n} \left(L_k(x_1,\ldots,x_n)\right)}{\det(L_1,\ldots,L_n)}$$

is the *Markov—Davenport form* for the MCRS-group $\mathfrak{A}$ and denote it by $\Phi_{\mathfrak{A}}$.

Remark 26.5. Notice that Markov—Davenport forms are closely related to MD-characteristics studied in the previous section. Namely, if $A \in \mathfrak{A}$ is an operator with distinct eigenvalues, then the Markov—Davenport form is proportional to the MD-characteristic Δ_A.

Let us continue with a particular example.

Example 26.6. Consider the MCRS-group corresponding to the Fibonacci operator

$$\begin{pmatrix} 1 & 1 \\ 1 & 0 \end{pmatrix}.$$

The Fibonacci operator has two eigenlines,

$$y = -\theta x \quad \text{and} \quad y = \theta^{-1}x,$$

where θ is the golden ration $\frac{1+\sqrt{5}}{2}$. So the Markov—Davenport form of the Fibonacci operator is

$$\frac{(y+\theta x)(y-\theta^{-1}x)}{\theta - \theta^{-1}} = \frac{1}{\sqrt{5}}(-x^2+xy+y^2).$$

Notice that we have already studied two-dimensional Markov—Davenport forms in Chapter 11 and Chapter 14. Let us formulate and prove the following two simple properties of Markov—Davenport forms.

Proposition 26.7. *The Markov—Davenport form of an MCRS-group $\mathfrak{A}$ is defined by $\mathfrak{A}$ up to a sign.*

Proof. All regular operators of $\mathfrak{A}$ have the same set of eigenlines. Hence the forms L_i are uniquely defined by the MCRS-group up to a multiplication by a scalar and permutations. Every scalar multiplication is normalized by the determinant in the denominator, while a sign of a permutation is not caught by the denominator.   □

Proposition 26.8. *All the coefficients of the Markov—Davenport form are either simultaneously real, if there is an even number of minimal invariant planes of the corresponding MCRS-group, or simultaneously complex.*

Proof. By definition, every MCRS-group contains a real operator with distinct eigenvalues. Linear forms related to real eigenvalues have real coefficients. Pairs of linear forms corresponding to pairs of complex conjugate eigenvalues are complex conjugate up to a multiplicative complex factor. Without loss of generality we choose this factor to be equal to 1. This implies that the product of two such linear forms is real. Due to the determinant in the denominator each pair of complex conjugate roots brings an additional multiplicative factor $I = \sqrt{-1}$.   □

Further details of the proof we leave to the reader.

26.1.3 Rational subgroups and their sizes

Recall some definitions related to complex geometry. A complex number is called a *Gaussian integer* if it is of the form $a + Ib$ with integers a and b. The set of all Gaussian integers forms a lattice with the operation of vector addition. We say that a vector is *Gaussian* if all its coordinates are Gaussian integers. A one-dimensional subspace of $\mathbb{C}^n$ is called *Gaussian* if it contains a nonzero Gaussian vector. A Gaussian vector is said to be *primitive* if all its coordinates are relatively prime with respect to the multiplication of Gaussian integers.

Proposition 26.9. *Every Gaussian one-dimensional subspace contains exactly four primitive Gaussian vectors. In addition, a vector v is primitive if and only if the vectors $-v$, Iv, and $-Iv$ are primitive as well.*   □

Notice that in the case of $\mathbb{C}^1$ there are four primitive vectors (or *units*): they are ± 1 and $\pm I$.

For a complex vector $v = (a_1 + Ib_1, \ldots, a_n + Ib_n)$, denote by $|v|$ the norm

$$\max_{i=1,\ldots,n} \left(\sqrt{a_i^2 + b_i^2} \right).$$

It is clear that the minimum of the norm $|*|$ among the Gaussian vectors in a Gaussian one-dimensional subspace is attained at a primitive Gaussian vector.

Now we are ready to define rational MCRS-groups.

Definition 26.10. An MCRS-group $\mathfrak{A}$ is called *rational* if every of its eigenlines contains a Gaussian nonzero vector.

Definition 26.11. Consider a rational MCRS-group $\mathfrak{A}$. Let $l_1, \ldots, l_n$ be the eigenlines of $\mathfrak{A}$. Let v_i be a primitive Gaussian vector in l_i for $i = 1, \ldots, n$. The *size* of $\mathfrak{A}$ is the real number

$$\min_{i=1,\ldots,n} (|v_1|, \ldots, |v_n|),$$

which we denote by $v(\mathfrak{A})$.

26.1.4 Discrepancy functional

Let us define a natural distance between MCRS-groups. Let $\mathfrak{A}_1$ and $\mathfrak{A}_2$ be two MCRS-groups. Consider the following two symmetric homogeneous forms of degree n each, the combination of two Davenport forms:

$$\Phi_{\mathfrak{A}_1} + \Phi_{\mathfrak{A}_2} \quad \text{and} \quad \Phi_{\mathfrak{A}_1} - \Phi_{\mathfrak{A}_2}.$$

Take two maximal absolute values of the coefficients of these forms (separately). The minimum of these two maximal values is considered the distance between $\mathfrak{A}_1$ and $\mathfrak{A}_2$, which we call the *discrepancy* and denote by $\rho(\mathfrak{A}_1, \mathfrak{A}_2)$.

Example 26.12. Consider the following two operators

$$\begin{pmatrix} 0 & -1 \\ 1 & 0 \end{pmatrix} \quad \text{with eigenvectors } (I, 1) \text{ and } (-I, 1),$$

$$\begin{pmatrix} 1 & 1 \\ 4 & 1 \end{pmatrix} \quad \text{with eigenvectors } (1, 2) \text{ and } (1, -2).$$

These operators have distinct eigenvalues, and therefore, they define MCRS-groups (denote them by $\mathfrak{A}_1$ and $\mathfrak{A}_2$ respectively). Both MCRS-groups are rational, where the operators of the first one have real eigenvalues, and where the operators of the second have pairs of complex conjugate eigenvalues. Direct calculations show that

$$v(\mathfrak{A}_1) = 1 \quad \text{and} \quad v(\mathfrak{A}_2) = 2.$$

Now let us calculate the discrepancy $\rho(\mathfrak{A}_1, \mathfrak{A}_2)$. We have

$$\left| \Phi_{\mathfrak{A}_1}(v) \pm \Phi_{\mathfrak{A}_2}(v) \right| = \left| I \frac{x^2 + y^2}{2} \pm \frac{y^2 - 4x^2}{4} \right|$$

and therefore $\rho(\mathfrak{A}_1, \mathfrak{A}_2) = \frac{\sqrt{3}}{2}$.

26.1.5 Approximation model

In the above notation the problem of approximation of MCRS-groups by rational MCRS-groups can be formulated as follows.

Definition 26.13. (The problem of best approximation.) For a given MCRS-group $\mathfrak{A}$ and a positive integer N, find a rational MCRS-group $\mathfrak{A}_N$ of size not exceeding N such that for every other rational MCRS-group $\mathfrak{A}'$ of size not exceeding N we have

$$\rho(\mathfrak{A}, \mathfrak{A}_N) \leq \rho(\mathfrak{A}, \mathfrak{A}').$$

Remark 26.14. There is another important class of MCRS-groups that would also be interesting to consider in this context. An MCRS-group is said to be *algebraic* if it contains regular operators of $GL(n, \mathbb{Z})$. It would be interesting to develop approximation techniques of MCRS-groups by algebraic MCRS-groups.

26.1.6 Diophantine approximation and MCRS-group approximation

The classical problem of approximating real numbers by rational numbers is a particular case of the problem of best approximations of MCRS-groups in the following sense.

For a nonzero real number α, we denote by $\mathfrak{A}[\alpha]$ the MCRS-group of $GL(2, \mathbb{R})$ with invariant lines $x = 0$ and $y = \alpha x$. Denote by $\Omega^{\mathbb{R}}_{[0,1]}$ the set of $\mathfrak{A}[\alpha]$ satisfying $1 > \alpha > 0$. Denote also by $\Omega^{\mathbb{Q}}_{[0,1]}$ the subset of $\Omega^{\mathbb{R}}_{[0,1]}$ defined by rational numbers α.

For every pair of relatively prime integers (p, q) satisfying $0 \leq \frac{p}{q} \leq 1$ we have

$$v\left(\mathfrak{A}\left[\frac{p}{q}\right]\right) = q.$$

Let us calculate the discrepancy between two MCRS-groups $\mathfrak{A}[\alpha_1]$ and $\mathfrak{A}[\alpha_2]$. for arbitrary positive real numbers α_1 and α_2,

$$\Phi_{\mathfrak{A}[\alpha_1]} - \Phi_{\mathfrak{A}[\alpha_2]} = \frac{x(y - \alpha_1 x)}{1} - \frac{x(y - \alpha_2 x)}{1} = (\alpha_2 - \alpha_1)x^2,$$

$$\Phi_{\mathfrak{A}[\alpha_1]} + \Phi_{\mathfrak{A}[\alpha_2]} = \frac{x(y - \alpha_1 x)}{1} + \frac{x(y - \alpha_2 x)}{1} = 2xy - (\alpha_2 + \alpha_1)x^2.$$

Since $\alpha_1 > 0$ and $\alpha_2 > 0$, we have

$$\rho(\mathfrak{A}[\alpha_1], \mathfrak{A}[\alpha_2]) = |\alpha_1 - \alpha_2|.$$

Now the classical problem of approximation of real numbers by rational numbers is reformulated in terms of MCRS-groups as follows: *for a given MCRS-group*

$\mathfrak{A}[\alpha] \in \Omega_{[0,1]}^{\mathbb{R}}$ *and a positive N, find an approximation* $\mathfrak{A}[p/q] \in \Omega_{[0,1]}^{\mathbb{Q}}$ *of size not exceeding N such that for every other MCRS-group* $\mathfrak{A}[p'/q'] \in \Omega_{[0,1]}^{\mathbb{Q}}$ *of size not exceeding N,*

$$\rho(\mathfrak{A}[\alpha],\mathfrak{A}[p/q]) \le \rho(\mathfrak{A}[\alpha],\mathfrak{A}[p'/q']).$$

26.2 Simultaneous approximation in $\mathbb{R}^3$ and MCRS-group approximation

The theory of simultaneous approximation of a real vector by vectors with rational coefficients is also to some extent a special case of MCRS-group approximation. We say a few words about this similarity in this section. After a brief description of the approximation model, we study two particular examples of simultaneous approximation in the framework of MCRS-group approximation.

26.2.1 General construction

Let (a,b,c) be a vector in $\mathbb{R}^3$. Denote by $\mathfrak{A}[a,b,c]$ the maximal commutative subgroup defined by an operator with the following three eigenvectors

$$(a,b,c), \quad (0,1,I), \quad (0,1,-I).$$

We state the problem of best approximation as follows: *for a given MCRS-group* $\mathfrak{A}[a,b,c]$ *and a positive number N, find an approximation* $\mathfrak{A}[a',b',c']$ *of size not exceeding N such that for every other MCRS-group* $\mathfrak{A}[a'',b'',c'']$ *of size not exceeding N, we have*

$$\rho(\mathfrak{A}[a,b,c],\mathfrak{A}[a',b',c']) \le \rho(\mathfrak{A}[a,b,c],\mathfrak{A}[a'',b'',c'']).$$

Notice that

$$v(\mathfrak{A}[a,b,c]) = \max(a,b,c)$$

and

$$\Phi_{\mathfrak{A}[a,b,c]}(x,y,z) = I\left(-\frac{b^2+c^2}{2a^2}x^3 + \frac{b}{a}x^2y + \frac{c}{a}x^2z - \frac{1}{2}xy^2 - \frac{1}{2}xz^2\right).$$

The discrepancy between the MCRS-groups $\mathfrak{A}[a,b,c]$ and $\mathfrak{A}[a',b',c']$ is as follows:

$$\rho\left(\mathfrak{A}[a,b,c],\mathfrak{A}[a',b',c']\right) = \min\left(\max\left(\left|\frac{b}{a}-\frac{b'}{a'}\right|,\left|\frac{c}{a}-\frac{c'}{a'}\right|,\left|\frac{b^2+c^2}{2a^2}-\frac{b'^2+c'^2}{2a'^2}\right|\right),\right.$$
$$\left.\max\left(\left|\frac{b}{a}+\frac{b'}{a'}\right|,\left|\frac{c}{a}+\frac{c'}{a'}\right|,\left|\frac{b^2+c^2}{2a^2}+\frac{b'^2+c'^2}{2a'^2}\right|\right)\right).$$

In the case of positive a, b, c, a', b', and c', we simply have

$$\rho\left(\mathfrak{A}[a,b,c],\mathfrak{A}[a',b',c']\right) = \max\left(\left|\frac{b}{a}-\frac{b'}{a'}\right|,\left|\frac{c}{a}-\frac{c'}{a'}\right|,\left|\frac{b^2+c^2}{2a^2}-\frac{b'^2+c'^2}{2a'^2}\right|\right).$$

Remark 26.15. For the case of classical simultaneous approximation one takes a slightly different distance between the vectors:

$$\rho\left(\mathfrak{A}[a,b,c],\mathfrak{A}[a',b',c']\right) = \max\left(\left|\frac{b}{a}-\frac{b'}{a'}\right|,\left|\frac{c}{a}-\frac{c'}{a'}\right|\right).$$

So the described approximation model is not exactly the classical model of simultaneous approximation.

Let us study two particular examples.

26.2.2 A ray of a nonreal spectrum operator

Consider the nonreal spectrum algebraic operator

$$B = \begin{pmatrix} 0 & 1 & 1 \\ 0 & 0 & 1 \\ 1 & 0 & 0 \end{pmatrix}.$$

This operator can be thought of as the simplest nonreal spectrum operator from a geometric point of view. Denote the real eigenvalue of the operator B by ξ_1 and its complex conjugate eigenvalues by ξ_2 and ξ_3. Notice that

$$|\xi_1| > |\xi_2| = |\xi_3|.$$

Let us approximate the eigenline corresponding to ξ_1. Let v_{ξ_1} be the vector in this eigenline having the first coordinate equal to 1. In this setting we have

$$\xi_1 \approx 1.3247179573 \quad \text{and} \quad v_{\xi_1} \approx (1, 0.5698402911, 0.7548776662).$$

The set of best approximations with size not exceeding 10^6 contains 48 elements. All these elements are of type $B^{n_i}(1,0,0)$, where (n_i) is the following sequence:

$$(4,6,7,8,9,10,\ldots,50,51,52).$$

We conjecture that the set of best approximations coincides with the set of points $B^k(1,0,0)$, where $k = 4$ or $k \geq 6$; the approximation rate in this case is $CN^{-3/2}$. The structure of the set of best approximations is usually closely related to the corresponding Dirichlet group. In this case the Dirichlet group $\Xi(B)$ is homeomorphic to $\mathbb{Z} \oplus \mathbb{Z}/2\mathbb{Z}$; the generators are B and $-\mathrm{Id}$.

26.2.3 Two-dimensional golden ratio

Consider now the algebraic real spectrum operator of the two-dimensional golden ratio

$$G = \begin{pmatrix} 3 & 2 & 1 \\ 2 & 2 & 1 \\ 1 & 1 & 1 \end{pmatrix}.$$

As we have already seen in Chapter 22, this operator has the simplest continued fraction in the sense of Klein. The Dirichlet group $\Xi(G)$ is generated by the following two operators:

$$H_1 = \begin{pmatrix} 1 & 1 & 1 \\ 1 & 1 & 0 \\ 1 & 0 & 0 \end{pmatrix} \quad \text{and} \quad H_2 = \begin{pmatrix} 0 & 1 & 0 \\ 1 & -1 & 1 \\ 0 & 1 & -1 \end{pmatrix}.$$

Notice that $G = H_1^2$ and $H_2 = (H_1 - \mathrm{Id})^{-1}$. The operator H_1 here is the three-dimensional Fibonacci operator. Denote the eigenvalues of H_1 by ξ_1, ξ_2, and ξ_3 such that

$$|\xi_1| > |\xi_2| > |\xi_3|.$$

Let us approximate the eigenline corresponding to ξ_1. Denote by v_{ξ_1} the vector of this eigenline having the last coordinate equal to 1. Notice that

$$\xi_1 \approx 2.2469796037 \quad \text{and} \quad v_{\xi_1} \approx (2.2469796037, 1.8019377358, 1).$$

The set of best approximations of size not exceeding 10^6 contains 41 elements; 40 of them (except for the third best approximation) are contained in the set

$$\left\{ H_1^m H_2^n (1,0,0) \;\middle|\; m,n \in \mathbb{Z} \right\}.$$

In the following table we show the consecutive pairs of powers (m_i, n_i) for the these 41 best approximations except for the third one. So in the column i we get $m = m_i$, $n = n_i$ for the approximation $H_1^{m_i} H_2^{n_i}(1,0,0)$.

i	1	2	4	5	6	7	8	9	10	11	12	13	14	15	16	17	18	19	20	21	22
m_i	1	2	3	3	4	4	5	5	6	6	6	7	7	8	8	9	9	10	10	11	11
n_i	1	1	2	1	2	1	3	2	3	2	1	3	2	3	2	4	3	4	3	5	4

i	23	24	25	26	27	28	29	30	31	32	33	34	35	36	37	38	39	40	41
m_i	11	12	12	13	13	14	14	15	15	15	16	16	17	17	18	18	19	19	19
n_i	3	4	3	5	4	5	4	6	5	4	5	4	6	5	6	5	7	6	5

The third best approximation is $\mathfrak{A}[3,2,1]$; it is not of the form $H_1^m H_2^n(1,0,0)$. We conjecture that the set of all best approximations except $\mathfrak{A}[3,2,1]$ is contained in the set of all points of type $H_1^m H_2^n(1,0,0)$, with the approximation rate in this case being $CN^{-3/2}$.

We conclude this chapter with the following general remark.

Remark 26.16. In the above two examples we have several algebraic artifacts (such as missing $B^5(1,0,0)$ as a best approximation for the first example and an additional best approximation $\mathfrak{A}[3,2,1]$ for the second example). This is not a surprise, since we do not approximate the triples of eigenvectors simultaneously but a certain eigenvector together with two vectors

$$(0,1,I), \quad (0,1,-I).$$

We mix vectors of different natures, and as a result we have irregularities. It is probable that in most common situations such artifacts may occur infinitely many times.

26.3 Exercises

Exercise 26.1. **(a).** Prove that if an operator A is regular, then the centralizer of $A \in GL(n,\mathbb{R})$ is commutative.
(b). Describe all matrices whose centralizers are commutative.

Exercise 26.2. For an arbitrary $A \in GL(n,\mathbb{R})$ describe all invariant subspaces of $C_{GL(n,\mathbb{R})}(A)$.

Exercise 26.3. Let G be a commutative subgroup of $GL(n,\mathbb{R})$. Suppose that for every matrix $A \in GL(n,\mathbb{R}) \setminus G$ there exists a matrix $\tilde{A} \in G$ such that A and $\tilde{A}$ do not commute. Is it true that there exists a matrix $B \in G$ such that $G = C_{GL(n,\mathbb{R})}(B)$?

Exercise 26.4. Prove that the space $Simpl_2$ is homeomorphic to the Möbius band.

Exercise 26.5. Find the first 10 best approximations $\mathfrak{A}[a,b,c]$ for the minimal and the middle eigenvectors of the three-dimensional golden ratio G.

Exercise 26.6. Using the theory of MCRS-groups formulate a problem of approximation of planes in $\mathbb{R}^4$ by integer planes.

Chapter 27
Other Generalizations of Continued Fractions

In this chapter we present some other generalizations of regular continued fractions to the multidimensional case. The main goal for us here is to give different geometric constructions related to such continued fractions (whenever possible). We say a few words about Minkowski—Voronoi continued fractions (Section 27.1), triangle sequences related to Farey addition (Section 27.2), O'Hara's algorithm related to decomposition of rectangular parallelepipeds (Section 27.3), geometric continued fractions (Section 27.4), and determinant generalizations of continued fractions (Section 27.5). Finally, in Section 27.6 we describe the relation of regular continued fractions to rational knots and links.

We do not pretend to give a complete list of generalizations of continued fractions. The idea is to show the diversity of generalizations. Let us give several references to some subjects that are beyond the scope of this book but that may be interesting and useful in the framework of geometry of continued fractions: p-adic continued fractions [144], [187], [87], [81], complex continued fractions [189], Rosen continued fractions and their expansions [186], [131], geodesic continued fraction approach based on Minkowski reduction [139]. Finally, we would like to mention a relation of continued fractions to tiling of a square by rectangles [137], [64], [183].

27.1 Relative minima

In this section we consider a geometric generalization of continued fractions in terms of local minima. This generalization appears for the first time in the works of H. Minkowski [151] and G.F. Voronoi [219] (see also in [220]) and is used to study units in algebraic fields. Several properties of local minima were later studied by G. Bullig in [34]. The main object of Minkowski—Voronoi continued fractions is an arbitrary complete lattice Γ in $\mathbb{R}^3$ with respect to the fixed coordinate lattice. Statistical properties of relative minima were studied by A.A. Illarionov in [89] and [90]. See also [37] and [78] for more information about three-dimensional rela-

© Springer-Verlag GmbH Germany, part of Springer Nature 2022
O. N. Karpenkov, *Geometry of Continued Fractions*,
Algorithms and Computation in Mathematics 26,
https://doi.org/10.1007/978-3-662-65277-0_27

tive minima. A description of the multidimensional case can be found in paper [36] by V.A. Bykovskii.

The main idea of the Minkowski approach is to define extremal nodes of a lattice Γ with respect to a family of convex sets that can be chosen with certain freedom (for instance, vertices of sails considered in previous chapters of this book can be considered tetrahedral local minima). In this section we describe the classical approach that works with coordinate parallelepipeds.

For simplicity, in this section we consider only sets in $\mathbb{R}^3$ in general position (in the sense that no two vertices of the lattice are in a plane parallel to some coordinate plane).

27.1.1 Relative minima and the Minkowski—Voronoi complex

We begin with a rather algebraic definition of the Minkowski—Voronoi complex. In later subsections, we will construct a tessellation of the plane showing the geometric nature of the complex.

Let T be an arbitrary subset of $\mathbb{R}^n_{\geq 0} = (\mathbb{R}_{\geq 0})^n$. For $i = 1,\ldots,n$ we set

$$T_i = \max\{x_i \mid (x_1,\ldots,x_n) \in T\}.$$

We associate to T the following parallelepiped:

$$\Pi(T) = \{(x_1,\ldots,x_n) \mid 0 \leq x_i \leq T_{x_i}, i = 1,\ldots,n\}.$$

Definition 27.1. Let S be an arbitrary subset of $\mathbb{R}^n_{\geq 0}$. An element $s \in S$ is called a *Voronoi relative minimum* with respect to S if the parallelepiped $\Pi(\{s\})$ contains no points of S except for the origin.

Definition 27.2. Let S be an arbitrary subset of $\mathbb{R}^n_{\geq 0}$ and let $\hat{S}$ be the set of all its Voronoi relative minima. A finite subset $T \subset \hat{S}$ is called *minimal* if the parallelepiped $\Pi(T)$ contains no points of $\hat{S} \setminus T$ except for the origin. We denote the set of all minimal k-element subsets of $\hat{S} \subset S$ by $\mathfrak{M}_k(S)$.

It is clear that any minimal subset of a minimal subset is also minimal.

Definition 27.3. A *Minkowski—Voronoi complex* $MV(S)$ is an $(n-1)$-dimensional complex such that
  (*i*) the k-dimensional faces of $MV(S)$ are enumerated by its minimal $(n-k)$-element subsets (i.e., by the elements of $\mathfrak{M}_{n-k}(S)$);
  (*ii*) a face with a minimal subset T_1 is adjacent to a face with a minimal subset $T_2 \neq T_1$ if and only if $T_1 \subset T_2$.

Remark 27.4. In the three-dimensional case it is also common to consider the Voronoi and Minkowski graphs that are subcomplexes of the Minkowski—Voronoi complex. They are defined as follows.

The *Voronoi graph* is the graph whose vertices and edges are respectively vertices and edges of the Minkowski—Voronoi complex.

The *Minkowski graph* is the graph whose vertices are edges are respectively faces and edges of the Minkowski—Voronoi complex (two vertices in the Minkowski graph are connected by an edge if and only if the corresponding faces in the Minkowski—Voronoi complex have a common edge.

So in some sense Minkowski and Voronoi graphs are dual to each other.

Example 27.5. Let us consider an example of a 6-element set $S_0 \subset \mathbb{R}^3$ defined as follows

$$S_0 = \{s_1, s_2, s_3, s_4, s_5, s_6\},$$

where

$$s_1 = (3,0,0), \ s_2 = (0,3,0), \ s_3 = (0,0,3),$$
$$s_4 = (2,1,2), \ s_5 = (1,2,1), \ s_6 = (2,3,4).$$

There are only five Voronoi relative minima for the set S_0, namely the vectors $s_1, \ldots, s_5$. The Minkowski—Voronoi complex contains 5 vertices, 6 edges, and 5 faces. Its vertices are

$$v_1 = \{s_1, s_3, s_4\}, \quad v_2 = \{s_3, s_4, s_5\}, \quad v_3 = \{s_1, s_4, s_5\},$$
$$v_4 = \{s_2, s_3, s_5\}, \quad v_5 = \{s_1, s_2, s_5\}.$$

Its edges are

$$e_1 = \{s_1, s_3\}, \ e_2 = \{s_3, s_2\}, \ e_3 = \{s_1, s_2\},$$
$$e_4 = \{s_3, s_4\}, \ e_5 = \{s_1, s_4\}, \ e_6 = \{s_4, s_5\},$$
$$e_7 = \{s_3, s_5\}, \ e_8 = \{s_1, s_5\}, \ e_9 = \{s_2, s_5\}.$$

Its faces are

$$f_1 = \{s_1\}, \quad f_2 = \{s_2\}, \quad f_3 = \{s_3\}, \quad f_4 = \{s_4\}, \quad f_5 = \{s_5\}.$$

Finally, we describe the complex $MV(S)$ as a tessellation of an open two-dimensional disk. We show vertices (on the left), edges (in the middle), and faces (on the right) separately:

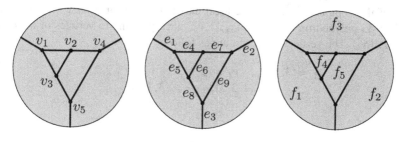

27.1.2 Minkowski—Voronoi tessellations of the plane

In this subsection we discuss some geometric images standing behind the notion of the Minkowski—Voronoi complex in the three-dimensional case.

Definition 27.6. Let S be an arbitrary subset of $\mathbb{R}^3_{\geq 0}$. The *Minkowski polyhedron* for S is the boundary of the set

$$S \oplus \mathbb{R}^3_{\geq 0} = \{s + r \mid s \in S, r \in \mathbb{R}^3_{\geq 0}\}.$$

In other words, the Minkowski polyhedron is the boundary of the union of copies of the positive octant shifted by vertices of the set S.

Definition 27.7. The *Minkowski—Voronoi tessellation* for a set $S \subset \mathbb{R}^3_{\geq 0}$ is a tessellation of the plane $x + y + z = 0$ obtained by the following three steps.

Step 1. Consider the Minkowski polyhedron for the set S and project it orthogonally to the plane $x + y + z = 0$. This projection induces a tessellation of the plane by edges of the Minkowski polyhedron.

Step 2. Remove from the tessellation of Step 1 all vertices corresponding to local minima of the function $x + y + z$ on the Minkowski polyhedron (this is exactly the relative minima of S). Remove also all edges coming from all the removed vertices.

Step 3. After Step 2 some of the vertices are of valence 1. For each vertex v of valence 1 and the only remaining edge wv with endpoint at v we replace the edge wv by the ray wv with vertex at w and passing through v.

Proposition 27.8. The Minkowski—Voronoi tessellation for a nice set S has the combinatorial structure of the Minkowski—Voronoi complex $MV(S)$.   □

Remark 27.9. Here we do not specify what the word "nice" means. We say only that it includes finite sets and complete lattices considered below.

Example 27.10. Consider the set S_0 as in Example 27.5. In Fig. 27.1 we show the Minkowski polyhedron (on the left) and the corresponding Minkowski—Voronoi tessellation (on the right). The local minima of the function $x + y + z$ on the Minkowski polyhedron for S_0 are the relative minima $f_1, \ldots, f_5$. They identify the faces of the complex $MV(S_0)$. The local maxima of the function $x + y + z$ on the Minkowski polyhedron for S_0 are $v_1, \ldots, v_5$, corresponding to vertices of the complex $MV(S_0)$. The vertices $v_1, \ldots, v_5$ are as follows:

$$v_1 = (3,1,3), \quad v_2 = (2,2,3), \quad v_3 = (3,2,2),$$
$$v_4 = (1,3,3), \quad v_5 = (3,3,1).$$

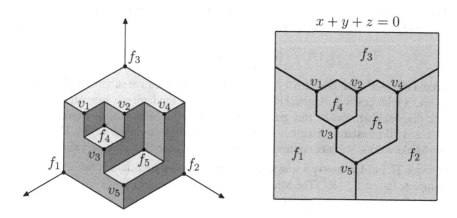

Fig. 27.1 The Minkowski polyhedron (on the left) and the corresponding Minkowski—Voronoi tessellation (on the right).

27.1.3 Minkowski—Voronoi continued fractions in $\mathbb{R}^3$

Now let us give a definition of Minkowski—Voronoi continued fractions for complete lattices in $\mathbb{R}^3$. Consider a lattice Γ defined as follows:

$$\Gamma = \{m_1 v_1 + m_2 v_2 + m_3 v_3 : m_1, m_2, m_3 \in \mathbb{Z}\},$$

where v_1, v_2, v_3 are linearly independent vectors in $\mathbb{R}^3$. Define

$$|\Gamma| = \{(|x|, |y|, |z|) \mid (x,y,z) \in \Gamma\} \setminus \{(0,0,0)\}.$$

Definition 27.11. We say that an edge $\{s_1, s_2\}$ of the complex $MV(|\Gamma|)$ with vertices $\{s_1, s_2, s_3\}$ and $\{s_1, s_2, s_4\}$ is *nondegenerate* if at least one of the triples (s_1, s_2, s_3) and (s_1, s_2, s_4) generates the lattice Γ.

Suppose that $\{s_1, s_2, s_3\}$ generates the lattice. Then we equip the edge $\{s_1, s_2\}$ with a matrix that sends s_1 and s_2 to themselves and takes s_3 to s_4.

So if both triples (s_1, s_2, s_3) and (s_1, s_2, s_4) generate the lattice Γ, then the corresponding edge is equipped with a pair of matrices inverse to each other.

Finally, we have all the tools that are needed to give a definition of the multidimensional Minkowski—Voronoi continued fraction.

Definition 27.12. Let Γ be an arbitrary complete lattice in $\mathbb{R}^3$. The *Minkowski—Voronoi continued fraction* for Γ is the Minkowski—Voronoi complex $MV(|\Gamma|)$ all of whose nondegenerate edges are equipped.

We refer to [151] and [82] for more details related to general minimal systems and to the papers [78] and [216] concerning the arbitrary case.

Let us give a small remark on the periodic algebraic case.

Remark 27.13. Let α, β, and γ be three roots of a cubic polynomial $x^3 + ax^2 + bx + 1$ with integer coefficients a and b. Consider the lattice generated by vectors

$$(1,1,1), \quad (\alpha,\beta,\gamma), \quad \text{and} \quad (\alpha^2,\beta^2,\gamma^2).$$

From Dirichlet's unit theorem it follows that the multiplicative subgroup of diagonal matrices with positive diagonal elements preserving the lattice is isomorphic to $\mathbb{Z}^2$. The diagonal matrices of this group preserve both the Minkowski—Voronoi complex and all translation matrices for the corresponding equipped edges. Therefore, the Minkowski—Voronoi continued fraction in this case is doubly periodic.

Remark 27.14. Let us say a few words about the two-dimensional case. Consider a complete lattice $\Gamma \subset \mathbb{R}^2$. The Minkowski—Voronoi complex for Γ is a broken line. Each edge e_i is equipped with a matrix

$$\begin{pmatrix} 0 & \pm 1 \\ 1 & a_i \end{pmatrix}.$$

Consider now sails of lattices in each of the four coordinate octants for the lattice Γ (i.e., we consider the boundaries of the convex hulls for points of Γ but not of $\mathbb{Z}^3$). Then the integers a_i correspond to the integer lengths of these sails.

27.1.4 Combinatorial properties of the Minkowski—Voronoi tessellation for integer sublattices

If Γ is a full-rank sublattice of $\mathbb{Z}^n$, then the Minkowski—Voronoi complex $MV(|\Gamma|)$ contains finitely many faces. Not much is known about its combinatorial structure. The first open problem here is as follows.

Problem 34. Give a combinatorial description (a complete invariant) of all realizable finite Minkowski—Voronoi complexes for integer sublattices.

In particular, it is interesting to consider the following problem.

Problem 35. How many different N-vertex Minkowski—Voronoi complexes for integer lattices do there exist for a fixed N?

Notice that the problem of a combinatorial description of the Minkowski—Voronoi complex is also open in the case of infinite complexes as well. Nevertheless it seems almost impossible to have a nice complete invariant for it, since there is no testing tool to check whether an infinite complex is realizable as $MV(|\Gamma|)$ for some lattice Γ. As in the case of sails in multidimensional continued fractions, the only infinite case in which it is possible to calculate some examples is the case of periodic Minkowski—Voronoi complexes corresponding to algebraic lattices (see Remark 27.13 above). So for the infinite case, we propose a simpler version of Problem 34.

Fig. 27.2 The Minkowski—Voronoi tessellation for $|\Gamma_{171}|$.

Problem 36. Find criteria of nonrealizability of the Minkowski—Voronoi complexes for arbitrary lattices.

In [225] C. Yannopoulos gives several examples of representations using only three directions for the edges of the Voronoi graph in the plane (which is the 1-skeleton of the Minkowski—Voronoi complex). In a paper [108] the authors extend such representations to the general case of integer sublattices. We announce this result here.

Theorem 27.15. *Let Γ be a sublattice of $\mathbb{Z}^n$. The Minkowski—Voronoi tessellation can be chosen in such a way that the following conditions hold:*

— *All finite edges are straight segments of one of the directions $\frac{k\pi}{3}$ (for $k \in \mathbb{Z}$).*

— *Each vertex that does not contain infinite edges is a vertex of the following 8 types:*

$\square$

Example 27.16. Consider the lattice Γ_{171} generated by the following three vectors:

$$(171,0,0), \quad (0,171,0), \quad \text{and} \quad (2,32,-1).$$

Notice that the first integer point on the z-axis is $(0,0,171)$. The diagram of the Minkowski—Voronoi tessellation for $|\Gamma_{171}|$ satisfying the conditions of Theorem 27.15 is shown in Fig. 27.2.

27.2 Farey addition, Farey tessellation, triangle sequences

27.2.1 Farey addition of rational numbers

We begin with a definition of Farey addition of rational numbers.

Definition 27.17. Let $\frac{p_1}{q_1}$ and $\frac{p_2}{q_2}$ be two rational numbers. We assume that p_1 is relatively prime to q_1 and p_2 is relatively prime to q_2. The rational number

$$\frac{p_1}{q_1} \oplus \frac{p_2}{q_2} = \frac{p_1 + p_2}{q_1 + q_2}$$

is called the *Farey sum* of $\frac{p_1}{q_1}$ and $\frac{p_2}{q_2}$. (Note that for every integer m we consider the fraction $\frac{m}{1}$.)

There is a nice way to construct the set of rational numbers using the operation of Farey addition. We do it inductively in countably many steps.

Construction of the rational numbers.

Base of construction. We start with the sequence of integer numbers $(a_{i,1})$, where $a_{i,1} = i$ for $i \in \mathbb{Z}$.

The kth step of the construction. Suppose that we have constructed a sequence $(a_{i,k})$ (which is infinite on both sides). Let us extend this sequence to the sequence $(a_{i,k+1})$. Set

$$a_{2i,k+1} = a_{i,k}, \quad \text{for } i \in \mathbb{Z};$$
$$a_{2i+1,k+1} = a_{i,k} \oplus a_{i+1,k}, \quad \text{for } i \in \mathbb{Z}.$$

In other words, we put between every two numbers of the sequence $(a_{i,k})$ their Farey sum.

Example 27.18. The sequences for the first three steps are as follows:

Base: $\dots, -4, -3, -2, -1, 0, 1, 2, 3, 4, \dots$

Step 1: $\dots, -2, -\frac{3}{2}, -1, -\frac{1}{2}, 0, \frac{1}{2}, 1, \frac{3}{2}, 2, \dots$

Step 2: $\dots, -1, -\frac{2}{3}, -\frac{1}{2}, -\frac{1}{3}, 0, \frac{1}{3}, \frac{1}{2}, \frac{2}{3}, 1, \dots$

$\dots$ $\dots$

Remark 27.19. The part of the sequence $(a_{i,k})$ contained between the elements 0 and 1 is sometimes called the *kth Farey sequence*.

Remark 27.20. The denominators appearing in the Farey sequences form *Stern's diatomic sequence*, which is defined as follows:

$$a_1 = 1;$$
$$a_{2i} = a_i, \quad \text{for } i \in \mathbb{Z}_+;$$
$$a_{2i+1} = a_i + a_{i+1}, \quad \text{for } i \in \mathbb{Z}_+.$$

The first few elements of this sequence are

$$1,1,2,1,3,2,3,1,4,3,5,2,5,3,4,1,\ldots$$

Further properties of Stern's diatomic sequence can be found in a recent survey [161] by S. Northshield. In this relation we would like to mention the work [123] by A. Knauf on ferromagnetic spin chains, where this sequence appears as Pascal's triangle with memory (see also [121] and [122]).

27.2.2 Farey tessellation

The above construction of rational numbers appears in hyperbolic geometry as *Farey tessellation*. In this section we work with the model of the hyperbolic plane in the upper half-plane of the real plane $\mathbb{R}^2$ with coordinates (x,y).

We add the point at infinity to the line $y = 0$ and call it the *absolute*. Considering the coordinate x as the coordinate on the absolute, we say that the coordinate of infinity is ∞. For every pair of points (a,b) in the absolute there exists a unique hyperbolic line passing through these points, denoted by $l(a,b)$. For every line l in the hyperbolic plane there exists a unique hyperbolic reflection with the axis l. A triangle all three of whose vertices are on the absolute is called an *ideal hyperbolic triangle*.

Definition 27.21. The *Farey tessellation* is the minimal possible decomposition of the hyperbolic plane into ideal triangles such that the following two conditions hold:
— it contains the ideal triangle with vertices 0, 1, and ∞;
— the tessellation is preserved by the group of isometries of the hyperbolic plane generated by all reflections whose axes are sides of ideal triangles in the tessellation.

There is a simple way to construct this tessellation. We start with the ideal triangle with vertices 0, 1, and ∞. Reflecting this triangle with respect to the axes $l(0,1)$, $l(0,\infty)$, and $l(1,\infty)$, we get three new triangles with vertex sets $\{1,2,\infty\}$, $\{-1,0,\infty\}$, and $\{0,\frac{1}{2},1\}$ respectively. Add all of them to the tessellation. Continue iteratively to reflect the obtained picture with respect to the edges of new triangles. In countably many steps we construct all ideal triangles of the tessellation. We show the Farey tessellation in Fig. 27.3.

Proposition 27.22. (*i*) *The set of vertices of all ideal triangles in the Farey tessellation coincides with the subset of the points on the absolute whose coordinates are rational or ∞.*
(*ii*) *An ideal triangle with vertices $p < q < r$ is in the Farey tessellation if and only if there exist i and k such that $p = a_{i,k}$, $r = a_{i+1,k}$, and $q = p \oplus r$ (i.e., $q = a_{2i+1,k+1}$).*   □

The continued fractions for the vertices of ideal triangles of the Farey tessellation has the following surprising regularity.

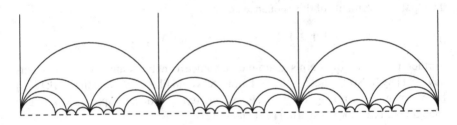

Fig. 27.3 Farey tessellation in the hyperbolic plane.

Proposition 27.23. *For every ideal triangle T in the Farey tessellation there exists a sequence of integers $a_0, a_1, \ldots, a_{n+1}$ such that the vertices of T have the coordinates p, q, and $p \oplus q$, where*

$$p = [0; a_0 : a_1 : \cdots : a_n], \quad q = [0; a_0 : a_1 : \cdots : a_n : a_{n+1}],$$
$$and \quad p \oplus q = [0; a_0 : a_1 : \cdots : a_n : a_{n+1} + 1].$$

$\square$

27.2.3 Descent toward the absolute

It turns out that the continued fraction of a real number itself can be interpreted in terms of the Farey tessellation. Let us explain this.

It is clear that the Farey tessellation is invariant with respect to the shift on the vector $(1, 0)$, so for simplicity we restrict ourselves to the case of real numbers α satisfying $0 \leq \alpha < 1$.

In the upper half-plane model one can imagine the Farey tessellation as a pyramid. We say one more time that we consider now only the part of the tessellation contained in the band $0 \leq x < 1$. On the top of the pyramid there is the ideal triangle with vertices 0, 1, and ∞. Suppose that we are at this triangle. It is permitted either to exit the pyramid at vertex 1 or to descend to the adjacent triangle with vertices 0, $\frac{1}{2}$, and 1. To fix notation, we say that we then *descend left*.

Suppose now that we are at some ideal triangle of the tessellation with vertices $x_1 < x_2 < x_3$. Then it is permitted either to exit the pyramid at vertex x_2 or to descend to one of the neighboring triangles in the tessellation: either to *descend left* to the triangle with vertices $x_1 < x_4 < x_2$ or to *descend right* to the triangle with vertices $x_2 < x_5 < x_3$ for the appropriate point x_4 or x_5 respectively (see Fig. 27.4).

In a finite or infinite number of steps we descend to some point α of the absolute. At each step of the descent we go either to the right or to the left, so it is natural to define the *descent sequence* whose elements are letters L (if we descend to the left) and R (if we descend to the right). If the sequence is nonempty, the first is always L.

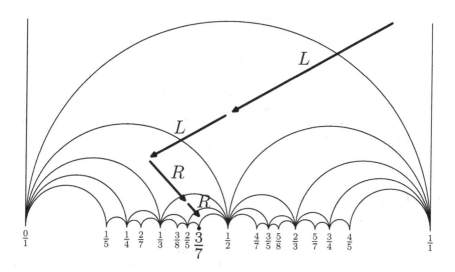

Fig. 27.4 Descent in the Farey pyramid. The descent sequence is *LLRR*, and hence the exit point is $\frac{3}{7} = [0; 2 : 3]$.

Let us denote by a_1 the number of letters L standing before the first letter R. Denote by a_2 the number of letters R standing before the next letter L, etc.

Example 27.24. For instance, for the sequence $(L, L, L, R, R, L, L, L, L, R, R, L, L)$ we get

$$a_1 = 3; \quad a_2 = 2; \quad a_3 = 4; \quad a_4 = 2; \quad a_5 = 2.$$

We have the following interesting theorem.

Theorem 27.25. *In the above notation the traveler will either exit the pyramid at the rational number*

$$[0; a_1 : a_2 : \cdots : a_{n-1} : a_n+1]$$

in a finite number of steps or descend to the irrational number

$$[0; a_1 : a_2 : \cdots]$$

on the absolute. □

For instance, in Example 27.24 we exit at point $\frac{69}{238} = [0; 3 : 2 : 4 : 2 : 3]$ in finitely many steps. In the example considered in Fig. 27.4 we have the descent sequence *LLRR* and the exit point $\frac{3}{7} = [0; 2 : 3]$.

Currently it is not clear what the analogue of the Farey tessellation in multidimensional hyperbolic geometry is, so we conclude this section with the following general open problem.

Problem 37. Find a natural generalization of the Farey tessellation to higher-dimensional hyperbolic geometry.

In this regard we would like to mention the works [158], [157], [159], [160] by V.V. Nikulin on discrete reflection groups in hyperbolic spaces.

Remark 27.26. It is also interesting to consider Farey tessellation in the de Sitter geometry, which is in some sense dual to the projective geometry. This is a new geometric approach to the classic theory of quadratic forms, developed by V.I. Arnold [7] and F. Aicardi [1] and [2]).

27.2.4 Triangle sequences

In this subsection we briefly observe several results related to *triangle sequences*. Triangle sequences generalize classical Farey addition to the two-dimensional case. They were introduced by T. Garrity in [67]. For multidimensional triangle sequences we refer to [51].

27.2.4.1 Definition of triangle sequence and its algebraic properties

We work with the triangle

$$\triangle = \{(x,y) \mid 1 \geq x \geq y > 0\}.$$

Consider the following partition of the triangle $\triangle$ into smaller triangles $\triangle_1, \triangle_2, \ldots,$ where

$$\triangle_k = \{(x,y) \in \triangle \mid 1-x-ky \geq 0 > 1-x-(k+1)y \geq 0\}$$

for $k \in \mathbb{Z}_+$. Now let us define the map T that maps each triangle $\triangle_n$ to the triangle $\triangle$. For every point (x,y) of the triangle $\triangle_n$, define

$$T(x,y) = \left(\frac{y}{x}, \frac{1-x-ky}{x}\right).$$

The point $T(x,y)$ is in the closure $\overline{\triangle} = \triangle \cup \partial\triangle$.

Definition 27.27. Consider an arbitrary point (x,y) of the triangle $\triangle$. Let $T^k(x,y) \in \triangle_n$. Then we put $a_k = n$, for $k = 1, 2, \ldots$. The sequence

$$(a_0, a_1, a_2, \ldots)$$

is called the *triangle sequence* for the pair (x,y).

Note that the triangle sequence is said to *terminate* at step k if $T^k(x,y) \in \overline{\triangle} \setminus \triangle$. In most cases the triangle sequence is infinite.

Remark 27.28. In [150] the authors proved that the triangle map $T : \triangle \to \triangle$ is ergodic with respect to the Lebesgue measure.

27.2.4.2 Multidimensional Farey addition

The definition of Farey addition is straightforward.

Definition 27.29. Consider two vectors in $\mathbb{R}^n$ whose coordinates are represented as ratios of numbers:

$$T = \left(\frac{p_1}{q_1}, \ldots, \frac{p_n}{q_n} \right) \quad and \quad S = \left(\frac{r_1}{s_1}, \ldots, \frac{r_n}{s_n} \right).$$

Then the *Farey sum* of T and S is the rational vector

$$T \oplus S = \left(\frac{p_1 + r_1}{q_1 + s_1}, \ldots, \frac{p_n + r_n}{q_n + s_n} \right).$$

We now use the following notation. Let $T = (p_1, \ldots, p_n)$ be an integer vector in $\mathbb{R}^n$. We denote by $\hat{X}$ the following representation of a rational vector in $\mathbb{R}^{n-1}$:

$$\hat{T} = \left(\frac{p_2}{p_1}, \frac{p_3}{p_1}, \ldots, \frac{p_n}{q_1} \right).$$

Finally, we put by definition

$$T \hat{+} S = \hat{T} \oplus \hat{S}.$$

27.2.4.3 Explicit formula for vertices of triangles

The map T defines a natural nested partition of the triangle $\triangle$ into smaller triangles. This partition contains triangles of the following type.

Definition 27.30. Let $(a_0, \ldots, a_n)$ be a sequence of nonnegative integers. Define

$$\triangle(a_0, \ldots, a_n) = \{(x, y) \mid T^k(x, y) \in \triangle_{a_k}, \text{ for all } k \leq n \}.$$

Our next goal is to write an explicit formula for such triangles. We begin a brief geometric description of triangle sequences. One can consider the triangle sequence for (α, β) as a method to construct a sequence of integer vectors (C_k) whose elements are "almost orthogonal" to the vector $(1, \alpha, \beta)$, i.e., the dot product of the vectors in the sequence with the point $(1, \alpha, \beta)$ in $\mathbb{R}^3$ is small. Set

$$C_{-3} = \begin{pmatrix} 1 \\ 0 \\ 0 \end{pmatrix}, \quad C_{-2} = \begin{pmatrix} 0 \\ 1 \\ 0 \end{pmatrix}, \quad C_{-1} = \begin{pmatrix} 0 \\ 0 \\ 1 \end{pmatrix}.$$

Suppose that the triangle sequence for (α, β) is (a_i). Then we set by definition

$$C_k = C_{k-3} - C_{k-2} - a_k C_{k-1}.$$

Define also

$$X_k = C_k \times C_{k+1},$$

where $v \times w$ denotes the cross product of two vectors v and w in $\mathbb{R}^3$.

Theorem 27.31. The vertices of the triangle $\triangle(a_0, \ldots, a_n)$ are the points $\hat{X}_{n-1}$, $\hat{X}_n$, and $X_n \hat{+} X_{n-2}$. $\qquad\square$

As is usual for continued fraction algorithms, the proof is obtained by induction (see the proof in [15]).

27.2.4.4 Convergence of triangle sequences

A triangle sequence does not always converge to a unique number. However, it does so under certain conditions.

Consider an infinite triangle sequence $(a_0, a_1, a_2, \ldots)$ of nonzero elements. Let the triangle $\triangle(a_0, \ldots, a_n)$ have vertices $\hat{X}_{n-1}$, $\hat{X}_n$, and $X_n \hat{+} X_{n-2}$. Set

$$\lambda_n = \frac{d(\hat{X}_{n-1}, \hat{X}_{n+1})}{d(\hat{X}_{n-1}, X_n \hat{+} X_{n-2})},$$

where $d(Y_1, Y_2)$ is the Euclidean distance between the points Y_1 and Y_2.

We have the following theorem.

Theorem 27.32. Assume that (a_n) is a triangle sequence of nonzero elements. Then the triangle sequence describes a unique pair (α, β) when

$$\prod_{n=N}^{\infty}(1 - \lambda_n) = 0.$$

$\qquad\square$

Let us now give a criterion of nonconvergence of a triangle sequence.

Theorem 27.33. Consider a sequence (a_n). Suppose that there exists N such that $a_n > 0$ for any $n > N$ and

$$\prod_{n=N}^{\infty}(1 - \lambda_n) > 0.$$

Then the triangle sequence (a_n) does not correspond to a unique point. $\qquad\square$

We refer to [15] for the proof of the above two theorems. Weak convergence is discussed in [150].

27.2.4.5 Algebraic periodicity

One of the most challenging properties that one thinks of while generalizing continued fractions to the multidimensional case is algebraic periodicity. If a triangle sequence is periodic, then the corresponding pair of real numbers correspond to certain solutions of a cubic equation with integer coefficients. Although the converse statement is not true, there is the following nice result announced in [50], which seems to improve the situation.

For simplicity we embed the triangle $\triangle$ in $\mathbb{R}^3$, simply by adding the first coordinate 1 and denoting the resulting triangle by $\tilde{\triangle}$. The vertices of $\tilde{\triangle}$ are

$$v_1 = \begin{pmatrix} 1 \\ 0 \\ 0 \end{pmatrix}, \quad v_2 = \begin{pmatrix} 1 \\ 1 \\ 0 \end{pmatrix}, \quad v_3 = \begin{pmatrix} 1 \\ 1 \\ 1 \end{pmatrix}.$$

Now we define three important matrices:

$$A_0 = \begin{pmatrix} 0 & 0 & 1 \\ 1 & 0 & 0 \\ 0 & 1 & 1 \end{pmatrix}, \quad A_1 = \begin{pmatrix} 1 & 0 & 1 \\ 0 & 1 & 0 \\ 0 & 0 & 1 \end{pmatrix}, \quad B = \begin{pmatrix} 1 & 1 & 1 \\ 0 & 1 & 1 \\ 0 & 0 & 1 \end{pmatrix}.$$

We embed the permutation group on three elements into $GL(3,\mathbb{Z})$ as a subgroup of matrices permuting the columns.

Definition 27.34. For an arbitrary triple of column permutations (σ, τ_0, τ_1) we define the following two matrices:

$$F_{0,\sigma,\tau_0} = \sigma A_0 \tau_0 \quad \text{and} \quad F_{1,\sigma,\tau_1} = \sigma A_1 \tau_1.$$

Denote by $\triangle_k(\sigma, \tau_0, \tau_1)$ the image of $\triangle$ under the map $F_{1,\sigma,\tau_1}^k F_{0,\sigma,\tau_0}$.

We use M^t to denote the transpose matrix to M.

Definition 27.35. Let us define the map T_{σ,τ_0,τ_1} for an arbitrary triple of column permutations σ, τ_0, τ_1. Consider a pair $(x,y) \in \triangle$. Suppose that $(x,y) \in \triangle_k(\sigma, \tau_0, \tau_1)$. Then we define

$$(a,b,c) = (1,x,y) \cdot (B F_{0,\sigma,\tau_0}^{-1} F_{1,\sigma,\tau_1}^{-k} B^{-1})^t.$$

We define $T_{\sigma,\tau_0,\tau_1}(x,y)$ as follows:

$$T_{\sigma,\tau_0,\tau_1}(x,y) = \left(\frac{b}{a}, \frac{c}{a} \right).$$

Remark 27.36. We have actually defined 216 different maps corresponding to different triples of permutations. Notice that the matrix $(B F_{0,\sigma,\tau_0}^{-1} F_{1,\sigma,\tau_1}^{-k} B^{-1})^t$ is in $SL(3,\mathbb{Z})$. The triangle partition maps T_{σ,τ_0,τ_1} are called *TRIP maps*.

Finally, we describe the following three classes of maps, where $n \in \mathbb{Z}_{\geq 0}$:

Class 1. $T_{e,e,e} \circ \left(T_{e,(123),(123)}^1 \right)^n$;

Class 2. $T_{e,(23),e} \circ \left(T^1_{e,(123),(123)}\right)^n$;

Class 3. $T_{(23),(23),e} \circ \left(T^0_{(13),(12),e}\right)^n$.

Notice that for each choice of a class and a parameter n we get a different algorithm defined by the composition $\hat{T}$ of the corresponding $n+1$ TRIP maps. Similar to Definition 27.35, the composed map $\hat{T}$ sends the triangle $\triangle$ to the set $\triangle_{(m_1,m_2,\ldots,m_n,m_e)}$, which is encoded by $n+1$ integer parameters appearing in the composition $\hat{T}$.

Definition 27.37. Let $(x,y) \in \triangle$. Fix a composed map $\hat{T}$ of one of Classes 1, 2, 3 with a fixed parameter n. For an arbitrary integer k we denote by $\bar{a}_k$ the $(n+1)$-dimensional vector of nonnegative integers such that $\hat{T}^k(x,y)$ is in $\triangle_{\bar{a}_n}$. We say that $(\bar{a}_i)$ is the *triangle sequence for* (x,y) *with respect to* $\hat{T}$.

Now we are ready to formulate an interesting theorem announced in [50] in which the authors in some sense cover all cubic irrationalities by periodic triangle sequences of several continued fraction algorithms in the following way.

Theorem 27.38. Let K be a cubic extension of $\mathbb{Q}$. Suppose that the element $u \in O_K$ satisfies $0 < u < 1$. Then either (u, u^2), (u^2, u^4), $(u, u^2 - u)$, $(u^2, u^2 - u^4)$, $(uu', (uu')^2 - uu')$, or $((uu')^2, (uu')^2 - (uu')^4)$, has a periodic triangle sequence under some composed map $\overline{T}$ in Class 1, 2, or 3 (where u' is one of the conjugates of u).   □

The following conjecture is given for the multidimensional case.

Conjecture 38. **(T. Garrity.)** For each positive integer d, there is a finite number of multidimensional continued fraction algorithms, so that for every d-tuple $(\xi_1,\ldots,\xi_d)$ with $\mathbb{Q}(\xi_1,\ldots,\xi_d)$ a degree-d algebraic number field, there is a multidimensional continued fraction algorithm in the family spanned by the initial algorithms such that it generates a periodic output.

Remark 27.39. Farey fractions give rise to a thermodynamic approach to real numbers (for more information we refer to [121], [62], [122], and [68]). Triangle sequences are used to generalize this approach to the case of pairs of real numbers [69].

Remark 27.40. There are several other generalizations in the spirit of triangle sequences. For instance, the classical Minkowski ?(x) function (see [152], [82], [212], [218], etc.) was generalized to the two-dimensional case by O.R. Beaver and T. Garrity in [21] (see also [171]).

27.3 Decompositions of coordinate rectangular bricks and O'Hara's algorithm

Surprisingly, the Euclidean algorithm arises in the theory of partitions related to O'Hara's algorithms, introduced by K.M. O'Hara in [164] (see also [165]). In this section we briefly describe the geometric approach to O'Hara's algorithm via special decompositions of parallelepipeds. In particular we describe their relation to regular continued fractions and their generalizations. For simplicity we consider here only the finite-dimensional case. For the general infinite-dimensional case for permutations of $\mathbb{Z}_+$ and further information on the finite-dimensional case, we refer to [125] and [169].

27.3.1 Π-congruence of coordinate rectangular bricks

We begin with several general definitions. If a polyhedron P is a disjoint union of a finite or countable number of convex polyhedra $P_1, P_2, \ldots$, then we say that $\{P_1, P_2, \ldots\}$ is a *decomposition* of P.

Definition 27.41. Let π be a linear subspace of codimension one in $\mathbb{R}^n$. We say that two convex polyhedra P and Q in $\mathbb{R}^n$ are π-*congruent* if there exist decompositions $\{P_1, \ldots, P_k\}$ and $\{Q_1, \ldots, Q_k\}$ of P and Q respectively such that for every $i \in \{1, \ldots, k\}$ there exists a translation T_i along a vector in π taking P_i to Q_i (i.e., $T_i(P_i) = Q_i$).

Two convex polyhedra are said to be Π-*congruent* if there exists a space π such that these polyhedra are π-congruent.

Definition 27.42. If instead of finite decompositions of polyhedra we have countable decompositions, then the corresponding polyhedra are called *asymptotically* π-*congruent* (respectively *asymptotically* Π-*congruent*).

In this section we are interested only in the coordinate parallelepipeds defined as

$$R(a_1, \ldots, a_n) = [0, a_1) \times \cdots \times [0, a_n).$$

We call them *coordinate bricks*.

Example 27.43. Let us consider two bricks $R(6, 11)$ and $R(22, 3)$. We show their decompositions in Fig. 27.5 simultaneously. Each of the bricks is subdivided into six smaller rectangles labeled by Roman numerals. The rectangle with label n is shifted to the rectangle with label n'. For instance, the gray rectangle labeled by V is taken to the gray rectangle labeled by V'. Here all the translation vectors are in the linear space $x + 2y = 0$.

In the next two subsections we answer the following two natural questions:
(a) *Which coordinate bricks are (asymptotically) Π-congruent?*
(b) *How does one construct the decompositions and translations showing the (asymptotic) Π-congruence of such boxes?*

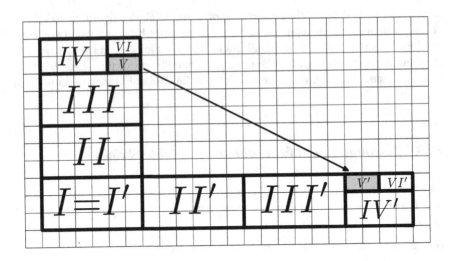

Fig. 27.5 A decomposition of bricks $R(6,11)$ and $R(22,3)$.

27.3.2 Criteron of Π-congruence of coordinate bricks

We begin with the following definition.

Definition 27.44. Two coordinate bricks $P = R(a_1,\dots,a_n)$ and $Q = R(b_1,\dots,b_n)$ are called *relatively rational* if there exists $\lambda \neq 0$ such that λP and λQ are integer regular parallelepipeds.

In an arbitrary dimension n the following general theorem holds.

Theorem 27.45. (*i*) *Two coordinate bricks of the same dimension are asymptotically Π-congruent if and only if they have the same volume.*
(*ii*) *Two relatively rational rectangular coordinate boxes are Π-congruent if and only if they have the same volume.* □

In addition we have the following simple formula to calculate the equation of the plane π.

Proposition 27.46. *Let $R(a_1,\dots,a_n)$ and $R(b_1,\dots,b_n)$ be π-congruent. Then the hyperspace π is uniquely defined by the equation*

$$x_1 + \frac{a_1}{b_2}x_2 + \frac{a_1 a_2}{b_2 b_3}x_3 + \cdots + \frac{a_1 \dots a_{n-1}}{b_2 \dots b_n}x_n = 0.$$

□

For a proof of Theorem 27.45 we refer to [125].

27.3.3 Geometric version of O'Hara's algorithm for partitions

The answer to the second question (on construction of decompositions and translations) is given by O'Hara's algorithm, which comes from the theory of general partitions (we refer the interested reader to [5], [165], [164], [168], [169]). In this subsection we consider only the geometric representation of the algorithm related to continued fractions. In some sense the idea of this algorithm is an algorithm-definition, giving a constructive description of all the translations T_i: for an arbitrary point of the first coordinate brick the algorithm constructs the image of this point in the second coordinate brick. The resulting decomposition is a decomposition into boxes, all of whose edges are parallel to coordinate vectors.

O'Hara's algorithm in a geometric form.

Predefined parameters of the algorithm. We are given two coordinate bricks $R(a_1, \ldots, a_n)$ and $R(b_1, \ldots, b_n)$.

Input data. A point $v \in R(a_1, \ldots, a_n)$.

Aim of the algorithm. To construct the point $w \in R(b_1, \ldots, b_n)$ corresponding to v.

Step 1. Put $v_1 = v$.

Recursive step k. We have constructed a point v_{k-1}. If v_{k-1} is contained in the brick $R(b_1, \ldots, b_n)$, then the algorithm terminates and we put $w = v_{k-1}$. Suppose now that v_{k-1} is not in the brick $R(b_1, \ldots, b_n)$. Then there exists an index j such that the jth coordinate of v_{k-1} is not less than b_j, allowing us to recursively define

$$v_k = v_{k-1} - b_j e_j + a_{j-1} e_{j-1}$$

(here e_j and e_{j-1} are the corresponding coordinate vectors whose indices are in the group $\mathbb{Z}/n\mathbb{Z}$).

Output. The algorithm returns a vector $w \in R(b_1, \ldots, b_n)$.

Remark 27.47. It turns out that the result w does not depend on the choice of the coordinate j in any recursive step, but entirely on the input data v (see [164] and [165] for more details).

Now we are ready to define a mapping between the coordinate bricks.

Definition 27.48. Let P and Q be two coordinate bricks of the same volume in $\mathbb{R}^n$. We define a bijection $\varphi_{P,Q} : P \to Q$ at every $v \in P$ as follows:

$$\varphi_{P,Q}(v) = w,$$

where w is the output of the algorithm with the input data v.

Observe the following property of $\varphi_{P,Q}$.

Proposition 27.49. The function $\varphi_{P,Q}$ is a piecewise linear bijective function between the coordinate bricks P and Q. $\square$

Due to piecewise linearity we have the following natural decomposition of the coordinate bricks.

Definition 27.50. Let P and Q be two coordinate bricks of the same volume in $\mathbb{R}^n$. Consider a natural decomposition of the coordinate brick P into regions on which the function $\varphi_{P,Q}$ is linear. Define a decomposition of Q as the image of the decomposition of P via the map $\varphi_{P,Q}$. We say that these decompositions of P and Q are *associated* to the bijection $\varphi_{P,Q}$.

In the general case the described decompositions have a countable number of boxes. However, when P and Q are relatively rational, the decompositions of P and Q are finite.

The following proposition gives an answer to the second question of this section: *how does one construct the decompositions and translations showing the (asymptotic) Π-congruence of such boxes?*

Proposition 27.51. *Let $P = R(a_1,\ldots,a_n)$ and $Q = R(a_1,\ldots,a_n)$ be asymptotically Π-congruent coordinate bricks. Then P and Q are asymptotically π-congruent, where the hyperplane π is defined by the equation*

$$x_1 + \frac{a_1}{b_2}x_2 + \frac{a_1 a_2}{b_2 b_3}x_3 + \cdots + \frac{a_1 \ldots a_{n-1}}{b_2 \ldots b_n}x_n = 0.$$

In addition, the decompositions of P and Q associated to the bijection $\varphi_{P,Q}$ establish the asymptotic π-congruence. The restriction of $\varphi_{P,Q}$ to parallelepipeds in the decomposition of P identifies all the translation vectors.   □

Here we notice again that in the case of relatively rational coordinate bricks P and Q, one has π-congruence instead of asymptotic π-congruence.

Remark 27.52. In the special case of rectangular bricks $R(a,b)$ and $R(b,a)$ with relatively prime positive integers a and b and the space $\pi = \{x + y = 0\}$, O'Hara's algorithm is a version of the Euclidean algorithm. Geometrically the elements of the continued fraction for b/a correspond to the numbers of equivalent squares in the corresponding layers of the decomposition of $R(a,b)$.

Example 27.53. We illustrate this with the example of $a = 9$ and $b = 25$. The associated decompositions of $R(9,25)$ and of $R(25,9)$ consist of two large squares, one average square, three small squares, and two very small squares. This can be read from the regular continued fraction: $\frac{25}{9} = [2; 1 : 3 : 2]$. In Fig. 27.6 we show the layers of equivalent squares as bold rectangles.

Remark 27.54. Notice that after a coordinate rescaling the decomposition remains combinatorially the same. Therefore, we can always choose the coordinate scale to work with $R(1,b)$ and $R(b,1)$, where $\pi = \{x + y = 0\}$ for some $b > 0$.

From the above two remarks it follows that the notion of partitions in the two-dimensional case correlates with the notion of regular continued fractions. So we conclude with the following natural problem.

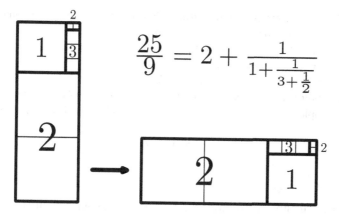

$$\frac{25}{9} = 2 + \cfrac{1}{1+\cfrac{1}{3+\frac{1}{2}}}$$

Fig. 27.6 A decomposition of bricks $R(9,25)$ and $R(25,9)$.

Problem 39. (**I. Pak [169].**) Find a relation between multidimensional continued fractions and the geometry of the bijection constructed by O'Hara's algorithm.

27.4 Algorithmic generalized continued fractions

We have already considered two algorithms that generalize the Euclidean algorithm: O'Hara's algorithm and triangle sequences. There are in fact many other different generalizations of the Euclidean algorithm in the spirit of the two mentioned above. No geometric interpretation is known for most of them. To give some impression of the algorithmic approaches to continued fractions we mention here several other algorithms of this nature. We are not planning to go into details of these approaches, since it is far away from the scope of this book. The best reference to the algorithmic approach to continued fractions is the book [195] by F. Schweiger (see also [28] and [210]). First we give a general algorithmic scheme in which such algorithms can be realized (for additional details we refer to the preprint [196] of F. Schweiger). We then show some examples of known algorithms. Finally, we briefly discuss some problems related to algorithmic generalizations of continued fractions.

27.4.1 General algorithmic scheme

Most of the generalized Euclidean algorithms work according to the following scheme.

Scheme of a generalized Euclidean algorithm.

Predefined parameters of the algorithm. Consider the space $\mathbb{R}^n$ for some integer n. The following data specifies each algorithm:

(i) the domain $\Delta \subset \mathbb{R}^n$ on which the algorithm is defined;

(ii) the operation of *subtraction* $S : \mathbb{R}^n \to \mathbb{R}^n$;

(iii) the operation of *inversion* $T : \mathbb{R}^n \to \mathbb{R}^n$;

(iv) the *termination domain* $\hat{\Delta} \subset \Delta$, at which the algorithm terminates.

We give examples of 4-tuples $(\Delta, S, T, \hat{\Delta})$ below. Notice that it is usually required that

$$T \circ S(\Delta) \subset \Delta.$$

Input data. A real vector $v \in \Delta$.

Aim of the algorithm. To construct the generalized continued fraction.

Step 0. Put $v_0 = v$.

Recursive step k. We have constructed v_{k-1}. If $v_k \in \hat{\Delta}$, then the algorithm terminates. If $v_k \notin \hat{\Delta}$, the recursive step consists of three substeps:

Substep $k.1$: perform an operation of subtraction S;

Substep $k.2$: invert the result via the operation of inversion T;

Substep $k.3$: write the $(k-1)$th element of the continued fraction based on Substep $k.1$ and Substep $k.2$.

We conclude the step by putting

$$v_k = T \circ S(v_{k-1}).$$

Output. If the algorithm stops (say at step $m+1$), then the last element v_m is considered a generalized greatest common divisor. The corresponding continued fraction is generated in the iterative steps; it is calculated by parameters of subtractions S and inversions T.

27.4.2 Examples of algorithms

First we write the classical Euclidean algorithm in the framework of the general algorithmic scheme. We will then give several examples of generalizations.

Example 27.55. **Euclidean algorithm.** Predefined parameters of the algorithm are as follows:

$$\Delta = \{(x,y) \mid x \geq y > 0; x,y \in \mathbb{R}\};$$
$$S : (x,y) \mapsto \left(x - \lfloor \tfrac{x}{y} \rfloor y, y\right);$$
$$T : (x,y) \mapsto (y,x);$$
$$\hat{\Delta} = \{(x,0) \mid x \in \mathbb{Z}_+\}.$$

In Substep $k.3$ we remember the element $a_{k-1} = \lfloor \frac{p_k}{q_k} \rfloor$, where $v_{k-1} = (p_{k-1}, q_{k-1})$ is the vector defined in Step $k-1$. The resulting continued fraction is

$$[a_0; a_1 : a_2 : \cdots].$$

Example 27.56. Jacobi—Perron algorithm ([92], [178], [177]). Historically, the first algorithmic generalization of regular continued fractions is the Jacobi—Perron algorithm (see also [191]). We put

$$\Delta = \{(x_0, x_1, \ldots, x_n) \mid x_0 \geq x_i \geq 0, 1 \leq i \leq n\};$$
$$S : (x_0, x_1, \ldots, x_n) \mapsto \left(x_0 - \lfloor \tfrac{x_0}{x_1} \rfloor, x_1, x_2 - \lfloor \tfrac{x_2}{x_1} \rfloor, \ldots, x_n - \lfloor \tfrac{x_n}{x_1} \rfloor\right);$$
$$T : (x_0, x_1, \ldots, x_n) \mapsto (x_1, \ldots, x_n, x_0);$$
$$\hat{\Delta} = \{(x_0, x_1, \ldots, x_n) \mid x_1 = 0\} \cap \Delta.$$

In Substep $k.3$ we remember the n-dimensional vector

$$a_{k-1} = \left(\left\lfloor \frac{x_{0,k-1}}{x_{1,k-1}} \right\rfloor, \left\lfloor \frac{x_{2,k-1}}{x_{1,k-1}} \right\rfloor, \ldots, \left\lfloor \frac{x_{n,k-1}}{x_{1,k-1}} \right\rfloor\right),$$

where $v_{k-1} = (x_{0,k-1}, x_{1,k-1}, \ldots x_{n,k-1})$ is the vector in $\mathbb{R}^{n+1}$ defined in Step $k-1$. The resulting generalized continued fraction here is the sequence of n-dimensional vectors

$$(a_0, a_1, a_2, \ldots).$$

Remark 27.57. If the Jacobi—Perron algorithm terminates, one actually can remove the zero coordinate $x_1 = 0$ and continue the algorithm in the real space whose dimension is one less (finally we end up at $n = 1$).

Example 27.58. Generalized subtractive algorithm ([192]).
  The generalized subtractive algorithm is defined by the following data. First, p is fixed for the algorithm. Second, we put

$$\Delta = \{(x_0, x_1, \ldots, x_n) \mid x_0 \geq x_1 \geq \cdots \geq x_n \geq 0\};$$
$$S : (x_0, x_1, \ldots, x_n) \mapsto (x_0 - x_p, x_1, \ldots, x_n);$$
$$T \text{ is a permutation of coordinates putting them in nonincreasing order;}$$
$$\hat{\Delta} = \{(x_0, x_1, \ldots, x_n) \mid x_p = 0\} \cap \Delta.$$

We remember the permutation that was used in Substep $k.2$. Indeed, we always exchange the first element with some element $a_{k-1} \in \{0, 1, \ldots, n\}$. So we can choose a sequence of integers

$$(a_0, a_1, a_2, \ldots)$$

as the generalized p-subtractive continued fraction.

Remark 27.59. The first subtractive algorithm was introduced in [30] by V. Brun, where he considered the case $p = 1$. Later, R. Mönkemeyer in [153] (and further E.S. Selmer in [198]) studied the subtractive algorithm for the case $p = n$.

Example 27.60. **Fully subtractive algorithm ([193], [194]).** In the fully subtractive algorithm we also fix p and consider

$$\Delta = \{(x_0, x_1, \ldots, x_n) \mid x_0 \geq x_1 \geq \cdots \geq x_n \geq 0\};$$
$$S : (x_0, x_1, \ldots, x_n) \mapsto (x_0 - x_p, x_1 - x_p, \ldots, x_{p-1} - x_p, x_p, \ldots, x_n);$$
T is a permutation of coordinates putting them in nonincreasing order;
$$\hat{\Delta} = \{(x_0, x_1, \ldots, x_n) \mid x_p = 0\} \cap \Delta.$$

We remember the permutation that was used in Substep $k.2$, denoting it by a_{k-1}. The generalized continued fraction is the sequence of permutations $(a_0, a_1, a_2, \ldots)$.

27.4.3 Algebraic periodicity

Let us mention here a famous conjecture known as *Jacobi's last theorem*.

Problem 40. **(Jacobi's last theorem.)** Let K be a totally real cubic number field. Consider arbitrary elements y and z of K satisfying $0 < y, z < 1$ such that 1, y, and z are independent over $\mathbb{Q}$. Is it true that the Jacobi—Perron algorithm generates an eventually periodic continued fraction with starting data $v = (1, y, z)$?

This problem is open not only for the Jacobi—Perron algorithm but also for other similar algorithms. The converse statement is true.

Proposition 27.61. *Periodic algorithmic continued fractions correspond to certain algebraic irrationalities.*   □

27.4.4 A few words about convergents

Let us briefly explain how to generate convergents for a vector v with a generalized continued fraction $(a_0, a_1 \ldots)$.

This is usually done by inverting the map $T \circ S$. First of all, notice that the values in the image of the map $(T \circ S)^{-1}$ are enumerated by the possible values of the elements of the corresponding continued fractions, so we can write $(T \circ S)_a^{-1}$ to indicate which of the images should be taken.

One starts with some vector w_0. It is natural to choose w_0 either $(1, 0, \ldots, 0)$ or the generalized greatest common divisor (in case of existence). Then the vectors

$$w_1 = (T \circ S)_{a_0}^{-1}(w_0), \quad w_2 = (T \circ S)_{a_1}^{-1}(w_1), \quad w_3 = (T \circ S)_{a_2}^{-1}(w_2), \quad \ldots$$

are called the *convergents* of v.

Remark 27.62. In questions of approximation it is natural to reduce projectively the first coordinates:

$$p = (x_0, \ldots, x_n) \quad \mapsto \quad \tilde{p} = \left(\frac{x_1}{x_0}, \ldots, \frac{x_n}{x_0}\right).$$

Now arises the question of *the quality of approximation of the vector $\tilde{v}$ by the vectors $\tilde{w}_i$*. For further information on approximation aspects we refer to [195].

27.5 Branching continued fractions

In this subsection we observe some interesting representations of real numbers as ratios of certain determinants of infinite matrices related to continued fractions or their special generalizations. It is nice to admit that for every algebraic number there exists a periodic representation defining the corresponding number. Matrix representations do not give a complete invariant of algebraic numbers in the following sense: any real number has many different representations, and it is quite hard to determine whether two of them define the same number. For further information and references we refer to the works of V.Ya. Skorobogatko [204], [205], V.I. Shmoïlov [201], [202], [203], I. Gelfand and V. Retakh [73], [71].

We start with a representation of real numbers by tridiagonal determinants whose elements are the elements of the following continued fractions (for further details we refer to [94]):

$$a_{11} + \cfrac{a_{12}}{a_{22} + \cfrac{a_{23}}{a_{33} + \cfrac{a_{34}}{a_{44} + \cfrac{a_{45}}{a_{55} + \ldots}}}} = \frac{\begin{vmatrix} a_{11} & a_{12} & 0 & 0 & 0 & \ldots \\ -1 & a_{22} & a_{23} & 0 & 0 & \ldots \\ 0 & -1 & a_{33} & a_{34} & 0 & \ldots \\ 0 & 0 & -1 & a_{44} & a_{45} & \ldots \\ 0 & 0 & 0 & -1 & a_{55} & \ldots \\ \cdot & \cdot & \cdot & \cdot & \cdot & \ldots \end{vmatrix}}{\begin{vmatrix} a_{22} & a_{23} & 0 & 0 & \ldots \\ -1 & a_{33} & a_{34} & 0 & \ldots \\ 0 & -1 & a_{44} & a_{45} & \ldots \\ 0 & 0 & -1 & a_{55} & \ldots \\ \cdot & \cdot & \cdot & \cdot & \ldots \end{vmatrix}}. \tag{27.1}$$

Equation (27.1) holds for any continued fraction whose sequence of convergents converges to a real number. Here we use dots in matrices to denote the limit of the ratios of determinants of the corresponding $(n+1) \times (n+1)$ matrix and $n \times n$ matrix as n tends to infinity. We will also do the same in the above formula. In the case of equation (27.1) we have an even stronger statement for n-convergents:

$$a_{11} + \cfrac{a_{12}}{a_{22} + \cfrac{a_{23}}{\cdots + \cfrac{a_{n-1,n-1}}{a_{n,n}}}} = \frac{\begin{vmatrix} a_{11} & a_{12} & 0 & \cdots & 0 & 0 \\ -1 & a_{22} & a_{23} & \cdots & 0 & 0 \\ 0 & -1 & a_{33} & \cdots & 0 & 0 \\ & & & \ddots & & \\ 0 & 0 & 0 & \cdots & a_{n-1,n-1} & a_{n-1,n} \\ 0 & 0 & 0 & \cdots & -1 & a_{n,n} \end{vmatrix}}{\begin{vmatrix} a_{22} & a_{23} & \cdots & 0 & 0 \\ -1 & a_{33} & \cdots & 0 & 0 \\ & & \ddots & & \\ 0 & 0 & \cdots & a_{n-1,n-1} & a_{n-1,n} \\ 0 & 0 & \cdots & -1 & a_{n,n} \end{vmatrix}}.$$

Let us give the definition of Hessenberg continued fractions.

Definition 27.63. Consider a real number x and a sequence of real numbers a_i. Suppose that

$$x = \frac{\begin{vmatrix} -a_1 & -a_2 & -a_3 & -a_4 & \cdots \\ -1 & -a_1 & -a_2 & -a_3 & \cdots \\ 0 & -1 & -a_1 & -a_2 & \cdots \\ 0 & 0 & -1 & -a_1 & \cdots \\ & & & & \cdots \end{vmatrix}}{\begin{vmatrix} -a_1 & -a_2 & -a_3 & \cdots \\ -1 & -a_1 & -a_2 & \cdots \\ 0 & -1 & -a_1 & \cdots \\ & & & \cdots \end{vmatrix}},$$

i.e., the limit from the right exists and equals x. Then the expression from the right is called the *Hessenberg continued fraction* for x.

The most interesting case here occurs when a sequence (a_i) has only finitely many nonzero entries.

Proposition 27.64. Let (a_i) be a sequence of real numbers with finitely many nonzero elements, assuming that the element a_n is the last nonzero element. Suppose that the Hessenberg continued fraction for the sequence (a_i) converges to some real number x. Then x satisfies the equation

$$x^n + a_1 x^{n-1} + a_2 x^{n-2} + \cdots + a_{n-1} x + a_n = 0.$$

$\square$

In the particular case of cubic equations we have an additional nice formula, which could be considered a generalized two-dimensional regular continued fraction.

Proposition 27.65. Suppose that a Hessenberg continued fraction for a real number x is defined by a sequence $(a_1, a_2, a_3, 0, 0, \ldots)$. Then we have

$$x = -a_1 + \cfrac{a_1 - \cfrac{a_2 - \cfrac{a_1 - \cfrac{a_1 - \cfrac{a_2 - \cfrac{a_3}{a_1 - \cdots}}{a_1 - \cdots}}{a_2 - \cfrac{a_3}{a_2 - \cdots}}}{a_2 - \cdots}}{\cdots}}{\cdots}.$$

$\Box$

Later A.Z. Nikiporetz gave the following generalization of Hessenberg continued fractions.

Definition 27.66. Consider a real number x_k and a two-sided sequence of real numbers (a_i). Suppose that

$$x_k = \frac{\begin{vmatrix} -a_k & -a_{k+1} & -a_{k+2} & -a_{k+3} & \cdots \\ -a_{k-1} & -a_k & -a_{k+1} & -a_{k+2} & \cdots \\ -a_{k-2} & -a_{k-1} & -a_k & -a_{k+1} & \cdots \\ -a_{k-3} & -a_{k-2} & -a_{k-1} & -a_k & \cdots \\ & & & & \cdots \end{vmatrix}}{\begin{vmatrix} -a_k & -a_{k+1} & -a_{k+2} & \cdots \\ -a_{k-1} & -a_k & -a_{k+1} & \cdots \\ -a_{k-2} & -a_{k-1} & -a_k & \cdots \\ -a_{k-3} & -a_{k-2} & -a_{k-1} & \cdots \\ & & & \cdots \end{vmatrix}} : \frac{\begin{vmatrix} -a_{k-1} & -a_k & -a_{k+1} & -a_{k+2} & \cdots \\ -a_{k-2} & -a_{k-1} & -a_k & -a_{k+1} & \cdots \\ -a_{k-3} & -a_{k-2} & -a_{k-1} & -a_k & \cdots \\ -a_{k-4} & -a_{k-3} & -a_{k-2} & -a_{k-1} & \cdots \\ & & & & \cdots \end{vmatrix}}{\begin{vmatrix} -a_{k-1} & -a_k & -a_{k+1} & \cdots \\ -a_{k-2} & -a_{k-1} & -a_k & \cdots \\ -a_{k-3} & -a_{k-2} & -a_{k-1} & \cdots \\ & & & \cdots \end{vmatrix}},$$

i.e., the limit from the right exists and equals x_k. Then the expression from the right is called the *Nikiporetz continued fraction* for x.

In analogy to Proposition 27.64 for Hessenberg continued fractions we get the following statement for Nikiporetz continued fractions.

Proposition 27.67. *Let (a_i) be a two-sided sequence of real numbers with finitely many nonzero elements. We assume that all elements with negative indices are zeros, $a_0 = 1$, and the element a_n is the last nonzero element. Suppose that the Nikiporetz continued fraction for the sequence (a_i) converges to some real number x_k. Then x_k satisfies the equation*

$$x_k^n + a_1 x_k^{n-1} + a_2 x_k^{n-2} + \cdots + a_{n-1} x_k + a_n = 0.$$

$\Box$

Notice that in the particular case $k = 1$ we have the statement of Proposition 27.64.

We continue with Skorobogatko continued fractions.

Definition 27.68. The expression

$$\alpha = b_0 + \cfrac{a_1}{b_1 + \cfrac{a_3}{b_3 + \cdots} + \cfrac{a_4}{b_4 + \cdots}} + \cfrac{a_2}{b_2 + \cfrac{a_5}{b_5 + \cdots} + \cfrac{a_6}{b_6 + \cdots}}$$

is called the *Skorobogatko continued fraction* for α.

Proposition 27.69. The Skorobogatko continued fraction for α as in the previous definition is written in determinant form as follows:

$$\alpha = \frac{\begin{vmatrix} b_0 & a_1 & a_2 & 0 & 0 & 0 & 0 & \cdots \\ -1 & b_1 & 0 & a_3 & a_4 & 0 & 0 & \cdots \\ -1 & 0 & b_2 & 0 & 0 & a_5 & a_6 & \cdots \\ 0 & -1 & 0 & b_3 & 0 & 0 & 0 & \cdots \\ 0 & -1 & 0 & 0 & b_4 & 0 & 0 & \cdots \\ 0 & 0 & -1 & 0 & 0 & b_5 & 0 & \cdots \\ 0 & 0 & -1 & 0 & 0 & 0 & b_6 & \cdots \\ & & \cdot & \cdot & \cdot & & \cdots \end{vmatrix}}{\begin{vmatrix} b_1 & 0 & a_3 & a_4 & 0 & 0 & \cdots \\ 0 & b_2 & 0 & 0 & a_5 & a_6 & \cdots \\ -1 & 0 & b_3 & 0 & 0 & 0 & \cdots \\ -1 & 0 & 0 & b_4 & 0 & 0 & \cdots \\ 0 & -1 & 0 & 0 & b_5 & 0 & \cdots \\ 0 & -1 & 0 & 0 & 0 & b_6 & \cdots \\ & \cdot & \cdot & \cdot & & \cdots \end{vmatrix}}.$$

$\square$

Finally, we would like to mention another generalization of continued fractions for the case of noncommuting elements and its expression in terms of quasideterminants studied by I. Gelfand and V. Retakh in [73], [71] (see also in [170]). Here we formulate only a consequence for the case of commutative elements (for the noncommutative case we refer the interested reader to the original papers mentioned above).

Definition 27.70. Consider a real number x and a collection of real numbers $a_{i,j}$. Suppose that

$$x = \frac{\begin{vmatrix} a_{11} & a_{12} & a_{13} & a_{14} & \cdots \\ -1 & a_{22} & a_{23} & a_{24} & \cdots \\ 0 & -1 & a_{33} & a_{34} & \cdots \\ 0 & 0 & -1 & a_{44} & \cdots \\ & \cdot & \cdot & \cdot & \cdots \end{vmatrix}}{\begin{vmatrix} a_{22} & a_{23} & a_{24} & \cdots \\ -1 & a_{33} & a_{34} & \cdots \\ 0 & -1 & a_{44} & \cdots \\ & \cdot & \cdot & \cdots \end{vmatrix}},$$

i.e., the limit from the right exists and equals x. Then the expression from the right is called the *quasideterminant* for x.

We mention once more that the notion of quasideterminant in the case of non-commutative elements $a_{i,j}$ is more complicated; see [72] and [73].

Proposition 27.71. *In the setting of the above definition, we have*

$$x = a_{11} + \sum_{j_1 \neq 1} a_{1j_1} \cfrac{1}{a_{2j_1} + \sum_{j_2 \neq 1, j_1} a_{2j_2} \cfrac{1}{a_{3j_2} + \dots}}.$$

$\square$

27.6 Continued fractions and rational knots and links

In this section we explain how continued fractions are used in topology for the classification of rational knots and links. For additional information and references we refer to [114] and [115].

27.6.1 Necessary definitions

Recall that a *knot* is a smooth embedding of a circle $\mathbb{R}/\mathbb{Z}$ into $\mathbb{R}^3$; a link is a smooth embedding of several circles into $\mathbb{R}^3$.

Definition 27.72. We say that a knot K has an *n-bridge* representation if K is isotopic to some knot having only n maxima and n minima as critical points of the natural height function on K given by the z-coordinate in $\mathbb{R}^3$.

The *bridge number* of a knot K is the minimal number n such that K has an n-bridge representation.

We say that a knot of bridge number n is an *n-bridge knot*.

A 2-bridge knot is said to be a *rational knot*.

For the study of n-bridge knots it is natural to consider n-tangles.

Definition 27.73. An *n-tangle* is a proper embedding of the disjoint union of n arcs into a three-dimensional ball with $2n$ marked points such that the endpoints of the arcs map to distinct marked boundary points.

An n-tangle is called *trivial* if its endpoints are connected by straight lines.

27.6.2 Rational tangles and operations on them

Definition 27.74. We say that a 2-tangle is a *rational tangle* if there exists a smooth deformation of the three-dimensional ball under which the 2-tangle evolves to the trivial 2-tangle.

Graphically, tangles are represented as special regular mappings of the arcs to the plane (i.e., having only finitely many singular points, and such that all these singularities are double crossings). Two pairs of endpoints are mapped to the lines $y = 0$ and $y = 1$ respectively, and the rest are mapped to the band bounded by these lines. At each crossing we indicate which of the two branches of the double crossing is above and which is below. (See an example of a tangle in Fig. 27.7.)

Basic tangles. First we define the following three tangles:

Let us now define the addition, multiplication by -1, and inversion operations on 2-tangles.

The sum operation. The *sum* of two tangles T_1 and T_2 is the tangle $T_1 + T_2$ defined as follows:

Multiplication by -1. If we change all the crossings in the tangle T, then we get the tangle called *opposite* to T and denoted by $-T$.

The inversion operation. Let T be an arbitrary 2-tangle. The *inverse* tangle T^i is defined as follows:

Definition 27.75. Let m be a positive integer. We define

$$T(m) = \underbrace{T(1) + \cdots + T(1)}_{m}.$$

Letting $[a_0; a_1 : a_2 : \cdots : a_n]$ be a continued fraction with integer coefficients, then set

$$T\big([a_0; a_1 : a_2 : \cdots : a_n]\big) = T(a_0) + \big(T(a_1) + \big(T(a_2) + \big(\cdots + T^i(a_n)\big)^i\big)^i\big)^i.$$

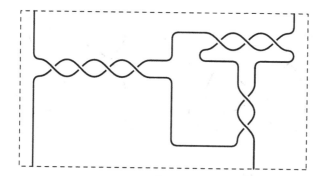

Fig. 27.7 A rational tangle $T([4;2:3])$.

As we will see later, in Theorem 27.76, the isotopy class of a tangle $T(\frac{p}{q})$ does not depend on the choice of the integer elements of the continued fractions representing the fraction $\frac{p}{q}$.

27.6.3 Main results on rational knots and tangles

The next theorem shows that the set of rational tangles is in a natural one-to-one correspondence with the rational numbers. This theorem is a reformulation of the Conway theorem from [42].

Theorem 27.76. (*i*) *Two tangles $T([a_0;a_1:a_2:\cdots:a_n])$ and $T([b_0;b_1:b_2:\cdots:b_m])$ for continued fractions with integer coefficients are isotopic if and only if*

$$[a_0;a_1:a_2:\cdots:a_n] = [b_0;b_1:b_2:\cdots:b_m].$$

(*ii*) *For every rational tangle T there exists a rational number $\frac{p}{q}$ such that T is isotopic to $T(\frac{p}{q})$.* □

To get a knot from a tangle one should use the following *closing operation*. Each tangle diagram has two endpoints on the line $y=0$. Connect them by some curve that does not intersect the diagram. Do the same for the two endpoints on the line $y=1$. See an example of a tangle $T(\frac{3}{2})$ and the corresponding trefoil knot in Fig. 27.8.

Theorem 27.77. (**H. Schubert [190]**) *Let $\frac{p_1}{q_1}$ and $\frac{p_2}{q_2}$ be two rational numbers satisfying $\gcd(p_1,q_1) = \gcd(p_2,q_2) = 1$. Then the corresponding knots or links $K(\frac{p_1}{q_1})$ and $K(\frac{p_2}{q_2})$ are isotopic if and only if the following two conditions hold:*
   (*a*) $p_1 = p_2$;
   (*b*) *either* $q_1 \equiv q_2 \pmod{p_1}$ *or* $q_1 q_2 \equiv 1 \pmod{p_1}$. □

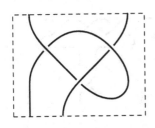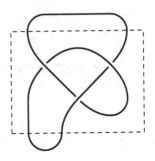

Fig. 27.8 A rational tangle and its closure.

Remark 27.78. The classification of rational knots arises in the topology of three-dimensional manifolds. Namely, the 2-fold branch coverings spaces of S^3 along the rational knots and links represent lens spaces; see the book [197] for more details.

So the problem of classification of rational knots is completely solved. The next step in this direction would be the classification of 3-bridge knots.

Problem 41. Describe all 3-bridge knots and links.

It is natural to suppose that one should use an appropriate generalization of continued fractions to investigate this question. So we have an additional question on a natural extension of continued fractions.

Problem 42. Find a generalization of continued fractions that classifies rational 3-tangles.

Rational 3-tangles can be defined similarly to *rational 2-tangles*. Nevertheless, we leave some freedom in the definition in order to try to achieve the goals of Problem 41.

References

1. F. Aicardi. Symmetries of quadratic form classes and of quadratic surd continued fractions. I. A Poincaré tiling of the de Sitter world. *Bull. Braz. Math. Soc. (N.S.)*, 40(3):301–340, 2009.
2. F. Aicardi. Symmetries of quadratic form classes and of quadratic surd continued fractions. II. Classification of the periods' palindromes. *Bull. Braz. Math. Soc. (N.S.)*, 41(1):83–124, 2010.
3. M. Aigner. *Markov's theorem and 100 years of the uniqueness conjecture. A mathematical journey from irrational numbers to perfect matchings.* Springer, Cham, 2013.
4. R. C. Alperin. $\mathrm{PSL}_2(\mathbb{Z}) = \mathbb{Z}_2 * \mathbb{Z}_3$. *The American Mathematical Monthly*, 100(4):385–386, 1993.
5. G. E. Andrews. *The theory of partitions.* Cambridge Mathematical Library. Cambridge University Press, Cambridge, 1998. Reprint of the 1976 original.
6. H. Appelgate and H. Onishi. The similarity problem for 3×3 integer matrices. *Linear Algebra Appl.*, 42:159–174, 1982.
7. V. Arnold. Arithmetics of binary quadratic forms, symmetry of their continued fractions and geometry of their de Sitter world. *Bull. Braz. Math. Soc. (N.S.)*, 34(1):1–42, 2003. Dedicated to the 50th anniversary of IMPA.
8. V. I. Arnold. Statistics of integral convex polygons. *Funct. Anal. Appl.*, 14(2):79–81, 1980. Russian version: Funkt. Anal. Prilozh. 14 (2), 1980, 1–3.
9. V. I. Arnold. A-graded algebras and continued fractions. *Comm. Pure Appl. Math.*, 42(7):993–1000, 1989.
10. V. I. Arnold. *Ordinary differential equations.* Springer Textbook. Springer-Verlag, Berlin, 1992. Translated from the third Russian edition by Roger Cooke.
11. V. I. Arnold. Higher-dimensional continued fractions. *Regul. Chaotic Dyn.*, 3(3):10–17, 1998. J. Moser at 70 (Russian).
12. V. I. Arnold. Preface. In *Pseudoperiodic topology*, volume 197 of *Amer. Math. Soc. Transl. Ser. 2*, pages ix–xii. Amer. Math. Soc., Providence, RI, 1999.
13. V. I. Arnold. *Continued fractions (In Russian).* Moscow: Moscow Center of Continuous Mathematical Education, 2002.
14. P. Arnoux and T. A. Schmidt. Natural extensions and Gauss measures for piecewise homographic continued fractions. *Bull. Soc. Math. France* 147(3): 515–544, 2019.
15. S. Assaf, L.-C. Chen, T. Cheslack-Postava, B. Cooper, A. Diesl, T. Garrity, M. Lepinski, and A. Schuyler. A dual approach to triangle sequences: a multidimensional continued fraction algorithm. *Integers*, 5(1):A8, 39, 2005.

© Springer-Verlag GmbH Germany, part of Springer Nature 2022
O. N. Karpenkov, *Geometry of Continued Fractions*,
Algorithms and Computation in Mathematics 26,
https://doi.org/10.1007/978-3-662-65277-0

16. M. Baake and J. A. G. Roberts. Reversing symmetry group of GL(2, ℤ) and PGL(2, ℤ) matrices with connections to cat maps and trace maps (English summary), *J. Phys. A*, 30(5):1549–1573, 1997.

17. J. Barajas and O. Serra. The lonely runner problem with seven runners. In *Fifth Conference on Discrete Mathematics and Computer Science (Spanish)*, volume 23 of *Ciencias (Valladolid)*, pages 111–116. Univ. Valladolid, Secr. Publ. Intercamb. Ed., Valladolid, 2006.

18. J. Barajas and O. Serra. The lonely runner with seven runners. *Electron. J. Combin.*, 15(1):Research paper 48, 18 pp., 2008.

19. I. Bárány and A. M. Vershik. On the number of convex lattice polytopes. *Geom. Funct. Anal.*, 2(4):381–393, 1992.

20. Margherita Barile, Dominique Bernardi, Alexander Borisov, and Jean-Michel Kantor. On empty lattice simplices in dimension 4. *Proc. Amer. Math. Soc.*, 139(12):4247–4253, 2011.

21. O. R. Beaver and T. Garrity. A two-dimensional Minkowski ?(x) function. *J. Number Theory*, 107(1):105–134, 2004.

22. M. Beck and S. Robins. *Computing the continuous discretely*. Undergraduate Texts in Mathematics. Springer, New York, 2007. Integer-point enumeration in polyhedra.

23. U. Betke and J. M. Wills. Untere Schranken für zwei diophantische Approximations-Funktionen. *Monatsh. Math.*, 76:214–217, 1972.

24. W. Bienia, L. Goddyn, P. Gvozdjak, A. Sebő, and M. Tarsi. Flows, view obstructions, and the lonely runner. *J. Combin. Theory Ser. B*, 72(1):1–9, 1998.

25. T. Bohman, R. Holzman, and D. Kleitman. Six lonely runners. *Electron. J. Combin.*, 8(2):Research Paper 3, 49 pp. (electronic), 2001. In honor of Aviezri Fraenkel on the occasion of his 70th birthday.

26. A. I. Borevich and I. R. Shafarevich. *Number theory*. Translated from the Russian by Newcomb Greenleaf. Pure and Applied Mathematics, Vol. 20. Academic Press, New York, 1966.

27. P. Borwein. *Computational excursions in analysis and number theory*. CMS Books in Mathematics/Ouvrages de Mathématiques de la SMC, 10. Springer-Verlag, New York, 2002.

28. A. J. Brentjes. *Multidimensional continued fraction algorithms*, volume 145 of *Mathematical Centre Tracts*. Mathematisch Centrum, Amsterdam, 1981.

29. K. Briggs. Klein polyhedra, http://keithbriggs.info/klein-polyhedra.html.

30. V. Brun. Algorithmes euclidiens pour trois et quatre nombres. In *Treizième congrès des mathématiciens scandinaves, tenu à Helsinki 18-23 août 1957*, pages 45–64. Mercators Tryckeri, Helsinki, 1958.

31. A. D. Bryuno and V. I. Parusnikov. Klein polyhedra for two Davenport cubic forms. *Math. Notes*, 56(3-4):994–1007, 1995. Russian version: Mat. Zametki, 56(4), 1994, 9–27.

32. J. Buchmann. A generalization of Voronoï's unit algorithm. I. *J. Number Theory*, 20(2):177–191, 1985.

33. J. Buchmann. A generalization of Voronoï's unit algorithm. II. *J. Number Theory*, 20(2):192–209, 1985.

34. G. Bullig. Zur Kettenbruchtheorie im Dreidimensionalen (Z 1). *Abh. math. Sem. Hansische Univ.*, 13:321–343, 1940.

35. R. T. Bumby. Hausdorff dimensions of Cantor sets. *J. Reine Angew. Math.*, 331:192–206, 1982.

36. Bykovskiĭ. Connectivity of Minkowskii graph for multidimensional complete lattices. *Vladivostok, Dalnauka*, Preprint 8, 2000.

37. V. A. Bykovskiĭ and O. A. Gorkusha. Minimal bases of three-dimensional complete lattices. *Sb. Math.*, 192(1-2):215–223, 2001. Russian version in Mat. Sb., 192(2):57–66, 2001.

38. J. T. Campbell and E. C. Trouy. When are two elements of GL(2, ℤ) similar? *Linear Algebra and its Applications*, 157:175–184, 1991.

39. J. W. S. Cassels and H. P. F. Swinnerton-Dyer. On the product of three homogeneous linear forms and the indefinite ternary quadratic forms. *Philos. Trans. Roy. Soc. London. Ser. A.*, 248:73–96, 1955.

40. S. Chowla, J. Cowles and M. Cowles. On the number of conjugacy classes in SL(2, ℤ), *Journal of Number Theory*, 12(3):372–377, 1980.

41. H. Cohen. *A course in computational algebraic number theory*, volume 138 of *Graduate Texts in Mathematics*. Springer-Verlag, Berlin, 1993.

42. J. H. Conway. An enumeration of knots and links, and some of their algebraic properties. In *Computational Problems in Abstract Algebra (Proc. Conf., Oxford, 1967)*, pages 329–358. Pergamon, Oxford, 1970.

43. T. W. Cusick. View-obstruction problems. *Aequationes Math.*, 9:165–170, 1973.

44. T. W. Cusick. The largest gaps in the lower Markoff spectrum. *Duke Math. J.*, 41:453–463, 1974.

45. T. W. Cusick. View-obstruction problems in n-dimensional geometry. *J. Combinatorial Theory Ser. A*, 16:1–11, 1974.

46. T. W. Cusick. View-obstruction problems. II. *Proc. Amer. Math. Soc.*, 84(1):25–28, 1982.

47. T. W. Cusick and M. E. Flahive. *The Markoff and Lagrange spectra*, volume 30 of *Mathematical Surveys and Monographs*. American Mathematical Society, Providence, RI, 1989.

48. T. W. Cusick and C. Pomerance. View-obstruction problems. III. *J. Number Theory*, 19(2):131–139, 1984.

49. V. I. Danilov. The geometry of toric varieties. *Russian Math. Surveys*, 33(2):97–154, 1978. Russian Version: Uspekhi Mat. Nauk, 33(1978), 2(200), 85–134.

50. K. Dasaratha, L. Flapan, T. Garrity, Ch. Lee, C. Mihaila, N. Neumann-Chun, S. Peluse, and M. Stroffregen. Cubic Irrationals and Periodicity via a Family of Multi-dimensional Continued Fraction Algorithms. *arXiv:1208.4244 [math.NT]*, page 14 pp., 2012.

51. K. Dasaratha, L. Flapan, T. Garrity, Ch. Lee, C. Mihaila, N. Neumann-Chun, S. Peluse, and M. Stroffregen. A Generalized Family of Multidimensional Continued Fractions: TRIP Maps. *arXiv:1206.7077 [math.NT]*, page 36 pp., 2012.

52. H. Davenport. On the product of three homogeneous linear forms. I. *Proc. London Math. Soc.*, 13:139–145, 1938.

53. H. Davenport. On the product of three homogeneous linear forms. II. *Proc. London Math. Soc.(2)*, 44:412–431, 1938.

54. H. Davenport. On the product of three homogeneous linear forms. III. *Proc. London Math. Soc.(2)*, 45:98–125, 1939.

55. H. Davenport. Note on the product of three homogeneous linear forms. *J. London Math. Soc.*, 16:98–101, 1941.

56. H. Davenport. On the product of three homogeneous linear forms. IV. *Proc. Cambridge Philos. Soc.*, 39:1–21, 1943.

57. H. Davenport. On the product of three non-homogeneous linear forms. *Proc. Cambridge Philos. Soc.*, 43:137–152, 1947.

58. B. N. Delone and D. K. Faddeev. *The theory of irrationalities of the third degree*. Translations of Mathematical Monographs, Vol. 10. American Mathematical Society, Providence, R.I., 1964.

59. J. W. Demmel. *Applied numerical linear algebra*. Society for Industrial and Applied Mathematics (SIAM), Philadelphia, PA, 1997.

60. E. Ehrhart. Sur les polyèdres rationnels homothétiques à n dimensions. *C. R. Acad. Sci. Paris*, 254:616–618, 1962.

61. G. Ewald. *Combinatorial convexity and algebraic geometry*, volume 168 of *Graduate Texts in Mathematics*. Springer-Verlag, New York, 1996.

62. J. Fiala and P. Kleban. Generalized number theoretic spin chain-connections to dynamical systems and expectation values. *J. Stat. Phys.*, 121(3-4):553–577, 2005.

63. David Fowler. *The mathematics of Plato's academy*. The Clarendon Press Oxford University Press, New York, second edition, 1999. A new reconstruction.

64. C. Freiling, M. Laczkovich, and D. Rinne. Rectangling a rectangle. *Discrete Comput. Geom.*, 17(2):217–225, 1997.

65. G. Frobenius. *Über die Markoffschen Zahlen*. Preuss. Akad. Wiss. Sitzungsberichte, 1913.

66. W. Fulton. *Introduction to toric varieties*, volume 131 of *Annals of Mathematics Studies*. Princeton University Press, Princeton, NJ, 1993. The William H. Roever Lectures in Geometry.

67. T. Garrity. On periodic sequences for algebraic numbers. *J. Number Theory*, 88(1):86–103, 2001.

68. T. Garrity. A thermodynamic classification of real numbers. *J. Number Theory*, 130(7):1537–1559, 2010.

69. T. Garrity. A thermodynamic classification of pairs of real numbers via the Triangle Multidimensional continued fraction. *arXiv:1205.5663 [math.NT]*, page 21 pp., 2012.

70. C. F. Gauss. *Recherches arithmétiques*. Blanchard, Paris, 1807.

71. I. Gelfand and V. Retakh. Quasideterminants. I. *Selecta Math. (N.S.)*, 3(4):517–546, 1997.

72. I. M. Gelfand and V. S. Retakh. Determinants of matrices over noncommutative rings. *Funct. Anal. Appl.*, 25(2):91–102, 1991. Russian version: Funktsional. Anal. i Prilozhen., 25(2), 1991, 13–25.

73. I. M. Gelfand and V. S. Retakh. Theory of noncommutative determinants, and characteristic functions of graphs. *Funct. Anal. Appl.*, 26(4):231–246 (1993), 1992. Russian Version: Funktsional. Anal. i Prilozhen., 26(4), 1992, 1–20.

74. O. N. German. Sails and Hilbert bases. *Proc. Steklov Inst. Math.*, 239(4):88–95, 2002. Russian version: Tr. Mat. Inst. Steklova, (4)239, 2002, 98–105.

75. O. N. German. Sails and norm minima of lattices. *Russian Acad. Sci. Sb. Math.*, 196(3):337–367, 2005. Russian version: Mat. Sb., 196 (3), 2005, 31–60.

76. O. N. German. Klein polyhedra and norm minima of lattices. *Dokl. Akad. Nauk*, 406(3):298–302, 2006.

77. O. N. German and E. L. Lakshtanov. On a multidimensional generalization of Lagrange's theorem for continued fractions. *Izv. Math.*, 72(1):47–61, 2008. Russian Version: Izv. Ross. Akad. Nauk Ser. Mat., 72(1), 2008, 51–66.

78. O. A. Gorkusha. Minimal bases of three-dimensional complete lattices. *Math. Notes*, 69(3-4):320–328, 2001. Russian version: Mat. Zametki, 69(3):353–362, 2001.

79. P. M. Gruber and C. G. Lekkerkerker. *Geometry of numbers*, volume 37 of *North-Holland Mathematical Library*. North-Holland Publishing Co., Amsterdam, second edition, 1987.

80. F. J. Grunewald. Solution of the conjugacy problem in certain arithmetic groups. In *Word problems, II (Conf. on Decision Problems in Algebra, Oxford, 1976)*, volume 95 of *Stud. Logic Foundations Math.*, pages 101–139. North-Holland, Amsterdam, 1980.

81. J. Hančl, A. Jaššová, P. Lertchoosakul, and R. Nair. On the metric theory of p-adic continued fractions. *Indag. Math. (N.S.)*, 24(1):42–56, 2013.

82. H. Hancock. *Development of the Minkowski geometry of numbers. Vols. One, Two*. Dover Publications Inc., New York, 1964.

83. J. P. Henniger. Factorization and similarity in $GL(2, \mathbb{Z})$, *Linear Algebra Appl.* 251:223–237, 1997.

84. C. Hermite. Letter to C. D. J. Jacobi. *J. Reine Angew. Math.*, 40:286, 1839.

85. K. Hessenberg. *Thesis*. Darmstadt, Germany: Technische Hochschule, 1942.

86. A. Higashitani and M. Masuda. Lattice multi-polygons. *arXiv:1204.0088 [math.CO]*, page 21 pp., 2012.

87. J. Hirsh and L. C. Washington. p-adic continued fractions. *Ramanujan J.*, 25(3):389–403, 2011.

88. F. Hirzebruch. Über vierdimensionale Riemannsche Flächen mehrdeutiger analytischer Funktionen von zwei komplexen Veränderlichen. *Math. Ann.*, 126:1–22, 1953.

89. A. A. Illarionov. Estimates for the number of relative minima of lattices. *Math. Notes*, 89(1-2):245–254, 2011. Russian version: Mat. Zametki, 89(2):249–259, 2011.

90. A. A. Illarionov. The average number of relative minima of three-dimensional integer lattices. *St. Petersburg Math. J.*, 23(3):551–570, 2012. Russian version: Algebra i Analiz, 23(3):189–215, 2011.

91. T. H. Jackson. Note on the minimum of an indefinite binary quadratic form. *J. London Math. Soc. (2)*, 5:209–214, 1972.

92. C. G. J. Jacobi. Allgemeine Theorie der Kettenbruchähnlichen Algorithmen, in welchen jede Zahl aus drei vorhergehenden gebildet wird (Aus den hinterlassenen Papieren von C. G. J. Jacobi mitgetheilt durch Herrn E. Heine). *Journal für die Reine und Angewandte Mathematik*, 69(1):29–64, 1868.

93. H. W. E. Jung. Darstellung der Funktionen eines algebraischen Körpers zweier unabhängigen Veränderlichen x,y in der Umgebung einer Stelle $x = a$, $y = b$. *J. Reine Angew. Math.*, 133:289–314, 1908.

94. V. F. Kagan. *Origins of determinant theory (In Russian)*. Gos. Izd. Ukraine, Odessa, 1922.

95. J.-M. Kantor and K. S. Sarkaria. On primitive subdivisions of an elementary tetrahedron. *Pacific J. Math.*, 211(1):123–155, 2003.

96. O. Karpenkov. On the triangulations of tori associated with two-dimensional continued fractions of cubic irrationalities. *Funct. Anal. Appl.*, 38(2):102–110, 2004. Russian version: Funkt. Anal. Prilozh. 38 (2), 2004, 28–37.

97. O. Karpenkov. On two-dimensional continued fractions of hyperbolic integer matrices with small norm. *Russian Math. Surveys*, 59(5(359)):959–960, 2004. Russian version: Uspekhi Mat. Nauk, 59(5), 2004, 149–150.

98. O. Karpenkov. Classification of three-dimensional multistoried completely hollow convex marked pyramids. *Russian Math. Surveys*, 60:165–166, 2005. Russian version: Uspekhi Mat. Nauk, 60, 1(361), 2005, 169–170.

99. O. Karpenkov. Three examples of three-dimensional continued fractions in the sense of Klein. *C. R. Math. Acad. Sci. Paris*, 343(1):5–7, 2006.

100. O. Karpenkov. Classification of lattice-regular lattice convex polytopes, *Funct. Anal. Other Math.*, 1(1):17–35, 2006.

101. O. Karpenkov. On an invariant Möbius measure and the Gauss-Kuz'min face distribution. *Proc. Steklov Inst.*, 258:74–86, 2007. Russian Version: Tr. Mat. Inst. Steklova 258 (2007), 79–92.

102. O. Karpenkov. Elementary notions of lattice trigonometry. *Math. Scand.*, 102(2):161–205, 2008.

103. O. Karpenkov. Constructing multidimensional periodic continued fractions in the sense of Klein. *Math. Comp.*, 78(267):1687–1711, 2009.

104. O. Karpenkov. On irrational lattice angles. *Funct. Anal. Other Math.*, 2(2-4):221–239, 2009.

105. O. Karpenkov. On determination of periods of geometric continued fractions for two-dimensional algebraic hyperbolic operators. *Math. Notes*, 88(1-2):28–38, 2010. Russian version: Mat. Zametki, 88(1), (2010), 30–42.

106. O. Karpenkov. Continued fractions and the second Kepler law. *Manuscripta Math.*, 134(1-2):157–169, 2011.

107. O. Karpenkov. Multidimensional Gauss Reduction Theory for conjugacy classes of SL(n,Z). *J. Théor. Nombres Bordeaux*, 25(1):99–109, 2013.

108. O. N. Karpenkov and A. V. Ustinov. Geometry and combinatorics of Minkowski-Voronoi 3-dimensional continued fractions. *J. Number Theor.*, 176:375–419, 2017.

109. O. N. Karpenkov and A. M. Vershik. Rational approximation of maximal commutative subgroups of GL($n, \mathbb{R}$). *J. Fixed Point Theory Appl.*, 7(1):241–263, 2010.

110. O. Karpenkov and M. Van-Son. Perron identity for arbitrary broken lines, *J. Théor. Nombres Bordeaux*, 31(1):131–144, 2019.

111. O. Karpenkov and M. Van-Son. Generalised Markov numbers, *J. Number Theor.*, 213:16–66, 2020.

112. O. Karpenkov, Continued Fraction approach to Gauss Reduction Theory. In Bell P.C., Totzke P., Potapov I. (eds) Reachability Problems. RP 2021. *Lecture Notes in Computer Science*, vol. 13035, pages 100–114, Springer, Cham, 2021.

113. S. Katok. Continued fractions, hyperbolic geometry and quadratic forms. In *MASS selecta*, pages 121–160. Amer. Math. Soc., Providence, RI, 2003.

114. L. H. Kauffman and S. Lambropoulou. On the classification of rational knots. *Enseign. Math. (2)*, 49(3-4):357–410, 2003.

115. L. H. Kauffman and S. Lambropoulou. On the classification of rational tangles. *Adv. in Appl. Math.*, 33(2):199–237, 2004.

116. A. Ya. Khinchin. *Continued fractions*. Moscow, FISMATGIS, 1961.

117. A. G. Khovanskiĭ. Newton polytopes, curves on toric surfaces, and inversion of Weil's theorem. *Russian Math. Surveys*, 52(6):1251–1279, 1997. Russian version: Uspekhi Mat. Nauk, 52(6(318)), 1997, 113–142.

118. V. Klee. Some characterizations of convex polyhedra. *Acta Math.*, 102:79–107, 1959.
119. F. Klein. Ueber eine geometrische Auffassung der gewöhnliche Kettenbruchentwicklung. *Nachr. Ges. Wiss. Göttingen Math-Phys. Kl.*, 3:352–357, 1895.
120. F. Klein. Sur une représentation géométrique de développement en fraction continue ordinaire. *Nouv. Ann. Math.*, 15(3):327–331, 1896.
121. A. Knauf. On a ferromagnetic spin chain. *Comm. Math. Phys.*, 153(1):77–115, 1993.
122. A. Knauf. On a ferromagnetic spin chain. II. Thermodynamic limit. *J. Math. Phys.*, 35(1):228–236, 1994.
123. Andreas Knauf. Number theory, dynamical systems and statistical mechanics. *Rev. Math. Phys.*, 11(8):1027–1060, 1999.
124. M. L. Kontsevich and Yu. M. Suhov. Statistics of Klein polyhedra and multidimensional continued fractions. In *Pseudoperiodic topology*, volume 197 of *Amer. Math. Soc. Transl. Ser. 2*, pages 9–27. Amer. Math. Soc., Providence, RI, 1999.
125. M. Konvalinka and I. Pak. Geometry and complexity of O'Hara's algorithm. *Adv. in Appl. Math.*, 42(2):157–175, 2009.
126. S. V. Konyagin and K. A. Sevastyanov. Estimate of the number of vertices of a convex integral polyhedron in terms of its volume. *Funktsional. Anal. i Prilozhen.*, 18(1):13–15, 1984.
127. E. Korkina. La périodicité des fractions continues multidimensionnelles. *C. R. Acad. Sci. Paris Sér. I Math.*, 319(8):777–780, 1994.
128. E. I. Korkina. The simplest 2-dimensional continued fraction. In *International Geometrical Colloquium, Moscow*, 1993.
129. E. I. Korkina. Two-dimensional continued fractions. The simplest examples. *Trudy Mat. Inst. Steklov*, 209(Osob. Gladkikh Otobrazh. s Dop. Strukt.):143–166, 1995.
130. E. I. Korkina. The simplest 2-dimensional continued fraction. *J. Math. Sci.*, 82(5):3680–3685, 1996. Topology, 3.
131. C. Kraaikamp and I. Smeets. Approximation results for α-Rosen fractions. *Unif. Distrib. Theory*, 5(2):15–53, 2010.
132. R. O. Kuzmin. On one problem of Gauss. *Dokl. Akad. Nauk SSSR Ser. A*, pages 375–380, 1928.
133. R. O. Kuzmin. On a problem of Gauss. *Atti del Congresso Internazionale dei Matematici, Bologna*, 6:83–89, 1932.
134. G. Lachaud. Polyèdre d'Arnol'd et voile d'un cône simplicial: analogues du théorème de Lagrange. *C. R. Acad. Sci. Paris Sér. I Math.*, 317(8):711–716, 1993.
135. G. Lachaud. Sails and Klein polyhedra. In *Number theory (Tiruchirapalli, 1996)*, volume 210 of *Contemp. Math.*, pages 373–385. Amer. Math. Soc., Providence, RI, 1998.
136. G. Lachaud. *Voiles et polyhedres de Klein*. Act. Sci. Ind., Hermann, 2002.
137. M. Laczkovich and G. Szekeres. Tilings of the square with similar rectangles. *Discrete Comput. Geom.*, 13(3-4):569–572, 1995.
138. J. C. Lagarias. Best simultaneous Diophantine approximations. I. Growth rates of best approximation denominators. *Trans. Amer. Math. Soc.*, 272(2):545–554, 1982.
139. J. C. Lagarias. Geodesic multidimensional continued fractions. *Proc. London Math. Soc. (3)*, 69(3):464–488, 1994.
140. J.-L. Lagrange. Solution d'un problème d'arithmétique. In *OEuvres de Lagrange*, volume 1, pages 671–732. Paris: Gauthier-Villars, 1867–1892.
141. A. K. Lenstra, H. W. Lenstra, Jr., and L. Lovász. Factoring polynomials with rational coefficients. *Math. Ann.*, 261(4):515–534, 1982.
142. P. Levy. Sur les lois de probabilité dont dependent les quotients complets et incomplets d'une fraction continue. *Bull. Soc. Math. France*, 57:178–194, 1929.
143. J. Lewis and D. Zagier. Period functions and the Selberg zeta function for the modular group. In *The mathematical beauty of physics (Saclay, 1996)*, volume 24 of *Adv. Ser. Math. Phys.*, pages 83–97. World Sci. Publ., River Edge, NJ, 1997.
144. K. Mahler. *Lectures on diophantine approximations. Part I: g-adic numbers and Roth's theorem.* Prepared from the notes by R. P. Bambah of my lectures given at the University of Notre Dame in the Fall of 1957. University of Notre Dame Press, Notre Dame, Ind, 1961.

145. Y. I. Manin and M. Marcolli. Continued fractions, modular symbols, and noncommutative geometry. *Selecta Math. (N.S.)*, 8(3):475–521, 2002.
146. A. Markoff. Sur les formes quadratiques binaires indéfinies. *Math. Ann.*, 15(3-4):381–406, 1879.
147. A. Markoff. Sur les formes quadratiques binaires indéfinies. *Math. Ann.*, 17(3):379–399, 1879.
148. P. Mattila. *Geometry of sets and measures in Euclidean spaces*, volume 44 of *Cambridge Studies in Advanced Mathematics*. Cambridge University Press, Cambridge, 1995. Fractals and rectifiability.
149. P. McMullen. Lattices compatible with regular polytopes, *European J. Combin.*, 29(8):1925–1932, 2008.
150. A. Messaoudi, A. Nogueira, and F. Schweiger. Ergodic properties of triangle partitions. *Monatsh. Math.*, 157(3):283–299, 2009.
151. H. Minkowski. Généralisation de la théorie des fractions continues. *Ann. Sci. École Norm. Sup. (3)*, 13:41–60, 1896.
152. H Minkowski. *Gesammelte Abhandlungen (pp. 293–315)*. AMS-Chelsea, 1967.
153. R. Mönkemeyer. Über Fareynetze in *n* Dimensionen. *Math. Nachr.*, 11:321–344, 1954.
154. P.-L. Montagard and N. Ressayre. Regular lattice polytopes and root systems, *Bull. Lond. Math. Soc.*, 41(2):227–241, 2009.
155. J.-O. Moussafir. Sails and Hilbert bases. *Funct. Anal. Appl.*, 34(2):114–118, 2000. Russian version: Funkt. Anal. Prilozh. 34, (2) 2000, 43–49.
156. J.-O. Moussafir. *Voiles et Polyédres de Klein: Geometrie, Algorithmes et Statistiques*. docteur en sciences thèse, Université Paris IX - Dauphine, 2000.
157. V. V. Nikulin. On the classification of arithmetic groups generated by reflections in Lobachevskiĭ spaces. *Izv. Akad. Nauk SSSR Ser. Mat.*, 45(1):113–142, 240, 1981.
158. V. V. Nikulin. Quotient-groups of groups of automorphisms of hyperbolic forms by subgroups generated by 2-reflections. Algebro-geometric applications. In *Current problems in mathematics, Vol. 18*, pages 3–114. Akad. Nauk SSSR, Vsesoyuz. Inst. Nauchn. i Tekhn. Informatsii, Moscow, 1981.
159. V. V. Nikulin. Discrete reflection groups in Lobachevsky spaces and algebraic surfaces. In *Proceedings of the International Congress of Mathematicians, Vol. 1, 2 (Berkeley, Calif., 1986)*, pages 654–671, Providence, RI, 1987. Amer. Math. Soc.
160. V. V. Nikulin. Finiteness of the number of arithmetic groups generated by reflections in Lobachevskiĭ spaces. *Izv. Math.*, 71(1):53–56, 2007. Russian version: Izv. Ross. Akad. Nauk Ser. Mat., 71(1), 2007, 55–60.
161. S. Northshield. Stern's diatomic sequence $0, 1, 1, 2, 1, 3, 2, 3, 1, 4, \ldots$. *Amer. Math. Monthly*, 117(7):581–598, 2010.
162. T. Oda. *Convex bodies and algebraic geometry*, volume 15 of *Ergebnisse der Mathematik und ihrer Grenzgebiete (3) [Results in Mathematics and Related Areas (3)]*. Springer-Verlag, Berlin, 1988. An introduction to the theory of toric varieties, translated from the Japanese.
163. J. O'Hara. *Energy of knots and conformal geometry*, volume 33 of *Series on Knots and Everything*. World Scientific Publishing Co. Inc., River Edge, NJ, 2003.
164. K. M. O'Hara. *Structure and Complexity of the Involution Principle for Partitions*. ProQuest LLC, Ann Arbor, MI, 1984. Thesis (Ph.D.)–University of California, Berkeley.
165. K. M. O'Hara. Bijections for partition identities. *J. Combin. Theory Ser. A*, 49(1):13–25, 1988.
166. R. Okazaki. On an effective determination of a Shintani's decomposition of the cone $\mathbf{R}^n_+$. *J. Math. Kyoto Univ.*, 33(4):1057–1070, 1993.
167. J. M. Ortega and H. F. Kaiser. The LL^T and QR methods for symmetric tridiagonal matrices. *Comput. J.*, 6:99–101, 1963/1964.
168. I. Pak. Partition identities and geometric bijections. *Proc. Amer. Math. Soc.*, 132(12):3457–3462 (electronic), 2004.
169. I. Pak. Partition bijections, a survey. *Ramanujan J.*, 12(1):5–75, 2006.
170. I. Pak, A. Postnikov, and V. Retakh. Noncommutative Lagrange Theorem and Inversion Polynomials (preprint). *www.math.ucla.edu/ pak/papers*, page 15 pp., 1995.

171. Giovanni Panti. Multidimensional continued fractions and a Minkowski function. *Monatsh. Math.*, 154(3):247–264, 2008.

172. V. I. Parusnikov. Klein's polyhedra for the thierd extremal ternary cubic form. Technical report, preprint 137 of Keldysh Institute of the RAS, Moscow, 1995.

173. V. I. Parusnikov. Klein's polyhedra for the fifth extremal cubic form. Technical report, preprint 69 of Keldysh Institute of the RAS, Moscow, 1998.

174. V. I. Parusnikov. Klein's polyhedra for the seventh extremal cubic form. Technical report, preprint 79 of Keldysh Institute of the RAS, Moscow, 1999.

175. V. I. Parusnikov. Klein polyhedra for the fourth extremal cubic form. *Math. Notes*, 67(1-2):87–102, 2000. Russian version: Mat. Zametki, 67(1),2000), 110–128.

176. V. I. Parusnikov. Klein polyhedra for three extremal cubic forms. *Math. Notes*, 77(4):566–583, 2005. Russian version: Mat. Zametki, 77(3-4),2000), 523–538.

177. O. Perron. Grundlagen für eine Theorie des Jacobischen Kettenbruchalgorithmus. *Math. Ann.*, 64(1):1–76, 1907.

178. O. Perron. Erweiterung eines Markoffschen Satzes über die Konvergenz gewisser Kettenbrüche. *Math. Ann.*, 74(4):545–554, 1913.

179. O. Perron. Über die Approximation irrationaler Zahlen durch rationale. *I S.-B. Heidelberg Akad. Wiss.*, Essay 4:17 pp., 1921.

180. O. Perron. Über die Approximation irrationaler Zahlen durch rationale. *II S.-B. Heidelberg Akad. Wiss.*, Essay 8:12 pp., 1921.

181. Bjorn Poonen and Fernando Rodriguez-Villegas. Lattice polygons and the number 12. *Amer. Math. Monthly*, 107(3):238–250, 2000.

182. P. Popescu-Pampu. The geometry of continued fractions and the topology of surface singularities. In *Singularities in Geometry and Topology 2004*, pages 119–195, 2007.

183. M. Prasolov and M. Skopenkov. Tiling by rectangles and alternating current. *J. Combin. Theory Ser. A*, 118(3):920–937, 2011.

184. R. Rankin. *Modular Forms and Functions*. Cambridge Univ. Press, Cambridge, 1977.

185. D. Repovsh, M. Skopenkov, and M. Tsentsel. An elementary proof of the twelve lattice point theorem. *Math. Notes*, 77(1-2):108–111, 2005. Russian Version: Mat. Zametki, 77(1), 2005, 117–120.

186. David Rosen. A class of continued fractions associated with certain properly discontinuous groups. *Duke Math. J.*, 21:549–563, 1954.

187. A. A. Ruban. Certain metric properties of the p-adic numbers. *Sibirsk. Mat. Ž.*, 11:222–227, 1970.

188. L. Schläfli. Theorie der vielfachen Kontinuität, written 1850-1852; *Zürcher und Furrer*, Zürich 1901; *Denkschriften der Schweizerischen naturforschenden Gesellschaft*, 38:1–237, 1901; reprinted in Ludwig Schläfli 1814–1895, *Gesammelte Mathematische Abhandlungen*, Vol. I, pp. 167–387 Birkhäuser, Basel 1950.

189. Asmus L. Schmidt. Ergodic theory of complex continued fractions. In *Number theory with an emphasis on the Markoff spectrum (Provo, UT, 1991)*, volume 147 of *Lecture Notes in Pure and Appl. Math.*, pages 215–226. Dekker, New York, 1993.

190. H. Schubert. Knoten mit zwei Brücken. *Math. Z.*, 65:133–170, 1956.

191. F. Schweiger. *The metrical theory of Jacobi-Perron algorithm*. Lecture Notes in Mathematics, Vol. 334. Springer-Verlag, Berlin, 1973.

192. F. Schweiger. Ergodic properties of multi-dimensional subtractive algorithms. In *New trends in probability and statistics, Vol. 2 (Palanga, 1991)*, pages 91–100. VSP, Utrecht, 1992.

193. F. Schweiger. Invariant measures for fully subtractive algorithms. *Anz. Österreich. Akad. Wiss. Math.-Natur. Kl.*, 131:25–30 (1995), 1994.

194. F. Schweiger. Fully subtractive algorithms. *Österreich. Akad. Wiss. Math.-Natur. Kl. Sitzungsber. II*, 204:23–32 (1996), 1995.

195. F. Schweiger. *Multidimensional continued fractions*. Oxford Science Publications. Oxford University Press, Oxford, 2000.

196. F. Schweiger. Multidimensional Continued Fractions – New Results and Old Problems (preprint). *http://www.cirm.univ-mrs.fr/videos/2008/exposes/287w2/Schweiger.pdf*, page 11 pp., 2008.

197. H. Seifert and W. Threlfall. *Seifert and Threlfall: a textbook of topology*, volume 89 of *Pure and Applied Mathematics*. Academic Press Inc. [Harcourt Brace Jovanovich Publishers], New York, 1980. Translated from the German edition of 1934 by Michael A. Goldman, With a preface by Joan S. Birman, With "Topology of 3-dimensional fibered spaces" by Seifert, Translated from the German by Wolfgang Heil.

198. E. S. Selmer. Continued fractions in several dimensions. *Nordisk Nat. Tidskr.*, 9:37–43, 95, 1961.

199. J.-P. Serre. *A course in arithmetic*. Graduate Texts in Mathematics, No. 7. Springer-Verlag, New York-Heidelberg, 1973.

200. T. Shintani. On evaluation of zeta functions of totally real algebraic number fields at non-positive integers. *J. Fac. Sci. Univ. Tokyo Sect. IA Math.*, 23(2):393–417, 1976.

201. V. I. Shmoïlov. *Nepreryvnye drobi. Tom I*. Merkator, Lviv, 2004. Periodicheskie nepreryvnye drobi. [Periodic continued fractions].

202. V. I. Shmoïlov. *Nepreryvnye drobi. Tom II*. Merkator, Lviv, 2004. Raskhodyashchiesya nepreryvnye drobi. [Diverging continued fractions].

203. V. I. Shmoïlov. *Nepreryvnye drobi. Tom III*. Merkator, Lviv, 2004. Iz istorii nepreryvnykh drobei. [On the history of continued fractions].

204. V. Ya. Skorobogatko. Ideas and results of the theory of branching continued fractions and their application to the solution of differential equations. In *General theory of boundary value problems*, pages 187–197. "Naukova Dumka", Kiev, 1983.

205. V. Ya. Skorobogatko. *Teoriya vetvyashchikhsya tsepnykh drobei i ee primenenie v vychislitelnoi matematike.* "Nauka", Moscow, 1983.

206. B. F. Skubenko. *Minima of a decomposable cubic form of three variables.* 168. Zapiski nauch. sem. LOMI, 1988.

207. B. F. Skubenko. *Minima of decomposable forms of degree n of n variables for $n \geq 3$.* 168. Zapiski nauch. sem. LOMI, 1990.

208. J. Stoer and R. Bulirsch. *Introduction to numerical analysis*, volume 12 of *Texts in Applied Mathematics*. Springer-Verlag, New York, third edition, 2002. Translated from the German by R. Bartels, W. Gautschi and C. Witzgall.

209. H. P. F. Swinnerton-Dyer. On the product of three homogeneous linear forms. *Acta Arith.*, 18:371–385, 1971.

210. G. Szekeres. Multidimensional continued fractions. *Ann. Univ. Sci. Budapest. Eötvös Sect. Math.*, 13:113–140 (1971), 1970.

211. E. Thomas and A. T. Vasquez. On the resolution of cusp singularities and the Shintani decomposition in totally real cubic number fields. *Math. Ann.*, 247(1):1–20, 1980.

212. R. F. Tichy and J. Uitz. An extension of Minkowski's singular function. *Appl. Math. Lett.*, 8(5):39–46, 1995.

213. L. N. Trefethen and D. Bau. *Numerical linear algebra*. Society for Industrial and Applied Mathematics (SIAM), Philadelphia, PA, 1997.

214. A. Trevisan. Lattice polytopes and toric varieties (http://www.math.leidenuniv.nl/scripties/trevisan.pdf). Master's thesis, Leiden University, 2007.

215. H. Tsuchihashi. Higher-dimensional analogues of periodic continued fractions and cusp singularities. *Tohoku Math. J. (2)*, 35(4):607–639, 1983.

216. A. V. Ustinov. Minimal Vector Systems in 3-Dimensional Lattices and Analog of Vahlen's Theorem for 3-Dimensional Minkowski's Continued Fractions. *Sovrem. Probl. Mat.*, 16:103–128, 2012.

217. A. M. Vershik. Statistical mechanics of combinatorial partitions, and their limit configurations. *Funct. Anal. Appl.*, 30(2):90–105, 1996.

218. P. Viader, J. Paradís, and L. Bibiloni. A new light on Minkowski's $?(x)$ function. *J. Number Theory*, 73(2):212–227, 1998.

219. G. Voronoï. *A generalization of the algorithm of continued fractions*. PhD thesis, Warsaw (in Russian), 1896.

220. G. F. Voronoï. *On a Generalization of the Algorithm of Continued Fraction. Collected works in three volumes (In Russian)*. USSR Ac. Sci., Kiev., 1952.

221. D. Wells. *The Penguin Dictionary of Curious and Interesting Numbers*. Middlesex, England: Penguin Books, 1986.

222. G. K. White. Lattice tetrahedra. *Canad. J. Math.*, 16:389–396, 1964.

223. J. M. Wills. Zwei Sätze über inhomogene diophantische Approximation von Irrationalzahlen. *Monatsh. Math.*, 71:263–269, 1967.

224. E. Wirsing. On the theorem of Gauss-Kusmin-Lévy and a Frobenius-type theorem for function spaces. *Acta Arith.*, 24:507–528, 1973/74. Collection of articles dedicated to Carl Ludwig Siegel on the occasion of his seventy-fifth birthday, V.

225. C. Yannopoulos. Zur Kettenbruchtheorie im Dreidimensionalen. *Math. Z.*, 47:105–110, 1940.

226. R. T. Živaljević. Rotation number of a unimodular cycle: an elementary approach. *arXiv:1209.4981 [math.CO]*, page 13 pp., 2012.

Index

© Springer-Verlag GmbH Germany, part of Springer Nature 2022
O. N. Karpenkov, *Geometry of Continued Fractions*,
Algorithms and Computation in Mathematics 26,
https://doi.org/10.1007/978-3-662-65277-0

Printed in the United States
by Baker & Taylor Publisher Services